# PRACTICAL DIGITAL
# DESIGN USING ICs

# NEW TITLES IN ELECTRONIC TECHNOLOGY

*Theodore Bogart*
**LAPLACE TRANSFORMS AND CONTROL SYSTEMS THEORY FOR TECHNOLOGY (1982)**

*Rodney B. Faber*
**APPLIED ELECTRICITY AND ELECTRONICS FOR TECHNOLOGY, 2nd edition (1982)**

*Luces M. Faulkenberry*
**AN INTRODUCTION TO OPERATIONAL AMPLIFIERS, 2nd edition (1982)**

*Joseph D. Greenfield*
**PRACTICAL DIGITAL DESIGN USING ICs, 2nd edition (1983)**

*Joseph D. Greenfield and William C. Wray*
**USING MICROPROCESSORS AND MICROCOMPUTERS: THE 6800 FAMILY (1981)**

*Curtis Johnson*
**PROCESS CONTROL INSTRUMENTATION TECHNOLOGY, 2nd edition (1982)**

*Larry Jones and A. Foster Chin*
**ELECTRONIC INSTRUMENTS AND MEASUREMENTS (1983)**

*Thomas Young*
**LINEAR INTEGRATED CIRCUITS (1981)**

*William Starr*
**ELECTRICAL WIRING AND DESIGN: A PRACTICAL APPROACH (1983)**

# PRACTICAL DIGITAL DESIGN USING ICs

*SECOND EDITION*

**Joseph D. Greenfield**
ROCHESTER INSTITUTE OF TECHNOLOGY

**JOHN WILEY & SONS**
*NEW YORK     CHICHESTER     BRISBANE     TORONTO     SINGAPORE*

MANUSCRIPT EDITOR: Dex Ott
PRODUCTION SUPERVISOR: Deborah Pokos
COVER DESIGN: Jules Perlmutter

*Library of Congress Cataloging in Publication Data:*

Greenfield, Joseph D., 1930–
    Practical digital design using ICs.

    (Electronic technology series, ISSN 0422-910X)
    Includes bibliographical references and index.
    1. Digital integrated circuits. 2. Digital electronics.
3. Logic circuits. I. Title. II. Title: Practical
digital design using I.C.s. III. Series.
TK7874.G72 1983      621.3815'3      82-10931
ISBN 0-471-05791-6

Printed in the United States of America

10 9 8 7 6 5 4 3 2

# PREFACE

The purpose of the second edition of *Practical Digital Design Using ICs* is to update the material in the first edition. The general thrust of the first edition—explaining the **7400** series of ICs—has been retained because it was well received, and because the **7400** is still the most popular and widely used IC series.

The major event in the digital field in the past five years has been the proliferation of microprocessors. I have accommodated this event in the second edition by providing a chapter on microprocessor interfacing and by emphasizing 3-state gates and line drivers and transceivers. Many new and more sophisticated ICs that work with microprocessors, such as UARTs and PIAs, have also been produced, and these are covered in as much detail as space and their importance allow.

Though the microprocessor seems to be ubiquitous, the use of small and medium scale ICs is even more widespread. Microprocessors will not soon replace ICs. Instead, they will work in conjunction with ICs to provide more sophisticated and powerful systems. As I point out in Chapter 1, the Apple computer uses about 50 ICs, most of which are discussed in this book, in addition to its **6502** microprocessor. Thus the study of ICs remains a necessary part of the curriculum for digital engineers and technologists.

Chapter 1 is an introduction to binary arithmetic. 2s complement and hexadecimal arithmetic are introduced in Chapter 14, where they can be presented in conjunction with the arithmetic circuits that actually use them.

An introduction to the basic IC gates has been moved into Chapter 2 to facilitate laboratory work with the Boolean algebra presented in the chapter. I have added a section on multiple outputs to Chapter 3 on reduction and Karnaugh maps. In Chapter 5 the material on 3-state gates has been expanded because of their widespread use.

Chapters 6, 7, 8, and 9 on flip-flops, one-shots, counters, and shift registers all have been modernized and some new material is introduced. Greater use of oscilloscope traces has been made where they can best illustrate a point.

In Chapter 10 on IC construction and testing, I have added a section on logic analyzers. The chapter on multiplexers has been moved to Chapter 11, so it can precede the chapters on multiplexing 7-segment displays and arithmetic circuits.

Chapter 15 has been updated to include more commonly used memories. The sections on dynamic RAMs and EPROMs have been expanded.

Chapter 16 is an entirely new chapter explaining how a digital computer works and how it can be built using the ICs introduced in the previous chapters. A very small computer, built by RIT students, is also introduced and explained. Making students build their own computer is, I feel, the best way to be sure they understand how computers work.

Chapter 17 is on minicomputer interfacing, and Chapter 19, which is new, examines microprocessor interfacing. In Chapter 17, the material on A/D converters has been expanded.

Chapter 18 discusses RS-232 and 20-mA current loop conversion. A detailed discussion of a commonly used UART has been included.

Chapter 20 is on the CRT Display Generator. It has been updated to conform to modern practice. I have added appendixes on the ASCII code and answers to selected problems. The latter has been requested by many people.

I would like to thank all the people who helped me with the second edition. I must mention my colleagues in the Department of Electronic Technology at Rochester Institute of Technology, and my students there, whose input kept my courses up to date. I must also thank the people at John Wiley, who kept pushing and prodding to get the manuscript done, and the pre-publication reviewers, who were very kind and enthusiastic, and provided many excellent suggestions for improvement.

I especially thank my wife Gladys. Without her love, encouragement, cooperation, and typing, this book would never have been written.

JOSEPH D. GREENFIELD
*Rochester, New York*

# CONTENTS

1  BINARY NUMBERS                                                          1

   1-1   Introduction to Digital Electronics                               1
   1-2   Instructional Objectives                                          3
   1-3   Self-Evaluation Questions                                         3
   1-4   Uses of Binary Bits                                               3
   1-5   Binary-to-Decimal Conversion                                      5
   1-6   Decimal-to-Binary Conversion                                      8
   1-7   Addition and Subtraction of Binary Numbers                       14
         Glossary                                                         16
         References                                                       17
         Problems                                                         17

2  BOOLEAN ALGEBRA                                                        19

   2-1   Instructional Objectives                                         19
   2-2   Self-Evaluation Questions                                        19
   2-3   Introduction to Boolean Algebra                                  20
   2-4   Operations with Boolean Variables                                23
   2-5   Theorems                                                         25
   2-6   Logical Manipulation                                             28
   2-7   IC Gates                                                         32
   2-8   Complementation of Functions                                     37
   2-9   Inverting Gates                                                  41
   2-10  Implementation of Logic Expressions Using Gates                  45
   2-11  Relays                                                           52
         Glossary                                                         55
         References                                                       56
         Problems                                                         56

3  SYSTEMATIC REDUCTION OF BOOLEAN EXPRESSIONS                            60

   3-1   Instructional Objectives                                         60
   3-2   Self-Evaluation Questions                                        60
   3-3   Standard Forms—Introduction                                      61
   3-4   Equivalence of SOP and POS Standard Forms                        72

3-5 Karnaugh Maps 74
3-6 Don't Cares 91
3-7 Five-Variable Maps 93
3-8 Practical Examples 95
Summary 101
Glossary 101
References 102
Problems 102

4 LOGIC FAMILIES AND THEIR CHARACTERISTICS 105

4-1 Instructional Objectives 105
4-2 Self-Evaluation Questions 105
4-3 Evaluation of IC Families 106
4-4 Transistor-Transistor Logic 109
4-5 Characteristics of TTL Gates 111
4-6 Emitter-Coupled Logic 116
4-7 Complementary Metal Oxide Semiconductor Gates 118
Summary 122
Glossary 123
References 123

5 BASIC TTL GATE 125

5-1 Instructional Objectives 125
5-2 Self-Evaluation Questions 125
5-3 Introduction 126
5-4 Schmitt Triggers 127
5-5 Open and Unused Inputs 130
5-6 Wire-ANDing and Open Collector Gates 132
5-7 Three-State Devices 139
5-8 Strobed Gates, Expandable Gates, and Expanders 143
5-9 AND-OR-INVERT Gates 147
5-10 The EXCLUSIVE-OR Gate 151
Summary 152
Glossary 153
References 154
Problems 154

6 FLIP-FLOPS 158

6-1 Instructional Objectives 158
6-2 Self-Evaluation Questions 158
6-3 Introduction 159
6-4 The Basic Flip-Flop 159
6-4 NOR gate Flip-Flops 160
6-6 NAND Gate Flip-Flops 162

6-7   D-type Flip-Flops                                    165
6-8   Bistable Latches                                     167
6-9   J-K Master-Slave Flip-Flops                          167
6-10  Edge-Triggered Flip-Flops                            172
6-11  Timing Charts                                        173
6-12  Direct SETS and Direct CLEARS                        175
6-13  Race Conditions                                      176
6-14  Flip-Flop Parameters                                 178
6-15  Uses of Flip-Flops                                   179
6-16  Synchronizing Flip-Flops                             186
6-17  Glitches                                             187
      Summary                                              189
      References                                           189
      Problems                                             190


7   ONE-SHOTS                                              195

7-1   Instructional Objectives                             195
7-2   Self-Evaluation Questions                            195
7-3   Introduction to One-Shots                            196
7-4   The 74121 One-Shot                                   197
7-5   Retriggerable One-Shots                              201
7-6   Integrated Circuit Oscillators                       207
7-7   Timing Generation Problems                           215
7-8   Switch Bounce                                        217
7-9   Debouncing Class A Switches                          221
7-10  The One-Shot Discriminator                           224
      Summary                                              225
      Glossary                                             225
      References                                           226
      Problems                                             226


8   COUNTERS                                               230

8-1   Instructional Objectives                             230
8-2   Self-Evaluation Questions                            230
8-3   Introduction                                         231
8-4   Divide-by-N Circuits                                 231
8-5   Ripple Counters                                      232
8-6   Synchronous Counters                                 234
8-7   The 3s Counters                                      238
8-8   Irregular and Truncated Count Sequences              239
8-9   IC Counters                                          245
8-10  UP-DOWN Counters                                     252
8-11  Divide-by-N Circuits Using Counters                  257
      Summary                                              262

Glossary 262
References 263
Problems 263

**9 SHIFT REGISTERS** 267

9-1 Instructional Objectives 267
9-2 Self-Evaluation Questions 267
9-3 The Basic Shift Register 268
9-4 LEFT-RIGHT Shift Registers 270
9-5 Serial Inputs and Parallel Loading of Shift Registers 273
9-6 Parallel Load and Parallel Output Shift Registers 275
9-7 Applications 283
9-8 MOS Shift Registers 288
Summary 293
Glossary 293
References 293
Problems 294

**10 CONSTRUCTION AND DEBUGGING OF IC CIRCUITS** 297

10-1 Instructional Objectives 297
10-2 Self-Evaluation Questions 297
10-3 Wire Wrapping 298
10-4 Printed Circuits 301
10-5 Construction of Wire-Wrap Circuits 303
10-6 Error Detection in Combinatorial Circuits 310
10-7 Error Detection in Sequential Circuits 312
10-8 Logic Analyzers 316
Summary 321
Glossary 322
References 322

**11 MULTIPLEXERS AND DEMULTIPLEXERS** 323

11-1 Instructional Objectives 323
11-2 Self-Evaluation Questions 323
11-3 Multiplexers 324
11-4 Demultiplexers 332
11-5 Practical Applications 336
Summary 344
Glossary 344
References 344
Problems 344

**12 BINARY CODED DECIMAL** 346

12-1 Instructional Objectives 346

12-2    Self-Evaluation  Questions                                    346
12-3    Expressing  Numbers  in  Binary  Coded  Decimal              347
12-4    Conversion  Using  Algorithms                                349
12-5    Conversion  Using  ICs                                       355
12-6    Indicating  Lights                                           362
12-7    Multiplexed  Displays                                        369
12-8    Liquid  Crystal  Displays                                    370
        Summary                                                      371
        Glossary                                                     371
        References                                                   371
        Problems                                                     372


13  EXCLUSIVE  OR  CIRCUITS                                          374

13-1    Instructional  Objectives                                    374
13-2    Self-Evaluation  Questions                                   374
13-3    Comparison  Circuits                                         375
13-4    Parity  Checking  and  Generation                            380
13-5    Parity  Checking  and  Generation  Using  the  **74180**     387
13-6    More  Sophisticated  Error-Correcting  Routines              390
13-7    The  Gray  Code                                              393
13-8    Liquid  Crystal  Displays                                    398
        Summary                                                      398
        Glossary                                                     399
        References                                                   399
        Problems                                                     400


14  ARITHMETIC  CIRCUITS                                             402

14-1    Instructional  Objectives                                    403
14-2    Self-Evaluation  Questions                                   403
14-3    The  Basic  Adder                                            403
14-4    Subtraction                                                  406
14-5    2s  Complement  Arithmetic                                   409
14-6    Hexadecimal  Arithmetic                                      414
14-7    The  **7483**  4-Bit  Adder                                  420
14-8    Overflow  and  Underflow  in  2s  Complement  Arithmetic     425
14-9    BCD  Arithmetic                                              427
14-10   Arithmetic/Logic  Units                                      431
14-11   Look-Ahead  Carry                                            435
14-12   Binary  Multiplication                                       438
14-13   Arithmetic  Processing  Units                                444
        Summary                                                      445
        Glossary                                                     445
        References                                                   446
        Problems                                                     446

15  MEMORIES                                                           452

    15-1   Instructional Objectives                                     452
    15-2   Self-Evaluation Questions                                    452
    15-3   Memory Concepts                                              453
    15-4   Core Memories                                                455
    15-5   Introduction to Semiconductor Memories                       459
    15-6   Bipolar RAMs                                                 461
    15-7   MOS Memories                                                 468
    1508   Dynamic RAMs                                                 472
    15-9   Read Only Memories                                           480
    15-10  IC ROMs                                                      483
           Summary                                                      487
           Glossary                                                     487
           References                                                   488
           Problems                                                     488


16  THE BASIC COMPUTER                                                 491

    16-1   Instructional Objectives                                     491
    16-2   Self-Evaluation Questions                                    491
    16-3   Introduction to the Computer                                 492
    16-4   Flowcharts                                                   496
    16-5   Branch Instructions and Loops                                498
    16-6   The Control Unit                                             505
    16-7   The Hardware Design of a Computer                            509
    16-8   The Complete Computer                                        514
    16-9   Building the Computer                                        517
           Summary                                                      529
           Glossary                                                     529
           References                                                   530
           Problems                                                     531


17  COMPUTER INTERFACES                                                532

    17-1   Instructional Objectives                                     532
    17-2   Self-Evaluation Questions                                    532
    17-3   Introduction to Computer Interfaces                          533
    17-4   The I-O Bus                                                  534
    17-5   Direct Memory Accesses                                       546
    17-6   Interrupts                                                   552
    17-7   Communication Between the Analog and Digital World           557
           Summary                                                      565
           Glossary                                                     566
           References                                                   567
           Problems                                                     568

18 MODEMS AND TELETYPES                                          570

   18-1  Instructional Objectives                       570
   18-2  Self-Evaluation Questions                     570
   18-3  Introduction to MODEMs                        571
   18-4  Low Speed MODEMs                              572
   18-5  High Speed MODEMs                             575
   18-6  Teletypes                                     581
   18-7  UARTs                                         586
         Summary                                   592
         Glossary                                  592
         References                                593
         Problems                                  593

19 MICROPROCESSOR INTERFACING                                    595

   19-1  Instructional Objectives                      595
   19-2  Self-Evaluation Questions                     595
   19-3  Introduction to Microprocessor Interfacing    596
   19-4  Input and Output Ports                        601
   19-5  Interrupts on the 8080 and 8085               605
   19-6  Memory Mapped I-O                             612
   19-7  The Motorola PIA                              613
   19-8  CA2 and CB2                                   617
   19-9  Interrupts on the 6800                        620
   19-10 Other Interface ICs                            622
   19-11 The SDK-85 Kit                                 623
   19-12 Other Microprocessor Busses                    625
         Summary                                   626
         Glossary                                  626
         References                                627
         Problems                                  627

20 DISPLAY GENERATORS                                            629

   20-1  Instructional Objectives                      629
   20-2  Self-Evaluation Questions                     630
   20-3  Introduction to the Display Generator         630
   20-4  The Raster-Scan Display Generator             632
   20-5  Components of the Display Generator           637
   20-6  Timing the Display Generator                  643
   20-7  The Cursor                                    649
   20-8  CRT Controllers                               652
         Summary                                   653
         Glossary                                  654
         References                                655
         Problems                                  655

APPENDIX A TABLE OF POWERS OF 2                                      659
APPENDIX B RESISTOR CALCULATIONS FOR OPEN
                  COLLECTOR ICs                                      660
APPENDIX C REDUCTION OF EQUATION 12-2 TO
                  EXCLUSIVE ORs                                      661
APPENDIX D PROOF OF EXCLUSIVE OR-PARITY RELATIONSHIP                 662
APPENDIX E THE ASCII CODE                                           663
APPENDIX F ANSWERS TO SELECTED PROBLEMS                              664
INDEX                                                                709
SUPPLEMENTARY INDEX                                                  715

# PRACTICAL DIGITAL DESIGN USING ICs

# CHAPTER 1

# BINARY NUMBERS

## 1-1 INTRODUCTION TO DIGITAL ELECTRONICS

This book introduces the reader to digital electronics and enables him or her to understand, design, and construct digital circuits. The book is primarily concerned with the functions and uses of digital integrated circuits (ICs) and emphasizes the 7400 TTL (transistor transistor logic) series. These ICs have experienced a phenomenal growth in the past several years and millions of them have been manufactured. They can be purchased from manufacturers, electronics distributors, electronics stores such as Radio Shack, many computer stores, and by mail or phone from discount houses.[1] Their low price and wide availability have caused them to be designed into almost all digital circuits produced in the last 10 years.

Microprocessors ($\mu$Ps) have also enjoyed phenomenal growth recently. Some engineers believed that $\mu$Ps would replace ICs, but this has not happened. Rather, $\mu$Ps work in conjunction with digital ICs to provide a high level of intelligence and sophistication to digital hardware. The Apple computer, for example, uses a 6502 $\mu$P as its main computing element. It also uses about 50 TTL ICs (mostly types described in this book), plus ROMs and RAMs. Thus even with a $\mu$P in the system, many small and medium scale ICs are required to support it.

[1]These discount houses often advertise in magazines such as *Popular Electronics* or *Byte*.

Both standard and special purpose ICs are also needed to enable a $\mu$P to interface or communicate with other devices such as disks and printers. Some of these special purpose ICs and interaction between $\mu$Ps and ICs is covered in the later chapters of this book.

Both ICs and $\mu$Ps are the results of advances in *digital electronics*. Modern digital electronics is the product of a marriage between switching theory (a discipline in its infancy in the 1950s) and integrated circuits (a family of devices that was not developed until the 1960s). However, digital circuits have proven to be the best way to construct many useful devices, including the digital computer; consequently, its growth has been rapid, and a significant percentage of the jobs currently available in electrical engineering require digital training.

The output of most electronic circuits is an *analog quantity*, typically a voltage. An analog quantity is a quantity that may assume any numerical value within the range of possible outputs. Therefore, an electronic circuit is capable of producing many outputs in response to different inputs. For example, 5.12 volts (V) might be one output, 3.76 V another; any voltage within the precision of the voltmeter is a legitimate output. *Digital circuits* depart from analog circuits by providing only two values as an output: an output can either be a one (1) or a zero (0), *and nothing else*. Of course, digital engineers are still dealing with electronic circuits whose outputs are voltages. What they have done is to *define* a certain range of voltages as a logic 1 and another range of voltages as a logic 0. Typically, the 1 and 0 ranges are separated by a *forbidden* range of voltages. TTL integrated circuits define a 1 output as any voltage between 2 and 5 V and a 0 as any voltage between ground (zero) and 0.8 V. But what if the actual output of the circuit is in a forbidden or undefined range, 1 V, for instance? "Then," says the digital engineer succinctly, "the circuit is malfunctioning. Fix it." Methods of diagnosing and repairing malfunctions in digital circuits are discussed in Chapter 10.

Advances in digital technology have been more spectacular than advances in switching theory. The first computers were built in the early 1950s using large vacuum tubes that had low reliability and consumed a great deal of power. Tubes were replaced by transistors as soon as the latter became available and reliable. Engineers quickly found ways to build small, efficient, and inexpensive digital circuits using a single transistor, a couple of diodes, and a few resistors. Now, however, these discrete components (the resistors, transistors, and diodes) have all disappeared inside the IC. Recently ICs have become slightly larger and far more complex. The microprocessor itself is an example of a very large scale (VLSI) integrated circuit containing several thousand gates.

The first three chapters of this book develop the required switching theory. The remaining chapters describe the behavior of ICs, predominantly the 7400 TTL series, and ways in which to design useful circuits with them. These circuits can then be used to build computers, or *controllers* and *interfaces* that allow other devices to communicate with a computer. Examples of computer interfaces and controller design are presented in the later chapters of this book.

## 1-2  INSTRUCTIONAL OBJECTIVES

This first chapter acquaints the student with the binary number system and gives some facility in handling binary numbers. After reading this chapter, the student should be able to:

1.  Convert binary numbers to decimal numbers.
2.  Convert decimal numbers to binary numbers.
3.  Find the sum and difference of two binary numbers.

## 1-3  SELF-EVALUATION QUESTIONS

As the student reads the chapter, he or she should be able to answer the following questions:

1.  What is the difference between digital and analog circuits?
2.  How are the outputs of a circuit defined to make the output digital?
3.  What are the advantages of digital circuits?
4.  What is the difference between a *bit* and a *decimal digit?* In what respects are they similar?
5.  In a *flow chart,* what is the function of the rectangular and diamond-shaped boxes?
6.  How are binary addition and subtraction different from decimal addition and subtraction? How are they similar?

## 1-4  USES OF BINARY BITS

The output of a digital circuit is a single binary digit (a 1 or a 0) commonly called a *bit.* There are two advantages gained by restricting the output of an electronic circuit to one of two possible values. First, it is rarely necessary to make fine distinctions. Whether an output is 3.67 or 3.68 V no longer matters; in both cases it is a logic 1. Since well-designed logic circuits produce voltages near the middle of the range defined for 1 or 0, there is no difficulty in distinguishing between them. In addition, a digital circuit is very tolerant of any drift in the output caused by component aging or changes. A change in a component would almost have to be catastrophic to cause the output voltage to drift from a 1 to a 0 or an undefined value. The second advantage is that it is far easier for electronic circuits to remember a 1 or a 0 than to remember an analog quantity like 3.67 V. Since all but the simplest digital circuits require the ability to remember the value of a voltage *after the conditions that caused that voltage have disappeared*, this is a very important consideration.

The output of a single digital circuit, a single bit, is enough to answer any question that has *only two* possible answers. For example, a typical job application might ask "What is your sex?" A 1 could arbitrarily be assigned to a male and a 0 to a female, so a single bit is enough to describe the answer to this question. A single bit is all the space a programmer needs to reserve in his computer for this answer.

However, another question on the job application might be "What is the color of your hair?"If the possible answers are black, brown, blonde, and red, a single bit cannot possibly describe them all. Now *several* bits are needed to describe all possible answers. We could assign one bit to each answer (i.e., brown = 0001, black = 0010, blonde = 0100, red = 1000), but if there are many possible answers to the given question many bits are required. The coding scheme presented above is not optimum; it requires more bits than are really necessary to answer the question.

It is most economical to use *as few bits as possible* to express the answer to a question, or a number, or a choice. The crucial question is:

"*What is the minimum number of bits required to distinguish between n different things?*"

Whether these $n$ things are objects, possible answers, or $n$ numbers is immaterial. To answer this question we realize that each bit has two possible values. Therefore, $k$ bits would have $2^k$ possible values. This fact is the basis for Theorem 1.

## Theorem 1

*The minimum number of bits required to express **n** different things is **k**, where **k** is the smallest number such that $2^k \geq n$.*

A few examples should make this clear.

---

### EXAMPLE 1-1

What is the minimum number of bits required to answer the hair color question, and how could they be coded to give distinct answers?

## SOLUTION

There are four possible answers to this question; therefore, $2^k = 4$. Since 2 is the smallest number such that $2^2 \geq 4$, $k = 2$ and 2 bits are needed. One way of coding the answers is $00 =$ brown, $01 =$ black, $10 =$ blonde, and $11 =$ red.

### EXAMPLE 1-2

How many bits are needed to express a single decimal digit?

## SOLUTION

There are 10 possible values for a single decimal digit (0 through 9); therefore, $2^k \geq 10$. Since $k = 4$ is the smallest integer such that $2^k \geq 10$, 4 bits are required.

### EXAMPLE 1-3

A computer must store the names of a group of people. If we assume that no name is longer than 20 letters, how many bits must the computer reserve for each name?

## SOLUTION

To express a name, we need only the 26 letters of the alphabet, plus a space and perhaps a period, or a total of 28 characters. Here $2^k \geq 28$ so that $k = 5$. Therefore, 5 bits are required for *each* character and, since space must be reserved for 20 such characters, 100 bits are needed for each name.

## 1-5  BINARY TO DECIMAL CONVERSION

In the early chapters of this book only two number systems, *binary* (base 2) and *decimal* (base 10) are considered. The *hexidecimal* (base 16) number system, used in arithmetic circuits and computers, is introduced in Chapter 14. To eliminate any possible confusion, a subscript is used to indicate which number system is employed. Thus, $101_{10}$ is the decimal number whose value is one hundred and one, while $101_2$ is a *binary* number whose decimal value is five. Of course, *any* number containing a digit from 2 to 9 is a decimal number.

The value of a decimal number depends on the magnitude of the decimal digits expressing it and on their *position*. A decimal number is equal to the sum $D_0 \times 10^0 + D_1 \times 10^1 + D_2 \times 10^2 + \cdots$, where $D_0$ is the least significant digit, $D_1$ the next significant, and so on.

---

**EXAMPLE** 1-4

Express the decimal number 7903 as a sum to the base 10.

**SOLUTION**

Here $D_0$, the least significant digit is 3, $D_1 = 0$, $D_2 = 9$, and $D_3 = 7$. Therefore, 7903 equals:

$$
\begin{array}{ll}
\phantom{+}3 \times 10^0 & 3 \\
+\phantom{} 0 \times 10^1 & 0 \\
+\phantom{} 9 \times 10^2 & 900 \\
+\phantom{} 7 \times 10^3 & \underline{7000} \\
& \overline{7903}
\end{array}
$$

---

Similarly a group of *binary* bits can represent a number in the *binary* system. The binary base is 2; therefore the digits can only be 0 or 1. However, a binary number is also equal to a sum, namely $B_0 \times 2^0 + B_1 \times 2^1 \cdots$, where $B_0$ is the least significant bit, $B_1$ the next significant bit, and so on. The powers of 2 are given in the "Binary Boat" or table of Appendix A. In this table, $n$ is the exponent and the corresponding positive and negative powers of 2 are listed to the left and right of $n$, respectively.

A binary number is a group of ones (1s) and zeros (0s). To find the equivalent decimal number, we simply add the powers of 2 that correspond to the 1s in the number and omit the powers of 2 that correspond to the 0s of the number.

---

**EXAMPLE** 1-5

Convert $100011011_2$ to a decimal number.

**SOLUTION**

The first bit to the left of the binary point corresponds to $n = 0$, and $n$ increases by one (increments) for each position further to the left. The number 100011011 has 1s in positions 0, 1, 3, 4, and 8, so the conversion is made by obtaining those powers of 2 corresponding to these $n$ values (using Appendix A, if necessary) and adding them:

| $n$ | $2^n$ |
|---|---|
| 0 | 1 |
| 1 | 2 |
| 3 | 8 |
| 4 | 16 |
| 8 | 256 |
|   | 283 |

Therefore, $100011011_2 = 283_{10}$.

## EXAMPLE 1-6

A *word* is a basic unit of computer information and consists of several bits. In the PDP-8 computer each word consists of 12 bits, that is, $k = 12$. How many numbers can be represented by a single PDP-8 word and what are they?

## SOLUTION

Since 12 bits are available, any one of 4096 ($2^{12}$) numbers can be expressed. These numbers range from a minimum of twelve 0s to a maximum of twelve 1s, which is the binary equivalent of 4095. Therefore, the 4096 different numbers that can be expressed by a single word are the decimal numbers 0 through 4095.

## 1-5.1   Conversion of Binary Fractions to Decimals

Decimal fractions can be expressed as a sum of digits times 10 to *negative* powers. For example, 0.3504 equals:

$$
\begin{aligned}
3 \times 10^{-1} &= 0.3 \\
+ 5 \times 10^{-2} &= 0.05 \\
+ 0 \times 10^{-3} &= 0 \\
+ 4 \times 10^{-4} &= 0.0004 \\
\hline
&\phantom{=} 0.3504
\end{aligned}
$$

Similarly, binary fractions can be expressed as sums of *negative* powers of two. The table of Appendix A can again be used if $n = 1$ is taken as the first position to the *right* of the decimal[2] point and $n$ increases as the position moves to the right. Here the *negative* powers of 2 are added up.

[2]Strictly speaking a "decimal point" in a binary number should be called a binary point.

---

**EXAMPLE 1-7**

Convert the binary fraction 0.11010001 to a decimal fraction.

**SOLUTION**

In this example the 1s appear in the 1, 2, 4, and 8 positions (reading toward the right). From Appendix A we find:

| $n$ | $2^{-n}$ |
| --- | --- |
| 1 | 0.5 |
| 2 | 0.25 |
| 4 | 0.0625 |
| 8 | 0.00390625 |
| | 0.81640625 |

Therefore, $0.11010001_2 = 0.81640625_{10}$.

---

## 1-6 DECIMAL TO BINARY CONVERSION

It is often necessary to convert decimal numbers to binary. Humans, for example, supply and receive decimal numbers from computers that work in binary; consequently, computers are continually making *binary-to-decimal* and *decimal-to-binary* conversions.

To convert a decimal number to its equivalent binary number, the following algorithm or procedure may be used, where **K** is the position of the bit:

1. Obtain **N.** (The decimal number to be converted.)
2. Determine if **N** is odd or even.
3. a. If **N** is odd, write 1 and subtract 1 from **N.** Go to step 4.
   b. If **N** is even, write 0.
4. Obtain a new value of **N** by dividing the **N** of step 3 by 2.
5. a. If **N** > 1, go back to step 2 and repeat the procedure.
   b. If **N** = 1, write 1. The number written is the binary equivalent of the original decimal number. The number written first is the least significant bit, and the number written last is the most significant bit.

This procedure can also be implemented by following the *flow chart* of Fig. 1-1. Computer programmers often use flow charts to describe their programs graphically. For the rudimentary flow charts drawn in this book, the *square* box is a *command*, which must be obeyed *unconditionally*. The *diamond-shaped* box is a *decision* box. If the answer to the question within the decision box is YES, the YES path must be followed; otherwise the NO path is followed. The flow chart of Fig. 1-1 starts with the given number N; since $K$ equals 0, initially we are writing $B_0$, the least significant digit. Note that equations in a flow chart are programmer's equations, not algebraic equations. The "equation" $N = N - 1$ makes no sense mathematically. What it means here is that $N$ is *replaced* by $N - 1$.

On the initial pass through the flow chart, $B_0$, the *least* significant bit, is written as 0 or 1, depending on whether $N$ is even or odd. Next $N$ is divided by 2 and $K$ is incremented so that on the following pass $B_1$, the second least significant digit, will be written. We continue looping through the flow chart and repeating the procedure until $N = 1$. Then the most significant bit is written as a 1, and the process stops. The bits written are the binary equivalent of the decimal number.

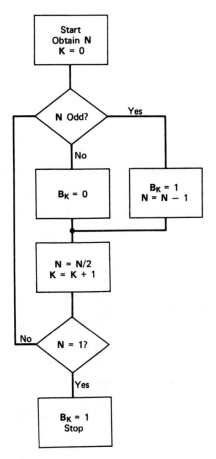

FIGURE 1-1 Flow chart for decimal-to-binary conversion of whole numbers.

## EXAMPLE 1-8

Find the binary equivalent of the decimal number 217.

## SOLUTION

The solution proceeds according to the algorithm or flow chart. When an odd number is encountered, a 1 is written as the binary digit and subtracted from the remaining number; when the remaining number is even, 0 is written as the binary digit. The number is then divided by 2. The process continues until the number is reduced to 1.

| Remaining Number | | Binary Digit or Bit |
|---|---|---|
| 217 | Odd—subtract 1 | 1 |
| 216 | Divide by 2 | |
| 108 | Even—divide by 2 | 0 |
| 54 | Even—divide by 2 | 0 |
| 27 | Odd—subtract 1 | 1 |
| 26 | Divide by 2 | |
| 13 | Odd—subtract 1 | 1 |
| 12 | Divide by 2 | |
| 6 | Even—divide by 2 | 0 |
| 3 | Odd—subtract 1 | 1 |
| 2 | Divide by 2 | |
| 1 | Finish | 1 |

Note that the least significant bit was written first. Therefore, $217_{10} = 11011001_2$.
This result can be checked by converting back from the binary to the decimal number. $11011001 = 128 + 64 + 16 + 8 + 1 = 217_{10}$.

## 1-6.1 Converting Decimal Fractions to Binary Fractions

Decimal fractions must also be converted to binary fractions in certain applications. Decimal fractions may not have an *exact* binary equivalent, and the decimal-to-binary conversion often produces a *repetitive sequence* of binary bits.[3] The decimal-to-binary conversion procedure starts with the *most significant* binary bit, the bit immediately to the right of the decimal point, and then proceeds to the right, one bit at a time. Each bit has only *half* the value of the preceding bit, and the engineer stops the conversion when he has a sufficiently accurate binary representation of the decimal fraction.

[3]This also happens with decimal fractions such as $\frac{1}{3}$ or $\frac{1}{9}$, etc.

The following procedure can be used to convert a decimal fraction to a binary fraction:

1. Obtain **N.**
2. Double **N.**
3.  a.  If the new value of **N** is greater than 1, write 1 as the next most significant bit, subtract 1 from **N,** and go back to step 2.
    b.  If the new value of **N** is less than 1, write 0 as the next most significant bit and go back to step 2.

An equivalent procedure is given by the flow chart of Fig. 1-2. Notice that we start with the most significant fractional bit, the bit immediately to the right of the decimal point, and proceed one bit to the right each time we loop through the flow chart.

---

## EXAMPLE 1-9

Convert $0.78125_{10}$ to binary.

## SOLUTION

Follow the procedure or the flow chart:

| Number | | Binary Bit |
|---|---|---|
| 0.78125 | | |
| | Double **N** | |
| 1.5625 | | |
| | Subtract 1 | 1 |
| 0.5625 | | |
| | Double **N** | |
| 1.125 | | |
| | Subtract 1 | 1 |
| 0.125 | | |
| | Double **N** | |
| 0.25 | | |
| | Double **N** | 0 |
| 0.5 | | |
| | Double **N** | 0 |
| 1.0 | | |
| | Subtract 1 | 1 |
| 0.0 | | |

Each binary bit is written immediately after **N** is doubled. If **N** is greater than or equal to 1, the bit is a 1 and 1 is subtracted from **N,** but if **N** is less than 1, the bit written is a 0 and no subtraction is performed.

Here we find that $0.78125_{10} = 0.11001_2$. This is one of the happy examples where the decimal fraction has an exact binary equivalent, as indicated by the fact that **N** is exactly 0 after the last subtraction.

**EXAMPLE 1-10**

Convert 0.85 to binary.

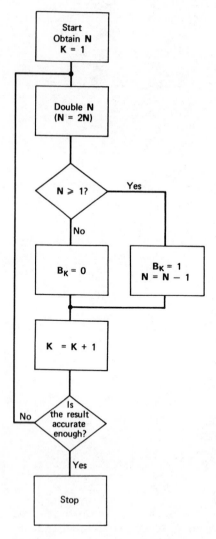

FIGURE 1-2  Flow chart for decimal-to-binary conversion of fractions.

**SOLUTION**

Again the flow chart or algorithm procedure is followed.

| Number | | Binary Bit |
|---|---|---|
| 0.85 | | |
| | Double **N** | |
| 1.7 | | 1 |
| | Subtract 1 | |
| 0.7 | | |
| | Double **N** | |
| 1.4 | | 1 |
| | Subtract 1 | |
| 0.4 | | |
| | Double **N** | |
| 0.8 | | 0 |
| | Double **N** | |
| 1.6 | | 1 |
| | Subtract 1 | |
| 0.6 | | Repetitive |
| | Double **N** | Sequence |
| 1.2 | | 1 |
| | Subtract 1 | |
| 0.2 | | |
| | Double **N** | |
| 0.4 | | 0 |

The results indicate that to 5 binary places $0.85 = 0.11011$. The repetitive sequence is 0110 because $N = 0.4$ at the beginning and at the end of this sequence. Therefore we can immediately write the equivalent of 0.85 to 9 or 13 bits by just appending repetitive sequences; that is,

$0.85 = 0.11011001100110011(0)$ etc.

Repetitive sequences

This result was checked with a hand calculator using the Binary Boat (Appendix A).

$0.11011 = 0.84375$
$0.110110011 = 0.84960937$
$0.1101100110011 = 0.8499755859$

Notice that each result is less than 0.85 because the bits of lesser significance were not included; however, as more bits are used the result approaches 0.85.

## 1-6.2  Mixed Numbers

Mixed numbers can be converted from binary to decimal or decimal to binary by working on the integer and fraction portions separately.

---

**EXAMPLE 1-11**

Convert $10011.101_2$ to a decimal number.

**SOLUTION**

The integer part, 10011, converts to 19, and the decimal portion converts to 0.625; therefore, $10011.101_2 = \mathbf{19.625}$.

**EXAMPLE 1-12**

Convert the decimal number 33.3 to binary.

**SOLUTION**

33 converts to 100001 and 0.3 becomes 0.0100110011, where the 0011 pattern is repetitive. Therefore, $33.3_{10} = \mathbf{100001.0100110011_2}$.

---

## 1-7 ADDITION AND SUBTRACTION OF BINARY NUMBERS

The binary number system is a valid mathematical system, and operations such as addition, subtraction, multiplication, and division can be performed on binary numbers. In this section, the most commonly performed arithmetic operations, addition and subtraction, will be discussed. The reader should consult more specialized texts (see the References at the end of the chapter) for multiplication, division, squares, square roots, and other arithmetic operations.

### 1-7.1 Addition of Binary Numbers

The addition of binary numbers is similar to the addition of decimal numbers except that $1 + 1 = 0$ with a carry out to the *next significant* place. A carry into a *more significant* position acts like an additional 1.

---

**EXAMPLE 1-13**

Add the binary numbers $A = 11101100$ and $B = 1100110$.

**SOLUTION**

| Column | 9 8 7 6 5 4 3 2 1 | (Decimal addition) |
|---|---|---|
| A | 1 1 1 0 1 1 0 0 | (236) |
| B | 1 1 0 0 1 1 0 | (102) |
| | 1 0 1 0 1 0 0 1 0 | (338) |

The above addition proceeded as follows:

| | | |
|---|---|---|
| 1. | Column 1 (least significant digit) | $0 + 0 = 0$. |
| 2. | Column 2 | $0 + 1 = 1$. |
| 3. | Column 3 | $1 + 1 = 0$ plus a carry output. |
| 4. | Column 4 | $0 + 1$ plus a carry input from Column 3 sums to a 0 and produces a carry out to Column 5. |
| 5. | Column 5 | $0 + 0 = 0$, but the carry input from Column 4 makes the sum 1. |
| 6. | Column 6 | $1 + 1 = 0$ and a carry output to Column 7. |
| 7. | Column 7 | $1 + 1$ plus a carry input results in a sum of 1 and a carry output. |
| 8. | Column 8 | $B$ does not have an eighth bit; therefore, a leading 0 can be assumed. Here $0 + 1$ plus a carry input yields a 0 sum plus a carry output. |
| 9. | Column 9 | Since neither $A$ nor $B$ has a ninth digit, leading 0s are written for both. In Column 9 we have $0 + 0$ plus a carry in from Column 8, which gives a sum of 1. Since there is no carry out of Column 9 the addition is complete. |

The sum of Ex. 1-13 can be checked by converting the numbers to their decimal equivalents. These numbers are shown in parentheses beside the sum.

## 1-7.2 Subtraction of Binary Numbers

The rules for the subtraction of binary numbers are:

1. $1 - 1 = 0$
2. $0 - 0 = 0$
3. $1 - 0 = 1$
4. $0 - 1 = 1$ with a borrow out

In order to borrow, change the next 1 in the minuend to a 0 and change all intervening 0s to 1s.

---

**EXAMPLE 1-14**

Subtract 101101001 from 100011010011.

## SOLUTION

Column      12 11 10 9 8 7 6 5 4 3 2 1    (Decimal subtraction)

```
        12 11 10 9 8 7 6 5 4 3 2 1   (Decimal subtraction)
         1  0  0 0 1 1 0 1 0 0 1 1   (2259)
                    1 0 1 1 0 1 0 0 1  −(361)
         ─────────────────────────
         1  1  1 0 1 1 0 1 0 1 0      (1898)
```

1. Column 1    $1 - 1 = 0$.
2. Column 2    $1 - 0 = 1$.
3. Column 3    $0 - 0 = 0$.
4. Column 4    $0 - 1 = 1$. The 1 in Column 5 is changed to a 0 because of the borrow out generated in Column 4.
5. Column 5    This is now $0 - 0 = 0$.
6. Column 6    $0 - 1 = 1$. The 1 in Column 7 is changed to a 0.
7. Column 7    Because of the borrow from Column 6 this now becomes $0 - 1$ or 1 with a borrow out that changes the 1 in Column 8.
8. Column 8    This becomes $0 - 0 = 0$.
9. Column 9    $0 - 1 = 1$. Since Columns 10 and 11 are 0, the borrow must be from Column 12. Columns 10 and 11 contain intervening 0s so they change to 1s and Column 12 changes to a 0.
10. Column 10    This is now $1 - 0 = 1$.
11. Column 11    This is now $1 - 0 = 1$.
12. Column 12    This is now $0 - 0 = 0$.

The results were checked by converting the binary numbers to their decimal equivalents, which are shown in parentheses beside the numbers.

---

## GLOSSARY

**Analog quantity.** A continuously variable quantity; one that may assume any value usually within a limited range.

**Binary system.** A number system with 2 as the base.

**Bit.** A single digital quantity: a 1 or a 0.

**Digital quantity.** A variable that has one of two possible values.

**Integrated circuit (IC).** A small electronic package usually containing several circuits.

**Microprocessor ($\mu$P).** A small computer in an IC package.

**Word.** A group of bits that constitutes the basic unit of information within a computer.

## REFERENCES

**Louis Nashelsky,** *Introduction to Digital Computer Technology,* Wiley, New York, 1972.

**Arpad Barna** and **Dan I. Porat,** *Integrated Circuits in Digital Electronics,* Wiley, New York, 1973.

**George K. Kostopoulos.** *Digital Engineering,* Wiley, New York, 1975.

**Frederick J. Hill** and **Gerald R. Peterson,** *Introduction to Switching Theory and Logical Design,* Second Edition, Wiley, New York, 1974.

**Ronald J. Tocci,** *Digital Systems,* Prentice Hall, Englewood Cliffs, N.J., 1980.

## PROBLEMS

1-1.    How many bits are required to distinguish between 100 different things?

1-2.    A major league baseball team plays 162 games a year. Before the season starts, how many bits must be reserved to express the number of games the team will win and how many bits to express the number of games the team will lose?

1-3.    A line printer is capable of printing 132 characters on a single line, and each character is 1 of 64 symbols (26 alphabetics plus 10 numbers plus punctuation). How many bits are needed to print an entire line?

1-4.    Express the following decimal numbers as a sum.
  (a) 4507
  (b) 137,659
  (c) 889.7061

1-5.    Convert the following binary numbers to decimal.
  (a) 10111
  (b) 110101
  (c) 110001011
  (d) 11011.10111
  (e) .00010011
  (f) 110001111.011101

1-6.    Convert the following decimal numbers to binary.
  (a) 66
  (b) 252
  (c) 5795
  (d) 106,503
  (e) 0.3
  (f) 0.635
  (g) 0.82
  (h) 67.65
  (i) 3,477.842

1-7. For each of the following pairs of numbers, find A + B and A − B, completing the third and fourth columns below:

| | A | B | A + B | A − B |
|---|---|---|---|---|
| (a) | 11011 | 1001 | | |
| (b) | 111001 | 101010 | | |
| (c) | 111000111 | 100101 | | |
| (d) | 101111011 | 1000101 | | |

1-8. Find A + B and A − B by converting each number to binary and doing the additions and subtractions in binary. Check the results by converting back to decimal.

| | A | B | A + B | A − B |
|---|---|---|---|---|
| (a) | 67 | 39 | | |
| (b) | 145 | 78 | | |
| (c) | 31,564 | 26,797 | | |

1-9. A PDP 11 is a minicomputer with a 16-bit word length. What range of numbers can a single word contain?

1-10. How high can you count using only your fingers?

Answers to selected problems are given in Appendix F.

# 2

# BOOLEAN ALGEBRA

## 2-1 INSTRUCTIONAL OBJECTIVES

This chapter presents the rules, theorems, and operations of Boolean algebra so that they may be applied to digital design. The reader will learn to:

1. Construct truth tables.
2. Simplify Boolean expressions.
3. Analyze electronic circuits using Boolean algebra.
4. Construct electronic circuits that physically implement Boolean expressions.
5. Take the complement of a given Boolean expression.
6. Analyze and construct relay circuits using Boolean algebra.

## 2-2 SELF-EVALUATION QUESTIONS

Watch for the answers to these questions as you read the chapter. They should help you to understand the material. When you have completed the chapter, return to this section and be sure you can answer the questions below:

1. Given N variables, how many entries must a truth table contain?
2. How are the outputs of a truth table determined?
3. How is addition similar to OR operation and multiplication to AND operation?

4.   How are the theorems of Boolean algebra used?

5.   What arithmetic operations are not allowed in Boolean algebra?

6.   What is the relationship between a Boolean expression and its complement?

7.   What is the significance of a bubble in a logic circuit?

8.   How do you obtain the alternate representation for a gate?

9.   Why should logical expressions be simplified before attempting to build them?

10.   What is the advantage of selecting gates so that bubbles are connected to bubbles?

11.   Are normally open (n.o.) and normally closed (n.c.) relay contacts complements of each other? Explain.

## 2-3   INTRODUCTION TO BOOLEAN ALGEBRA

Boolean algebra has proven to be a mathematical discipline directly applicable to switching theory and digital design. The basic Boolean equation is simply:

$$G = f(w, x, y, z, \cdots) \tag{2-1}$$

Equation (2-1) states that the output, $G$, is a *function* of the input variables $w$, $x$, $y$, $z$, and so on. Stated another way, the value of $G$ is *determined by the values of the input variables*. The unique feature of Boolean algebra is that all the variables in a Boolean equation, *outputs as well as inputs*, may only assume the values 0 or 1—that is, true or false. Thus, there is a direct correspondence between digital circuits, also restricted to 0s and 1s on both their inputs and outputs, and Boolean equations.

### 2-3.1   Word Problems

When a supervisor first presents a design problem to an engineer, he does so by giving the engineer a verbal description of what he *feels* he wants designed. All too often, the descriptions are vague in many respects and fail to define the output desired in response to many significant input combinations. Boolean algebra can help clarify problems stated in words as well as electronic problems. To apply Boolean algebra to a word problem, a two-step procedure must be followed:

1.   The number of variables implied by the definition of the problem must be determined.

2.   The significance of each of the two values of each variable must be defined.

An example should make this clear.

## EXAMPLE 2-1

Given the following statement, identify the variables and assign a value to each:

The president of a company and 3 of his assistants are voting on whether to accept a contract. If the president votes for it, then 2 yes votes (including the president's) are enough; but if the president votes no, all 3 of his assistants must vote yes in order for the contract to be accepted.

## SOLUTION

After some thought we realize that the result or output of this statement is the decision that determines the acceptance or rejection of the contract. Let us call the output C, and assign C a value of 1 if the contract is accepted, and a value of 0 if it is rejected.

There are 4 people whose votes determine the result. Let us call the president $P$, and his 3 assistants $A_1$, $A_2$, and $A_3$. We can assign a 1 to each individual if he votes to accept the contract and a 0 if he votes to reject it. This statement may be described by the equation:

$$C = f(P, A_1, A_2, A_3) \qquad (2\text{-}2)$$

This is a Boolean equation; all input variables and the output, C, have a value of either 0 or 1. Only the function itself remains to be defined.

## 2-3.2  Truth Tables

Truth tables are often constructed to relate the output of a circuit to its inputs. *A truth table lists every possible combination of inputs and their corresponding outputs.* Therefore, if a function has N inputs, there are $2^N$ possible combinations of these inputs and there will be $2^N$ entries in the truth table. If the $2^N$ entries are listed in a haphazard or arbitrary manner, one might inadvertently list some entries twice and omit others. To avoid this the following entry-listing procedure is recommended:

1.  Select one variable as the least significant variable. List this in a column of alternate 1s and 0s.
2.  Select a second variable as the next least significant variable. List this in a second column as pairs of alternating 1s and 0s (00110011 $\cdots$).
3.  List the third variable in a column of alternating groups of 4 (0000 1111 0000 $\cdots$).
4.  List additional variables in alternating groups of 8, 16, 32, and so on, until all possible input combinations have been used.
5.  List the ouput corresponding to each input combination. The output must be determined by examining the statement of the problem and the value of each input variable. A clear problem statement allows the designer to determine the proper output for each combination of input variables.

## EXAMPLE 2-2

Construct the truth table for Example 2-1.

## SOLUTION

There are four input variables in this example so we must have $2^4 = 16$ lines in the truth table. Arbitrarily, $A_3$ is chosen as the least significant variable, $A_2$ next, then $A_1$, and the president, $P$, as the most significant variable. The truth table is shown in Fig. 2-1. First the columns for the 4 input variables are listed in accordance with the above procedure, giving 16 lines in the table, each with a different set of inputs. The output, $C$, for each line is determined by examining the inputs and the statement of the problem. For example, the first line of the truth table reads 0000, which means that everyone votes against the contract. Of course it cannot be accepted and $C = 0$.

The sixth line of the truth table is 0101, which indicates that the president and his second assistant are against the contract but the first and third assistants are for it. An examination of the word statement indicates the contract is not accepted and again $C = 0$.

The tenth line, 1001, indicates the president and the third assistant are for the contract, which is enough to have the contract accepted, and $C = 1$.

| $P$ | $A_1$ | $A_2$ | $A_3$ | $C$ |
|---|---|---|---|---|
| 0 | 0 | 0 | 0 | 0 |
| 0 | 0 | 0 | 1 | 0 |
| 0 | 0 | 1 | 0 | 0 |
| 0 | 0 | 1 | 1 | 0 |
| 0 | 1 | 0 | 0 | 0 |
| 0 | 1 | 0 | 1 | 0 |
| 0 | 1 | 1 | 0 | 0 |
| 0 | 1 | 1 | 1 | 1 |
| 1 | 0 | 0 | 0 | 0 |
| 1 | 0 | 0 | 1 | 1 |
| 1 | 0 | 1 | 0 | 1 |
| 1 | 0 | 1 | 1 | 1 |
| 1 | 1 | 0 | 0 | 1 |
| 1 | 1 | 0 | 1 | 1 |
| 1 | 1 | 1 | 0 | 1 |
| 1 | 1 | 1 | 1 | 1 |

FIGURE 2-1 Truth table for Ex. 2-2.

From the truth table we can see whether the contract is accepted for all possible combinations of votes. Thus the truth table implicitly specifies the function, $f$, in Eq. (2-2).

## 2-4 OPERATIONS WITH BOOLEAN VARIABLES

Truth tables are a very precise and comprehensive way of describing a function. Unfortunately, they may quickly become large and unwieldy. For example, 8 input variables need a 256 line truth table, which we might have to write on a scroll. In Boolean algebra, certain operations and manipulations are permissible and they allow us to simplify expressions without the use of truth tables.

### 2-4.1 Complementation

To *complement* a variable is to *reverse* its value. A complemented variable is represented by placing a bar over the variable. Thus, if $x = 1$, $\bar{x} = 0$; conversely, if $x = 0$, $\bar{x} = 1$.[1] An electronic gate whose function is to change logic 1s to 0s (and vice versa) is called an *inverter*. (See Sec. 2-7.3)

### 2-4.2 Addition—The OR Operation

Boolean addition is equivalent to a logical OR. The *plus* symbol $(+)$ is the symbol used to indicate addition or ORing. As in ordinary arithmetic, $0 + 0 = 0$ and $0 + 1 = 1$. The difficult question is "How much is $1 + 1$ in a system where the number 2 is not allowed?" By definition, $1 + 1 = 1$. Thus, the expression $s = x_1 + x_2$ means that $s = 0$ only if *both* $x_1$ and $x_2$ are 0, but $s = 1$ if *either* $x_1$ or $x_2$ **or** both equal 1. In general, the equation

$$s = x_1 + x_2 + \cdots + x_n$$

means $s = 1$ *if any x is 1 and* $s = 0$ *only if all the x's are 0.*

---

**EXAMPLE 2-3**

You want to go to the movies but refuse to go alone. So you send a note to each of your four girl friends, Alice, Betty, Cindy, and Doris, asking them to join you. You will go to the movies if *one or more* of them say yes. Express this situation as a Boolean equation.

---

[1]Some authors use $x'$ instead of $\bar{x}$ to denote the complement of $x$.

**SOLUTION**

Let us choose the variables $A$, $B$, $C$, and $D$ to represent the girls, and assign a value of 1 to each variable if that girl says yes. $M$ is chosen as the variable "movies," and $M = 1$ means you will go. Equation (2-3) represents the situation.

$$M = A + B + C + D \qquad (2-3)$$

Arithmetically, if any one or more of the girls say yes, $M$ is a 1 and you go to the movies. Logically we can say you will go to the movies if Alice OR Betty OR Cindy OR Doris OR any combination of the girls say yes. This demonstrates the equivalence of Boolean algebra and the logical OR.

## 2-4.3  Multiplication—The AND Operation

Boolean multiplication is equivalent to a logical AND operation. The rules for multiplication are the same as in ordinary arithmetic. If $s = x \cdot y$,[2] $s = 1$ only if *both x and y are* 1, and $s = 0$ *if either or both x and y are* 0. In general the equation

$$s = x_1 \, x_2 \cdots x_n$$

means that $s = 1$ *only if all the x values are* 1 or that $s = 0$ *if **any** x has a value of* 0.

---

**EXAMPLE 2-4**

The girls (named in Ex. 2-3) are thinking about going to Rochester. They each have \$3.00 and their car needs \$11.00 worth of gas to get there. Let $R$ be a variable that equals 1 if they do go to Rochester and express this situation as a Boolean equation.

**SOLUTION**

They can go to Rochester only if they all agree to go and chip in for the gas. Thus the Boolean equation is:

$$R = ABCD \qquad (2-4)$$

Logically, Eq. (2-4) says they will go to Rochester only if Alice AND Betty AND Cindy AND Doris all agree to go and pay for the gas.

---

[2]As in ordinary algebra, the dot signifying a product is usually omitted. Thus $s = x \cdot y$ is the same as $s = xy$; the *latter* form is *preferred* and used throughout this book.

## 2-4.4 Other Permissible Boolean Operations

The Boolean operations of *addition* and *multiplication* (ANDing and OR-ing) are both *commutative* and *associative*. Therefore,

1. $xy = yx$
2. $x + y = y + x$
3. $(x + y) + z = x + (y + z) = x + y + z$
4. $(xy) z = x (yz) = xyz$

This allows us to manipulate Boolean expressions *without* concern for the order in which variables appear.

*Factoring* is another permissible operation in Boolean algebra. For example, $xy + xz = x (y + z)$.

## 2-4.5 Prohibited Operations

The operations of *subtraction* and *division* are *not* permitted in Boolean algebra. Consequently if identical terms appear on either side of an equation they *cannot* be cancelled (by subtraction) as in ordinary algebra (see Ex. 2-15). It also means both sides of an equation cannot be divided by a common variable. Fractions such as $x/y$ do *not* occur in Boolean algebra.

## 2-5 THEOREMS

Several *basic theorems* in Boolean algebra are used to *simplify* expressions and equations. These are listed here along with pertinent explanations and derivations as necessary.

### 2-5.1 Addition Theorems

1. $x + 0 = x$

Simply adding 0 to a variable does *not* change the value of that variable.

2. $x + 1 = 1$

This theorem should be interpreted as 1 plus anything (a variable or a combination of several variables forming an expression) equals 1. This is extremely useful; it often allows the engineer to delete some variables from an expression (see Ex. 2-10).

### 2-5.2 Multiplication Theorems

3. $x \cdot 0 = 0$
4. $x \cdot 1 = x$

As in ordinary algebra, multiplication by 0 gives a 0 result and the multiplication of a variable by 1 does not change the value of the variable.

### 2-5.3 Other Theorems Involving a Single Variable

5. $x \cdot x = x$
6. $x + x = x$
7. $x \cdot \bar{x} = 0$
8. $x + \bar{x} = 1$

These theorems are best demonstrated by simply substituting 0s and 1s for $x$. Theorems 7 and 8 are apparent when we remember that if $x = 1$, $\bar{x} = 0$ and vice versa.

---

**EXAMPLE 2-5**

Simplify $x + \bar{x} + y$.

**SOLUTION**

We note that $x + \bar{x} = 1$ by Theorem 8. Therefore,

$$x + \bar{x} + y = 1 + y = 1 \qquad \text{(Theorem 2)}$$

**EXAMPLE 2-6**

Simplify $x(y + z)\,\bar{x} + w$.

**SOLUTION**

This expression can be rewritten as

$$x \cdot \bar{x} \cdot (y + z) + w$$

but $x \cdot \bar{x} = 0$, which eliminates the first term entirely. Therefore,

$$x(y + z)\,\bar{x} + w = w$$

---

### 2-5.4 Theorems Involving More Than One Variable

9. $x + xy = x$

> *Proof*
> The expression $x + xy$ can be factored
>
> $$x + xy = x(1 + y) = x$$
>
> Since $1 + y = 1$ by Theorem 2.

After a theorem has been proven, it is wise to go back and examine the original equality. This verifies the theorem and gives one a feeling for it and its uses. When the expression $x + xy$ is examined, we note that it equals 1 if $x = 1$ and it equals 0 if $x = 0$, regardless of the value of $y$.

10. $x(x + y) = x$

**Proof**
By multiplying, the left side becomes

$$x(x + y) = x \cdot x + xy = x + xy$$

This is identical to the left side of Theorem **9**.

11. $x + \bar{x}y = x + y$

**Proof**
Here a bit of imagination helps. Consider the expression $(x + \bar{x})(x + y)$.

$$(x + \bar{x})(x + y) = x \cdot x + xy + \bar{x}x + \bar{x}y$$
$$= x + xy + \bar{x}y$$
$$= x + \bar{x}y$$

But $(x + \bar{x}) = 1$ so it is also apparent that

$$(x + \bar{x})(x + y) = x + y$$

This proves Theorem **11**.

Examining both sides of the original equality we see that they are surely 1 if $x = 1$, but if $x = 0$ both sides of the expression are equal to $y$.

12. $xy + \bar{y}z + xz = xy + \bar{y}z$

**Proof**
This theorem can be proved by properly expanding the right side.

$$xy + \bar{y}z = xy(1 + z) + \bar{y}z(1 + x)$$
$$= xy + xyz + \bar{y}z + x\bar{y}z$$
$$= xy + \bar{y}z + xz(y + \bar{y})$$

The situations where Theorem **12** can be applied are not always obvious. One must generally look for a *variable multiplied by a second variable*, and the *complement of the first variable multiplied by a third variable*. Then, *if the product of the second and third variables is present, it can be eliminated.*

---

**EXAMPLE 2-7**

Simplify $f(w, x, y, z) = \bar{x}y + wy\bar{z} + xw\bar{z}$.

## SOLUTION

In this problem one must be careful. The $x$, $y$, and $z$ of the specified function do not correspond directly to the $x$, $y$, and $z$ variables in Theorem 12. Some engineers find it clearer to change the variables to $a$, $b$, $c$, and $d$.

$$f(w, x, y, z) = f(a, b, c, d) = \bar{b}c + ac\bar{d} + ba\bar{d}$$

Examining this expression, we find a variable ($b$) times a second variable ($a\bar{d}$) and its complement ($\bar{b}$) times a third variable ($c$). Looking at Theorem 12, the following correspondence can be established:

$$y = b, \quad x = a\bar{d} \quad \text{and} \quad z = c$$

Therefore,

$$\bar{b}c + ac\bar{d} + ba\bar{d} = \bar{b}c + ba\bar{d}$$

or returning to the original function:

$$f(w, x, y, z) = \bar{x}y + wy\bar{z} + xw\bar{z} = \bar{x}y + xw\bar{z}$$

The term $wy\bar{z}$ can be deleted without changing the value of the expression.

---

13.  $(x + y)(x + \bar{y}) = x$

   **Proof**

$$
\begin{aligned}
(x + y)(x + \bar{y}) &= x + xy + x\bar{y} + \bar{y}y \\
&= x + x(y + \bar{y}) + 0 \\
&= x + x \\
&= x
\end{aligned}
$$

14.  $(y + z)(\bar{y} + x) = xy + \bar{y}z$

   **Proof**
   Direct expansion gives

$$(y + z)(\bar{y} + x) = xy + \bar{y}z + xz$$

By Theorem 12 this reduces to $xy + \bar{y}z$.

## 2-6   LOGICAL MANIPULATION

In the practical world, each logic expression is translated into electronic gates or relays and their interconnections. Therefore, the *simplest* and *most compact* Boolean statement results in the *fewest* gates and wires. Gates themselves are relatively inexpensive. But when the additional costs, in both time and money, of mounting extra circuits, debugging them, and repairing them when they fail are considered, appreciable savings accrue to people who make an effort to eliminate unneeded circuits.

## 2-6.1   Reduction of Logical Expression

Often Boolean expressions an engineer sees in industry (and these expressions may describe actual circuits as built) are *not* written in their simplest form. By using the theorems of Sec. 2-5, many of these expressions can be reduced, which leads to the elimination of useless circuitry. The following examples demonstrate some of the reduction techniques used in simplifying and reducing logical expressions.

### EXAMPLE 2-8

Simplify the expression

$$f(x, y, z) = \bar{x}(\bar{y} + \bar{z}) + yz + x\bar{z}$$

### SOLUTION

It is usually best to begin by breaking parentheses open.

$$f(x, y, z) = \bar{x}(\bar{y} + \bar{z}) + yz + x\bar{z} = \bar{x}\bar{y} + \bar{x}\bar{z} + yz + x\bar{z}$$

Examining this expression, we find that

$$\bar{x}\bar{z} + x\bar{z} = \bar{z}(x + \bar{x}) = \bar{z}$$

Therefore,

$$
\begin{aligned}
f(x, y, z) &= \bar{x}\bar{y} + \bar{x}\bar{z} + yz + x\bar{z} \\
&= \bar{x}\bar{y} + yz + \bar{z} \\
&= \bar{x}\bar{y} + y + \bar{z} \qquad \text{(Theorem 11)} \\
&= \bar{x} + y + \bar{z} \qquad \text{(Theorem 11)}
\end{aligned}
$$

The last two simplifications $(yz + \bar{z} = y + \bar{z})$ and $(\bar{x}\bar{y} + y = y + \bar{x})$ used Theorem 11. No further simplifications can be made.

### EXAMPLE 2-9

Simplify the expression

$$f(A, B, C) = (A + B)(A + BC) + \overline{A}\overline{B} + \overline{A}\overline{C}$$

### SOLUTION

$$
\begin{aligned}
f(A,B,C) &= (A + B)(A + BC) + \overline{A}\overline{B} + \overline{A}\overline{C} \\
&= A + AB + ABC + BC + \overline{A}\overline{B} + \overline{A}\overline{C} \\
&= A(1 + B + BC) + BC + \overline{A}\overline{B} + \overline{A}\overline{C}
\end{aligned}
$$

The term in parentheses equals 1 and can be dropped:

$$f(A,B,C) = A + \overline{A}\overline{B} + \overline{A}\overline{C} + BC$$

Since A is a term already in the expression, another identical term (another A) can be added without changing the value of the expression (by Theorem 6). Therefore,

$$f(A,B,C) = (A + \overline{A}\overline{B}) + (A + \overline{A}\overline{C}) + BC$$
$$= A + \overline{B} + A + \overline{C} + BC$$
$$= A + \overline{B} + \overline{C} + B$$

But $B + \overline{B} = 1$. Therefore, the value of the function $f(A,B,C) = 1$.

Logically this means there is no combination of variables A, B, and C one can choose that makes $f(A,B,C) = 0$. Physically it means no gates are actually needed to implement $f$, other than a conductor. If $f$ is needed it can be wired directly to a voltage in the logic 1 range.

### EXAMPLE 2-10

Simplify

$$f(x,y,z) = x + \bar{y}z + (x + \bar{y}z) \text{ times } Q$$

where $Q$ = any expression.

### SOLUTION

The expression $x + \bar{y}z$ can be factored from the given expression.

$$f(x,y,z) = x + \bar{y}z + (x + \bar{y}z) Q$$
$$= (x + \bar{y}z)(1 + Q)$$
$$= x + \bar{y}z \qquad\qquad \text{(Theorem 2)}$$

since the last term in parentheses equals 1.

### EXAMPLE 2-11

Simplify

$$f(W,X,Y,Z) = \overline{W}XY\overline{Z} + \overline{W}XYZ + W\overline{X}\overline{Y}Z + W\overline{X}YZ + WX\overline{Y}Z + WXY\overline{Z}$$

### SOLUTION

$$f(W,X,Y,Z) = \overline{W}XY\overline{Z} + \overline{W}XYZ + W\overline{X}\overline{Y}Z$$
$$+ W\overline{X}YZ + WX\overline{Y}Z + WXY\overline{Z}$$
$$= \overline{W}XY(\overline{Z} + Z) + W\overline{X}Z(\overline{Y} + Y)$$
$$+ WX\overline{Y}Z + WXY\overline{Z}$$

Before the indicated simplifications are performed it is best to rewrite the third and first terms near the last two terms. This allows us to simplify the last two terms.

$$f(W,X,Y,Z) = \overline{W}XY(\overline{Z} + Z) + W\overline{X}Z(\overline{Y} + Y) + WX\overline{Y}Z$$
$$+ WX\overline{Y}Z + WXY\overline{Z} + \overline{W}XY\overline{Z}$$
$$= \overline{W}XY(\overline{Z} + Z) + W\overline{X}Z(\overline{Y} + Y)$$
$$+ W\overline{Y}Z(\overline{X} + X) + (W + \overline{W})XY\overline{Z}$$
$$= \overline{W}XY + W\overline{X}Z + W\overline{Y}Z + XY\overline{Z}$$

## EXAMPLE 2-12

Show that the following expression is an equality by adding a term to each side:

$$ab + \bar{a}\bar{b} + bc = ab + \bar{a}\bar{b} + \bar{a}c$$

## SOLUTION

By using Theorem 12 we can expand the last two terms on the left side to

$$\bar{a}\bar{b} + bc = \bar{a}\bar{b} + bc + \bar{a}c$$

By using Theorem 12 on the first and third terms of the right side, we find that

$$ab + \bar{a}c = ab + \bar{a}c + bc$$

When we make these substitutions

$$ab + \bar{a}\bar{b} + bc + \bar{a}c = ab + \bar{a}\bar{b} + \bar{a}c + bc$$

A term-by-term comparison now shows that the two sides are equal.

## 2-6.2 Logical Equivalence

Two Boolean expressions are equivalent *only* if they are equal for *all possible values of the variables* in *both* expressions. *To prove two expressions unequal, it is only necessary to find a single set of values for the variables that makes the expressions unequal.* But if two expressions are equal, this must be proved by logical manipulation or, as a last resort, by truth tables.

## EXAMPLE 2-13

Does $X + WZ = WX + \bar{W}X\bar{Y} + W\bar{X}Z + \bar{W}Y$?

## SOLUTION

We start by manipulating the right-hand side in an attempt to reduce it.

$$
\begin{aligned}
WX &+ \bar{W}X\bar{Y} + W\bar{X}Z + \bar{W}Y \\
&= WX + \bar{W}(Y + X\bar{Y}) + W\bar{X}Z \\
&= WX + \bar{W}(Y + X) + W\bar{X}Z \qquad \text{(Theorem 11)} \\
&= (W + \bar{W})X + \bar{W}Y + W\bar{X}Z \qquad \text{(Theorem 8)} \\
&= X + W\bar{X}Z + \bar{W}Y \qquad \text{(Theorem 11)} \\
&= X + WZ + \bar{W}Y
\end{aligned}
$$

Since no further reductions seem possible, the question is now:

$$\text{Does } X + WZ \stackrel{?}{=} X + WZ + \bar{W}Y$$

The two sides do not look alike; but to prove it, a set of values must be found to make the sides unequal. The right side of the equation contains all the terms in the left side, so if either term (X or WZ) equals 1 both sides must be 1. However, if we choose values such that the left side is 0, it may be possible to make the right side equal to 1. If we choose X = 0, W = 0, and Y = 1, we get $\overline{W}Y = 1$. Now the left side equals 0 and the right side equals 1. This proves the terms are unequal.

## EXAMPLE 2-14

Does $bc + abd + a\bar{c} \stackrel{?}{=} bc + a\bar{c}$

## SOLUTION

First we try to make the term $abd$ equal 1 while making the right side equal 0. Unfortunately, we do not succeed (if $a$ and $b$ both equal 1, the right side equals 1), so perhaps the terms are equal. Manipulating the left side we obtain

$$
\begin{aligned}
bc &+ abd + a\bar{c} \\
&= bc + abd + a\bar{c} + ab \qquad &\text{(Theorem 12)} \\
&= bc + a\bar{c} + ab \qquad &\text{(Theorem 9)} \\
&= bc + a\bar{c} \qquad &\text{(Theorem 12)}
\end{aligned}
$$

Now we have demonstrated that the two sides *are* equal. This explains why it was *not* possible to find a set of values to make them different.

## EXAMPLE 2-15

A misguided mathematician would like to subtract the term $a\bar{c}$ from both sides of the *equality*:

$$bc + abd + a\bar{c} = bc + a\bar{c}$$

Would they still be equal if he did so?

## SOLUTION

With the term $a\bar{c}$ removed the expressions would be

$$bc + abd = bc$$

Now we find that if $a = 1$, $b = 1$, $c = 0$, and $d = 1$, the left side equals 1 and the right side equals 0; therefore, the terms are unequal. This example demonstrates why subtraction is *not* allowed in Boolean algebra.

## 2-7   IC GATES

The Boolean expressions of this chapter can be implemented physically using electronic gates or relays. *An electronic gate is a circuit that has one or more inputs and only one output.* All the inputs and the output are voltages at either the 0 or 1 levels. The output therefore is a logical function of the inputs.

Relays are used for switching in *high* power circuits. For most electronic and computer circuits, however, the logic expression itself is paramount, and these circuits use very little power. In low power circuits, relays have been superseded by modern integrated circuits (ICs). A typical IC comes in a *dual in-line* (DIP) package (Fig. 2-2) and is composed of several logic gates within the same package. ICs are faster, smaller, cheaper, more reliable, and consume less power than relays. Modern designers use electronic gates almost exclusively.

## 2-7.1 AND Gates

The first basic gate to be considered is the AND gate, which conceptually may have any number of inputs. Its output is the logical AND of the inputs (i.e., all inputs must be 1 in order for the output to be a 1). The standard symbol for an AND gate is shown in Fig. 2-3, where both 2- and 4-input AND gates are shown along with their logical equations.

The basic TTL AND gate in the standard TTL series is the 7408 quad 2-input AND gate. Triple 3- and dual 4-input AND gates are available in the low power Schottky family, but they may not be as widely available as

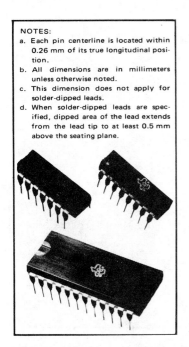

NOTES:
a. Each pin centerline is located within 0.26 mm of its true longitudinal position.
b. All dimensions are in millimeters unless otherwise noted.
c. This dimension does not apply for solder-dipped leads.
d. When solder-dipped leads are specified, dipped area of the lead extends from the lead tip to at least 0.5 mm above the seating plane.

FIGURE 2-2 ICs in dual in-line packages. (From the *TTL Data Book for Design Engineers*, 2nd ed., Texas Instruments, Inc. Courtesy of Texas Instruments, copyright 1976.)

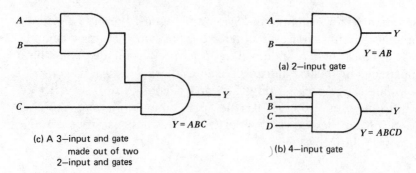

(c) A 3—input and gate
made out of two
2—input and gates

(a) 2—input gate

$Y = AB$

(b) 4—input gate

$Y = ABCD$

$Y = ABC$

FIGURE 2-3 AND gates.

the 2-input **7408.** If 3- or 4-input AND gates are needed, they can be constructed from **7408** gates as shown in Fig. 2-3c.

The manufacturer's literature must be consulted to properly use any TTL gate. Figure 2-4 is taken from the TTL data book, second edition, published by Texas Instruments, Inc. (1976), which describes the **7408** AND gate. It is labelled a **quad 2-input AND gate.** *Quad* means there are four AND gates within the package. The bold black number in the upper left corner, **08,** means the IC is a **7408.** The prefix, **74,** is understood. The drawing shows the 4 gates in each IC package and shows the pinout or pin arrangement for each gate (gate 1, for example, has its inputs on pins 1 and 2 and its output on pin 3). As with most gates, power ($V_{CC}$) is on pin 14 and ground is on pin 7. For TTL gates, $V_{CC}$ is always 5 V.

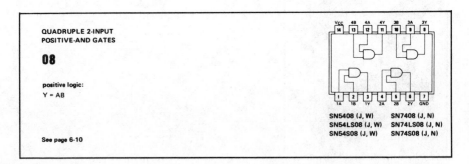

FIGURE 2-4 The **7408** Quad 2-input AND gate. (From the *TTL Data Book for Design Engineers*, 2nd ed., Texas Instruments, Inc. Courtesy of Texas Instruments, copyright 1976.)

## 2-7.2  OR Gates

As its name indicates, the output of an OR gate is the logical OR of its inputs. The standard symbol for an OR gate is shown in Fig. 2-5, where a 2-input OR gate and a 3-input OR gate made of two OR gates are shown. Note that the OR symbol is much like the AND symbol except that the front is curved rather than straight, and the output is pointed rather than curved.

In the TTL series only a quad 2-input OR gate, the **7432,** is available. Again the manufacturer's specifications must be consulted for further information, such as the pinout of the gates.

---

**EXAMPLE  2-16**

Use AND and OR gates to implement the function $Y = AB + BCD$.

**SOLUTION**

The first term requires $A$ and $B$ to be ANDed, and the second term requires that $B$, $C$, and $D$ be ANDed. These two partial results must then be ORed together. The circuit using only 2-input gates is shown in Fig. 2-6.

---

## 2-7.3  Inverters and Bubbles

The function of an *inverter* is either to change a 1 input to a 0 output, or a 0 input to a 1 output. Hence it *inverts* or *complements* the signal. An

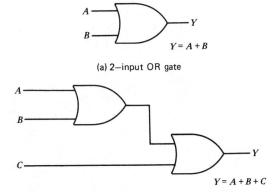

(a) 2–input OR gate

$Y = A + B$

(b) A 3–input OR gate made up of two 2–input OR gates

$Y = A + B + C$

FIGURE 2-5  OR gates. (a) 2-input OR gate. (b) A 3-input OR gate made up of two 2-input OR gates.

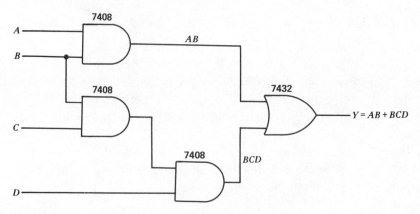

FIGURE 2-6   Circuit to produce the function $Y = AB + BCD$.

inverter has only one input and one output. The symbol for an inverter, a small triangle with a bubble, is shown in Fig. 2-7. Both Figs. 2-7*a* and 2-7*b* are valid representations of an inverter. The standard TTL inverter is the 7404 IC.

The small circle shown at the *output* of Fig. 2-7*a* (or the input of Fig. 2-7*b*) is called a *bubble*, and is used to indicate the parts of the circuit where the asserted or true level of a signal is LOW. If a signal is actually present (asserted), *we expect those points in the signal path that are not connected to bubbles to be logic 1s, and those points that are connected to bubbles to be logic 0s.* Conversely, if the same signal is not present or asserted (logic 0) the points connected to bubbles should be 1, and the points without bubbles should be 0. Thus, in Fig. 2-7*a* if the input signal to the inverter is asserted (1), the output will be 0. This is indicated by the bubble on the output. If the signal is not asserted, the input, A, will be LOW and the output ($\overline{A}$) will be HIGH. In Fig. 2-7*b* the asserted level of the signal on

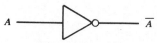

(a) Inverter-bubble on output

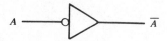

(b) Inverter-bubble on input

FIGURE 2-7   Two equivalent representations for an inverter.

the input is LOW and the asserted level of the output is HIGH, as indicated by the absence of a bubble. If the input A is asserted or present, the input to the gate of Fig. 2-7b should be a 0.

## 2-8 COMPLEMENTATION OF FUNCTIONS

Two functions, $F$ and $\bar{F}$, are complementary if they depend on the same set of input variables and if, *for every combination of values of the input variables*, the values of $F$ and $\bar{F}$ are *inverse*. One way to determine the complement of a function, $F$, is to draw up a *truth table*. $\bar{F}$ is then a 1 whenever $F$ is a 0, and 0 whenever $F$ is a 1.

---

### EXAMPLE 2-17

Draw up the truth table for $F = X + \bar{Y}Z$. From it find $F$ and $\bar{F}$ and simplify $\bar{F}$ if possible.

### SOLUTION

The truth table of Fig. 2-8 was constructed according to the procedure of Sec. 2-3.2. The value of $F$ is determined from the equation and $\bar{F}$ is then written simply as the inverse of $F$. Since upon examining the truth table we find $\bar{F}$ to be 1 on lines 0, 2, and 3, we can write the following equation:

$$\begin{aligned}
\bar{F} &= \bar{X}\bar{Y}\bar{Z} + \bar{X}Y\bar{Z} + \bar{X}YZ \\
&= \bar{X}(\bar{Y}\bar{Z} + Y\bar{Z} + YZ) \\
&= \bar{X}(\bar{Y}\bar{Z} + Y) & \text{(Theorem } 8) \\
&= \bar{X}(\bar{Z} + Y) & \text{(Theorem } 11)
\end{aligned}$$

---

| X | Y | Z | F | $\bar{F}$ |
|---|---|---|---|---|
| 0 | 0 | 0 | 0 | 1 |
| 0 | 0 | 1 | 1 | 0 |
| 0 | 1 | 0 | 0 | 1 |
| 0 | 1 | 1 | 0 | 1 |
| 1 | 0 | 0 | 1 | 0 |
| 1 | 0 | 1 | 1 | 0 |
| 1 | 1 | 0 | 1 | 0 |
| 1 | 1 | 1 | 1 | 0 |

FIGURE 2-8 Truth table for $F = X + \bar{Y}Z$.

## 2-8.1 DeMorgan's Theorem

Fortunately the complement of a Boolean expression can be found without using truth tables by **DeMorgan's theorem,** which states the following.

*The complement of a function f can be found by*:

1. *Replacing each variable by its complement.*
2. *Interchanging all the* **AND** *and* **OR** *signs.*

To apply DeMorgan's theorem the given expression is first investigated to find all the AND and OR signs. These are changed to OR and AND signs, respectively, and then the variables are complemented.

---

**EXAMPLE 2-18**

If $F = X + \bar{Y}Z$, as in Ex. 2-17, find $\bar{F}$.

**SOLUTION**

The $+$ sign between X and $\bar{Y}$ must be changed to a mulitplication sign, and the implied multiplication sign between Y and Z must be changed to a $+$ sign. The variables must also be complemented. This procedure is illustrated below.

$$F = X + \bar{Y} \cdot Z$$
$$\bar{F} = X(Y + \bar{Z})$$

The result, $\bar{F} = \bar{X}(Y + \bar{Z})$, agrees with the results obtained from the truth table.

---

Two important corrolaries of DeMorgan's theorem are:

$$\overline{(X_1 + X_2 + \cdots + X_N)} = \bar{X}_1\bar{X}_2 \cdots \bar{X}_n \qquad (2\text{-}5)$$

$$\overline{(X_1X_2X_3 \cdots X_n)} = \bar{X}_1 + \bar{X}_2 + \bar{X}_3 + \cdots + \bar{X}_n \qquad (2\text{-}6)$$

---

**EXAMPLE 2-19**

For Exs. 2-3 and 2-4 find $\bar{M}$ and $\bar{R}$ and explain their meaning physically.

**SOLUTION**

In Ex. 2-3 we found that

$$M = A + B + C + D$$

Using Eq. (2-5) we obtain

$$\bar{M} = \bar{A}\bar{B}\bar{C}\bar{D}$$

Physically this means you are *not* going to the movies ($\bar{M} = 1$) if all the girls say no ($\bar{A} = \bar{B} = \bar{C} = \bar{D} = 1$).

In Ex. 2-4 we found that

$$R = ABCD$$

Here using Eq. (2-6) we obtain

$$\bar{R} = \bar{A} + \bar{B} + \bar{C} + \bar{D}$$

This means that the girls are *not* going to Rochester ($\bar{R} = 1$) if any one of them doesn't want to go. The rest of the girls, collectively, cannot afford the cost of the gasoline.

---

Occasionally, it is necessary to simplify expressions that already contain *complemented subexpressions*. The best procedure:

1. Take the complement of that portion of the expression that is complemented.
2. Simplify the remainder of the expression.

---

### EXAMPLE 2-20

Simplify the expression:

$$F = (X + \bar{Z})(\overline{Z + WY}) + (VZ + W\bar{X})(\overline{Y + Z})$$

**SOLUTION**

A complement sign outside a parentheses refers to everything within the parentheses. By DeMorgan's theorem:

$$(\overline{Z + WY}) = \bar{Z}(\bar{W} + \bar{Y})$$

While

$$(\overline{Y + Z}) = \bar{Y}\bar{Z}$$

Then

$$\begin{aligned}
F &= (X + \bar{Z}) \cdot \bar{Z} \cdot (\bar{W} + \bar{Y}) + (VZ + W\bar{X})\bar{Y}\bar{Z} \\
&= (X\bar{Z} + \bar{Z})(\bar{W} + \bar{Y}) + WX\bar{Y}\bar{Z} \\
&= \bar{Z}(\bar{W} + \bar{Y}) + WX\bar{Y}\bar{Z} \\
&= \bar{Z}\bar{W} + \bar{Z}\bar{Y} + WX\bar{Y}\bar{Z} \\
&= \bar{Z}\bar{W} + \bar{Z}\bar{Y}(1 + WX) \\
&= \bar{Z}\bar{W} + \bar{Z}\bar{Y}
\end{aligned}$$

---

When expressions to be complemented become complex, the place to substitute signs is not always obvious. Any possible confusion can be eliminated by breaking the function up into subfunctions and simplifying them.

## EXAMPLE 2-21

(a) Find the complement of the expression of Ex. 2-20 without simplifying.
(b) Simplify the complement.

## SOLUTION

The function

$$F = (X + \bar{Z})(\overline{Z + WY}) + (VZ + W\bar{X})(\overline{Y + Z})$$

looks formidable. Let us write

$$F = F_1F_2 + F_3F_4$$
$$F = F_5 + F_6$$

where

$$F_1 = X + \bar{Z}$$
$$F_2 = \overline{(Z + WY)}$$
$$F_3 = (VZ + W\bar{X})$$
$$F_4 = \overline{(Y + Z)}$$
$$F_5 = F_1F_2$$
$$F_6 = F_3F_4$$

Now proceeding a step at a time we obtain:

$$F = F_5 + F_6$$
$$\bar{F} = \bar{F}_5 \cdot \bar{F}_6$$
$$\bar{F}_5 = \bar{F}_1 + \bar{F}_2$$
$$\bar{F}_6 = \bar{F}_3 + \bar{F}_4$$

then

$$\bar{F} = (\bar{F}_1 + \bar{F}_2)(\bar{F}_3 + \bar{F}_4)$$

and

$$\bar{F}_1 = \bar{X}Z$$
$$\bar{F}_2 = [\overline{(\overline{Z + WY})}] = Z + WY$$
$$\bar{F}_3 = (\bar{V} + \bar{Z})(\bar{W} + X)$$
$$\bar{F}_4 = Y + Z$$

Finally,

$$\bar{F} = (\bar{X}Z + Z + WY)[(\bar{V} + \bar{Z})(\bar{W} + X) + Y + Z]$$
$$= (Z + WY)(\bar{F}_3 + Y + Z)$$
$$= Z + ZY + Z\bar{F}_3 + WY\bar{F}_3 + WYZ + WY$$
$$= Z(1 + Y + \bar{F}_3) + WY(1 + Z + \bar{F}_3)$$
$$= Z + WY$$

We note that the simplified expression for $\overline{F}$ is indeed the complement of the simplified expression for $F$ found in Ex. 2-20. Thus we can be confident that both of these difficult examples were done correctly.

---

## 2-9 INVERTING GATES

Besides inverters, many gates perform inversion as a normal part of their function. The simplest inverting gates are NAND and NOR gates.

A NAND gate takes the AND of its inputs, and, in the process, inverts the output. Because common-emitter transistor circuits normally invert, NAND gates are easier to build and more commonly used than noninverting gates, such as AND gates. The standard symbol for a 3-input NAND gate is shown in Fig. 2-9. Note that it is simply an AND gate symbol with a bubble on the output to indicate inversion. The process is AND first, then invert; therefore, the output equation for the NAND gate of Fig. 2-9 is

$$Y = \overline{(ABC)}$$
$$= \overline{A} + \overline{B} + \overline{C} \qquad \text{(by DeMorgan's theorem, Sec. 2-8)}$$

A NOR gate similarly takes the OR of its inputs and then inverts it. The symbol for a NOR gate is shown in Fig. 2-10. Note that it is an OR gate with a bubble on the output. The logic equation for the NOR gate is

$$Y = \overline{(A + B + C)}$$
$$Y = \overline{A}\,\overline{B}\,\overline{C}$$

### 2-9.1 NAND Gates

The standard NAND gate in the 7400 series is the 7400 IC. It is a quad 2-input NAND gate.

In addition to the 7400, other NAND gates are available as shown in Fig. 2-11:

1. The 7404 hex inverter (an inverter can be viewed as a 1-input NAND gate).
2. The 7410 triple 3-input NAND gate.
3. The 7420 dual 4-input NAND gate.
4. The 7430 8-input NAND gate.

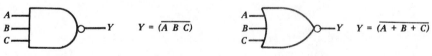

FIGURE 2-9 A 3-input NAND gate.          FIGURE 2-10 A 3-input NOR gate.

| Chip Number | Number of Identical Gates per Package | High Input Representation | Low Input Representation | Equation |
|---|---|---|---|---|
| 7404 | 6 | 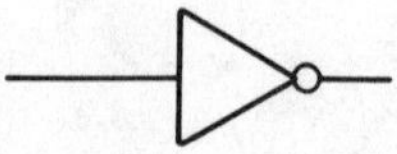 | 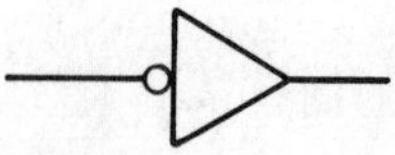 | $Y = \overline{A}$ |
| 7400 | 4 | 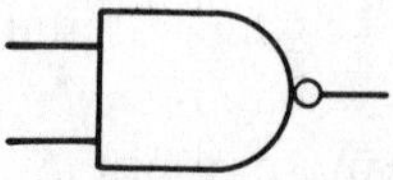 | 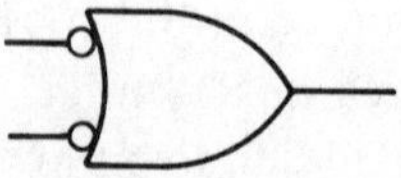 | $Y = \overline{(AB)}$ |
| 7410 | 3 | 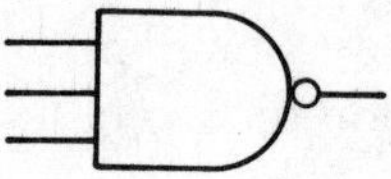 | 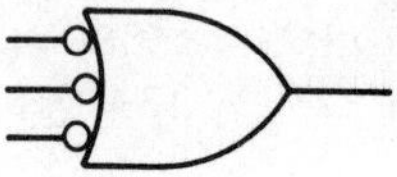 | $Y = \overline{(ABC)}$ |
| 7420 | 2 | 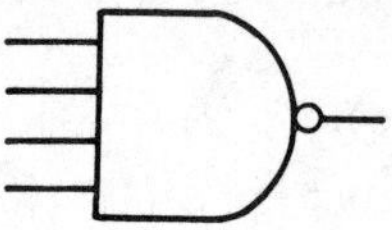 | 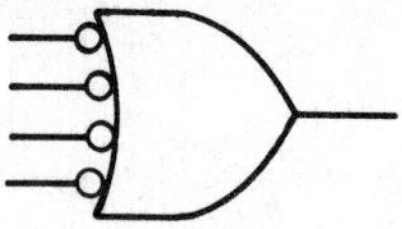 | $Y = \overline{(ABCD)}$ |
| 7430 | 1 | 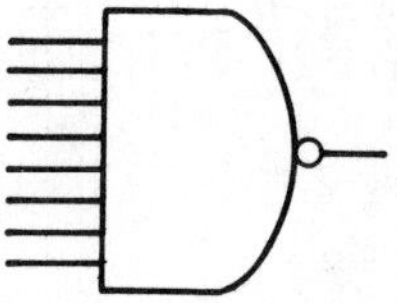 | 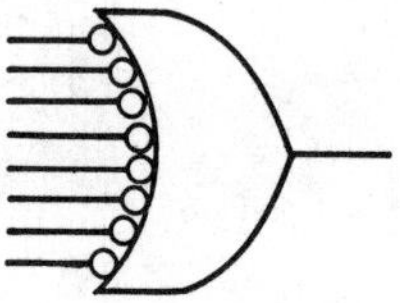 | $Y = \overline{(ABCDEFGH)}$ |

FIGURE 2-11   Common NAND gates of the **7400** series.

All of the above NAND gates are in 14 pin packages ($V_{CC}$ is on pin 14, ground is on pin 7), and are readily available from most suppliers.

---

**EXAMPLE 2-22**

Design a 30 input NAND gate using only the NAND gates of Fig. 2-11.

**SOLUTION**

There are many possible designs that satisfy the problem requirement. One solution, shown in Fig. 2-12, is to connect the inputs to 7430s in groups of 8. The last 7430 only requires 6 inputs so that the unused inputs on the 7430 are tied to a working

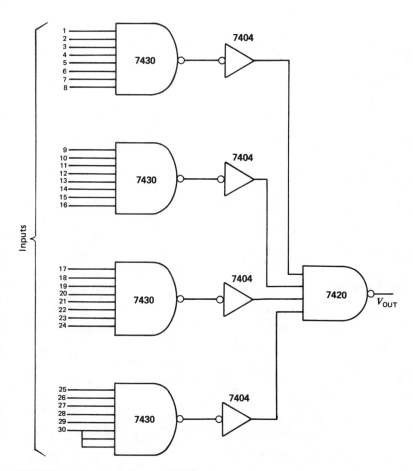

FIGURE 2-12 A 30-input NAND gate.

input (see Sec. 5-6). The outputs of the **7430** are then inverted and go to a **7420**, 4-input NAND gate. The final output will be low only if *all* 30 inputs are HIGH.

## 2-9.2 TTL NOR Gates

The **7402** is the basic quad 2-input NOR gate. Other NOR gates available in the standard TTL line are the **7427** triple 3-input NOR gate and the **7425** dual 4-input NOR gate. The **7425** is a strobed gate (see Sec. 5-8.1).

## 2-9.3 Equivalence of Gates and Inversion of Gates

Any gate of the AND-OR-NAND-NOR group may be represented in one of two ways. The two ways to symbolize an inverter were shown in Fig. 2-7. The alternate symbol for a gate is obtained if we:

1.  Change the function of the gate (from AND to OR or OR to AND).
2.  Change all the bubbles on both the inputs and outputs. (That is, delete bubbles when present and add them when absent.)

Both a symbol and its alternate represent the *same physical gate*. Alternate symbols are often used to clarify signal flow in a circuit (Sec. 2-10.1).

---

**EXAMPLE 2-23**

Find the alternate symbol for the 3-input NAND gate of Fig. 2-13a.

**SOLUTION**

The alternate symbol is shown in Fig. 2-13b. It was obtained by:

1.  Changing the function of the gate from AND to OR.
2.  Adding bubbles to the input.
3.  Removing the bubble from output.

Figure 2-13a indicates the output is LOW only if *all* (logic AND) *the inputs are HIGH*. Figure 2-13b indicates the output is HIGH if *any input* (logic OR) *is LOW*. With a little thought, we realize that these are two different ways of saying the same thing!

**EXAMPLE 2-24**

Find the alternate representation of the 2-input OR gate of Fig. 2-14a.

**SOLUTION**

The alternate representation, shown in Fig. 2-14b, is obtained by replacing the OR symbol with an AND symbol and adding bubbles on both the inputs and the output.

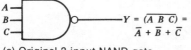

(a) Original 3-input NAND gate

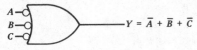

(b) Alternate representation

FIGURE 2-13 Equivalent representations of a 3-input NAND gate.

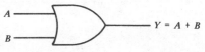

(a) 2-input OR gate

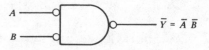

(b) Equivalent representation

FIGURE 2-14 Equivalent representations of a 2-input OR gate.

Figure 2-14$a$ indicates the output is HIGH if $A$ $or$ $B$ is high, and Fig. 2-14$b$ indicates the output is LOW only if $A$ $and$ $B$ are low. Again, these two statements are equivalent. DeMorgan's theorem shows that the output equations for each gate of Fig. 2-14 are equivalent, that is, $A + B = \overline{\overline{A}\overline{B}}$.

---

NAND and NOR gates are basically inverting gates with bubbles on $either$ the inputs $or$ the output, $but\ not\ both$ (Fig. 2-13). AND and OR gates are basically noninverting and either have bubbles on both inputs and outputs, or have no bubbles at all, as Fig. 2-14 shows.

## 2-10 IMPLEMENTATION OF LOGIC EXPRESSIONS USING GATES

All the logic expressions discussed previously can be implemented using AND gates, OR gates, and inverters. AND gates are simply placed wherever two variables are $multiplied$ together, and OR gates are placed wherever two variables are $added$. The resulting circuit is the hardware implementation of the given Boolean expression, as shown by Exs. 2-25 through 2-27 below.

---

### EXAMPLE 2-25

Implement the expression $Y = (A + \overline{B})C + \overline{A}B\overline{C}$ if:
(a) Both complemented and uncomplemented input variables are available.
(b) Only uncomplemented input variables are available.

### SOLUTION
(a) The term $A + \overline{B}$ suggests an OR gate. The output of this OR gate must be ANDed with $C$ to get the first term in the expression. To obtain the term $\overline{A}B\overline{C}$ a 3-input AND gate is required. Finally, since the output is a sum, the two terms must be ORed together to produce $Y$. The results are shown in Fig. 2-15$a$.

(b) If only $uncomplemented$ inputs are available, complemented variables can be obtained through the use of inverters. Figure 2-15$b$ is essentially the same as Fig. 2-15$a$, except that inverters have been added to convert uncomplemented inputs $A$, $B$, and $C$ to $\overline{A}$, $\overline{B}$, and $\overline{C}$ where needed.

---

### EXAMPLE 2-26

Using only the gates discussed previously, implement the expression:

$$Y = AB(\overline{\overline{C}D}) + \overline{B}CD + (\overline{A} + \overline{C})(B + D)$$

(a) Do not simplify the above expression before implementing.
(b) Simplify first, then implement the expression.

### SOLUTION

For part (a) the expression is implemented just as stated. The term $(\overline{CD})$ implies an AND gate and an inverter, or more simply a NAND gate. The output of this NAND gate is ANDed with A and B to produce the first term in the expression. The term $\overline{B}CD$ is a 3-input AND gate, and the term $(\overline{A} + \overline{C})(B + D)$ is the AND of two OR gates. Finally, the three terms have to be ORed together because the output is the sum of the terms. The solution is shown in Fig. 2-16a. For simplicity it was assumed that both complemented and uncomplemented variables were available.

The solution of Ex. 2-26a demonstrates that any combinatorial expression can be implemented using only ANDs, ORs, and so forth.

A better circuit is obtained by *simplifying the expression before* implementing it. The first step in any simplification is to use DeMorgan's theorem to eliminate the complemented term.

$$Y = AB(\overline{\overline{CD}}) + \overline{B}CD + (\overline{A} + \overline{C})(B + D)$$
$$= AB(C + D) + \overline{B}CD + (\overline{A} + \overline{C})(B + D)$$

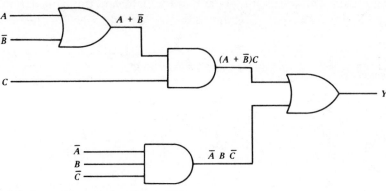

(a) Both complemented and uncomplemented variables available

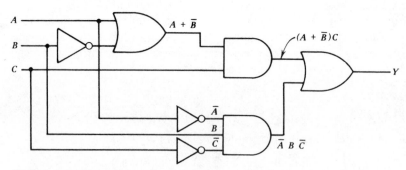

(b) Only uncomplemented inputs are
available

FIGURE 2-15   Two implementations of the expression $Y = (A + \overline{B})C + \overline{A}B\overline{C}$.

This expression can be simplified further either by drawing Karnaugh maps (see Sec. 3-5) or by algebraic manipulation. Trying algebraic manipulation, we obtain:

$$\begin{aligned}
Y &= ABC + AB\bar{D} + \bar{B}CD + \bar{A}B + \bar{A}D + B\bar{C} + \bar{C}D \\
&= B(AC + \bar{A}) + AB\bar{D} + D(\bar{B}C + \bar{C}) + \bar{A}D + B\bar{C} \\
&= B(C + \bar{A}) + AB\bar{D} + D(\bar{C} + \bar{B}) + \bar{A}D + B\bar{C} \\
&= BC + B\bar{C} + B\bar{A} + AB\bar{D} + D\bar{C} + D\bar{B} + \bar{A}D \\
&= B + D\bar{B} + D\bar{C} + \bar{A}D \\
&= B + D
\end{aligned}$$

Thus the expression which originally required 7 gates to build (in Fig. 2-16a) reduces to a single OR gate as shown in Fig. 2-16b. This example clearly demonstrates why it is wise to first simplify expressions, instead of immediately plunging in and building the circuit. The circuit of Fig. 2-16a would have been even more complicated if only uncomplemented variables were available. In that case four additional inverters would have been required while the circuit of Fig. 2-16b would still remain unchanged.

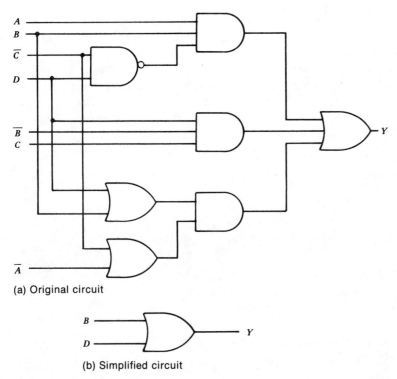

(a) Original circuit

(b) Simplified circuit

FIGURE 2-16 Two circuits to implement the expression $Y = AB(\overline{CD}) + \bar{B}CD + (\bar{A} + \bar{C})(B + D)$.

We should also be able to reverse the procedure and analyze a circuit that already exists to determine its logic equation. The equation is needed in order to predict its behavior for all combinations of input variables, because occasionally such poor designs as the circuit of Fig. 2-16a actually are built and incorporated into working equipment. Naturally they should be reduced and replaced with a simpler equivalent circuit, if at all possible.

To analyze a given circuit, the logic equations for each gate are written first. Then gates are chained together in accordance with the interconnecting wires until an expression for the entire circuit is achieved, as shown in Ex. 2-27.

---

**EXAMPLE 2-27**

Find the expression for the output of Fig. 2-17. Design a simpler circuit if possible.

**SOLUTION**

The circuit of Fig. 2-17a is redrawn in Fig. 2-17b with the gates and outputs labeled for clarity. In order to prevent confusion by introducing too many complementation signs, complemented expressions are immediately reduced using DeMorgan's theorem. The output is obtained by following the steps listed below:

1.  The output of AND gates 1 and 2 are $AB$ and $BC$. These form inputs to gates 7 and 4, respectively.
2.  Gate 4 is a NAND gate with inputs $A$ and $BC$. Since NAND gate inverts, the output of gate 4 is the complement of the AND of $A$ and $BC$ or $\overline{(ABC)} = \overline{A} + \overline{B} + \overline{C}$.
3.  Gate 5 is a NOR gate with inputs $BC$ and $\overline{C}$. The output is therefore $\overline{(BC + \overline{C})} = \overline{C}(\overline{B} + \overline{C}) = \overline{C}\overline{B}$.
4.  Gate 6 NORs together the output of gates 4 and 5. Its output is therefore $\overline{[C\overline{B} + \overline{A} + \overline{B} + \overline{C}]} = \overline{(\overline{A} + \overline{B} + \overline{C})} = ABC$.
5.  The output of gate 7 is the NAND of the outputs of gates 1 and 6. Thus the final output is:

$$Y = \overline{[(ABC)(AB)]}$$
$$= \overline{(ABC)}$$
$$= \overline{A} + \overline{B} + \overline{C}$$

Figure 2-17c shows that the output, $Y$, can be produced by a single 3-input NAND gate if uncomplemented variables are available, or by a 3-input OR gate if complemented variables are available.

## 2-10.1  Placement of Bubbles

With two alternate representations for each gate, the question arises: Which representation should be used? Many engineers solve the problem by placing bubbles on outputs only, and never on inputs. However, we feel it is easier

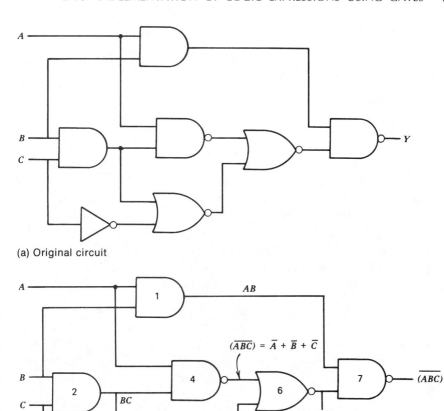

(a) Original circuit

(b) Circuit used for analysis

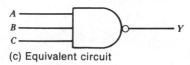

(c) Equivalent circuit

FIGURE 2-17   Circuit for Ex. 2-27

to trace the signal flow, and the operation of a circuit becomes clearer, if we can choose the gate representations so that:

1.   Output bubbles are connected to input bubbles.
2.   Outputs without bubbles are connected to inputs without bubbles.

Gate representations that satisfy these requirements cannot be selected in all cases, but they can in most cases. They help the engineer "see" the signal flow and facilitate trouble-shooting for the technician.

Consider the problem of finding an expression for the circuit of Fig. 2-18a where bubbles are placed only on the outputs. The output of gates 1 and 2 are $(\overline{AB})$ and $(\overline{CD})$, respectively. Here $Y$ is the output of a NAND gate:

$$Y = [(\overline{AB})(\overline{CD})] = AB + CD$$

Even for this simple circuit, visualization of the signal flow is somewhat difficult, and the answer was achieved by algebraic manipulation.

Now consider the circuit of Fig. 2-18b, which is the same as Fig. 2-18a, but NAND gate 3 has been replaced by its equivalent representation with negative inputs. In Fig. 2-18b, we can see that if $A$ and $B$ *or* $C$ and $D$ are *both* HIGH, the output of gates 1 or 2 (or both) is LOW. This LOW output is fed to the input of gate 3, causing its output to be HIGH. Therefore,

$$Y = AB + CD$$

The signal flow is traced by noting simply that two HIGH inputs cause a LOW output between the gates, and this LOW level causes the output of gate 3 to be HIGH.

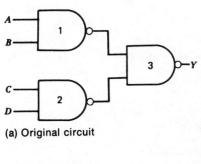

**(a) Original circuit**

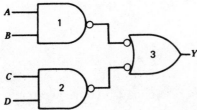

**(b) Circuit with alternate representation for gate 3**

FIGURE 2-18  Two identical circuits with different representations.

## EXAMPLE 2-28

Find an expression for the circuit of Fig. 2-19a. Change the bubbles to make the circuit clearer.

## SOLUTION

The representation of the 2-input NAND gate can be changed so that all inputs to the second stage NAND gate are positive, as shown in Fig. 2-19b. Now, however, the circuit output is negative. It is the product of the 3 inputs, or:

$$\overline{Y} = (\overline{A} + \overline{B})(CD)(E + F)$$

If a positive output is needed, an alternate solution is to change the 3-input NAND gate to its alternate representation, as shown in Fig. 2-19c. Since in this case it is convenient to have all the inputs to this NAND gate negative, the first level AND and OR gates are changed to their alternate representations. From Fig.

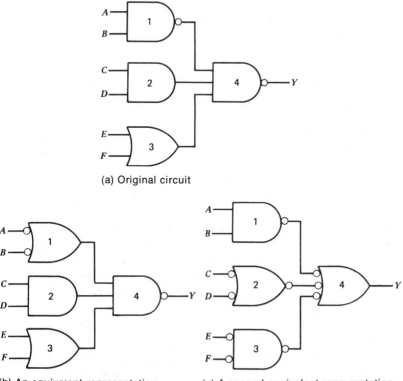

(a) Original circuit

(b) An equivalent representation    (c) A second equivalent representation

FIGURE 2-19   Circuit for Ex. 2-28.

2-19c it is easy to see that the output is HIGH for *any* combination of inputs that makes the output of one or more of the first level gates LOW. Therefore,

$$Y = AB + \bar{C} + \bar{D} + \overline{EF}$$

Note that the outputs of Fig. 2-19b and Fig. 2-19c can be shown to be identical by DeMorgan's theorem.

## 2-11   RELAYS

Algebraic expressions may be implemented using relays. Relays are still widely used in innumerable devices (the pinball machine is one common example of a device using relays), and the functions they perform are described by Boolean equations.

Physically a relay consists of a coil and a set of switch contacts or closures. A specified current in the coil causes the switch contacts to open or close and make or break another electric circuit in which the contacts are connected.

There are two types of relay contacts; *normally open* and *normally closed*. For a relay, the normal condition means its deenergized state, that is, *no* current flow in the relay coil. When current does flow the relay is said to be *energized*; normally open contacts close and normally closed contacts open. Figure 2-20 shows the circuit of a relay and the symbols for normally open and normally closed contacts. We see that a current in the coil magnetically attracts the armature of the normally open contact, causing it to close. The same current and mechanical motion pull down the normally closed contact, causing it to open.

Circuits involving *several* relays quickly become complicated. It is advantageous to use Boolean logic to simplify and reduce them. When Boolean

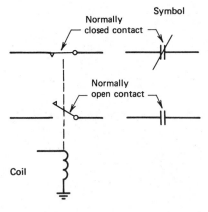

FIGURE 2-20   Diagram of a relay.

algebra is applied to a circuit, a 1 for the output indicates there is a current path through the circuit. But if the output = 0, the circuit is *open*. For a single relay let us define a closed contact as a 1 and an *energized* relay as a 1. Therefore, normally open contacts are written as unprimed variables. For example, a set of normally open contacts on relay A are written as A. If relay A is not energized, there is *no* current path through the contacts and A = 0. Conversely, a set of normally closed contacts is written as $\bar{A}$ because when A = 0, there *is* a path through the contacts, but when the relay is energized, A = 1 and the path opens ($\bar{A}$ = 0).

Typical relay problems involve determining whether a path exists between two points in a relay network, and whether that network can be simplified.

---

### EXAMPLE 2-29

Write the expression for the circuit of Fig. 2-21.

### SOLUTION

Because there is more than one possible path between two points in a circuit, as between points A and B in Fig. 2-21, an OR symbol is appropriate since there may be continuity through one path *or* another. Whenever there is *only one possible path*, as between point B and the output, an AND symbol is correct. In the circuit of Fig. 2-21, relay Z must also be closed to provide continuity between the input and the output.

Examining the circuit with these facts in mind, we write the final expression as:

$$\text{Output} = (W + XY)Z$$

In Fig. 2-21, either W or XY provides a path from A to B, but contact Z must provide the path from B to the output.

### EXAMPLE 2-30

Design a relay circuit to implement the function

$$\text{Output} = A(B + C) + \bar{B}(\bar{C} + DE)$$

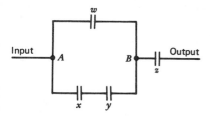

FIGURE 2-21   Relay circuit for Ex. 2-29.

**SOLUTION**

The solution is shown in Fig. 2-22. The equation $A(B + C) + \overline{B}(\overline{C} + DE)$ consists of two terms ORed together. This indicates that there are two major paths between the input and output. The term $A(B + C)$ is implemented by the top path and the term $\overline{B}(\overline{C} + DE)$ is implemented by the lower path.

**EXAMPLE 2-31**

Write the expression for the relay circuit of Fig. 2-23$a$. Simplify it and find a simpler relay circuit that implements the *same* function.

**SOLUTION**

1.   The expression is written by following every possible path from input to output.
2.   Then, the expression is simplified. Thus:

1. Output $= AB + A\overline{C}\overline{B} + A\overline{C}D + CD + C\overline{B} + C\overline{C}B$
2.         $= AB + \overline{B}(A\overline{C} + C) + A\overline{C}D + CD$
           $= AB + \overline{B}A + \overline{B}C + A\overline{C}D + CD$          (Theorem 11)
           $= A + \overline{B}C + A\overline{C}D + CD$                  (Theorem 8)
           $= A + \overline{B}C + CD$                       (Theorem 9)
           $= A + C(\overline{B} + D)$

   The logical equations indicate that if relay A is closed, there is always a path between the input and the output. This is correct, but it is *not* obvious from the illustration. A simpler relay circuit that only uses four contacts and conforms to the simplified final equation is shown in Fig. 2-23$b$.

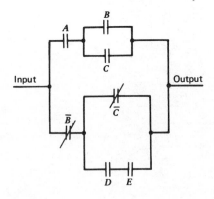

FIGURE 2-22   Circuit for Ex. 2-30.

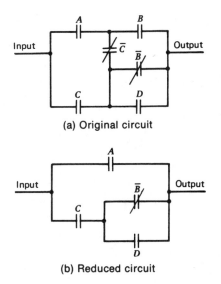

(a) Original circuit

(b) Reduced circuit

FIGURE 2-23   Relay circuit for Ex. 2-31.

## SUMMARY

In this chapter the techniques of manipulating, complementing, and reducing Boolean expressions were explained and demonstrated. DeMorgan's theorem for complementing functions was also discussed. These techniques are often used by design engineers to eliminate unnecessary circuitry.

The AND, OR, NAND, and NOR gates available in the **7400** series were introduced, and methods of producing the Boolean functions using those gates were explained. Relay logic, still used in some applications, was also explained.

## GLOSSARY

**Boolean algebra.** An algebra of two valued variables.

**Bubble.** A symbol (a small circle) indicating where the asserted level of the signal is LOW.

**Complement.** Invert the value of a variable or expression.

**Inverter.** An electronic gate whose output is the complement of its input.

**NAND gate.** A gate combining the functions of ANDing followed by inversion.

**NOR gate.** A gate combining the functions of ORing followed by inversion.

**Normally open contact.** A relay contact that is open when there is no current in the relay coil (relay deenergized).

**Normally closed contact.** A relay contact that is closed when there is no current in the relay coil and opens when current is applied (relay energized).

## REFERENCES

**Louis Nashelsky,** *Introduction to Digital Computer Technology,* Wiley, New York, 2nd ed., 1977.

**Arpad Barna** and **Dan I. Porat,** *Integrated Circuits in Digital Electronics,* Wiley, New York, 1973.

**George K. Kostopoulos,** *Digital Engineering,* Wiley, New York, 1975.

**Frederick J. Hill and Gerald R. Peterson,** *Introduction to Switching Theory and Logical Design,* 2nd ed., Wiley, New York, 1974.

**Alan B. Marcovitz** and **James H. Pugsley,** *An Introduction to Switching System Design,* Wiley, New York, 1971.

**Swaminathan Madhu,** *Theory and Design of Switching Circuits,* Unpublished work, Rochester Institute of Technology, Rochester, New York, 1972.

## PROBLEMS

2-1.   An engine (cooled by water and lubricated by oil under pressure) has a warning signal light that turns ON when one or both the following conditions are present: (1) engine temperature is *high*; (2) engine temperature is *low* but both the water level and the oil pressure are inadequate.

Let $x$, $y$, and $z$ denote, respectively, engine temperature, water level, and oil pressure. Assume that these are measured by sensors that put out either a 0 or a 1 signal. That is, $x = 0$ means temperature is low, $x = 1$ temperature high, and so on.

(a) Set up a truth table with $x$, $y$, and $z$ as the inputs and $T(x, y, z)$ as the output of the warning light control circuit. $T = 1$ indicates that the light goes on.

(b) Obtain a minimal form of $T$ using algebraic manipulation.

2-2.   Refer to Ex. 2-4. If it only cost \$9.00 for gas to get to Rochester, draw up a truth table to show when the girls can go.

2-3.   A machine operates with four essential variables controlling its operation. For the machine to be operating properly at least two of these control variables must be present at the same time. However, when the machine is not operating correctly we wish to have some signal to alert us to the problem. Draw up a truth table and find an expression for the alarm signal.

2-4.   The conditions under which an insurance company will issue a policy are (1) a married female 25 years old or older, or (2) a female under 25 years, or (3) a married male under 25 with no accident record, or (4) a married male with an accident record, or (5) a married male 25 years or older with no accident record. Obtain a simplified logic expression stating to whom a policy can be issued.

2-5.   Simplify each of the following expressions:

(a) $XY(X + Y\bar{Z})$

(b) $T(x,y,z) = (x + \bar{x} + \bar{y})x\bar{z} + x\bar{z}(y + \bar{y})$

(c) $T(x,y,z) = xy + x\bar{y} + \bar{x}z$

(d) $T(w,x,y,z) = (w + x)(x + y)(w + \bar{x} + y + \bar{z}) + \bar{x} + \bar{y} + \bar{w}$

(e) $T(a,b,c,d) = \bar{a}\bar{d}(\bar{b} + \bar{c}) + (\bar{b} + c)(b + \bar{c})$

2-6.    Show that each of the following identities is true:
  (a) $\bar{a}b + ac = ac + bc + \bar{a}\bar{c}b + b\bar{a}\bar{c}d$
  (b) $a\bar{b} + \bar{a}\bar{c} + \bar{a}b + \bar{c}b = \bar{c} + a\bar{b} + \bar{a}b$
  (c) $(X + \bar{Z})(X + Y + \bar{Z})(Y + \bar{Z}) = \bar{Z} + XY$
  (d) $AB + (\overline{AB + \bar{A}\bar{B}}) = A + B$
  (e) $XY\bar{Z} + (XZ + \bar{X}YZ + \bar{Y}Z) = \bar{Z}$
  2-7.    Determine whether each of the following equations is true. If any are false, find a set of values that makes the two sides unequal.
  (a) $(X + \bar{Y} + XY)(X + \bar{Y})\bar{X}Y = 0$
  (b) $xyz + w\bar{y}\bar{z} + wxz = xyz + w\bar{y}z + wx\bar{y}$
  (c) $ab + \bar{a}b + a\bar{b}c = ac + \bar{a}b + \bar{a}bc$
  (d) $X + YZ = XY + \bar{X}\bar{Y}W + \bar{X}YZ + \bar{Y}W$
2-8.    For the circuits of Fig. P2-8, find the logic equations without simplifying; then simplify the circuits.
2-9.    Use relays to implement the following logic expressions. Do not simplify.
  (a) $BD + \bar{B}(A\bar{C} + \bar{A}C)$
  (b) $XY\bar{Z} + XZ(YW + \bar{X}V)$
  (c) $AB(C\bar{B}D + ACD + BDE)$
2-10.    Complement the following expressions.
  (a) $T(a,b,c,d) = \bar{a}d(\bar{b} + c) + \bar{a}\bar{d}(b + c) + (\bar{b} + \bar{c})$
  (b) $T(x,y,z) = (x + y)(\bar{x} + z)(y + z)$
  (c) $T(a,b,c,d,e) = a\bar{b}c + (\bar{a} + b + d)(ab\bar{d} + \bar{e})$
  (d) $(a + b\bar{c})(c + \bar{d}(e + f))$

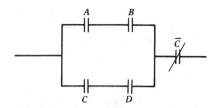

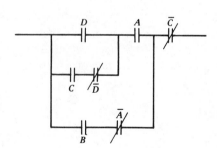

FIGURE P2-8   Circuits for Problem 2-8

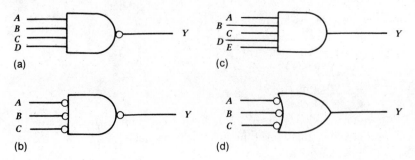

FIGURE P2-12

2-11.   Given $X\bar{Y} + \bar{X}Y = Z$, show that $X\bar{Z} + \bar{X}Z = Y$.

2-12.   Find the alternate representation for the gates of Fig. P2-12.

2-13.   Using only NOR gates, design:
   (a) A 2-input NAND gate.
   (b) A 2-input AND gate.
   (c) A 2-input OR gate.

2-14.   If only inverters and 3-input NAND gates are available, design:
   (a) A 3-input AND gate.
   (b) A 3-input OR gate.
   (c) A 3-input NOR gate.

2-15.   Implement the following expressions. Do not simplify or take complements.
   (a) $(a + b\bar{c})(c + \bar{d}(e + f))$
   (b) $AB + C(\bar{B} + CA) + \overline{AB\bar{D}}$
   (c) $WXY(Z + \overline{WX}) + WX(\bar{Y} + \bar{Z})$
   (d) $[(X + \bar{Y}Z) + W(XY + \bar{Z})]$
   (e) $[(A + B\bar{C})(D + \overline{\bar{A}C})] + CD$
   (f) $[X(\bar{Y} + Z) + (\overline{WX}\bar{Z})]$

2-16.   (a) Find the logic expression for the circuits of Fig. P2-16 without simplifying.
   (b) See if you can draw a simpler circuit.

2-17.   When approaching an intersection, an automobile driver may make a right turn if the traffic light is green, or if it is red and no car is approaching on the intersecting road. Put this statement into Boolean algebra, simplify it, and restate it more simply. (*Hint:* Red = $\overline{\text{Green}}$.)

To be sure you understand this chapter, return to Sec. 2-2 and review the questions. If you cannot answer certain questions, review the appropriate sections of the text to find the answers.

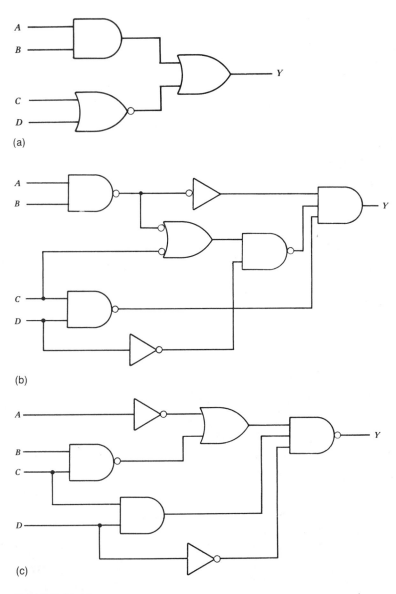

(a)

(b)

(c)

FIGURE  P2-16

# 3
# SYSTEMATIC REDUCTION OF BOOLEAN EXPRESSIONS

## 3-1 INSTRUCTIONAL OBJECTIVES

This chapter introduces the Karnaugh map method of simplifying logic expressions. After reading this chapter the student should be able to:

1. Express a given Boolean function in standard SOP and POS form.
2. Find the numerical equivalent of SOP and POS terms.
3. Construct Karnaugh maps for given functions.
4. Use Karnaugh maps to obtain the minimal form for given expressions.
5. Implement SOP and POS expressions using logic gates.

## 3-2 SELF-EVALUATION QUESTIONS

Watch for the answers to the following questions as you read the chapter. They should help you understand the material presented. When you have finished the chapter, return to this section and be sure you can answer all of the questions.

1. Define an SOP term. What is the difference between an SOP and a standard SOP term?
2. Define a POS term. What is the difference between a POS and a standard POS term?
3. If a term has $k$ literals missing, how many terms does its expansion to a standard POS or SOP form contain?

4.  If $k$ literals are missing from an SOP or POS expression, what specific terms does the expression represent?

5.  Define subcubes. How are they used in minimizing functions?

6.  Why is the largest essential subcube selected in minimizing functions?

7.  How many literals are missing when a 2, 4, or 8 cell subcube is written?

8.  How do you check a simplified expression to be sure all the specified terms are covered?

9.  Why is it wise to start simplifying Karnaugh maps by selecting essential subcubes?

10.  How do you plot a POS Karnaugh map?

11.  When do "don't cares" typically arise? Why are they advantageous to the designer?

## 3-3  STANDARD FORMS—INTRODUCTION

The expressions considered in Chapter 2 were simplified by applying the rules and theorems we had developed. It requires a generous helping of inspiration and experience, however, to decide *which* theorems may be applied usefully to a particular expression. An ounce of inspiration is often worth a pound of logic, but unfortunately there is no formula for inspiration. Consequently, a more systematic way of reducing Boolean expressions must be examined so that the pound of logic is there when the ounce of inspiration is missing! Furthermore, a systematic approach prevents the designer from deluding himself into thinking he has the minimal form of an expression when he has only achieved a partial simplification.

For example, if we try to simplify the expression

$$f(W,X,Y,Z) = \overline{W}\overline{X} + \overline{X}\overline{Y}\overline{Z} + \overline{W}\overline{Z} + YZ$$

we can try a few theorems, but we will achieve no reduction. Does this mean the expression cannot be simplified at all or merely that we have not learned to simplify it? The application of systematic logic answers this type of question.

In order to develop a systematic method of reducing Boolean expressions the following definitions are required:

□  **Literal.** The occurrence of a variable in either its complemented or uncomplemented state in an expression. The term $A\overline{B}C$ *consists of three literals:* A, $\overline{B}$, and C.

□  **Product term.** A product term consists solely of the product of literals. $A\overline{B}C$ is a product term, while $A(\overline{B} + C)$ is not a product term because of the + sign.

- **Sum term.** A sum term consists solely of the sum of literals. $A + \bar{B} + C$ is a sum term; similarly $A(B + C)$ is not a sum term because of the implied multiplication.
- **Domain.** The domain of a function is the set of variables for which that function exists. This may be defined *explicitly* or *implicitly*. Explicitly, the expression $f(x,y,z)$ means the function exists for the variables $x$, $y$, and $z$. Implicitly, domains are implied by word statements of a problem. Such statements must be examined to determine what the pertinent variables are; these form the domain of the function. In Ex. 2-1 the domain of the contract problem proved to be the president and his three assistants, or the variables $P$, $A_1$, $A_2$, $A_3$.
- **Standard product term.** A product term that contains *one* literal from *every* variable in the domain of the function. For example, if the domain is A, B, C, and D, the term $A\bar{B}CD$ is a standard product term while the term ACD is not because the literal B is missing.
- **Standard sum term.** A *sum* term that contains *one* literal from every variable in the domain. $\bar{A} + \bar{B} + \bar{C} + \bar{D}$ is a standard sum term, but $A + B + D$ is not.
- **Sum of products form.** An expression is in a sum of products (SOP) form if it is a sum of product terms. For example, $A\bar{B} + ABC$ is in SOP form. An expression is in *standard SOP form* if *each term* in the sum is a *standard product term*.
- **Product of sums form.** An expression is in the product of sums (POS) form if it consists only of the product of a group of sums. The expression $(A + B)(A + B + \bar{C})$ is in POS form. If the expression is the product of a group of *standard sum terms* it is in *standard POS form*

## 3-3.1  Conversion of Product Terms to Standard SOP Form

A product term that is not a standard product term because it has several literals missing can be expanded into the sum of several standard product terms. Consider a product term from a domain of $n$ variables. If the term has $k$ literals missing and $n-k$ literals specified, the term can be expanded into the standard SOP form where each standard product term has the $n-k$ literals that are specified and the terms consist of all combinations of the $k$ literals that are missing. Therefore, a product term with $k$ variables missing expands into $2^k$ standard terms to include all possible combinations of the $k$ variables. To demonstrate how the conversion to standard SOP form works, consider:

$$f(V,W,X,Y) = V\bar{W}\bar{X} \tag{3-1}$$

Note in Eq. (3-1) the Y term is missing. Equation (3-1) can be multiplied by $Y + \bar{Y}$ (which equals 1), however, to give:

$$V\bar{W}\bar{X}(Y + \bar{Y}) = V\bar{W}\bar{X}Y + V\bar{W}\bar{X}\bar{Y} \tag{3-2}$$

The function is now in standard SOP form because each term contains one literal for each variable in the domain.

If another variable, Z for instance, were included in the domain, Eq. (3-2) could be multiplied by $Z + \bar{Z}$. The result would be 4 terms. The terms would include all possible combinations of the missing variables (Y and Z) and each term would include $VW\bar{X}$.

---

### EXAMPLE 3-1

Given $f(a,b,c,d,e) = \bar{a}b\bar{d}$, write $f$ in standard SOP form.

#### SOLUTION

The missing variables are $c$ and $e$. Because 2 variables are missing the solution will contain 4 terms. Each term will contain 1 combination of $c$ and $e(\bar{c}\bar{e}, \ c\bar{e}, \ \bar{c}e, \ \text{or} \ ce)$ and the specified variables $\bar{a}b\bar{d}$. Therefore,

$$f(a,b,c,d,e) = \bar{a}b\bar{d}$$
$$= \bar{a}b\bar{c}\bar{d}\bar{e} + \bar{a}bc\bar{d}\bar{e} + \bar{a}b\bar{c}\bar{d}e + \bar{a}bc\bar{d}e$$

---

### EXAMPLE 3-2

Given $f(a,b,c,d) = a\bar{b} + ac\bar{d}$, write it in standard SOP form.

#### SOLUTION

The given function consists of 2 product terms. Each can be expanded into standard SOP form. The first term has 2 literals missing and will result in 4 product terms. The second term has only 1 literal missing and will result in 2 SOP terms. Thus the answer is assumed to consist of 6 SOP terms. However some of the 6 terms may occur more than once. In that case, the *redundant* terms may be eliminated (Theorem 6).[1]

$$f(a,b,c,d) = a\bar{b} + ac\bar{d}$$
$$= a\bar{b}\bar{c}\bar{d} + a\bar{b}\bar{c}d + a\bar{b}c\bar{d} + a\bar{b}cd + abc\bar{d} + a\bar{b}c\bar{d}$$

$$\underbrace{\phantom{a\bar{b}\bar{c}\bar{d} + a\bar{b}\bar{c}d + a\bar{b}c\bar{d} + a\bar{b}cd}}_{\text{Expansion of the term } a\bar{b}} \quad \underbrace{\phantom{abc\bar{d} + a\bar{b}c\bar{d}}}_{\substack{\text{Expansion of} \\ \text{the term } ac\bar{d}}}$$

$$= a\bar{b}\bar{c}\bar{d} + a\bar{b}\bar{c}d + a\bar{b}c\bar{d} + a\bar{b}cd + abc\bar{d}$$

The term $a\bar{b}c\bar{d}$ occurred in both expansions, but it need only be written once, as the final answer shows.

---

[1]See Sec. 2-5.

## 3-3.2 Expansion of Sum Terms to Standard POS Form

Sum terms can be expanded into standard POS form in a manner similar to the expansion of product terms into standard SOP form. Expanding a sum term with $k$ missing literals results in the product of $2^k$ POS terms and the terms contain all the $2^k$ possible combinations of the unspecified terms.

To demonstrate the expansion, consider:

$$
\begin{aligned}
f(X,Y,Z) &= X + \bar{Y} \\
&= X + \bar{Y} + \bar{Z}Z && \text{(Theorem \quad 5)}^2 \\
&= (X + \bar{Y} + Z)(X + \bar{Y} + \bar{Z}) && \text{(Theorem \quad 13)}^2
\end{aligned}
$$

Since the term $\bar{Z}Z = 0$, it may be added to an expression whenever it is helpful and it does not change the value of the expression.

---

**EXAMPLE 3-3**

Express $f(A,B,C,D) = (A + \bar{C} + \bar{D})(A + \bar{B})$ in standard POS form.

**SOLUTION**

The first term in the product, $A + \bar{C} + \bar{D}$, has the variable $B$ missing and expands to $(A + B + \bar{C} + \bar{D})(A + \bar{B} + \bar{C} + \bar{D})$. The second term has the variables $C$ and $D$ missing and expands to

$$A + \bar{B} = (A + \bar{B} + \bar{C} + \bar{D})(A + \bar{B} + \bar{C} + D)$$
$$\cdot (A + \bar{B} + C + \bar{D})(A + \bar{B} + C + D)$$

Therefore,

$$
\begin{aligned}
f(A,B,C,D) &= (A + \bar{C} + \bar{D})(A + \bar{B}) \\
&= (A + B + \bar{C} + \bar{D})(A + \bar{B} + \bar{C} + \bar{D}) \\
&\quad \cdot (A + \bar{B} + \bar{C} + \bar{D})(A + \bar{B} + \bar{C} + D) \\
&\quad \cdot (A + \bar{B} + C + \bar{D})(A + \bar{B} + C + D) \\
&= (A + B + \bar{C} + \bar{D})(A + \bar{B} + \bar{C} + \bar{D}) \\
&\quad \cdot (A + \bar{B} + \bar{C} + D)(A + \bar{B} + C + \bar{D}) \\
&\quad \cdot (A + \bar{B} + C + D)
\end{aligned}
$$

As before, the term $A + \bar{B} + \bar{C} + \bar{D}$ appeared twice and is written only once.

---

## 3-3.3 Numerical Representation of SOP Forms

Functions in standard SOP form are often represented by the sum ($\Sigma$) of a group of numbers, where each number corresponds to a particular SOP

---

[2]See Sec. 2-5.

term. The numeric form is found by assuming that the *first* variable in the domain is the *most significant* bit of binary number and proceeding until the *last* variable, which is the *least significant* bit of the number. In standard SOP form uncomplemented literals are assigned the value of 1 and complemented literals are assigned a value of 0. We arrive at a numerical result simply by reading each term as a binary number and writing its decimal equivalent.

---

**EXAMPLE 3-4**

Write the numerical form of the standard SOP function:

$$f(a,b,c,d) = \bar{a}\bar{b}\bar{c}\bar{d} + \bar{a}\bar{b}cd + \bar{a}b\bar{c}d + ab\bar{c}d + abc\bar{d}$$

**SOLUTION**

We can attach numbers to each term as described above.

$$f(a,b,c,d) = \bar{a}\bar{b}\bar{c}\bar{d} + \bar{a}\bar{b}cd + \bar{a}b\bar{c}d + ab\bar{c}d + abc\bar{d}$$

| 0000 | 0011 | 0101 | 1101 | 1110 |
|------|------|------|------|------|
| 0 | 3 | 5 | 13 | 14 |

Therefore,

$$f(a,b,c,d) = \Sigma(0,3,5,13,14)$$

**EXAMPLE 3-5**

If $f(W,X,Y,Z) = \Sigma(1,5,13,15)$, write $f$ in standard SOP form.

**SOLUTION**

In this example, we must write the literals to correspond to the given numbers.

$$\begin{aligned} f(W,X,Y,Z) &= \Sigma(1,5,13,15) \\ &= 0001 + 0101 + 1101 + 1111 \\ &= \bar{W}\bar{X}\bar{Y}Z + \bar{W}X\bar{Y}Z + WX\bar{Y}Z + WXYZ \end{aligned}$$

---

With a little practice one can go directly from the numerical representation to the SOP form and back without need of the intervening step.

If a product term is not in standard form, it becomes the sum of several terms or several numbers. It is possible to go from the literal form directly to the numbers by placing an X wherever a literal is missing and then writing numbers corresponding to the given literals and all possible combinations of X values.

**EXAMPLE 3-6**

If $f(a,b,c,d) = a\bar{c}\bar{d}$, find the numeric SOP representation of the function.

**SOLUTION**

The function $a\bar{c}\bar{d}$ corresponds to 1X00, where X is used in place of the missing literal $b$. Substituting 0 and 1 for X, we get the numbers 8 and 12 as a result. Therefore,

$$f(a,b,c,d) = a\bar{c}\bar{d} = \Sigma(8,12)$$

Indeed 8 and 12 are the only two 4-bit numbers where the most significant bit is 1 and the least significant bits are both 0.

**EXAMPLE 3-7**

If $f(a,b,c,d,e) = \bar{b}e$, express $f$ in numerical form.

**SOLUTION**

Since 3 variables are missing, we expect the result to contain $2^3 = 8$ terms. We can write

$$f(a,b,c,d,e) = \bar{b}e = \text{X0XX1}$$

where the Xs correspond to the missing variables $a$, $c$, and $d$, the 0 corresponds to $\bar{b}$, and the 1 in the least significant bit position corresponds to $e$. The results are shown in Fig. 3-1. First the $X_1$, $X_2$, and $X_3$ values are listed simply as the binary numbers from 0 to 7. This assures us 8 different combinations for the Xs. Then the $X_1$, $X_2$, and $X_3$ values are substituted into the term X0XX1 and the numerical results are simply the decimal equivalents of the resulting binary numbers. Therefore,

$$f(a,b,c,d,e) = \bar{b}e = \Sigma(1,3,5,7,17,19,21,23)$$

## 3-3.4 Numerical Representation of POS Forms

At this point an important distinction between SOP and POS forms must be made:

*If a function is expressed in SOP form, it equals 1 if any term in the sum equals 1 and 0 otherwise. But if a function is expressed in POS form, it equals 0 if any term in the product is 0 and 1 otherwise.*

| $X_1$ $X_2$ $X_3$ | X 0 X X 1 | Numerical Result |
|---|---|---|
| 0 0 0 | 0 0 0 0 1 | 1 |
| 0 0 1 | 0 0 0 1 1 | 3 |
| 0 1 0 | 0 0 1 0 1 | 5 |
| 0 1 1 | 0 0 1 1 1 | 7 |
| 1 0 0 | 1 0 0 0 1 | 17 |
| 1 0 1 | 1 0 0 1 1 | 19 |
| 1 1 0 | 1 0 1 0 1 | 21 |
| 1 1 1 | 1 0 1 1 1 | 23 |

FIGURE 3-1   Expansion of the function $be$.

Functions in the standard POS form are listed as a product ($\Pi$) of a group of numbers. *These numbers correspond to input values that cause the value of the function to be 0. A sum term can be 0 only if all the literals making up that sum are 0.* Therefore, a term such as $W + \bar{X} + Z$ equals 0 only if $W = 0$, $X = 1$, and $Z = 0$. For sum terms a numerical representation is obtained by assigning a *value of 0 to each uncomplemented variable and a value of 1 to each complemented variable.* This is the reverse of the procedure for obtaining a numerical representation of a product term.

---

**EXAMPLE 3-8**

What is the numerical representation of the sum term $\bar{W} + X + Y + \bar{Z}$?

**SOLUTION**

By assigning a 0 to the uncomplemented variables and a 1 to the complemented variables, we get

$$\bar{W} + X + Y + \bar{Z}$$
$$1 \quad\;\; 0 \quad 0 \quad\;\; 1$$

Since this is the binary equivalent of 9, the numerical representation of $\bar{W} + X + Y + \bar{Z}$ is 9.

Note that if $f(W,X,Y,Z) = \bar{W} + X + Y + \bar{Z}$, f is 0 only if $W = 1$, $X = 0$, $Y = 0$, and $Z = 1$, which corresponds to a numeric value of 9. For any other combination of input variables, at least one of the literals has a value of 1 and $f$ equals 1.

**EXAMPLE 3-9**

(a) Explain the meaning of

$$f(a,b,c,d) = \Pi(0,4,7,11,14,15)$$

(b)   Express it as literals in standard POS form.

**SOLUTION**

(a) Because of the product ($\Pi$) sign the function equals 0 when the numeric value of the input is 0, 4, 7, 11, 14, or 15. By using the rules developed in this section we can translate $f$ into literals:

(b) $f(a,b,c,d) = \Pi(0,4,7,11,14,15)$
$= (a + b + c + d)(a + \bar{b} + c + d)(a + \bar{b} + \bar{c} + \bar{d})$
$\cdot (\bar{a} + b + \bar{c} + \bar{d})(\bar{a} + \bar{b} + \bar{c} + d)(\bar{a} + \bar{b} + \bar{c} + \bar{d})$

where each sum term corresponds to one of the numbers in the original expression.

---

The numeric representation of sum terms that have one or more literals missing can be obtained by using the given literals and every combination of the unspecified literals. This is similar to finding the numeric value for SOP terms.

---

**EXAMPLE 3-10**

If $f(a,b,c,d) = a + \bar{c}$, find its numeric representation in POS form.

**SOLUTION**

The term $a + \bar{c}$ can be written as 0X1X, where the 0 in the most significant position represents $a$, the first X represents the missing $b$ literal, the 1 represents $\bar{c}$, and the second X represents the missing $d$ literal. All four combinations of Xs will be included in the numerical representation. Therefore,

$$f(a,b,c,d) = \Pi(0010, 0011, 0110, 0111)$$
$$= \Pi(2,3,6,7)$$

**EXAMPLE 3-11**

If

$$f(W,X,Y,Z) = (\bar{W} + Y)(W + \bar{Y} + Z)(X + Y + \bar{Z})$$

find the numeric representation of $f$.

SOLUTION

The first term, $\overline{W} + Y$, expands into 1X0X or 8, 9, 12, and 13. The term $W + \overline{Y} + Z$ expands to 0X10 or 2 and 6, and the term $X + Y + \overline{Z}$ expands to X001 or 1 and 9. Therefore,

$$f(W,X,Y,Z) = \Pi(1,2,6,8,9,12,13)$$

The number 9 occurred twice but should only be written once.

## 3-3.5  AND-OR and NAND Implementation of SOP Forms

Any expression in SOP form can be implemented by using AND gates followed by OR gates. This AND-OR implementation is accomplished by ANDing together the literals that comprise each term and ORing the output of each AND gate.

SOP expressions can also be implemented by using two levels of NAND gates. The first level NANDs the literals of each term producing a low output and a second level NAND gate, serving as an OR for negative inputs, ORs together the outputs of the first level NAND gates. In practice, the two level NAND gate implementation is used more often because it requires only one type of gate.

### EXAMPLE 3-12

The expression for segment A of a 7-segment display (see Sec. 3-8.3) is:

$$\text{Segment A} = A + BD + \overline{B}\,\overline{D} + CD$$

Implement this expression:
(a) Using AND-OR logic.
(b) Using NAND logic.

SOLUTION

(a) The simplest solution for the AND-OR circuit is to combine the last three terms using AND gates and then OR the four terms together as shown in Fig. 3-2a.[3] If only uncomplemented variables are available, $\overline{B}$ and $\overline{D}$ can be obtained from inverters. An alternate way to obtain $\overline{B}\,\overline{D}$ is to use a NOR gate (if available) in place of the AND gate since $\overline{B}\,\overline{D} = \overline{(B + D)}$. This circuit is shown in 3-2b.

(b) The solution using only NAND gates is shown in Fig. 3-2c. The last three terms are combined using NAND gates instead of AND gates. Note for this solution the A variable must be inverted because a LOW input to the second level NAND gate is needed to produce a HIGH output.

---

[3] The 4-input OR gate shown functionally in Fig. 3-2 is not available in a TTL package. It can be constructed from three 2-input OR gates, as explained in Chapter 2.

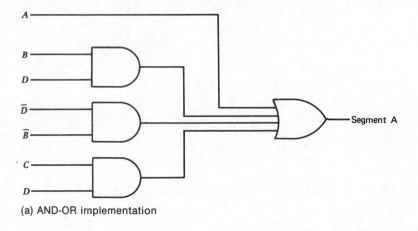

(a) AND-OR implementation

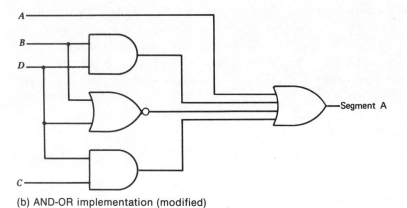

(b) AND-OR implementation (modified)

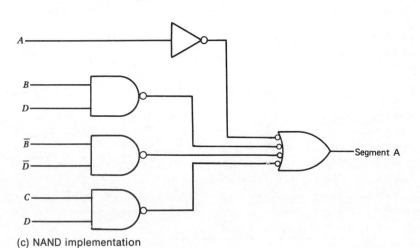

(c) NAND implementation

Figure 3-2  Three ways of building segment A of a 7-segment decoder.

## 3-3.6   NOR Implementation of POS Expressions

Expressions in POS form can be implemented by either a two level OR-AND circuit or a two level NOR circuit. For the two level OR-AND circuit, the first level gates OR together the literals comprising each term. The second level gate ANDs together the outputs of the OR gates. In the NOR implementation, the first level gates NOR together the literals for each term and the second level gate functions as an AND gate for LOW inputs to AND together the outputs of the first level gates.

---

**EXAMPLE 3-13**

The POS expression for segment A of a 7-segment decoder is:

$$\text{Segment A} = (\bar{B} + D)(A + B + C + \bar{D})$$

Implement this expression:
  (a) Using OR-AND logic.
  (b) Using NOR logic.

**SOLUTION**

For the OR-AND implementation, the first term becomes a 2-input OR gate and the second term becomes a 4-input OR gate. The outputs of the two OR gates are then ANDed to produce the final output as shown in Fig. 3-3a.

  For the NOR implementation the first level gates consist of a 2-input NOR and a 4-input NOR, and the second level gate is another 2-input NOR gate. The three NOR gates that implement the function are shown in Fig. 3-3b.

---

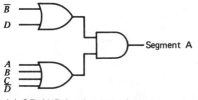

(a) OR-AND implementation

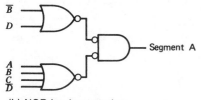

(b) NOR implementation

FIGURE 3-3   Two implementations of the POS form for segment A of a 7-segment decoder.

## 3-4 EQUIVALENCE OF SOP AND POS STANDARD FORMS

The SOP and POS numeric representations for a function are complementary. *The numbers that appear in the SOP representation are not in the POS representation and all numbers that are not in the SOP representation appear in the POS representation.*

---

### EXAMPLE 3-14

If $f(a,b,c,d) = \Sigma(1,2,5,6,7,10,11,12,13)$, find its numeric and literal POS representation.

### SOLUTION

The domain of the function contains 4 variables, which can be described by the numbers 0 through 15. All the numbers not in the SOP form appear in the POS representation. Therefore,

$$f(a,b,c,d) = \Pi(0,3,4,8,9,14,15)$$
$$= (a + b + c + d)(a + b + \bar{c} + \bar{d})(a + \bar{b} + c + d)$$
$$\cdot (\bar{a} + b + c + d)(\bar{a} + b + c + \bar{d})$$
$$\cdot (\bar{a} + \bar{b} + \bar{c} + d)(\bar{a} + \bar{b} + \bar{c} + \bar{d})$$

---

## 3-4.1 Truth Table Verification

To verify the equivalence of the SOP and POS numeric forms, consider the function defined by the truth table of Fig. 3-4. In SOP form:

$$f(X,Y,Z) = \Sigma(0,2,4,5,7)$$

| X | Y | Z | F(X, Y, Z) |
|---|---|---|:---:|
| 0 | 0 | 0 | 1 |
| 0 | 0 | 1 | 0 |
| 0 | 1 | 0 | 1 |
| 0 | 1 | 1 | 0 |
| 1 | 0 | 0 | 1 |
| 1 | 0 | 1 | 1 |
| 1 | 1 | 0 | 0 |
| 1 | 1 | 1 | 1 |

FIGURE 3-4  Truth table for the function considered in Sec. 3-4.1.

The SOP form lists the numbers for which the value of the function is 1.

To express the same function in POS form, the numbers where the function is 0 are used:

$$f(X,Y,Z) = \Pi(1,3,6)$$

---

**EXAMPLE 3-15**

Show that the two forms for $f(X,Y,Z)$ are identical using algebraic manipulation.

**SOLUTION**

Starting with the SOP form and simplifying, we find that

$$\begin{aligned}
f(X,Y,Z) &= \Sigma(0,2,4,5,7) \\
&= \overline{X}\overline{Y}\overline{Z} + \overline{X}Y\overline{Z} + X\overline{Y}\overline{Z} + X\overline{Y}Z + XYZ \\
&= \overline{X}\overline{Z}(\overline{Y} + Y) + X\overline{Y}(\overline{Z} + Z) + XZ(Y + \overline{Y}) \\
&= \overline{X}\overline{Z} + X\overline{Y} + XZ
\end{aligned}$$

Starting with the POS form and simplifying, we find that

$$\begin{aligned}
f(X,Y,Z) &= \Pi(1,3,6) \\
&= (X + Y + \overline{Z})(X + \overline{Y} + \overline{Z})(\overline{X} + \overline{Y} + Z) \\
&= (X + \overline{Z})(\overline{X} + \overline{Y} + Z) \qquad\qquad \text{(Theorem 13)}[4] \\
&= X\overline{Y} + XZ + \overline{X}\overline{Z} + \overline{Y}\overline{Z} \\
&= X\overline{Y} + XZ + \overline{X}\overline{Z} \qquad\qquad \text{(Theorem 12)}[4]
\end{aligned}$$

Thus the POS and SOP forms reduce to identical terms.

---

## 3-4.2  Verification Using DeMorgan's Theorem

For a function with a domain of N variables there are $2^N$ standard SOP terms. *For any given combination of input variables, however, only one of the $2^N$ terms can equal 1.* If an SOP function contains all $2^N$ standard product terms, it will equal 1 because one of the terms must equal 1 and the rest must equal 0 for *any and every* combination of the input variables. Therefore, if a function $f$ is defined in numeric SOP form, all the standard SOP terms that were not included in $f$ must together comprise $\overline{f}$ because if the term that equals 1 is in $f$, it is not in $\overline{f}$, and vice versa. In Ex. 3-15,

$$f(X,Y,Z) = \Sigma(0,2,4,5,7)$$

[4]See Sec. 2-5.

Therefore, the SOP form of

$$\bar{f}(X,Y,Z) = \Sigma(1,3,6)$$
$$= \overline{X}\overline{Y}Z + \overline{X}YZ + XY\overline{Z}$$

If, for example, $\overline{X}\overline{Y}Z = 1$, then $X = 0$, $Y = 0$, and $Z = 1$. This combination of variables makes $\bar{f} = 1$ and $f = 0$ since for these variables every term in $f$ will be 0. To verify the above statements, let us take the complement of $\bar{f}$.

$$\overline{\overline{f}(X,Y,Z)} = f(X,Y,Z,)$$
$$\overline{\overline{X}\overline{Y}Z + \overline{X}YZ + XY\overline{Z}} = (X + Y + \overline{Z})(X + \overline{Y} + \overline{Z})(\overline{X} + \overline{Y} + Z)$$

But this is exactly $f(X,Y,Z)$ in POS form, which verifies our statement.

Similarly, *only one of all the possible POS terms can equal 0; the rest must equal 1. Therefore, $\bar{f}$ can also be expressed in numeric POS form as the product of all the numbers not included in f.* Again referring to Ex. 3-15:

$$f(X,Y,Z) = \Pi(1,3,6)$$

Therefore,

$$\bar{f}(X,Y,Z) = \Pi(0,2,4,5,7)$$
$$= (X + Y + Z)(X + \overline{Y} + Z)(\overline{X} + Y + Z)$$
$$\cdot (\overline{X} + Y + \overline{Z})(\overline{X} + \overline{Y} + \overline{Z})$$

If the complement of $\bar{f}$ is taken, the result will be $f(X,Y,Z)$ in SOP form.

## 3-5  KARNAUGH MAPS

Karnaugh mapping is a graphic technique for reducing functions to their simplest terms. Once a function is expressed in standard SOP or POS form, it can be plotted on a Karnaugh map, which leads to a systematic minimization of the function. Karnaugh mapping is a viable reduction technique for functions of 3, 4, and 5 variables. For functions of more than 5 variables, more advanced techniques such as the Quine–McClusky algorithm are used.[5] However, these are more elaborate and cumbersome techniques and only Karnaugh mapping will be discussed in this book.

[5]For a discussion of the Quine–McClusky techniques, see F.J. Hill and G.R. Peterson, *Introduction to Switching Theory and Logical Design*, 3rd ed. (New York: Wiley, 1981), Chapter 7.

### 3-5.1 Construction of the 3-Variable Karnaugh Maps

A Karnaugh map for a function of 3 variables, $f(X,Y,Z)$, is shown in Fig. 3-5a. There are 8 squares within the map, one for each possible number in the decimal representation of the equation. The squares are arranged in a 4-by-2 matrix as shown. The two possible values for the most significant variable (X in this case) head the two columns. The four possible values for the two remaining variables are listed alongside the map. Note that the 11 coordinate is listed above the 10 coordinate. This is deliberately done so that *no coordinate differs from its adjacent coordinate in more than one bit position*.

Figure 3-5b shows the decimal value for each box. This value corresponds to the X, Y, and Z coordinates for that box. If a function is specified in standard SOP form, a 1 is entered into the Karnaugh map for each number in the decimal specification.

---

**EXAMPLE 3-16**

Construct the Karnaugh map for the function

$$f(a,b,c) = \bar{a}b\bar{c} + a\bar{b}c + b\bar{c}$$

**SOLUTION**

The solution is shown in Fig. 3-5c. First the 3-variable map is drawn as in Fig. 3-5a. The given function is not in standard SOP form so it is expanded using the methods of Sec. 3-3.

$$
\begin{aligned}
f(a,b,c) &= \bar{a}b\bar{c} + a\bar{b}c + b\bar{c} \\
&= \bar{a}b\bar{c} + a\bar{b}c + ab\bar{c} + \bar{a}b\bar{c} \\
&= \Sigma(0,5,6,2)
\end{aligned}
$$

1s are then placed in the Karnaugh map at the 0, 2, 5, and 6 locations and the map is complete, as shown.

---

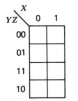

(a) Three-variable map

(b) Numeric placement on a three-variable map

(c) Map for the function $f(a,b,c) = \Sigma(0,2,5,6)$

FIGURE 3-5 Three-variable Karnaugh maps.

## 3-5.2 Simplification of 3-Variable SOP Expressions

A Boolean expression can be simplified using Karnaugh maps by the following procedure:

1. Write the expression in standard SOP form.
2. Plot the Karnaugh map.
3. Form subcubes that cover all the 1s in the expression.
4. Write the resulting expression.

If the subcubes are properly chosen, the result is the expression in minimum SOP form.

   *Subcubes are groups of 1s that are adjacent to each other on a Karnaugh map.* A 2-cell subcube consists of two adjacent 1s, which means the cells containing the 1s border each other in either a vertical or horizontal line. Cells next to each other diagonally are not adjacent. For any 2-cell subcube we find that one variable changes values, but the rest of them have a common value. *The SOP expression for the subcube consists of those variables with a common value. The variable that changes values is not needed in the expression.*

---

### EXAMPLE 3-17

Given $f(a,b,c) = \Sigma(0,2,6)$ draw the Karnaugh map and simplify the function.

### SOLUTION

The Karnaugh map for the given function is shown in Fig. 3-6, where three 1s are written in the squares whose coordinates are 0, 2, and 6. The 1s in the 2 and 6 squares are adjacent and form a 2-cell subcube. The coordinates corresponding to 2 and 6 are 010 and 110, so that $a$ (the most significant variable) changes, while $b = 1$ and $c = 0$ for both squares. Therefore, the subcube is written as $b\bar{c}$.

   The 1 in the 0 square seems all alone. Anytime a 1 is indeed all alone and cannot be combined with any other 1 to form a subcube, it is a *1-cell subcube* and is represented by a literal for each variable in the domain. The 1 in the 0 square is written as $\bar{a}\bar{b}\bar{c}$ and

$$f(a,b,c) = \bar{a}\bar{b}\bar{c} + b\bar{c}$$

---

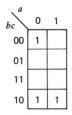

FIGURE 3-6   Karnaugh map for Ex. 3-17.

After examining the solution of Ex. 3-17, we realize it can be simplified.

$$f(a,b,c) = \bar{c}(\bar{a}\bar{b} + b)$$
$$= \bar{c}(\bar{a} + b) \qquad \text{(Theorem 11)}[6]$$
$$= \bar{a}\bar{c} + b\bar{c}$$

It appears the Karnaugh map didn't yield the simplest expression after all, but this is incorrect. Heretofore we have ignored the fact that *Karnaugh maps are spherical*; that is, *they wrap around themselves*. For an 8-cell map this means that *the top row is adjacent to the bottom row*. In this example a second subcube can be formed by the 0 and 2 cells. The 0 cell is numerically 000 while the 2 cell is 010. Therefore only the *b* variable changes; it is excluded and the subcube is represented by $\bar{a}\bar{c}$.

---

**EXAMPLE 3-18**

Find the minimal expression for the function

$$f(a,b,c) = \Sigma(0,2,5,6)$$

**SOLUTION**

This function was plotted in Fig. 3-5c. As in Ex. 3-17, the 0, 2, and 6 cells can be combined into 2-cell subcubes. Since the 1 in the 5 cell is truly alone and cannot be combined, it is expressed as $a\bar{b}c$. Therefore,

$$f(a,b,c) = \Sigma(0,2,5,6)$$
$$= \bar{a}\bar{c} + b\bar{c} + a\bar{b}c$$

Note that this expression is simpler than the original expression given in Ex. 3-16, from which the Karnaugh map was drawn.

---

## 3-5.3 Four-Cell Subcubes

The size of a subcube must be a *power* of 2 (i.e., only 2, 4, 8, 16, etc. subcubes exist). For example, a 3-cell subcube is not allowed. If a row of three adjacent 1s appears on a map, they must be combined into two 2-cell subcubes.

Four-cell subcubes are also allowed. A 4-cell subcube appears on a map as either a row of four 1s or a two-by-two square of 1s. For a 4-cell subcube, 2 variables will change and they are omitted. A 4-cell subcube is expressed by 2 fewer literals than the domain of the expression.

[6]See Sec. 2–5.

**EXAMPLE 3-19**

Find the minimal expression for the function,

$$f(X,Y,Z) = \Sigma(0,1,2,3,4,6)$$

**SOLUTION**

The Karnaugh map for the function is shown in Fig. 3-7. There is a column of four 1s in the X = 0 column. This forms a 4-cell subcube that is described by $\overline{X}$. (Note that for this 4-cell subcube the variables Y and Z changed and became the omitted literals.) The two other 1s at 100 and 110 are adjacent to each other because the top row is adjacent to the bottom row, and can be covered by a 2-cell subcube. However, they can also be covered by a 4-cell subcube consisting of the 000, 100, 010, and 110 cells. These form a two-by-two square because the top and bottom rows are adjacent. When a choice exists, it is wisest to use the *largest* subcube possible, because this will result in an expression with the fewest literals. The 4-cell subcube here is written as $\overline{Z}$ since each cell in the subcube has a Z value of 0. Therefore,

$$f(X,Y,Z) = \Sigma(0,1,2,3,4,6)$$
$$= \overline{X} + \overline{Z}$$

As a check, $\overline{X} = \Sigma(0,1,2,3)$ and $\overline{Z} = \Sigma(0,2,4,6)$. The total sum covers all the terms in $f$, and no terms that are not in $f$.

## 3-5.4 Construction of a 4-Variable Map

Functions of four variables require a map with 16 entries. A 4-variable Karnaugh map is shown in Fig. 3-8a. The four possible combinations for the two most significant variables (W and X) are listed along the top of the map and the four combinations for the least significant variables (Y and Z) are listed along the side. Again the order of listing is 00, 01, 11, and 10, for both the top and side variables. The numeric terms for each of the 16 squares in the map is shown in Fig. 3-8b.

|  YZ \ X | 0 | 1 |
|---|---|---|
| 00 | 1 | 1 |
| 01 | 1 |  |
| 11 | 1 |  |
| 10 | 1 | 1 |

FIGURE 3-7  Karnaugh map for Ex. 3-19.

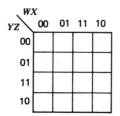

(a) The general 4-variable map

| WX YZ | 00 | 01 | 11 | 10 |
|---|---|---|---|---|
| 00 | 0 | 4 | 12 | 8 |
| 01 | 1 | 5 | 13 | 9 |
| 11 | 3 | 7 | 15 | 11 |
| 10 | 2 | 6 | 14 | 10 |

(b) The map with numeric terms entered

FIGURE 3-8  The 4-variable Karnaugh map.

## EXAMPLE 3-20

Construct the Karnaugh map for the function,

$$f(W,X,Y,Z) = \Sigma(0,3,5,9,12,15)$$

### SOLUTION

The solution is shown in Fig. 3-9 where 1s are written in the Karnaugh map for each term in the specification. For example, the 1 for the number 12 (binary 1100) is written in the square that is the intersection of 11 for the most significant variables and 00 for the least significant variables.

## EXAMPLE 3-21

What combination of variables does the square with the letter A in it in Fig. 3-9 represent?

### SOLUTION

The letter A is in the square at the intersection of the 01 and 10 coordinates. Therefore, this square is designated by the number 0110 or decimal 6.

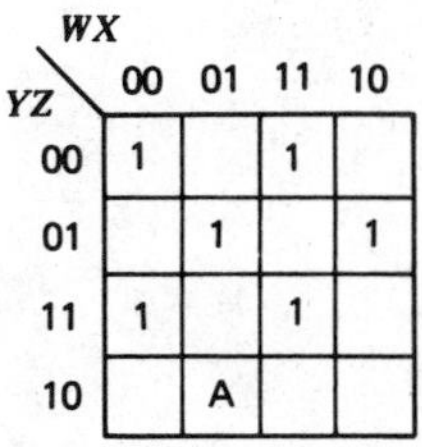

| WX YZ | 00 | 01 | 11 | 10 |
|---|---|---|---|---|
| 00 | 1 | | 1 | |
| 01 | | 1 | | 1 |
| 11 | 1 | | 1 | |
| 10 | | A | | |

FIGURE 3-9  Karnaugh map for Exs. 3-20 and 3-21.

## 3-5.5 Simplification of 4-Variable Maps for SOP Expressions

Four-variable maps are also simplified by following the procedure developed for 3-variable maps in Sec. 3-5.2. Four-variable maps are spherical also. This means the top row is adjacent to the bottom row and the leftmost column is adjacent to the rightmost column. Eight-cell subcubes may exist in a 4-variable map; if so, they must form a 2-by-4 matrix, containing any 2 complete adjacent rows or columns.

---

**EXAMPLE 3-22**

Find the minimal expression for

$$f(W,X,Y,Z) = \Sigma(0,1,2,3,4,8,9,10,15)$$

**SOLUTION**

First the Karnaugh map is plotted as shown in Fig. 3-10. The column $W = 0$, $X = 0$ is completely filled and forms a 4-cell subcube, $\overline{W}\overline{X}$. The 1 in the 1001 cell forms an obvious 2-cell subcube with the 1 in the 1000 cell. This 2-cell subcube can be expressed as $W\overline{X}\overline{Y}$; however, it is wisest to cover a cell by the largest subcube possible, and in this case the 1 in the 1001 subcube can be combined into a 4-cell subcube with 1000, 0000, and 0001, because the left column is adjacent to the right column. This yields the expression $\overline{X}\overline{Y}$, which is certainly simpler and easier to build than the expression $W\overline{X}\overline{Y}$.

The 1 in the 1010 is part of an interesting subcube. Each corner cell is adjacent to two other corner cells (but not to the corner cell diagonally opposite it). When all 4 corner cells are filled, however, they form a 4-cell subcube. Their coordinates are 0000, 1000, 1010, 0010. Note that the second and fourth variables are always 0; thus the subcube is described as $\overline{X}\overline{Z}$.

The 1 in the 0100 cell can only be combined with the 1 in the 0000 cell. A 2-cell subcube eliminates 1 literal, and the expression for this particular subcube is $\overline{W}\overline{Y}\overline{Z}$. The X variable changed; consequently it is not included in the expression.

Finally, since the 1 in the 1111 cell cannot be combined with any other 1 on the map, it must be described by using all 4 literals as WXYZ. Therefore,

$$\begin{aligned} f(W,X,Y,Z) &= \Sigma(0,1,2,3,4,8,9,10,15) \\ &= \overline{W}\overline{X} + \overline{X}\overline{Y} + \overline{X}\overline{Z} + \overline{W}\overline{Y}\overline{Z} + WXYZ \end{aligned}$$

This covers all the 1s on the map and is the minimal SOP expression for the function.

---

**EXAMPLE 3-23**

Find the minimal expression for

$$f(W,X,Y,Z) = \Sigma(0,1,2,3,4,6,8,9,10,11,15)$$

| YZ \ WX | 00 | 01 | 11 | 10 |
|---|---|---|---|---|
| 00 | 1 | 1 | | 1 |
| 01 | 1 | | | 1 |
| 11 | 1 | | 1 | |
| 10 | 1 | | | 1 |

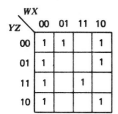

FIGURE 3-10 Karnaugh map for Ex. 3-22.

| YZ \ WX | 00 | 01 | 11 | 10 |
|---|---|---|---|---|
| 00 | 1 | 1 | | 1 |
| 01 | 1 | | | 1 |
| 11 | 1 | | 1 | 1 |
| 10 | 1 | 1 | | 1 |

FIGURE 3-11 Karnaugh map for Ex. 3-23.

**SOLUTION**

The Karnaugh map is shown in Fig. 3-11. The four 1s in the left-most column are adjacent to the four 1s in the right-most column. Together they form an 8-cell subcube, and 3 literals must disappear. In this case, X = 0 for each cell in the subcube while W, Y, and Z all change. Therefore, this subcube is designated as $\bar{X}$. The 2 cells in the upper left row are adjacent to the 2 cells in the lower left row, and form a 4-cell subcube, $\bar{W}\bar{Z}$. The remaining 1 in cell 1111 can be combined with the 1 in cell 1011 to form a 2-cell subcube. WYZ. This covers all the 1s on the map. Therefore,

$$\begin{aligned} f(W,X,Y,Z) &= \Sigma(0,1,2,3,4,6,8,9,10,11,15) \\ &= \bar{X} + \bar{W}\bar{Z} + WYZ \end{aligned}$$

The easiest way to check expressions of this kind is to make a check table as shown below:

| Term | Expression | Numbers Covered |
|---|---|---|
| $\bar{X}$ | X 0 X X | 0,1,2,3,8,9,10,11 |
| $\bar{W}\bar{Z}$ | 0 X X 0 | 0,2,4,6 |
| WYZ | 1 X 1 1 | 11,15 |

All the terms of the original specifications have been covered. This checks the solution.

### 3-5.6 Essential Subcubes

The simplest circuit realization results from choosing the best set of subcubes to cover all the 1s on a given Karnaugh map. To select the best set of subcubes, we first select all the *essential subcubes*, rather than the largest one. *An essential subcube is one that must be chosen to cover a certain 1 on the map because no other subcubes can cover that particular 1.* After the essential subcubes have been chosen, the remaining 1s on the map should be covered as simply as possible.

In Fig. 3-10, for example, we should not start out by deciding how to cover the 1 in cell 0000. This cell can be part of three different 4-cell subcubes. Since a *choice* exists, this cell is *not* essential. The 2-cell subcube, $\overline{W}\overline{Y}\overline{Z}$ which covers the 0000 and 0100 cell, is essential because there is no other way to cover the 0100 cell. The 0001 and 0010 cells are each part of two 4-cell subcubes and are therefore not essential, but the 1 in the 0011 cell can only be covered by a 4-cell subcube, $\overline{W}\overline{X}$; therefore this subcube is essential. (Essential subcubes are not necessarily 2-cell subcubes.) Similarly, the subcube $\overline{X}\overline{Z}$ is essential because it is the only way to cover the 1010 cell and the subcube $\overline{X}\overline{Y}$ is essential because it is the only way to cover the 1001 cell. Since all the terms needed to cover this map, as found in Ex. 3-22, are essential, the solution is minimal.

---

**EXAMPLE 3-24**

Find the minimal expression for $f$ if

$$F(W,X,Y,Z) = \Sigma(1,4,5,6,8,12,13,15)$$

**SOLUTION**

The Karnaugh map for the function is shown in Fig. 3-12*a*. It is tempting to grab the big 4-cell subcube in the center, $X\overline{Y}$. However, this would leave four 1s

(a) Karnaugh map

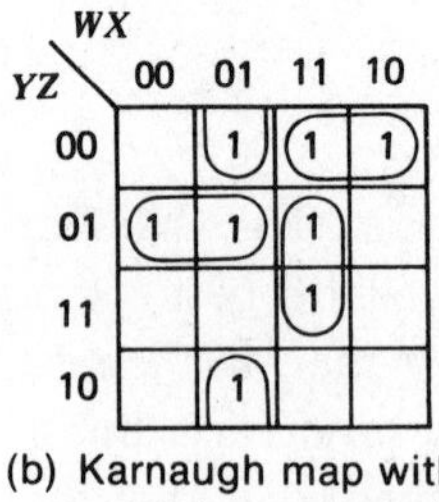

(b) Karnaugh map with the essential subcubes selected

FIGURE 3-12 Karnaugh maps for Ex. 3-24.

uncovered, and they must be covered by four 2-cell subcubes. The resulting expression is

$$F(W,X,Y,Z) = X\bar{Y} + W\bar{Y}\bar{Z} + \bar{W}\bar{Y}Z + \bar{W}X\bar{Z} + WXZ$$

Unfortunately this procedure does not result in a minimum expression for the function. This is because none of the cells in the 4-cell subcube are essential, whereas all the 2-cell subcubes are essential. If we start correctly by taking the four essential 2-cell subcubes, as shown in Fig. 3-12b, we find that each essential subcube covers a 1 in the 4-cell subcube, and together they cover all the 1s. Consequently, the term $X\bar{Y}$ for the 4-cell subcube covers no new 1s and should be omitted. Therefore,

$$F(W,X,Y,Z) = W\bar{Y}\bar{Z} + \bar{W}\bar{Y}Z + WXZ + \bar{W}X\bar{Z}$$

---

Some functions do not contain any essential cells. Consider $F(W,X,Y,Z) = \Sigma(0,4,5,8,10,11,13,15)$ whose Karnaugh map is shown in Fig. 3-13a. Since each cell in this map can be part of two 2-cell subcubes, no cells are essential. When this occurs the following general rules are helpful.

1.  For each new subcube, try to cover as many previously uncovered 1s as possible.
2.  Try to select subcubes so that no 1s are isolated.
3.  If each 1 in the map has been covered only once, a minimal expression can be obtained from this cover. It is not always possible to achieve this result.

If we started to cover the function by selecting subcubes as shown in Fig. 3-13b, rule 2 is violated because the 1s in cells 0101 and 1000 are isolated. Two additional subcubes would be needed to cover them and the resulting expression would contain five 2-cell subcubes.

A minimal cover is shown in 3-13c, where the eight 1s in the map are covered by four 2-cell subcubes and each 1 is covered only once. Therefore,

$$F(W,X,Y,Z) = \Sigma(0,4,5,8,10,11,13,15)$$
$$= \bar{W}\bar{Y}\bar{Z} + X\bar{Y}Z + WYZ + W\bar{X}\bar{Z}$$

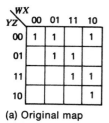

(a) Original map

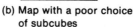

(b) Map with a poor choice of subcubes

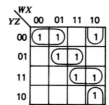

(c) Map with a good choice of subcubes

FIGURE 3-13   A Karnaugh map containing no essential subcubes.

is a minimal cover for the expression. Note that this minimal cover is not unique. The expression,

$$\overline{X}\overline{Y}\overline{Z} + \overline{W}X\overline{Y} + WXZ + W\overline{X}Y$$

also covers the function with four 2-cell subcubes. (See Problem 3-9.)

---

**EXAMPLE 3-25**

Find a minimal expression for the Karnaugh map of Fig. 3-14a.

**SOLUTION**

Inspection of the Karnaugh map reveals that each 1 is part of two 4-cell subcubes and there are no essential subcubes. If we select the two columns as 4-cell subcubes, as shown in Fig. 3-14b., we isolate the remaining 1s; this leads to a solution where four 4-cell subcubes are needed to cover the map. Selecting the subcubes as shown in Fig. 3-14c does not isolate any 1s and results in covering the twelve 1s with three 4-cell subcubes. This results in a minimal cover of

$$\overline{A}\overline{B} + BC + A\overline{C}$$

---

## 3-5.7 Simplification of 4-Variable Maps for POS Expressions

Karnaugh maps for functions in POS form can be plotted by placing a 0 in each cell where the value of the function is 0. Adjacent cells can then be combined into subcubes in a manner similar to that for SOP expressions. It must be remembered that the coordinates for POS maps are inverted because complemented variables represent 1s and uncomplemented variables represent 0s in POS expressions.

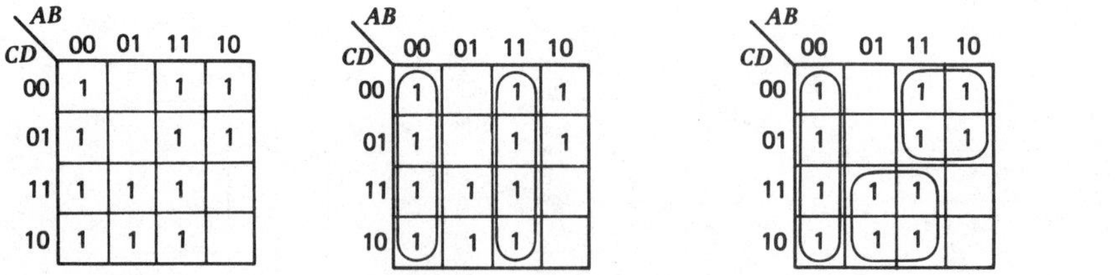

(a) Original map

(b) Map with a poor choice of 4-cell subcubes

(c) Map with a good choice of 4-cell subcubes

FIGURE 3-14  Karnaugh maps for Ex. 3-25.

---

**EXAMPLE 3-26**

Plot the Karnaugh map for the function

$$f(W,X,Y,Z) = (W + \bar{Z})(\bar{W} + \bar{X} + \bar{Y} + \bar{Z})$$

and find a simpler expression from the map.

**SOLUTION**

The term $W + \bar{Z}$ can be plotted by placing a 0 on the map at all cells where $W = 0$ and $Z = 1$, or, using the methods of Sec. 3-3.4,

$$W + \bar{Z} = 0XX1 = \Pi(1,3,5,7)$$

so that 0s are placed at the cells in locations 1, 3, 5, and 7. Similarly the term $\bar{W} + \bar{X} + \bar{Y} + \bar{Z}$ causes us to place a 0 in the 1111 cell. The resulting Karnaugh map is shown in Fig. 3-15. From it we see that the 0 in the 1111 cell can be combined as a 2-cell subcube with the 0 in the 0111 cell. For this 2-cell subcube, the W variable changes so it is not in the final expression, while X, Y, and Z are 1. Since 1s are represented by complemented variables for POS functions, this subcube is $\bar{X} + \bar{Y} + \bar{Z}$. Therefore, the function is covered by a 4-cell subcube and a 2-cell subcube and

$$f(W,X,Y,Z) = (W + \bar{Z})(\bar{X} + \bar{Y} + \bar{Z})$$

This expression is simpler than the original expression because it contains one less literal.

---

**EXAMPLE 3-27**

(a) Plot the Karnaugh map and obtain the POS expression for

$$f(a,b,c,d) = \Pi(1,5,6,7,10,12,13,15)$$

(b) Verify your solution using a check table.

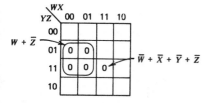

FIGURE 3-15   Karnaugh map for Ex. 3-26.

## SOLUTION

Since the function is specified by a product, 0s are written into each cell of the map that corresponds to a number in the specification, as shown in Fig. 3-16$a$. Next the essential subcubes are selected as shown in Fig. 3-16$b$. The 0 in cell 1010 cannot be combined with any other cell. Therefore it must be represented by a term containing all 4 literals, $\bar{a} + b + \bar{c} + d$. The three 2-cell subcubes are all essential, and each covers a term in the 4-cell subcube. The 1111 cell, however, can only be covered by the 4-cell subcube; therefore, the 4-cell subcube is also essential. Writing the proper expression for each subcube in the cover, we obtain

$$f(a,b,c,d) = \Pi(1,5,6,7,10,12,13,15)$$
$$= (\bar{a} + b + \bar{c} + d)(\bar{b} + \bar{d})(a + c + \bar{d})$$
$$(a + \bar{b} + \bar{c})(\bar{a} + \bar{b} + c)$$

A check table can be made for POS as well as SOP expressions.

| Term | Expression | Numbers Covered |
|---|---|---|
| $\bar{a} + b + \bar{c} + d$ | 1 0 1 0 | 10 |
| $\bar{b} + \bar{d}$ | X 1 X 1 | 5,7,13,15 |
| $a + c + \bar{d}$ | 0 X 0 1 | 1,5 |
| $a + \bar{b} + \bar{c}$ | 0 1 1 X | 6,7 |
| $\bar{a} + \bar{b} + c$ | 1 1 0 X | 12,13 |

Together they cover all the numbers (0s) in the original specification and no other numbers. This checks the solution.

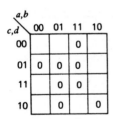

(a) Original map

(b) Map with subcubes selected

FIGURE 3-16   Karnaugh maps for Ex. 3-26.

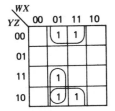

(a) SOP map

(b) POS map

FIGURE 3-17   Karnaugh maps for $f(w,x,y,z) = \Sigma (4,6,7,12,14)$.

**EXAMPLE 3-28**

Given the function

$$f(W,X,Y,Z) = \Sigma(4,6,7,12,14)$$

(a) Find the minimal SOP expression.
(b) Find the minimal POS expression.
(c) Show by algebraic manipulation that the two expressions are equal.

**SOLUTION**

The SOP Karnaugh map is drawn as shown in Fig. 3-17$a$. It simplifies to a 4-cell subcube and a 2-cell subcube. Therefore,

$$f(W,X,Y,Z) = X\bar{Z} + \bar{W}XY$$

The POS map is drawn simply by placing 0s in the map wherever there is not a 1, as shown in Fig. 3-17$b$. It simplifies into an 8-cell subcube and two 4-cell subcubes. In POS form,

$$f(W,X,Y,Z) = X(Y + \bar{Z})(\bar{W} + \bar{Z})$$

To show the two functions are identical, we operate on the POS form:

$$
\begin{aligned}
X(Y + \bar{Z})(\bar{W} + \bar{Z}) &= X(\bar{Z} + \bar{Z}Y + \bar{Z}\bar{W} + \bar{W}Y) \\
&= X(\bar{Z} + \bar{W}Y) \qquad \text{(Theorem 9)[7]} \\
&= X\bar{Z} + X\bar{W}Y
\end{aligned}
$$

---

[7]See Sec. 2-5.

## 3-5.8  Optimum Selection of Gates

Heretofore we have developed logic that minimizes the number of literals in a Boolean expression, and we have also developed a procedure for implementing any expression in POS or SOP form. For some logic expressions the SOP form is simpler and easier to implement. For other expressions, the POS form is preferable. There is no a priori way of knowing which form yields the simpler circuit. Consequently, the wise designer develops both the SOP and POS expressions before building the circuit. Implementation then depends on:

1.  Which form requires fewer gates.
2.  The availability of the required gates. (It makes no sense to use a gate that is not available from the manufacturer, or a gate that is not in stock and may require a long time for delivery when, as is typical in industry, the circuit is needed immediately.)
3.  The number of wires required.

Often a mixture of ANDs, ORs, NANDs, and so on produces the optimum circuit. As in other areas of engineering, experience, inspiration, and, above all, a little thought help the engineer make the best decision.

---

**EXAMPLE 3-29**

Given:

$$f(W,X,Y,Z) = \Sigma(5,6,7,9,10,11,13,14,15)$$

(a) How would the SOP and POS functions be implemented?
(b) Which form is preferable?

**SOLUTION**

(a) The Karnaugh map, shown in Fig. 3-18, is a 3-by-3 rectangle that is covered by four 4-cell subcubes. The SOP representation is

$$f(W,X,Y,Z) = WY + WZ + XY + XZ$$

| YZ\WX | 00 | 01 | 11 | 10 |
|-------|----|----|----|----|
| 00    |    |    |    |    |
| 01    |    | 1  | 1  | 1  |
| 11    |    | 1  | 1  | 1  |
| 10    |    | 1  | 1  | 1  |

FIGURE 3-18   Karnaugh map for the function of Ex. 3-29.

For NAND implementation, this requires four 2-input gates and a 4-input gate.

The POS representation is obtained by using the top row and left column as two 4-cell subcubes. This gives

$$f(W,X,Y,Z) = (W + X)(Y + Z)$$

(b) The POS form can be implemented with three 2-input NOR gates. Here the POS form has the clear advantage of requiring fewer and less complex gates.

## 3-5.9  Multiple Outputs

Many circuits require several outputs from the same set of inputs. One common example is 7-segment display drivers where the outputs for each of the 7-segments are derived from the same four inputs (see Sec. 3-8.3).

Karnaugh maps can and should be drawn for each output. Unfortunately, *minimizing the map for each output individually does not necessarily minimize the circuit as a whole.* As a simple example, consider two functions that must be generated on the same set of inputs:

$$f_1 = \Sigma \ (4,5,12,13,15)$$
$$f_2 = \Sigma \ (6,14, \ 15)$$

Their Karnaugh maps are shown in Fig. 3-19. Simplifying each individually yields:

$$f_1 = X\bar{Y} + WXZ$$
$$f_2 = XY\bar{Z} + WXY$$

The functions can also be expressed as:

$$f_1 = X\bar{Y} + WXYZ$$
$$f_2 = XY\bar{Z} + WXYZ$$

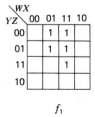

$f_1$

$f_2$

FIGURE 3-19  Karnaugh maps for $f_1$ and $f_2$.

This expression requires one more literal in each output, but the same term, WXYZ, can be used to contribute to *both* outputs. The NAND or AND-OR implementation would require one 4-input gate instead of two 3-input gates and is generally preferred (see Problem 3-23).

The theory of minimizing multiple output circuits is complex and not totally developed. We offer the following suggestions.

1.  Start by covering all the subcubes of one function that have *no common points* with the other function. For the above example these were the subcube $X\overline{Y}$ for $f_1$ and $XY\overline{Z}$ for $f_2$.

2.  Where common points exist examine the maps carefully to try to determine where gates used for one function can also be used for the other function.

---

**EXAMPLE 3-30**

Given:

$$f_1(WXYZ) = \Sigma\ (0,1,8,9,10,13)$$
$$f_2(WXYZ) = \Sigma\ (0,1,5,7,9,13,15)$$

Implement $f_1$ and $f_2$ using the minimum number of gates.

**SOLUTION**

The Karnaugh maps are shown in Fig. 3-20. If we were minimizing each function individually, we would certainly use the subcubes $X\overline{Y}$ for $f_1$ and $\overline{Y}Z$ and $XZ$ for $f_2$. Individual simplification would lead to three terms for each function.

To minimize the functions together, we start by observing that the subcube $W\overline{X}\overline{Z}$ (the right side corners) is essential for $f_1$ and has no commonality with $f_2$. We also see that the subcube $XZ$ is essential for $f_2$. The subcube $\overline{W}\overline{X}\overline{Y}$ is essential for $f_2$ but can be also used to cover two terms in $f_1$. At this point we have:

$$f_1 = W\overline{X}\overline{Z} + \overline{W}\overline{X}\overline{Y}$$
$$f_2 = XZ + \overline{W}\overline{X}\overline{Y}$$

We still must cover the terms 13 and 9 in $f_1$ and 9 in $f_2$. For $f_1$ this requires $W\overline{Y}Z$,

| WX\YZ | 00 | 01 | 11 | 10 |
|---|---|---|---|---|
| 00 | 1 | | | 1 |
| 01 | 1 | | 1 | 1 |
| 11 | | | | |
| 10 | | | | 1 |

$f_1$

| WX\YZ | 00 | 01 | 11 | 10 |
|---|---|---|---|---|
| 00 | 1 | | | |
| 01 | 1 | 1 | 1 | 1 |
| 11 | | 1 | 1 | |
| 10 | | | | |

$f_2$

FIGURE 3-20  Karnaugh maps for Ex. 3-30.

but this term can also be used in $f_2$ instead of the subcube $\overline{Y}Z$, and eliminates the need for that subcube. Therefore, the final expressions become:

$$f_1 = W\overline{X}\overline{Z} + \overline{W}X\overline{Y} + W\overline{Y}Z$$
$$f_2 = XZ + \overline{W}\overline{X}\overline{Y} + W\overline{Y}Z$$

As with the minimal expression, there are three terms for each expression, but now two of them are common so the expressions can be implemented with four first level gates instead of six. The implementation is shown in Fig. 3-21.

## 3.6 DON'T CARES

For some functions the outputs corresponding to certain combinations of input variables do not matter. This usually occurs because some combinations of the input variables should not exist. A common example is the use of 4 binary bits to represent a single-decimal digit. Usually these binary bits are in the binary-coded-decimal (BCD) form where the decimal digit is equivalent to the binary value of the 4 bits. The BCD code conversion table is shown in Fig. 3-22. Therefore, bit combinations with a value of 10 through 15 should not occur. On a truth table or Karnaugh map, outputs corresponding to these inputs are listed as *don't cares*.

*Don't cares are written as small ds* on Karnaugh maps, and increase the designer's versatility. As on any map, the 1s must be covered and the 0s must be left uncovered. The don't cares, however, may or may not be covered. *Don't cares should be covered if their inclusion simplifies the final expression by creating larger subcubes*, which are expressed by terms with fewer literals. Otherwise there is no need to cover them. Don't cares can be used on an "as needed" basis on both SOP and POS maps.

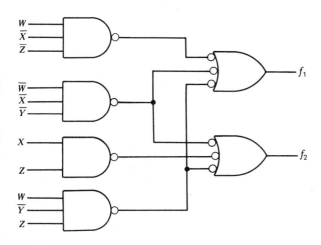

FIGURE 3-21  Circuit for Ex. 3-30.

| Binary – Coded Decimal Representation | Decimal Digit Equivalent |
|---|---|
| 0  0  0  0 | 0 |
| 0  0  0  1 | 1 |
| 0  0  1  0 | 2 |
| 0  0  1  1 | 3 |
| 0  1  0  0 | 4 |
| 0  1  0  1 | 5 |
| 0  1  1  0 | 6 |
| 0  1  1  1 | 7 |
| 1  0  0  0 | 8 |
| 1  0  0  1 | 9 |

FIGURE 3-22  The BCD code conversion table.

---

## EXAMPLE 3-31

Given:

$$F(W,X,Y,Z) = \Sigma(0,1,2,5,8,14) + d(4,10,13)$$

Find:

(a) The minimum SOP expression.

(b) The minimum POS expression.

(c) Whether the SOP and POS expressions can be shown to be equal by algebraic manipulation?

## SOLUTION

(a) The SOP Karnaugh map is shown in Fig. 3-23a. The map may be covered by two 4-cell subcubes (consisting of the upper left-hand corner and the four corners) and a 2-cell subcube. All three subcubes are essential and the minimal SOP expression is

$$f(W,X,Y,Z) = \overline{W}\overline{Y} + \overline{X}\overline{Z} + WY\overline{Z}$$

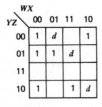

(a) SOP map          (b) POS map          (c) POS map with subcubes selected

FIGURE 3-23  Karnaugh map for Ex. 3-31.

Note that the don't care in the 1101 cell added nothing to the SOP expression and was not included. Note also that the minimal SOP expression would have been longer if the don't cares in 0100 and 1010 were not available.

(b) The POS Karnaugh map for the function is shown in Fig. 3-23b. It was drawn simply by writing 0s in all the vacant cells on the SOP map and leaving the don't cares as they were. On the POS map the two 4-cell subcubes $\bar{Y} + \bar{Z}$ and $\bar{W} + \bar{Z}$ are essential. The two remaining 0s must be covered by 2-cell subcubes that are not essential. One particular cover is shown in Fig. 3-23c. This leads to the expression:

$$F(W,X,Y,Z) = (\bar{W} + \bar{Z})(\bar{Y} + \bar{Z})(\bar{X} + Y + Z)(W + \bar{X} + \bar{Y})$$

In the POS map the $d$ in cell 1010 did not result in any simplification and consequently was not included.

(c) The SOP and POS expressions are *not* equivalent because they both covered the same subcube (0100); for $W = Y = Z = 0$ and $X = 1$ the SOP expression is 1 while the POS expression equals 0. Nevertheless, they are both valid solutions to the problem because we *don't care* what the result is in response to an input of 0100.

## 3-7   FIVE-VARIABLE MAPS

Karnaugh maps can be used to simplify functions of 5 and 6 variables, although the problem of selecting the proper subcubes becomes more difficult as the number of variables increase. Figure 3-24 shows a 5-variable map with its numeric representation for $f(V,W,X,Y,Z)$. Actually two 4-variable Karnaugh maps are drawn side-by-side; one for $V = 0$ and one for $V = 1$. SOP terms from 0 through 15 are plotted on the $V = 0$ map, and SOP terms from 16 through 31 are plotted on the $V = 1$ map. However, the $V = 0$ map must be considered to be above the $V = 1$ map so that a cell on the $V = 0$ map is adjacent to the corresponding cell on the $V = 1$ map (i.e., the 3 and 19 cells are adjacent, as are the 13 and 29 cells, etc.). This leads to the formation of three-dimensional subcubes and tests a person's powers of visualization.

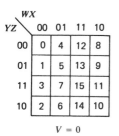

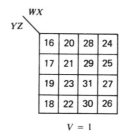

FIGURE 3-24   Numeric representation of a 5-variable Karnaugh map.

**EXAMPLE 3-32**

Simplify

$$f(V,W,X,Y,Z) = \Sigma(2,3,4,7,10,11,13,16,18,19,23,24,26,27,29,31)$$

using 5-variable Karnaugh maps.

**SOLUTION**

The Karnaugh map for this function is plotted in Fig. 3-25. The 1-cell 00010 is essential and is covered by the 8-cell subcube consisting of the 2, 3, 10, 11, 18, 19, 26, and 27 squares. An 8-cell subcube deletes 3 literals and leaves 2. The expression for this particular subcube is $\overline{X}Y$ since the variables V, W, and Z change. The 1 in the 00100 cell has no adjacent 1s and must be described by an SOP term containing all 5 literals, $\overline{V}\overline{W}X\overline{Y}\overline{Z}$. The 1 in the 00111 cell forms part of an essential 4-cell subcube consisting of the 3, 7, 19, and 23 cells. For this subcube V and X change values; the subcube is therefore $\overline{W}YZ$. The 1 in the 1101 cell forms an essential 2-cell subcube with the corresponding cell on the V = 1 map, $WX\overline{Y}Z$. This completes the cover for all the cells on the V = 0 map.

Many of the 1s on the V = 1 map were covered when the V = 0 map was covered, but the two in the upper corners were not. They can best be covered by the 4-cell subcube $V\overline{X}\overline{Z}$. The only remaining 1 not yet covered is in the 11111 cell. This can be covered either by the 2-cell subcube VWXZ or the 4-cell subcube VYZ. (This subcube is not drawn on the Karnaugh map for clarity.) We prefer using the 4-cell subcube since it contains fewer literals and would be easier to implement.

Gathering our subcubes together we find that

$$F(V,W,X,Y,Z) = \Sigma(2,3,4,7,10,11,13,16,18,19,23,24,26,27,29,31)$$
$$= \overline{X}Y + \overline{V}\overline{W}X\overline{Y}\overline{Z} + \overline{W}YZ + WX\overline{Y}Z + V\overline{X}\overline{Z} + VYZ$$

The answer can be checked by using the check table below:

| Term | Expression | Numbers Covered |
|------|------------|-----------------|
| $\overline{X}Y$ | X X 0 1 X | 2,3,10,11,18,19,26,27 |
| $\overline{V}\overline{W}X\overline{Y}\overline{Z}$ | 0 0 1 0 0 | 4 |
| $\overline{W}YZ$ | X 0 X 1 1 | 3,7,19,23 |
| $WX\overline{Y}Z$ | X 1 1 0 1 | 13,29 |
| $V\overline{X}\overline{Z}$ | 1 X 0 X 0 | 16,18,24,26 |
| $VYZ$ | 1 X X 1 1 | 19,23,27,31 |

Together these terms cover all the numbers in the specification, and no other terms. Therefore, we can be sure that our answer is correct.

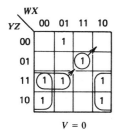
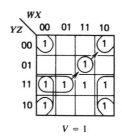

FIGURE 3-25 Karnaugh map for Ex. 3-32.

Six-variable Karnaugh maps can also be constructed using four 4-variable maps. Of course, it becomes more difficult to select the subcubes correctly as the number of variables increase.

### 3-7.1 Other Methods of Implementing Logic Expressions

Designers often use other methods besides small scale ICs for implementing logic expressions, especially complex logic expressions. They generally use larger scale ICs, which have more logic in the IC and simplify the designer's role. They will be described briefly. More detailed descriptions can be found in some of the references.

1. **ROMs or PROMs.** Read Only memories (discussed in Chapter 15) can be used to implement logic expressions. The input variables simply become memory addresses, and the output corresponding to each set of inputs is burned into the ROMs. If the logic requires several outputs based on the same set of input variables (a topic not discussed here), they can easily be accommodated by a ROM.

2. **Multiplexers.** Multiplexers in IC packages can be used to implement logic functions. A single 16-pin IC can handle any function of 4 variables and a 24-pin IC can handle any function of 5 variables. Multiplexer logic is discussed in Sec. 11-3.

3. **Programmable Logic Arrays (PLAs).** PLAs are ICs that contain a group of AND gates that can be connected to a group of OR gates. The user can determine the connections. By making the proper connections, the user can set up the output in SOP (AND-OR) form. Like ROMs, multiple outputs can be easily implemented using PLAs.

### 3-8 PRACTICAL EXAMPLES

The techniques of logical simplification discussed in this chapter can be applied to many practical problems. Once the simplest logical expression is obtained it can be implemented using relays, (Sec. 2-11) or using ICs,

(Sec. 2-7). Four practical simplification examples are presented in this section.

Here is the usual technique for solving a practical problem.

1. Draw the truth table.
2. Determine the numeric SOP function.
3. Draw the Karnaugh map.
4. Minimize the Karnaugh map for both the SOP and POS forms of the function.
5. Select the form of implementation most suitable for hardware implementation, as discussed in Sec. 3-5.8.

## 3-8.1   Control Problems

Logic circuits are often used to control machinery or processes. The logic inputs come from sensors (e.g., voltage, temperature, and pressure) that monitor the process, and the output of the logic circuit determines the next step in the procedure.

---

### EXAMPLE 3-33

A step in a space vehicle checkout depends on four sensors. Every circuit is functioning if sensor 1 and at least two of the other 3 sensors are also 1s.

(a) Find the minimal SOP expression for the output if the circuit is working properly.

(b) Find the minimal POS expression for the output if the circuit is working properly.

(c) If the circuit is not working properly, an alarm should sound. Find the minimal SOP expression needed to activate the alarm.

### SOLUTION

The truth table for the example is shown in Fig. 3-26a, where $S_1$ is sensor 1, $S_2$ is sensor 2, and so on. From it we find that

$$f(S_1S_2S_3S_4) = \Sigma(11,13,14,15)$$

The Karnaugh map is drawn in Fig. 3-26b and simplifies into three 2-cell subcubes.

$$f(S_1S_2S_3S_4) = S_1S_2S_3 + S_1S_3S_4 + S_1S_2S_4 = S_1(S_2S_3 + S_3S_4 + S_2S_4)$$

The POS map of Fig. 3-26c simplifies into an 8-cell subcube and three 4-cell subcubes.

$$f(S_1S_2S_3S_4) = S_1(S_3 + S_4)(S_2 + S_4)(S_2 + S_3)$$

| Sensors $S_1$ | $S_2$ | $S_3$ | $S_4$ | |
|---|---|---|---|---|
| 0 | 0 | 0 | 0 | 0 |
| 0 | 0 | 0 | 1 | 0 |
| 0 | 0 | 1 | 0 | 0 |
| 0 | 0 | 1 | 1 | 0 |
| 0 | 1 | 0 | 0 | 0 |
| 0 | 1 | 0 | 1 | 0 |
| 0 | 1 | 1 | 0 | 0 |
| 0 | 1 | 1 | 1 | 0 |
| 1 | 0 | 0 | 0 | 0 |
| 1 | 0 | 0 | 1 | 0 |
| 1 | 0 | 1 | 0 | 0 |
| 1 | 0 | 1 | 1 | 1 |
| 1 | 1 | 0 | 0 | 0 |
| 1 | 1 | 0 | 1 | 1 |
| 1 | 1 | 1 | 0 | 1 |
| 1 | 1 | 1 | 1 | 1 |

**(a) Truth table**

| $S_3S_4$ \ $S_1S_2$ | 00 | 01 | 11 | 10 |
|---|---|---|---|---|
| 00 | | | | |
| 01 | | | 1 | |
| 11 | | | 1 | 1 |
| 10 | | | 1 | |

**(b) SOP Karnaugh map**

| $S_3S_4$ \ $S_1S_2$ | 00 | 01 | 11 | 10 |
|---|---|---|---|---|
| 00 | 0 | 0 | 0 | 0 |
| 01 | 0 | 0 | | 0 |
| 11 | 0 | 0 | | |
| 10 | 0 | 0 | | 0 |

**(c) POS Karnaugh map**

FIGURE 3-26 Truth tables and maps for the space vehicle check-out system of Ex. 3-33.

The alarm must sound whenever there is a failure. The Karnaugh map for the alarm can be plotted simply by putting 1s in whenever there are 0s on the success map. This simplifies to

$$\bar{f}(S_1S_2S_3S_4) = \bar{S}_1 + \bar{S}_2\bar{S}_3 + \bar{S}_2\bar{S}_4 + \bar{S}_3\bar{S}_4$$

Note that $f$ could also be obtained by complementing the POS form of $f$.

## 3-8.2 Adders

An adder is the most common arithmetic circuit built. Usually an adder is built of many identical stages. A typical stage of an adder that sums two binary numbers, A and $B$, has inputs $A_n$, $B_n$ and $C_{n-1}$, where $A_n$ and $B_n$ are the Nth bits of A and $B$ (the augend and addend) and $C_{n-1}$ is the

| $C_{IN}$ | $A$ | $B$ | $S$ | $C_{OUT}$ |
|---|---|---|---|---|
| 0 | 0 | 0 | 0 | 0 |
| 0 | 0 | 1 | 1 | 0 |
| 0 | 1 | 0 | 1 | 0 |
| 0 | 1 | 1 | 0 | 1 |
| 1 | 0 | 0 | 1 | 0 |
| 1 | 0 | 1 | 0 | 1 |
| 1 | 1 | 0 | 0 | 1 |
| 1 | 1 | 1 | 1 | 1 |

(a) Truth table

S

(b) Karnaugh map for the sum

$C_{OUT}$

(c) Karnaugh map for the carry-out

FIGURE 3-27   Truth tables and Karnaugh maps for a full-adder.

carryout of the N - 1 stage. (See Sec. 1-7.1.) The outputs required are the sum and carry. Such a stage is called a *full-adder*, to distinguish it from a *half-adder*, which has A and B inputs but no carry input. A half-adder can be used to add the two least significant digits, $A_0$ and $B_0$, where there is no carry in.

---

## EXAMPLE 3-34

Design a full-adder.

## SOLUTION

First the truth table is plotted as shown in Fig. 3-27a. Since two outputs, sum and carry, must be generated, two 3-variable Karnaugh maps are drawn as shown in Fig. 3-27b and 3-27c. The Karnaugh map for the sum resembles a checkerboard. There are no 1s that can be combined with others; therefore no simplifications are possible[8] and the expression for the sum is

$$\text{Sum} = \bar{A}B\bar{C}_{IN} + A\bar{B}\bar{C}_{IN} + \bar{A}\bar{B}C_{IN} + ABC_{IN}$$

The Karnaugh map for the carry out can be simplified into three 2-cell subcubes. Therefore,

$$\text{Carry Out} = AB + AC_{IN} + BC_{IN}$$

---

[8]No further simplifications are possible using AND-OR logic. When EXCLUSIVE-OR gates are used, they greatly simplify checkerboard Karnaugh maps (see Chapter 13). The adder is discussed further in Chapter 14.

### 3-8.3 Seven-Segment Displays

Seven-segment displays are used to convert a 4-bit BCD number into a visible readout. A 7-segment display consists of 7 light-emitting diodes (LEDs) arranged as shown in Fig. 3-28a. Modern hand calculators use 7-segment displays for their readouts.

Logically we require that those segments turn on or light up that most closely approximate the shape of the decimal digit equivalent to the binary value of the input. For example, a 0 input turns on all segments except g to give the appearance of a 0, a 1 input turns on segments b and c, and a 7 input turns on segments a, b, and c. Since the numbers to be represented must be between 0 and 9, numbers greater than 9 should not appear on the inputs, and the outputs corresponding to these prohibited inputs are

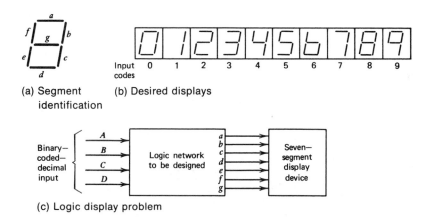

(a) Segment identification

(b) Desired displays

(c) Logic display problem

| Decimal Displayed | Inputs | | | | Outputs | | | | | | |
|---|---|---|---|---|---|---|---|---|---|---|---|
| | A | B | C | D | a | b | c | d | e | f | g |
| 0 | 0 | 0 | 0 | 0 | 1 | 1 | 1 | 1 | 1 | 1 | 0 |
| 1 | 0 | 0 | 0 | 1 | 0 | 1 | 1 | 0 | 0 | 0 | 0 |
| 2 | 0 | 0 | 1 | 0 | 1 | 1 | 0 | 1 | 1 | 0 | 1 |
| 3 | 0 | 0 | 1 | 1 | 1 | 1 | 1 | 1 | 0 | 0 | 1 |
| 4 | 0 | 1 | 0 | 0 | 0 | 1 | 1 | 0 | 0 | 1 | 1 |
| 5 | 0 | 1 | 0 | 1 | 1 | 0 | 1 | 1 | 0 | 1 | 1 |
| 6 | 0 | 1 | 1 | 0 | 0 | 0 | 1 | 1 | 1 | 1 | 1 |
| 7 | 0 | 1 | 1 | 1 | 1 | 1 | 1 | 0 | 0 | 0 | 0 |
| 8 | 1 | 0 | 0 | 0 | 1 | 1 | 1 | 1 | 1 | 1 | 1 |
| 9 | 1 | 0 | 0 | 1 | 1 | 1 | 1 | 0 | 0 | 1 | 1 |

(d) Truth table

FIGURE 3-28 Seven-segment displays (From Thomas Blakeslee. *Digital Design With Standard MSI & LSI.* Copywright John Wiley & Sons, Inc., 1975. Reprinted by permission of John Wiley & Sons, Inc.)

written as don't cares. The truth table for a 7-segment display is shown in Fig. 3-28d. Note that there are 7 outputs, one for each segment based on the 4 inputs.

---

**EXAMPLE 3-35**

Design the logic to display segment a on a 7-segment display.

**SOLUTION**

Segment a is ON for the numbers 0, 2, 3, 5, 7, 8, 9. The SOP Karnaugh map is shown in Fig. 3-29a. This leads to the simplification:

$$\text{Segment a} = A + BD + \bar{B}\bar{D} + CD$$

Note also that the POS Karnaugh map of Fig. 3-29b contains three 0s and two of them can be combined with don't cares to form a 4-cell subcube. This may be simpler to analyze and gives

$$\text{Segment a} = (\bar{B} + D)(A + B + C + \bar{D})$$

## 3-8.4   Code Conversion

Logic circuits can be used to convert from one code or representation to another. In general, an N-bit input can be converted into an M-bit output. (M is often, but not necessarily, equal to N.) To accomplish this, an N-variable truth table is set up, and M Karnaugh maps (one for each output variable) are drawn. We can then write the equations necessary to construct the code converter.

---

**EXAMPLE 3-36**

In the *Excess-3 code* each decimal digit is represented by its binary equivalent plus 3. The truth table is shown in Fig. 3-30. Design a code converter to convert from BCD to Excess-3 representation.

| CD \ AB | 00 | 01 | 11 | 10 |
|---------|----|----|----|----|
| 00 | 1 |  | d | 1 |
| 01 |  | 1 | d | 1 |
| 11 | 1 | 1 | d | d |
| 10 | 1 |  | d | d |

(a) SOP Karnaugh map

| CD \ AB | 00 | 01 | 11 | 10 |
|---------|----|----|----|----|
| 00 |  | 0 | d |  |
| 01 | 0 |  | d |  |
| 11 |  |  | d | d |
| 10 |  | 0 | d | d |

(b) POS Karnaugh map

FIGURE 3-29   Karnaugh maps for segment a of a 7-segment display.

| Decimal Digit | BCD Representation $A_3$ $A_2$ $A_1$ $A_0$ | | | | Excess – 3 Representation $B_3$ $B_2$ $B_1$ $B_0$ | | | |
|---|---|---|---|---|---|---|---|---|
| 0 | 0 | 0 | 0 | 0 | 0 | 0 | 1 | 1 |
| 1 | 0 | 0 | 0 | 1 | 0 | 1 | 0 | 0 |
| 2 | 0 | 0 | 1 | 0 | 0 | 1 | 0 | 1 |
| 3 | 0 | 0 | 1 | 1 | 0 | 1 | 1 | 0 |
| 4 | 0 | 1 | 0 | 0 | 0 | 1 | 1 | 1 |
| 5 | 0 | 1 | 0 | 1 | 1 | 0 | 0 | 0 |
| 6 | 0 | 1 | 1 | 0 | 1 | 0 | 0 | 1 |
| 7 | 0 | 1 | 1 | 1 | 1 | 0 | 1 | 0 |
| 8 | 1 | 0 | 0 | 0 | 1 | 0 | 1 | 1 |
| 9 | 1 | 0 | 0 | 1 | 1 | 1 | 0 | 0 |
| | 1 | 0 | 1 | 0 | d | d | d | d |
| | 1 | 0 | 1 | 1 | d | d | d | d |
| | 1 | 1 | 0 | 0 | d | d | d | d |
| | 1 | 1 | 0 | 1 | d | d | d | d |
| | 1 | 1 | 1 | 0 | d | d | d | d |
| | 1 | 1 | 1 | 1 | d | d | d | d |

(a) Truth table

$B_3$

| $A_1A_0$ \ $A_3A_2$ | 00 | 01 | 11 | 10 |
|---|---|---|---|---|
| 00 | | | d | 1 |
| 01 | | 1 | d | 1 |
| 11 | | 1 | d | d |
| 10 | | 1 | d | d |

$$B_3 = A_3 + A_2A_1 + A_2A_0$$

$B_2$

| $A_1A_0$ \ $A_3A_2$ | 00 | 01 | 11 | 10 |
|---|---|---|---|---|
| 00 | 1 | d | | |
| 01 | 1 | | d | 1 |
| 11 | 1 | | d | d |
| 10 | 1 | | d | d |

$$B_2 = A_2\bar{A}_1\bar{A}_0 + \bar{A}_2A_1 + \bar{A}_2A_0$$

$B_1$

| $A_1A_0$ \ $A_3A_2$ | 00 | 01 | 11 | 10 |
|---|---|---|---|---|
| 00 | 1 | 1 | d | 1 |
| 01 | | | d | |
| 11 | 1 | 1 | d | d |
| 10 | | | d | d |

$$B_1 = A_1A_0 + \bar{A}_1\bar{A}_0$$

$B_0$

| $A_1A_0$ \ $A_3A_2$ | 00 | 01 | 11 | 10 |
|---|---|---|---|---|
| 00 | 1 | 1 | d | 1 |
| 01 | | | d | |
| 11 | | | d | d |
| 10 | 1 | 1 | d | d |

$$B_0 = \bar{A}_0$$

(b) Karnaugh maps

FIGURE 3-30   Design of an Excess-3 code converter.

## SOLUTION

This code converter has 4 inputs (the BCD representation of a decimal digit) and 4 outputs. The outputs for inputs greater than 9 are not specified and are assumed to be don't cares. Four Karnaugh maps, one for each output bit are drawn up as shown in Fig. 3-30b. The simplifying subcubes are also shown and the SOP form of the results are listed beneath the maps. A circuit for each output now can be built.

## SUMMARY

In this chapter we discussed:

1. The method of expanding functions into standard SOP and POS forms.
2. Plotting the POS and SOP forms on Karnaugh maps.
3. Using Karnaugh maps to obtain the minimal expression for a function.
4. Implementing SOP and POS forms using AND-OR, NAND, and NOR logic.

Finally, several practical examples using these techniques to achieve a minimal expression were presented. The minimal expression is useful because it leads to a minimal hardware implementation.

## GLOSSARY

**Domain.** The variables for which a function is defined.
**Literal.** The occurrence of a variable in either its primed or unprimed state.

**SOP.** Sum of product terms.
**POS.** Product of sum terms.
**Subcube.** A group of two or more adjacent 1s on a Karnaugh map.
**BCD.** Binary coded decimal representation.

## REFERENCES

**Louis Nashelsky,** *Introduction to Digital Computer Technology,* Wiley, New York, 1977.
**George K. Kostopoulos.** *Digital Engineering,* Wiley, New York, 1975.
**Frederick J. Hill** and **Gerald R. Peterson,** *Introduction to Switching Theory and Logical Design,* 2nd ed., Wiley, New York, 1974.
**Alan B. Marcovitz** and **James K. Pugsley,** *An Introduction to Switching System Design,* Wiley, New York, 1971.
**Swaminathan Madhu,** *Theory and Design of Switching Circuits,* Unpublished Work, Rochester Institute of Technology, Rochester, New York, 1972.
**Dan I. Porat** and **Arpad Barna,** *Introduction to Digital Techniques,* Wiley, New York, 1979.
**Thomas R. Blakeslee,** *Digital Design With Standard MSI & LSI,* Wiley, New York, 1975.

## PROBLEMS

3-1.   Write each expression below in standard SOP form.
  (a) $f(a,b,c) = a\bar{b} + c$
  (b) $f(w,x,y,z) = w\bar{x} + xyz + \bar{w}\bar{y}z$
  (c) $f(a,b,c,d) = b + \bar{a}\bar{c} + a\bar{b}cd$
  (d) $f(v,w,x,y,z) = \bar{x}\bar{z} + wxy + v\bar{w}yz$
3-2.   Write the expressions of Problem 1 in standard POS form.
3-3.   Write the numerical expressions for Problem 1 in both SOP and POS form.
3-4.   Given:

$$f(W,X,Y,Z)$$

Which numbers do the following expressions represent?
  (a) $\bar{X}\bar{Z}$                    (d) $\bar{W}XY$
  (b) $W\bar{X}$                    (e) $(\bar{W} + X + \bar{Y} + Z)$
  (c) $W\bar{X}Z$                    (f) $(W + \bar{Y})$
3-5.   Find the expression for each subcube shown in Fig. P3-5.
3-6.   Find the minimum expression for

$$f(W,X,Y,Z) = \Sigma(0,2,3,4,8,9,10,11,15)$$

Is the subcube composed of the four corner squares essential?

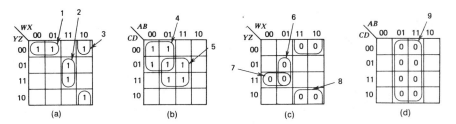

FIGURE P3-5   Karnaugh maps for Problem 3-5.

3-7.   Find the minimum expression for

$$f(W,X,Y,Z) = \Sigma\ (0,2,3,5,6,7,11,14,15)$$

Find a 4-cell subcube in your map that is not essential and should not be included in the minimal expression.

3-8.   Show by algebraic manipulation that

$$x\bar{y} + w\bar{y}\bar{z} + \bar{w}\bar{y}z + wxz + \bar{w}x\bar{z} = w\bar{y}\bar{z} + \bar{w}\bar{y}z + wxz + \bar{w}x\bar{z}$$

(*Hint:*   Multiply $x\bar{y}$ by 1 in the form of $(z + \bar{z})$.)

3-9.   Show how

$$\overline{X}\overline{Y}\overline{Z} + \overline{W}X\overline{Y} + WXZ + W\overline{X}Y$$

covers the Karnaugh map of Fig. 3-13c.

3-10.   Find a minimal cover for the Karnaugh map of Fig. 3-14 that is different from the result found in Ex. 3-25.

3-11.   Given

$$f(W,X,Y,Z) = \Pi(1,5,6,7,10,12,13,14,15)$$

Draw the Karnaugh map and find the minimal POS expression.

3-12.   Find the minimal SOP expression for the function of Fig. 3-15 by placing 1s in the map wherever there are no 0s. Show by algebraic manipulation that your answer is the same as the answer given in Ex. 3-26.

3-13.   Simplify the Karnaugh maps of Fig. P3-13.

3-14.   Find the POS form for the functions whose Karnaugh maps are given in Figs. P3-13a and P3-13b. Show that the POS forms are equivalent to the SOP forms by algebraic manipulation.

3-15.   Obtain the SOP form for the Karnaugh map of Fig. P3-13c. Show that it is equivalent to the POS form found in Prob. 3-13.

3-16.   For the functions given below, draw the Karnaugh maps and express them in POS and SOP form.

$$f(x,y,z) = \Sigma(1,2,3,6) + d(0)$$
$$f(w,x,y,z) = \Sigma(1,10,12,13,14) + d(8)$$
$$f(a,b,c,d) = \Pi(2,3,9,10,12,15) + d(7,11)$$

| X / YZ | 0 | 1 |
|---|---|---|
| 00 | 1 | 1 |
| 01 | | |
| 11 | 1 | |
| 10 | 1 | 1 |

| WX / YZ | 00 | 01 | 11 | 10 |
|---|---|---|---|---|
| 00 | | | | |
| 01 | 1 | | 1 | 1 |
| 11 | 1 | | | 1 |
| 10 | | | | 1 |

| AB / CD | 00 | 01 | 11 | 10 |
|---|---|---|---|---|
| 00 | 0 | | 0 | 0 |
| 01 | | | 0 | |
| 11 | | | | |
| 10 | 0 | | 0 | 0 |

FIGURE P3-13

3-17. Given

$$f(V,W,X,Y,Z) = \Sigma(0,1,2,8,9,10,14,15,16,17,22,24,25,28,30,31)$$

Find the minimal SOP expression for the function.

3-18. Find the minimal POS and SOP expressions for segments $c$ and $f$ of a 7-segment display.

3-19. Given inputs $A_3A_2A_1A_0$, we are required to produce outputs $B_3B_2B_1B_0$, where $B = A$ for $A < 10$ and $B = A - 4$ for $A \geq 10$. Show your truth tables and Karnaugh maps.

3-20. A hotel has four rooms (W,X,Y,Z) and two bellboys. Each room is capable of ringing for room service and a request will be answered by any one of the two bellboys. Unfortunately, the occupants of rooms W and X dislike the first bellboy and will not accept his services. Show your truth tables and Karnaugh maps for a circuit to sound an alarm whenever a request cannot be answered.

3-21. Design the circuits for the excess-3 code converter of Ex. 3-36

3-22. Implement the following expressions:
  (a) Using OR-AND logic.
  (b) Using NOR logic.

  1.  $(W + \bar{X})(\bar{W} + \bar{Y} + Z)$
  2.  $(W + X + Y + Z)(\bar{X} + \bar{Y})(\bar{W} + X + \bar{Z})$

3-23. Show the NAND gate implementation for $f_1$ and $f_2$ in Sec. 3-5.9.

3-24. Given

$$f_1(a,b,c,d) = \Sigma(3,7,9,10,11,14,15)$$
$$f_2(a,b,c,d) = \Sigma(0,3,4,7,8,9,12,14,15)$$

Minimize $f_1$ and $f_2$ together and draw the minimal circuit.

To be sure you understand this chapter, return to Sec. 3-2 and review the questions. If there is any question you cannot answer, review the appropriate section of the text to find the answers.

# 4

# LOGIC FAMILIES AND THEIR CHARACTERISTICS

## 4-1   INSTRUCTIONAL OBJECTIVES

This chapter describes the basic logic families in use today. It also describes the five types of TTL available. After reading the chapter, the student should be able to:

1.   Decide whether to use TTL, CMOS, or ECL logic for the particular application.
2.   Select the best series of TTL for the application if TTL is the choice.
3.   Interface between TTL and CMOS.
4.   Calculate the speed-power product of an IC gate.

## 4-2   SELF-EVALUATION QUESTIONS

Watch for the answers to the following questions as you read the chapter. They should help you understand the material presented. When you have finished the chapter, return to this section and be sure you can answer all of the questions.

1.   Why should more than one logical form be investigated before building a logic circuit?
2.   Why is power dissipation an important consideration in the design of circuits?

3.   What overhead cost is associated with an IC?

4.   What precautions must be observed when mixing ICs of different families?

5.   What are the speeds of the three families and the speed of each TTL series?

## 4-3   EVALUATION OF IC FAMILIES

The simplest way to implement a logic function such as AND or OR is with a *diode gate,* which is the proper connection of several diodes and a resistor. This was tried in the 1960s and was quickly found to be unsatisfactory because voltage levels in diode gates deteriorate and amplification is needed.

Diode gates were replaced by the RTL (Resistor Transistor Logic) and then the DTL (Diode Transistor Logic) series of logic gates. These are now obsolete and modern designers generally use one of three families of logic: Transistor Transistor Logic (TTL), Complementary Metal-Oxide Semiconductor (CMOS), and Emitter Coupled Logic (ECL). Each has its advantages, disadvantages, and uses in special applications. TTL (including its LS series) is the most popular series and will be emphasized in this book, but many MOS and CMOS gates will be discussed in those applications where they dominate (such as RAMs, ROMs, etc.).

Designers usually select a logic family on the basis of following criteria:

1.   Speed
2.   Power dissipation
3.   Cost
4.   Fanout
5.   Availability

### 4-3.1   Speed

The *speed* of a logic family is measured by the propagation delay time of its basic inverter or NAND gate. A square wave is applied to the input and the output observed. The difference in nanoseconds from when the input has completed 50 percent of its transition to the point where the output has completed 50 percent of its transition is the *propagation delay.* Actually there are two propagation delays, as shown in Fig. 4-1. One, called $t_{PHL}$, is the delay when the input causes the *output* to change from a HIGH to a LOW level, and the other, $t_{PLH}$, is the delay when the *output* goes from a LOW to a HIGH level. Usually $t_{PHL}$ and $t_{PLH}$ are quite close and the delay of the family is the average of the two.[1]

---

[1]Some designers prefer to specify the speed of a family as the maximum clock rate of its flip-flops.

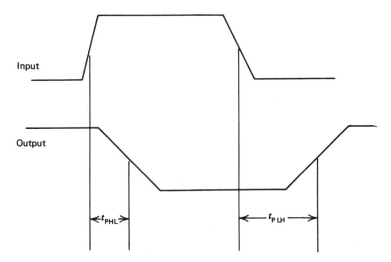

FIGURE 4-1   Response at an inverter to an input pulse, illustrating $t_{PHL}$ and $t_{PLH}$.

## 4-3.2   Power Dissipation

Power dissipation is the amount of power (in milliwatts) that an IC drains from its power supply. Generally each IC contains several circuits. Most typical power measurements are made when half the circuits on the chip are in the 1 state and the others are in the 0 state. Power measurements are also affected by the input frequency—that is, the rate at which the IC changes state. CMOS has almost no power dissipation at low frequencies, but its dissipation increases as the frequency increases (see Sec. 4-7).

Power dissipation is important not only because it drives up the cost of the power supplies, but because *it increases the heat in the vicinity of the ICs and within the electronic package. Excess heat is a prime cause of failure in electronic circuits.* If the cost of repairs and field maintenance are considered, we can see why a designer must pay careful attention to the power dissipation in the circuit.

## 4-3.3   Cost

The cost of ICs should be a relatively minor factor affecting the choice of a family. Generally the cost of an IC depends on the quantity manufactured. Standard TTL has been mass produced in very large quantities and enjoys a price advantage over all other families. A 7400 IC, which contains four NAND gates, can be purchased for less than 20 cents. This reduces the cost per NAND gate to less than 5 cents! More complex ICs cost considerably more, primarily because they are not manufactured in as large quantities as NAND gates.

For inexpensive ICs, the *overhead cost* per IC (overhead cost per IC is obtained by taking the total system cost—the cost of sockets, racks, power supplies, cooling, checkout, repair, and so on—and dividing it by the number of ICs in the system) is far more than the cost of the IC. You can buy an IC for 20 cents but you cannot buy a socket to plug it into for that price. Blakeslee, in a detailed analysis of a system,[2] estimates the overhead cost per IC to be $3.31. Although the overhead for any other system will vary somewhat, if $3.31 is accepted as a "ball-park" figure, it is obvious that the cost of most ICs is low compared to the overhead cost.

### 4-3.4 Fanout

*Fanout* is defined as *the number of loads that can be driven from a single source.* Usually fanout is determined for a standard gate in a family driving other gates in the *same* family. High fanout is an advantage because additional drivers are not needed to supply many loads connected to the same source, but it is a relatively minor factor affecting the choice of logic circuits. CMOS, because of its high input impedance, has the largest fanout capability of any logic family.

### 4-3.5 Availability

Availability is an important factor in determining the selection of a logic family. All too often progress on digital systems is delayed for weeks, or even months, awaiting the delivery of a particular part. Availability can be considered in two ways:

1. The popularity of the series.
2. The "breadth" of the series.

The popularity of a series among the customers helps availability. If several million ICs of a particular type are manufactured, a supply is generally available.

The breadth of the series refers to the number of different types of chips available. In a wide series a complex function may be available that would have to be built up from less complex chips in another family. For both popularity and breadth, standard TTL has a distinct advantage over other logic families.

Other factors such as temperature range, noise immunity, and so on, are important for special pupose circuits, but usually the choice of an IC family depends on the factors listed above.

[2]Thomas R. Blakeslee, *Digital Design With Standard MSI & LSI* (see Sec. 4-10).

## 4-4 TRANSISTOR-TRANSISTOR LOGIC

At the present time the three families of IC logic in general use are TTL, ECL, and CMOS. Transistor-Transistor Logic (TTL) is available in five series: Standard, Low Power, High Speed, Schottky, and Low Power Schottky. TTL has the lowest price, widest breadth, and best availability.

### 4-4.1 Standard TTL

The standard TTL series is the most popular logic series ever produced. It is still the most widely used and most commonly available TTL series. All the other TTL series are modifications of the circuit used in the standard series.

The basic TTL 2-input NAND gate is shown in Fig. 4-2. This is the circuit for the type 7400 NAND gate. The inputs, A and B, come into $Q_1$, which is a *multiple-emitter transistor*. A unique feature of TTL is that the inputs enter via a transistor that has several emitters.

The TTL output voltage is determined by $Q_2$, the *phasesplitter* transistor, which drives $Q_3$ and $Q_4$, the so-called *totem-pole* transistors. They operate as follows:

1. If $Q_2$ is ON, it supplies current to $Q_4$, saturating it.
2. At the same time the LOW collector voltage holds $Q_3$ OFF, and the result is a 0 output.
3. If $Q_2$ is OFF, no current enters the base of $Q_4$ and it turns OFF.
4. Now the high collector voltage turns $Q_3$ ON, which provides a HIGH output ($\approx 3.4$ V) and a low impedance path to $V_{CC}$.

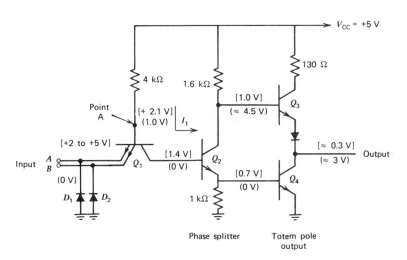

FIGURE 4-2 A TTL NAND gate.

There are two sets of numbers indicating the circuit voltages on Fig. 4-2. The upper set applies only when both inputs are HIGH, reverse-biasing both emitter-base junctions. Then the gate operates as follows.

1.   Current $I_1$ flows through the 4-kΩ resistor, the base-to-collector junction of $Q_1$, and the base-to-emitter junction of $Q_2$.
2.   The voltage at point A is the sum of three forward-biased *pn* junction drops. At 0.7 V per junction this is 2.1 V.
3.   The current through $Q_2$ saturates it and produces a LOW voltage on the base of $Q_3$, cutting it off. (The function of diode $D_3$ is to assure that $Q_3$ is cut off when the phase-splitter is on.)
4.   The current through $Q_2$ enters the base of $Q_4$, saturating it.
5.   Therefore when inputs are HIGH the output is LOW because $Q_4$, the bottom transistor of the totem-pole pair, is ON and saturated and NAND action has been achieved.

Note that any collector current that flows through $Q_4$ when it is saturated comes not through $Q_3$, but from the loads (commonly multiple-emitter transistor inputs to other gates) that are connected to it.

If either A or B (or both) inputs are LOW, the circuit operates as follows.

1.   Current through the 4 kΩ resistor flows through one or both of the multiple-emitters to the low input.
2.   The voltage at point A is only 1 V (0.7 V for the base-to-emitter drop and 0.3 V for $V_{CE(sat)}$ of the transistor that is turned ON and causing the input to be LOW).
3.   Current $I_1$ does not flow and $Q_2$ remains OFF because of the lack of current.
4.   $Q_4$ also receives no base current and remains OFF.
5.   There is a high voltage at the collector of $Q_2$ that turns $Q_3$ on and effectively connects the output to a high voltage.
6.   Therefore, with either input LOW the output is HIGH and again NAND action has been achieved.

### 4-4.2   Low Power TTL

Low Power TTL has been developed to reduce the power consumption of TTL ICs. The circuit for Low Power TTL is the same as the Standard TTL circuit except that the value of each resistor has been increased by a factor of 10. This reduces the current and power consumption, but increases the propagation delay time. Low Power TTL is not used very often for new designs.

### 4-4.3   High Speed TTL

High Speed TTL is constructed by adding an extra transistor to give the upper totem pole output transistor more drive. The speed improvement is less than that obtained by using Schottky TTL and the power dissipation is high. As a result, High Speed TTL is rarely used.

### 4-4.4   Schottky TTL

Schottky TTL is the fastest (shortest propagation delay) TTL series. It is formed by placing a Schottky (metal-silicon) diode between the base and collector at the transistors. This diode holds the transistors out of saturation and speeds up the circuit so that the propagation delay is about 3 ns. Because high speed circuits such as Schottky and ECL generate very sharp wave fronts that cause transmission and coupling problems, it is unwise to use these circuits except where the high speed is essential to the application.

### 4-4.5   Low Power Schottky TTL

Low Power Schottky TTL is formed by using Schottky diodes to speed up the TTL circuit and then raising the values of the internal resistors, which increase the delay. The net result is a circuit with a speed approximately equal to the standard TTL gate but, because of the high resistances, a Low Power Schottky gate consumes only about one fifth the power of a standard TTL gate. Many new designs use Low Power Schottky TTL, and it is a very popular series. Some newer ICs, such as the **74LS 257**, are manufactured only in the LS (Low Power Schottky) series.

## 4-5   CHARACTERISTICS OF TTL GATES

A table of the important characteristics of each of the TTL series is given in Table 4-1. An explanation of the entries follows:

1.  $V_{CC}$, power supply voltage. The nominal power supply voltage for all series is 5 V. For the **7400** series $V_{CC}$ can vary from 4.75 to 5.25 V.
2.  $V_{IH}$, input voltage when the input is HIGH. The *minimum* voltage *guaranteed* to be recognized as a 1 at the IC input is 2 V. There is also an absolute maximum value of $V_{IH}$ that may be applied to the input of an IC. For the **7400** series this is specified as 5.5 V.
3.  $V_{IL}$, input voltage when the input is LOW. The maximum voltage *guaranteed* to be *recognized as a* 0 at the input is 0.8 V.
4.  $V_{OH}$, output voltage when the output is HIGH. The minimum HIGH output voltage for the standard series is 2.4 V. The Schottky and Low Power Schottky series will provide at least 2.7 V output when its output is a 1.
5.  $V_{OL}$, output voltage when the output is LOW. A 7400 IC will produce no more than 0.4 V when its output is a 0. (The bottom transistor of the totem-pole pair must be ON and saturated, producing a $V_{CE(sat)}$ of no more than 0.4 V.)
6.  $I_{OL}$, output current when the output is LOW. This is the minimum amount of current the bottom transistor of a totem-pole pair can "sink" or absorb *when its output is* LOW.
7. . $I_{IL}$, 0 level input current. This is the maximum current an IC will source through the multi-emitter transistor when the input is connected to a 0.

TABLE 4-1 Typical TTL Characteristics

| | | Series | | | | |
|---|---|---|---|---|---|---|
| | Standard | Low Power | High Speed | Schottky | Low Power Schottky | Units |
| $V_{CC}$ | +5 | +5 | +5 | +5 | +5 | Volts |
| $V_{IH}$ (min) | 2 | 2 | 2 | 2 | 2 | Volts |
| $V_{IL}$ (max) | 0.8 | 0.7 | 0.8 | 0.8 | 0.8 | Volts |
| $V_{OH}$ (min) | 2.4 | 2.4 | 2.4 | 2.7 | 2.7 | Volts |
| $V_{OL}$ (max) | 0.4 | 0.4 | 0.4 | 0.5 | 0.5 | Volts |
| $I_{OL}$ (min) | 16 | 3.6 | 20 | 20 | 8 | mA |
| $I_{IL}$ (max) | −1.6 | −.18 | −2 | −2 | −0.4 | mA |
| Propagation Delay | 10 | 33 | 6 | 3 | 9.5 | nS |
| Power dissipation. | 10 | 1 | 22 | 19 | 2 | mW |

8.  **Propagation delay.** This is the average of $T_{PLH}$ and $T_{PHL}$ for each series under typical operating conditions.

9.  **Power dissipation.** This is the average power dissipation for each gate in each series.

Since all series operate with the same +5-V power supply and have approximately the same voltage levels for a logic 1 and 0, they can be interchanged without difficulty. The compatibility between series is another advantage of TTL.

## 4-5.1  Noise Margin

*Noise margin is the difference between the voltage produced at the output of a logic circuit and the voltage that will be recognized as that same logic level on its input.* This difference, or margin, allows for the rejection of external noise picked up between gates and assures proper operation of the circuit.

For standard TTL, noise margin is illustrated by Fig. 4-3 and can be calculated from the data given in Table 4-1. They show that the minimum HIGH voltage an IC will *generate* is 2.4 V, but the minimum voltage it will *recognize* as a 1 is 2.0 V. This difference, 0.4 V, is the noise margin. It assures us that *any 1 generated by an output will be recognized as a 1 by any input it is connected to.* Similarly, a low level output can be no more than 0.4 V, but any input up to 0.8 V *must be recognized* as a 0. Therefore, we also have a *low level noise margin* of 0.4V.

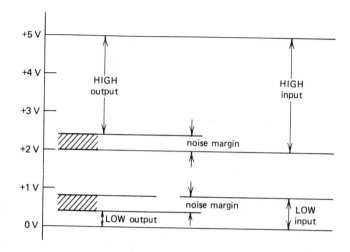

FIGURE 4-3 Band diagram showing the noise margins of TTL circuits. (From Porat and Barna. *Introduction to Digital Technique.* Copyright John Wiley & Sons, Inc. 1979. Reprinted by permission of John Wiley & Sons, Inc.

---

**EXAMPLE 4-1**

Calculate the high level noise margins for Low Power Schottky TTL.

**SOLUTION**

Table 4-1 gives $V_{OH}$, the minimum output an LSTTL gate will generate, as 2.7 V. Similarly, $V_{IH}$ is found as 2 V, so the high level noise margin for Low Power Schottky TTL is 0.7 V.

---

## 4-5.2 Speed-Power Product

The ideal gate is infinitely fast and consumes no power. Logic series are often compared on the basis of their *speed-power product*, which is the product of propagation delay times power consumption. A low speed-power product is a definite advantage.

---

**EXAMPLE 4-2**

Find the speed-power product of:
(a) Standard TTL
(b) Low Power Schottky TTL

**SOLUTION**

(a) For standard TTL, Table 4-1 gives a power dissipation of 10 mW per gate and a speed of 10 ns. So the speed-power product is 100.

(b) For Low Power Schottky TTL, the product is 9.5 ns × 2 mW = 19. Because Low Power Schottky has the lowest speed-power product of any TTL series, it is becoming very popular.

### 4-5.3 Fanout

*Fanout is the number of loads that can be driven by a single source.* If a source is required to drive more loads than its fanout allows, multiple sources should be used and the loads should be divided.

In TTL each source is capable of sinking a current of at least $I_{OL}$, and *each load can drive no more than* $I_{IL}$ into the source. Thus the fanout is $I_{OL}$ divided by $I_{IL}$. For most families the fanout is 10. Fanout can change, however, if one series is driving loads of another series. The worst case is where Low Power is driving Standard, High Speed, or Schottky TTL.

**EXAMPLE 4-3**

How many standard loads can a single Low Power gate drive?

**SOLUTION**

Table 4-1 shows that $I_{OL}$ for Low Power TTL is 3.6 mA and $I_{IL}$ for standard TTL is 1.6 mA.[3] Therefore, the fanout is

$$\frac{3.6}{1.6} = 2$$

If more than 2 loads must be driven, it is often best to buffer the Low Power source by letting it drive a single standard TTL gate or inverter. The standard TTL gate can then drive 10 loads.

### 4-5.4 Using Manufacturer's Data

The circuit and characteristics of most TTL gates are specified by the manufacturer in commonly available literature. Figure 4-4, for example, is taken from the *TTL Data Book*, published by the Texas Instruments Corporation, a leading manufacturer of TTL. It shows the pinouts for the 7400 quad 2-input NAND gate.

---

[3]This 1.6 mA is sometimes called *one standard TTL load.*

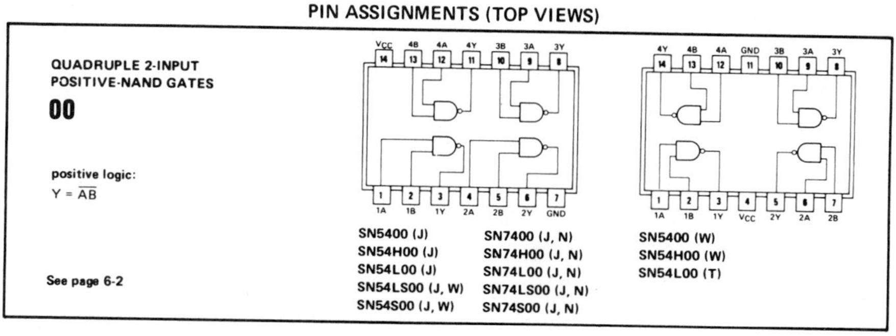

PIN ASSIGNMENTS (TOP VIEWS)

FIGURE 4-4 Pinouts for the **7400** NAND gate. (From the *TTL Data Book for Design Engineers*, 2nd ed., Texas Instruments, Inc. Courtesy of Texas Instruments, copyright 1976.)

Figure 4-4 shows that all five logic families are listed beneath the left pinout. This indicates that the **7400** is manufactured in all families. This is only true for a small number of ICs. Even a common IC, such as the **7402**, is not manufactured in the High Speed series.

The figure also shows the package type. J means a ceramic dual-in-line (DIP) package, N means a plastic DIP, and W indicates a flat package used only for ICs that are to be soldered instead of being mounted in sockets. The type N plastic package is the least expensive, most readily available, and most commonly used. Pinouts for the **5400** TTL series are also given. The **5400** series has a wide temperature range and is used primarily in military applications.

The figure also directs the reader to another page (6-2 in this case) which gives more information on the ICs. Pages 6-2 and 6-3 of the *TTL Data Book* are shown in Appendix E and give the following.

1. The operating characteristics ($V_{IH}$, $V_{OH}$, etc.).
2. The temperature range for the IC. **7400** series operate from 0° to 70°C, which is adequate for laboratory and office environments. **5400** series ICs have a wider temperature range.
3. Power supply current drain per IC for each series. The typical current per gate is generally used to calculate a system's current drain.
4. Propagation delays for various conditions of loading.
5. The circuits for each family.

These two pages thoroughly cover the specifications for TTL NAND gates with totem-pole outputs. Other types of gates are similarly described by the Data Book.

## 4-6  EMITTER-COUPLED LOGIC

Emitter-coupled logic (ECL) is faster than TTL and is used in applications where very high speed is essential. ECL has high switching speeds because the transistors act as difference-amplifier emitter followers, and are never in saturation. Because of this high speed, more attention must be paid to the construction of the boards and the interconnecting wiring, or else noise generated by the fast wavefronts may degrade performance.

The basic ECL gate is shown in Fig. 4-5. It is essentially an OR/NOR gate, and typically both OR and NOR outputs are available. $V_{CC1}$ and $V_{CC2}$ are both connected to ground. Two grounds are provided to eliminate crosstalk within the package (another precaution against high speed noise effects). The supply voltage, $V_{EE}$, is $-5.2$ V. A logic 1 output is typically $-0.9$ V and a logic 0 output is typically $-1.75$V.

In the circuit of Fig. 4-5 the input transistors ($Q_1$ through $Q_4$) operate as a difference amplifier in conjunction with $Q_5$. At any time, either $Q_5$ is ON or one or more of the input transistors is ON. The circuit functions as follows.

1.  Resistors $R_1$ and $R_2$ and the diodes clamp the base of $Q_6$ so that its emitter, which is also the base of $Q_5$, is always at $-1.29$ V.

2.  If all inputs are LOW ($-1.75$ V), $Q_5$ has the highest base voltage and turns ON.

3.  This clamps the emitters to $-2$ V, and each input transistor has a base-to-emitter voltage of $0.25$ V, which is a forward voltage, but not large enough to turn it ON.

4.  The OR output is LOW ($\approx -1.75$ V). It equals the IR drop across $R_5$ plus its base to emitter voltage drop.

5.  The NOR output is HIGH ($\approx -0.9$ V) because there is only a small voltage drop across $R_4$ with $Q_1$ through $Q_4$ OFF.

6.  Consequently, the OR output is a 0 and the NOR output is a 1. This is proper OR gate action when all inputs are LOW.

7.  If any input is a 1 ($-0.9$ V), the base of its transistor is higher than the base of $Q_5$ and it will turn ON, turning $Q_5$ OFF.

8.  If $Q_4$ is ON and $Q_5$ is OFF, the voltage at the base of $Q_7$ will be close to ground and the OR output will be one base-to-emitter drop below ground or $-0.9$ V.

9.  The current through $Q_4$ flows through $R_4$ causing the voltage on the base of $Q_8$ to be approximately $-1$ V so the NOR output is approximately $-1.75$ V.

10.  OR action is achieved since with any input HIGH, $Q_5$ turns OFF and $Q_4$ turns ON, the OR output is HIGH and the NOR output is LOW.

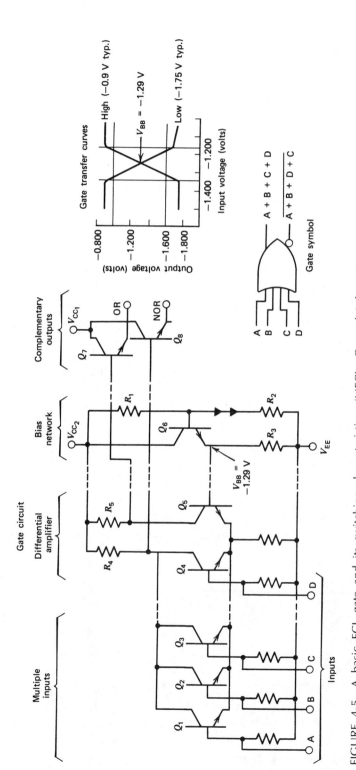

FIGURE 4-5  A basic ECL gate and its switching characteristics, (MECL, General Information Manual, Motorola, Inc. Copyright 1974. Courtesy of Motorola Integerated Circuits Division.)

## 4-7 COMPLEMENTARY METAL OXIDE SEMICONDUCTOR GATES

Complementary metal oxide semiconductor gates (CMOS) are currently proving to be popular because of their low power dissipation. A CMOS gate is composed of two *metal-oxide-semiconductor* (MOS) gates. There are two types of MOS gates, *n*-channel and *p*-channel. Each gate consists of *a drain, a source, a gate, and a substrate.* For logic circuits, *enhancement* mode gates are used. The symbols for MOS gates and a cross section of an *n*-channel transistor are shown in Fig. 4-6.

An *n*-channel MOS transistor is similar to a field-effect transistor (FET) and is set up so that the drain is positive with respect to the source. With no enhancement the *pn* junction between the drain and the substrate is reverse-biased and no current flows. Enhancement occurs if the gate is made positive with respect to the substrate. Note that the gate is separated from the substrate by an oxide insulator, which causes the input impedance of an MOS transistor to be very high. When the gate voltage is positive, electrons in the substrate are attracted toward the gate and the migration of electrons to the gate area effectively changes the *p*-doped silicon to *n* silicon. Thus a low impedance *n*-channel is formed between source and drain and current flows from drain to source. P-channel MOS transistors work in a similar manner as the table below shows.

|  | *n*-Channel | *p*-Channel |
| --- | --- | --- |
| Substrate material | *p* | *n* |
| Source material | *n* | *p* |
| Drain material | *n* | *p* |
| Gate effects (Enhancement) | Current flows when gate is positive with respect to substrate | Current flows when gate is negative with respect to substrate |
| Direction of conventional current | Drain to source | Source to drain |

MOS transistors are used in large scale memories (see Chapter 15) but are not used as ordinary gates. CMOS, however, does have the advantage of dissipating almost no power. Consequently, it provides an attractive alternative to TTL.

The basic CMOS inverter is shown in Fig. 4-7. As in all CMOS gates it consists of an *n*-channel transistor whose source and substrate are tied to ground, and a *p*-channel transistor whose source and substrate are tied to $V_{DD}$. $V_{DD}$ can be any voltage from $+3$ to $+15$ V. The output voltage is either ground or $V_{DD}$, denoting a 0 and a 1, respectively.

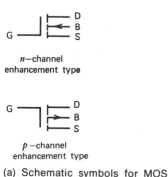

(a) Schematic symbols for MOS transistor (G: gate, D: drain, B: active bulk or substrate, S: source)

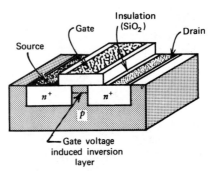

(b) Cross-sectional view of an n-channel enhancement-type MOS transistor

FIGURE 4-6   MOS transistors. *(RCA COS/MOS Integrated Circuits Manual, CMS-271, copyright 1971. Courtesy of RCA Solid State Division.)*

The unique feature of a CMOS inverter is that no current flows through it in either the 0 or 1 state. If the input is HIGH, the $n$-channel transistor is enhanced, but the $p$-channel is not. The output is disconnected from $V_{CC}$ by the open $p$-channel transistor. This makes the output LOW and inversion has taken place. Also, since the open $p$-channel blocks any current flow from $V_{CC}$, no power is dissipated.

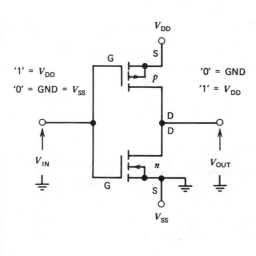

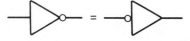

FIGURE 4-7   COS/MOS inverter circuit. *(RCA COS/MOS Integrated Circuits Manual, CMS-271, copyright 1971. Courtesy of RCA Solid State Division.)*

If the input is LOW, the $n$-channel transistor is effectively open, while the $p$-channel transistor is enhanced because its gate voltage is negative with respect to its substrate, which is tied to $V_{CC}$. The output is now shorted to $V_{CC}$ and is HIGH, but there is still no path for current to flow and again the power dissipation is very small. Power is only dissipated in a CMOS gate when it is in the process of *switching states*. Consequently, power dissipation is proportional to the frequency at which the gate is switched, but CMOS power is still much smaller than that of standard TTL. This is very attractive to engineers whose circuits must operate on small amounts of power, or who want to keep their circuits cool. On the other hand, CMOS is slower, more costly, and does not offer all the circuits available in standard TTL.

Logic gates are built from CMOS circuits by adding additional gates. A 2-input NOR gate and a 2-input NAND gate are shown in Fig. 4-8. The tables show how they operate. For the NOR gate, for example, note that if either input is HIGH, at least one of the $n$-channel transistors is shorted to ground and at least one of the $p$-channel transistors is open, so that the output is LOW. A HIGH output occurs only if both inputs are LOW, enhancing the two $p$-channel transistors. In that case, both $n$-channel transistors are open, isolating the output from ground.

### 4-7.1 CMOS Power Considerations

While CMOS absorbs practically no power in the *static state*, if a CMOS gate is switched frequently, its power dissipation increases because it must charge and discharge whatever capacity is connected to its gates. A chart showing the power dissipation for the three series is given in Fig. 4-9. It shows that a CMOS gate operating at 5 V and switching a million times a second dissipates as much power as an LS gate. CMOS gates are ideal for battery driven circuits that switch slowly because they take almost no current.

Figure 4-9 also shows that at higher voltages, CMOS dissipates more power, as we would expect. A second disadvantage of using 10- or 15-V CMOS is that it is not as compatible with TTL as 5-V CMOS. The advantage of high voltage is that it speeds up the CMOS circuits. Typical propagation delays are 40 ns for CMOS driven by a 5-V supply, but only 20 ns for 10-V CMOS. This speed-power trade off is an engineering decision.

### 4-7.2 Interfacing TTL to CMOS

TTL can be interfaced to 5-V CMOS almost directly. There is no problem when the TTL output is LOW. When the TTL output is HIGH, however,

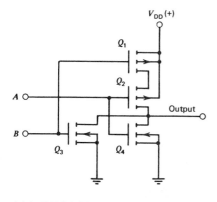

| A B | $Q_1$ $Q_2$ $Q_3$ $Q_4$ | Output |
|-----|-----------|--------|
| 0 0 | S S O O | H |
| 0 1 | O S S O | L |
| 1 0 | S O O S | L |
| 1 1 | O O S S | L |

S = Short
O = Open

(a) A CMOS NOR gate

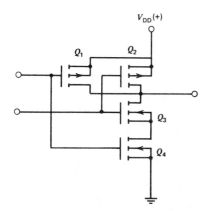

| A B | $Q_1$ $Q_2$ $Q_3$ $Q_4$ | Output |
|-----|-----------|--------|
| 0 0 | S S O O | H |
| 0 1 | S O S O | H |
| 1 0 | O S O S | H |
| 1 1 | O O S S | L |

S = Short
O = Open

(b) A CMOS NAND gate

FIGURE 4-8 CMOS gates. (From George K. Kostopoulos. *Digital Engineering.* Copyright John Wiley & Sons, Inc., 1975. Reprinted by permission of John Wiley & Sons, Inc.)

its voltage is only 3.5 V. A pull-up resistor (see Sec. 5-5.1) should be added to the output of any TTL gate that drives CMOS to raise the high level output voltage to 5 V for reliable operation.

Because CMOS gates are high impedance and source very little current, fanout from TTL to CMOS is very high and does not generally present a problem. To interface TTL to higher voltage CMOS, buffer/drivers such as the 7406 and 7407 should be used. (See Sec. 5-6.2.)

### 4-7.3 Interfacing CMOS to TTL

When CMOS drives TTL, it cannot reliably sink the 1.6 mA ($I_{IL}$) of the TTL gate. Manufacturers recommend that a single 74LS inverter be placed

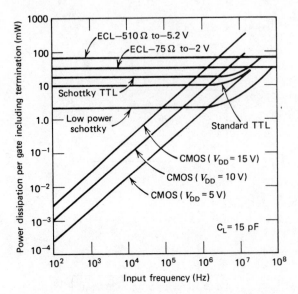

FIGURE 4-9 Typical power dissipation versus input frequency for several popular logic families. (From the Fairchild CMOS Data Book, 1977. Reprinted courtesy of Fairchild Camera and Instrument Corporation.)

at the CMOS interface. The LS gate inputs only source 0.4 mA (see Fig. 4-3) and CMOS can absorb this.

## 4-7.4 CMOS Availability

CMOS is widely available in two series; the 4000 series and the 74C series. Both series contain the standard gates and more complex ICs, such as shift registers and counters. The 4000 was the original series. The 74C series was brought out to emulate TTL. All 74C ICs are pin compatible and functionally the same as their TTL counterparts. Therefore, 74C CMOS of 5 V can be readily used with the 7400 TTL series as long as the interfacing rules of the two preceding paragraphs are followed.

### SUMMARY

In this chapter the three types of digital logic in current usage were described. As a result of these discussions, we make the following suggestions.

1.  ECL has the highest speed of any family, but use ECL for high speed circuits only. For low speed circuits there is no need to take on the extra expense, power, and care required to eliminate high speed noise.
2.  CMOS is very attractive where low power dissipation or battery operation is required. It is two to four times slower than TTL, but the additional speed is often

unnecessary. Fortunately, it is very easy to interface or combine TTL and 5 V CMOS.

3.   Standard TTL has the broadest line and can easily be mixed with other types of TTL for special requirements. It offers a good compromise between speed and power dissipation and is a good choice for most applications.

4.   For new designs, Low Power Schottky should be seriously considered. Its comparable speed and lower power dissipation make it an attractive alternate to standard TTL.

## GLOSSARY

**RTL.** Resistor-Transistor Logic.

**DTL.** Diode-Transistor Logic.

**TTL.** Transistor-Transistor Logic.

**ECL.** Emitter-Coupled Logic.

**MOS.** Metal-Oxide-Semiconductor Transistor.

**LSI.** Large scale integration.

**CMOS.** Complimentary Metal-Oxide-Semiconductor circuit, consisting, at least, of a $p$-channel and an $n$-channel MOS transistor.

**Propagation delay.** The time required for the output of a gate to respond to a change in the inputs.

**Fanout.** The number of loads that can be driven by a single source or input.

**Breadth.** The number of different types of gates and circuits available on an IC family.

**Saturation.** The state of a transistor when it draws maximum current. In saturation $V_{CE}$ is very low ($\approx 0$).

**Enhancement.** Providing the proper voltage at the gate of an MOS transistor to turn it on.

## REFERENCES

Louis Nashelsky, *Introduction to Digital Computer Technology*, Wiley, New York, 2nd ed., 1977.

Dan I. Porat and Arpad Barna, *Introduction to Digital Techniques*, Wiley, New York, 1979.

George K. Kostopoulos, *Digital Engineering*, Wiley, New York, 1975.

Thomas R. Blakeslee, *Digital Design with Standard LSI and MSI*, Wiley, New York, 2nd ed., 1978.

*The TTL Data Book for Design Engineers*, Texas Instruments, Inc., 2nd ed., 1976.

Morris and Miller, *Designing with TTL Integrated Circuits*, McGraw-Hill, New York, 1971.

**Lane S. Garrett,** "Integrated Circuit Digital Logic Families," *IEEE Spectrum,* Oct. 1970 and Nov. 1970.

*RCA COS/MOS Integrated Circuits Manual,* RCA, Sommerville, New Jersey, 1972.

*McMOS '72,* Motorola Semiconductor Products Division, Phoenix, Arizona, 1972.

*MECL 10,000,* Motorola Semiconductor Products Division, Phoenix, Arizona, 1972.

*CMOS Data Book,* Fairchild Camera and Instrument Corporation, Mountain View, California, 1977.

# 5

# BASIC TTL GATES

## 5-1 INSTRUCTIONAL OBJECTIVES

In this chapter the basic TTL gates are introduced. After reading this chapter the student should be able to:

1. Use Schmitt triggers to smooth input waveforms.
2. Properly connect, pull-up, or clamp any unused gate inputs.
3. Use open collector and 3-state gates, where necessary.
4. Use strobed gates, expandable gates, and expanders.
5. Determine where buffer/drivers are necessary and use them.
6. Build complementers and comparators using EXCLUSIVE-OR circuits.

## 5-2 SELF-EVALUATION QUESTIONS

Watch for the answers to the following questions as you read the chapter. They should help you understand the material presented. When you have finished the chapter, return to this section and be sure you can answer all of the questions.

1. What is the difference between a Schmitt trigger and an ordinary NAND gate?
2. Describe the three methods of handling unused inputs. State their advantages and disadvantages.

3.   Why is wire-ANDing of totem-pole gates prohibited?
4.   Why is a large $V_{OH}$ an advantage in an open-collector gate?
5.   When several 3-state gates are connected together, why must only 1 gate be enabled at any time?
6.   How does the strobe input affect the operation of a 7425?
7.   How can a 7460 affect the operation of a 7423 or a 7450?

## 5-3   INTRODUCTION

The most basic IC gates (ANDs, ORs, NANDs, and NORs) were introduced in Chapter 2. We will now begin to consider other ICs and the circuits that can be built from them. Henceforth, all ICs discussed in this book are identified by part number, and can be purchased from the manufacturer, a local distributor, or an electronics discount house. Any circuit described can easily be set up in the laboratory if the reader wants to investigate its behavior.

Most of the ICs considered will be taken from the 7400 TTL series. MOS and CMOS ICs will be discussed in those applications, such as memories, where they dominate. Because the emphasis is on circuit design, the choice of a particular IC series matters little. The knowledge and experience gained by studying the TTL circuits presented here are applicable to circuits built from other families, with relatively minor adjustments needed if built with CMOS or ECL. Indeed, the National Semiconductor Corporation has come out with a series of CMOS ICs that are compatable (functionally and pin-for-pin) with the 7400 TTL series. The circuits described in this book can be directly translated to CMOS using this series.

## 5-4   SCHMITT TRIGGERS

The basic NAND gates in the 7400 series were introduced in Sec. 2-9.1 The Schmitt trigger is a special type of NAND gate that is often used to smooth out or square-up noisy or irregular voltage inputs. It has the following special property: a Schmitt trigger will not turn on unless the input voltage is *greater* than a certain voltage called the *positive-going threshold voltage*, and will not turn off unless the input voltage is *less* than another voltage called the *negative-going threshold voltage*.

The positive-going threshold voltage ($V_{T+}$) is greater than the negative-going threshold voltage ($V_{T-}$). The difference between $V_{T+}$ and $V_{T-}$ is called *hysteresis*, and gives the Schmitt trigger the ability to square up slow and jagged waveforms. Sometimes a hysteresis symbol ($\sqcup$) is placed within the gate to distinguish a Schmitt trigger from a NAND gate.

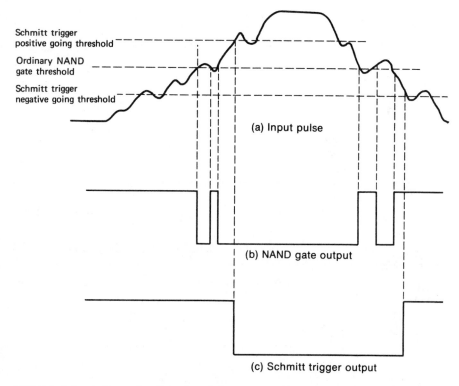

Schmitt trigger
positive going threshold

Ordinary NAND
gate threshold

Schmitt trigger
negative going threshold

(a) Input pulse

(b) NAND gate output

(c) Schmitt trigger output

FIGURE 5-1 Action of a Schmitt trigger compared to the action of a NAND gate.

The action of a Schmitt trigger is illustrated in Fig. 5-1, where it is compared with the action of an ordinary NAND gate. The output of the NAND gate is assumed to be LOW if its input is *above* the threshold, and HIGH if its input is *below* the threshold. If the input pulse is slow, noisy, or uneven, spikes appear on the output, as shown in Fig. 5-1*b*, because the input crosses and recrosses the NAND gate threshold.

The Schmitt trigger output, a clean pulse, is shown in Fig. 5-1*c*. The Schmitt trigger output does not go LOW until the input waveform crosses the positive-going threshold. Once triggered, however, the Schmitt trigger will not turn off merely because the input recrosses the positive-going threshold. The input must be sufficiently negative to cross the negative-going threshold for the Schmitt trigger to turn OFF.

There are three Schmitt trigger ICs available; the **7413** dual 4-input NAND gate, the **7414** hex Schmitt trigger inverters, and the **74132** quad 2-input NAND gates. For all gates the typical $V_{T+}$ is 1.7 V and $V_{T-}$ is 0.9 V.

## EXAMPLE 5-1

A 1.8-V, 1-MHz sine wave is applied to a threshold detector that is to produce an output pulse if the input exceeds 1.7 V. How long does the output pulse last if the input connected to:

    (a) A NAND gate whose output changes when the input crosses 1.7 V?

    (b) A **7413** Schmitt trigger whose $V_{T+} = 1.7$ V and $V_{T-} = 0.9$ V?

## SOLUTION

The solution for both parts is shown in Fig. 5-2.

    (a) The NAND gate turns on and off at an angle such that:

$$\theta = \sin^{-1}\frac{17}{18} = 70.8 \text{ degrees (turn ON) and } 109.2 \text{ degrees (turn OFF)}$$

    Thus the output is ON for 38.4/360 or 0.107 of a cycle. Since a cycle takes 1 $\mu$s, the output pulse will be 107 ns long.

    (b) The Schmitt trigger turns on when $\theta = \sin^{-1} 17/18 = 70.8$ degrees, but does not turn off until:

$$\theta = \sin^{-1}\frac{9}{18} = 150 \text{ degrees}$$

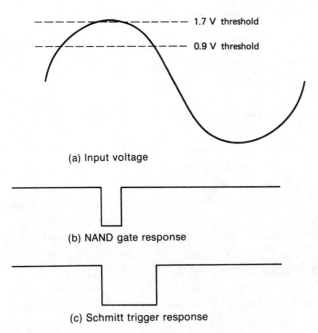

(a) Input voltage

(b) NAND gate response

(c) Schmitt trigger response

FIGURE 5-2   Response of threshold detectors to a 1.8 V sine wave.

The Schmitt trigger is ON for 79.2 degrees out of 360 degrees or 220 nanoseconds. Clearly the Schmitt trigger has improved the detector by delivering a longer output pulse.

---

Certain flip-flops and one-shots (multivibrators) in the TTL family have trigger inputs that have a voltage transition instead of a voltage level. They are known as "edge-triggered" devices. Normally the edges must change faster than 1 V per microsecond ($dv/dt \geq 1$ V/$\mu$s) to be effective. When a typical TTL gate changes state, it goes from 0.4 to 3.4 V in about 3 ns; therefore, it changes at the rate of 1 V/ns = 1000 V/$\mu$s and there is no problem. If the input is too slow, however, Schmitt triggers are usually used to square up and speed up the output, which can then be used as a trigger.

---

## EXAMPLE 5-2

The 120-V, 60-Hz "house lines" are to provide a series of triggers for a TTL circuit. In a practical circuit these triggers could be used to monitor commercial power. The absence of a trigger would then be an early warning that power has failed. The triggers must have edges faster than 1 V/$\mu$s. Design a circuit to produce the output pulses.

## SOLUTION

A solution is shown in Fig. 5-3. The circuit is designed as follows:

1.   The first problem is to limit the TTL input signals to approximately 0 to 5 V so that the maximum value of $V_{IH}$ will not be exceeded.
2.   The transformer is used to reduce the input ac voltage to a more manageable level. Here a transformer with a peak output voltage of 10 V is selected.
3.   When the transformer voltage is greater than 5 V diode $D_1$ turns on, clamping the input of the 7413 to 5 V. The 1-k$\Omega$ resistor prevents excessive current through the diode.

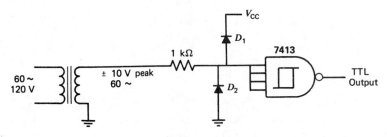

FIGURE 5-3   A Schmitt trigger used to square up a 60-Hz input.

4. When the transformer voltage is negative, diode $D_2$ turns on and clamps the input of the 7413 to ground. The 7413 is protected because the diodes hold its input voltage between ground and + 5 V.

5. The transformer output is:

$$V = 10 \sin 377t$$

and

$$\frac{dv}{dt} = 3770 \cos 377t$$

6. The maximum rate of change of the input is 0.00377 V/$\mu$s. This is too slow to be used as a trigger.

7. The Schmitt trigger 7413 can accept this slow input and produce a TTL output. The 7413 output is a square wave at a frequency of 60 Hz. The trigger output switches at TTL speeds, which is much faster than required in this problem.

## 5-5 OPEN AND UNUSED INPUTS

Occasionally, some of the inputs to a TTL gate are *not* used. This occurs most often in one-shots or flip-flops where the devices have features that are not needed for the particular circuit, but it can also occur with simple gates. Suppose, for example, an engineer needs two 3-input NAND gates and an inverter for his circuit. It is most economical to use a single **7410** IC for all three circuits, but this means transforming one 3-input NAND gate to an inverter, leaving two unused inputs.

In TTL, *open inputs* almost invariably *act as a logic 1*. It requires a *current* through a multiple-emitter input to ground (or $V_{CE(sat)}$) to produce a 0, and an open input provides no such current path. A **7410** will function as an inverter if two of its inputs are left open. The output is the complement of the signal on the remaining input.

**EXAMPLE 5-3**

One of the inputs to a **7402** 2-input NOR gate (see sec. 2-9.2) is left open. How does the output behave?

**SOLUTION**

The open input behaves as a 1. Since the **7402** sees a 1 on one of its inputs, it produces a 0 output regardless of the state of the connected input. One must be especially careful *not* to leave OR or NOR gate inputs unconnected.

TTL manufacturers advise against leaving inputs open. Open inputs are susceptible to noise and may occasionally provide an erroneous signal or spike. Also, while open inputs act like a logic 1, they appear as a 0 (or a voltage in the prohibited region) when viewed on an oscilloscope. This can add to the confusion when attempting to debug a circuit.

For MOS or CMOS gates the situation is far worse. If their inputs are floating, the outputs of MOS or CMOS gates drift and are *unpredictable*. Consequently, *unused MOS or CMOS inputs must be pulled up or tied to* $V_{CC}$ *for a 1 or grounded for a 0.*

## 5-5.1 Pull-up Resistors

Unused inputs can be "pulled up" by connecting them to $V_{CC}$ via a resistor as in Fig. 5-4a, where a **7420** with two unused inputs is shown. This decreases the noise susceptibility. Texas Instruments recommends that a

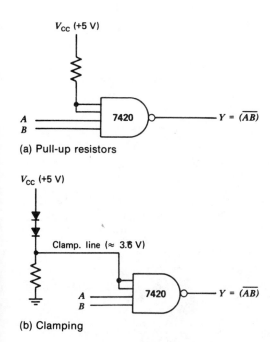

(a) Pull-up resistors

(b) Clamping

(c) Tying unused inputs to used inputs

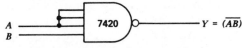

FIGURE 5-4 Methods of handling unused inputs.

1-kΩ resistor be used, and that no more than 25 unused inputs be connected to the same resistor.

## 5-5.2   Clamping

Another way to handle unused inputs is to clamp them, which effectively ties them to a constant voltage of 3.6 V. A typical clamp circuit, shown in Fig. 5-4*b*, consists merely of two diodes and a resistor. The voltage drop across the diodes lowers the output voltage to $V_{CC} - 1.4$ V. The advantage of clamping is that it reduces the IC inputs to 1.4 V less than the power supply. Therefore, any voltage spikes that occur on the power supply are unlikely to drive the clamp voltage above 5 V, where the IC input may be damaged.

## 5-5.3   Tying Used Inputs to Unused Inputs

If gate inputs are unused, they can be tied to used inputs on the same gate as shown in Fig. 5-4*c*. This is the most popular way of handling unused gate inputs, but often leads to mistakes in determining fanout. Engineers usually count the number of inputs connected to an output to determine fanout. If two or more inputs are connected to the same gate of NAND or AND gates they should only be counted as one load because a single gate cannot supply more than 1.6 mA through its 4-kΩ resistor regardless of how many inputs are connected to the multiple emitters.

TTL OR and NOR gates, however, absorb one standard load for each connected input. In general the input circuit of each load may have to be examined if a precise determination of fanout is required.

Because clamps and pull-up resistors tie unused inputs to a logic 1, they cannot be used on OR or NOR gates, although unused OR and NOR inputs can be tied to ground. Engineers tend to eliminate the whole problem by tying unused gate inputs to used inputs on the same gate. Flip-flops and one-shots, however, contain certain inputs (direct sets and direct clears, etc.) that must be HIGH if unused, and cannot be tied to varying inputs. These unused inputs are usually clamped. Pull-up resistors are used to terminate open collector gates (Sec. 5-6) and cables (Sec. 17-4).

## 5-6   WIRE-ANDING AND OPEN COLLECTOR GATES

For gates where the 0 logic level is caused by saturating a transistor, an additional level of logic can be obtained by connecting the outputs or collectors of several gates together. This is called *wire-ANDing*. Figure 5-5 shows the outputs of three DTL gates wire-ANDed together. If any one of the three transistors is saturated, it causes the output to be LOW. Therefore, the output is HIGH only if *all* three transistors are cut off, which corresponds to a 1 output for each individual gate. AND action occurs because the outputs of each of the three individual circuits must be a 1 in order that the wire-AND output be a 1.

If only one of the gates in Figure 5-5 is LOW, the transistor draws additional current because it is connected to $V_{CC}$ through three load resistors instead of one. This extra current decreases the fanout of the gate. If fanout is not a problem, however, there is no objection to wire-ANDing DTL and RTL circuits.

Unfortunately, TTL circuits with *totem-pole outputs* are *not* amenable to wire-ANDing. The wire-ANDing of totem-pole outputs is shown in Fig. 5-6. If the inputs on gate 1 cause its output to be a 1 and the inputs to gate 2 cause its output to be a 0, the top transistor of gate 1's totem-pole pair turns ON and the bottom transistor of gate 2's totem-pole pair turns ON.

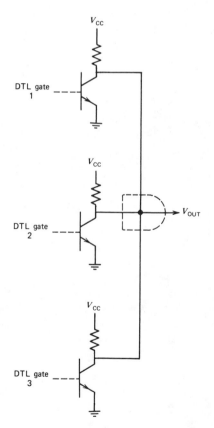

FIGURE 5-5 Three DTL gates with their outputs wire-ANDed together. Note the dashed AND symbol that is sometimes used to indicate wire-ANDing.

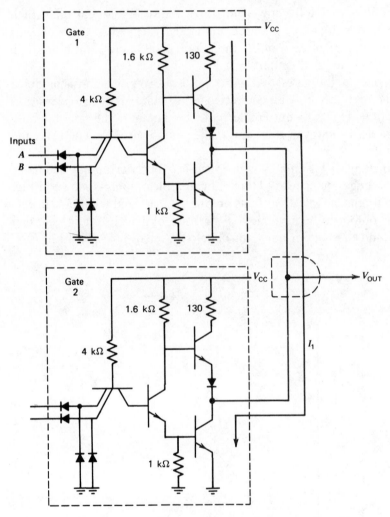

FIGURE 5-6 Wire ANDing two **7400s** together. Note that this is *not recommended*. If A OR B is low and C AND D are high, current $I_1$ will flow, which may damage the transistors.

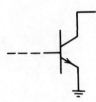

FIGURE 5-7 Output circuit of an open-collector gate.

Consequently, current $I_1$ flows. Note that there is only 130 $\Omega$ in the current path, and the current $I_1$ may be excessive and damage one of the gates. Even if the gates remain undamaged, the output voltage may enter the prohibited region because of the heavy current. Therefore, wire-ANDing of TTL gates is very poor design.

## 5-6.1   Open-Collector Gates

To allow wire-ANDing, TTL manufacturers have produced a series of *open-collector* gates. The output of an open-collector gate is shown in Fig. 5-7. The collector is tied only to a pin on the IC package and a load must be tied to the collector at this point. If the input conditions at the gate cause base current to flow, the open-collector output transistor saturates and pulls its output voltage to ground. If no base current flows, the output transistor acts like an open circuit.

Figure 5-8 is a table of the open-collector gates available in the standard TTL series. They accomplish the function of wire-ANDing by being tied to each other. To function properly open-collector gates must be tied to $V_{CC}$ through a pull-up resistor. The value of the pull-up resistor depends on the number of sources being wire-ANDed and the number of loads to be driven. The method of calculating this value is given in Appendix B.[1] As a practical rule, we recommend:

1.   Unless power dissipation is crucial, use a 1-k$\Omega$ pull-up resistor. This works for any configuration of up to 7 sources with a fanout of 7. When the number of sources or fanout exceeds 7 you must be very careful. (See Morris and Miller.)
2.   If power dissipation is extremely critical, calculate the maximum possible value for the pull-up resistor and use it.

---

**EXAMPLE 5-4**

Find an expression for the output of Fig. 5-9a.

**SOLUTION**

In Fig. 5-9a, two **7403** open-collector NAND gates and a **7409** open-collector AND gate are wire-ANDed together, causing $V_{OUT}$ to be LOW unless the output of each gate is HIGH. Therefore,

$$V_{OUT} = (\bar{A} + \bar{B})(\bar{C} + \bar{D})\, EF$$

The same circuit is shown in Fig. 5-9b, where the alternate gate representation is used for clarity.

---

[1]A table of the proper resistance values is given in Morris and Miller, *Designing with TTL Circuits*, p. 50.

| Open collector<br>I C | Description | Totem Pole<br>Equivalent | Diagram |
|---|---|---|---|
| **7401** | Quad 2–input<br>NAND gates | 7400 | 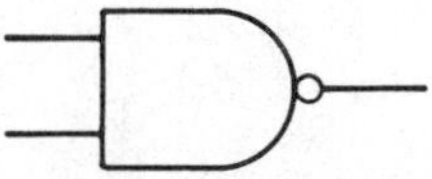 |
| **7403\*** | Quad 2–input<br>NAND gates | 7400 | 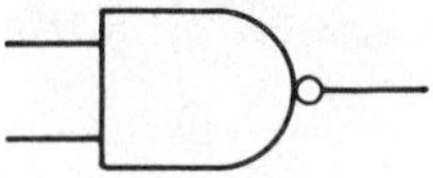 |
| **7405** | Hex Inverter | 7404 | 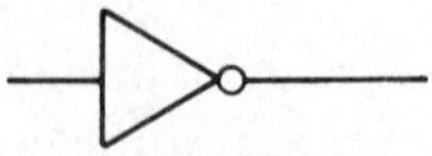 |
| **7409** | Quad 2–input<br>AND gates | 7408 | 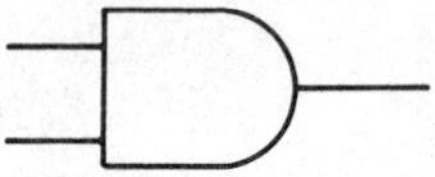 |
| **7412** | Triple 3–input<br>NAND gates | 7410 | 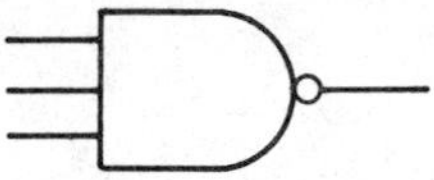 |
| **7422** | Dual 4–input<br>NAND gates | 7420 | 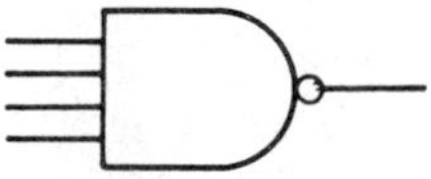 |
| **7433** | Quad 2–input<br>NOR gates | 7402 | 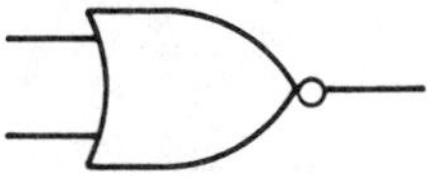 |

\* Pin for pin compatible with the 7400

FIGURE 5-8  A table of the most common open-collector gates.

## 5-6.2  Open-Collector Buffer/Drivers

Buffer/drivers differ from ordinary gates because they have a larger current sinking capability and a larger fanout. They are used to drive many loads or loads that require high current.

Two of the most popular open-collector buffer/drivers are the 7406 and 7407. The 7406 is a hex-inverting buffer and the 7407 is a hex-noninverting buffer/driver. The schematics and circuit diagrams are shown in Fig. 5-10.

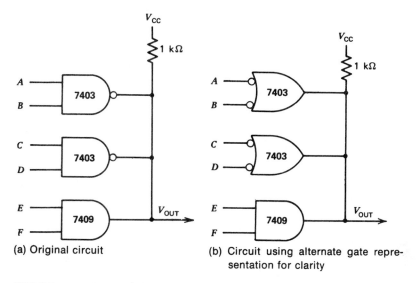

(a) Original circuit

(b) Circuit using alternate gate representation for clarity

FIGURE 5-9  Circuit for Ex. 5-4.

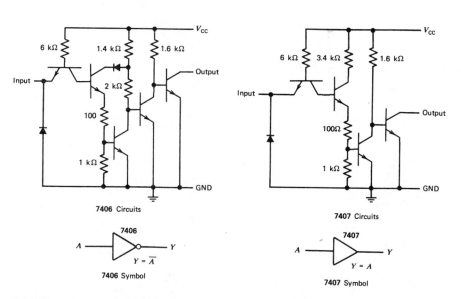

7406 Circuits

7406

$Y = \bar{A}$

7406 Symbol

7407 Circuits

7407

$Y = A$

7407 Symbol

FIGURE 5-10  **7406** and **7407** circuits and symbols. Note that there are six gates on each IC. (From *The TTL Data Book for Design Engineers*, 2nd ed.; Texas Instruments, Inc. Courtesy of Texas Instruments, copyright 1976.)

They can absorb a current of 40 mA (compared to an $I_{OL}$ of 16 mA for an ordinary gate) and have the additional advantage that $V_{OH}$ is 30 V for these ICs. Ordinary open-collector circuits like the **7403** have a $V_{OH}$ of 5 V, which means their collectors may not be connected to a supply voltage greater than 5 V. The **7406** and **7407** are often used to interface from TTL to circuits requiring higher voltages or currents than ordinary TTL gates can handle.

---

**EXAMPLE 5-5**

Small lamps are often used to indicate the state of a digital circuit. If a 10-V, 40-mA incandescent lamp is to indicate the level of a point in a digital circuit, and the lamp is to be lit when the point is a 1, how should the lamp be connected?

**SOLUTION**

Because of both the current and voltage requirements, the lamp cannot be connected directly to an ordinary TTL gate. The solution is to connect the lamp to the output of a **7406**, as shown in Fig. 5-11. When the point in the circuit is HIGH, the output of the **7406** is LOW, allowing current to flow through the lamp, turning it ON. When the point is LOW, the output of the **7406** is actually an open-collector transistor that is OFF. This prevents any current flow and keeps the lamp OFF. Note that when the **7406** gate is OFF, 10 V appear at the open collector, which would damage an ordinary open-collector gate with a $V_{OH}$ of 5 V. However, since $V_{OH}$ is 30 V for the **7406**, it is well within specifications.

---

Designers using incandescent indicators usually prefer 5-V lamps to eliminate the need for another power supply. Buffer/drivers are still used, however, to satisfy the current requirements. Pull-up resistors are not required in this circuit, because the lamp itself acts as the load.

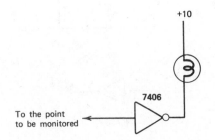

FIGURE 5-11   Driving an incandescent lamp with a **7406** open-collector inverter.

### 5-6.3 Other Buffer/Driver Gates

Three buffer/driver gates with totem-pole outputs are available in the 7400 series:

1. The 7428 quad 2-input NOR buffer, which is logically equivalent to the 7402.
2. The 7437 quad 2-input NAND buffer, which is logically equivalent to the 7400.
3. The 7440 dual 4-input positive NAND buffer, which is logically equivalent to the 7420.

The main advantage of these buffer/drivers is that they have an $I_{OL}$ of 40 mA instead of 16 mA for an ordinary gate. This means they have a fanout of 25 instead of 10, or they can drive a 40-mA lamp directly. Usually buffer/drivers are only used where the requirements for large fanout or heavy output current exists.

## 5-7 THREE-STATE DEVICES

Recently 3-state devices that have 3 *output states* have been developed for both TTL and CMOS. All modern microprocessors and their peripherals use 3-state outputs, which allow the user to directly connect outputs in parallel and to add or remove ICs connected to the output line without affecting circuit operation.

The three possible output states are:

1. Logic 1 (low impedance to $V_{CC}$).
2. Logic 0 (low impedance to ground).
3. Disabled (in the disabled state the device presents a high impedance to both $V_{CC}$ and ground).

Three-state devices have an enable/disable input in addition to the normal inputs. If enabled, the gate functions normally, but it presents a *very high output impedance if it is disabled*. The principle of 3-state operation is illustrated by the basic inverter of Fig. 5-12. Here, as with most 3-state gates, the gate is enabled if the inhibit input is LOW. This causes the output of the internal inverter ($\overline{\text{INHIBIT}}$) to be HIGH, reverse-biasing both its input to the multiple-emitter transistor and the diode to the base of $Q_3$. Consequently, the circuit functions as though there were no inhibit input and the output is the inverse of its input.

If the INHIBIT input is HIGH, $\overline{\text{INHIBIT}}$ is LOW. Now current flows through the base-to-emitter junction of $Q_1$ and deprives $Q_2$ of base current, cutting it OFF. The low voltage at $\overline{\text{INHIBIT}}$ drags the base of $Q_3$ down and cuts $Q_3$ off. As a result, both $Q_3$ and $Q_4$ are cut off, which causes a high output impedance to both $V_{CC}$ and ground. CMOS devices operate

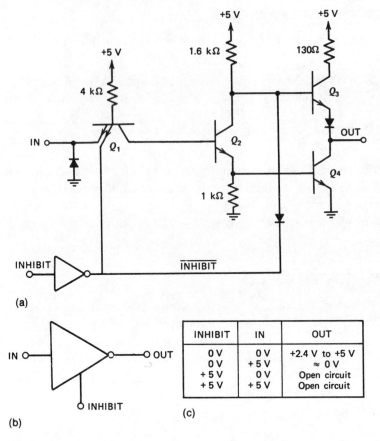

FIGURE 5-12 A 3-state TTL inverter circuit. (From Porat and Barna, *Introduction to Digital Techniques*. Copyright John Wiley & Sons, Inc., 1979. Reprinted by permission of John Wiley & Sons, Inc.)

similarly; when the device is inhibited, both the $p$ and the $n$-channel MOS transistors are open.

Three-state devices are currently being used in more complex devices, such as memories, shift registers, and multiplexers. They are designed for parallel operation and *one and only one of the parallel gates may be enabled at any time*. If no gates are enabled, the output presents a high impedance and its voltage may be in the prohibited region. If more than one gate is enabled, excessive current may flow and damage the ICs.

An oscilloscope trace of a 3-state microprocessor line is shown in Fig. 5-13. The top trace is the 3-state line and the bottom trace is the enabling pulse line. When the enabling pulses are LOW, the output is driven HIGH or LOW depending on the data. When not enabled, the output line is high

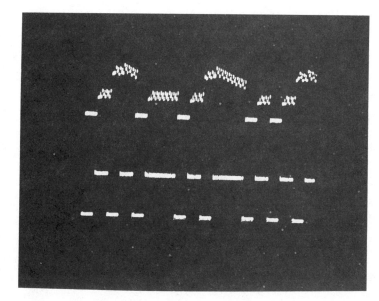

FIGURE 5-13  Signals on a 3-state bus line. (Courtesy of Ed Pickett and People's Cable TV Company, Rochester, N.Y.)

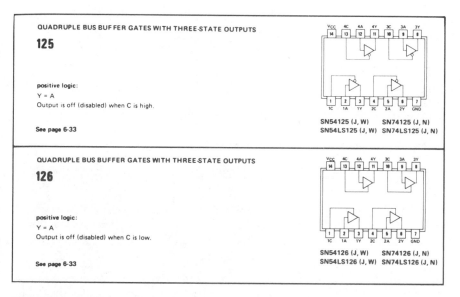

FIGURE 5-14  The **74125** and **74126** 3-state gates. (From the *TTL Data Book for Design Engineers*, 2nd ed., Texas Instruments, Inc. Courtesy of Texas Instruments, copyright 1976.)

impedance and floats to an intermediate level. The data shown on this line are 010-01-001.

### 5-7.1 The 74125 and 74126

A 3-state driver is a gate used to drive a 3-state bus. Two commonly used drivers are the **74125** and **74126**. Their pinouts are shown in Fig. 5-14. Each is a straight-through gate; the output is the same as the input when the IC is enabled. The **74125** is enabled by a LOW on its enable line, while the **74126** needs a HIGH enable line to function.

---

**EXAMPLE 5-6**

For the circuit of Fig. 5-15, find the output for each combination of inputs V and W.

**SOLUTION**

The operation of the circuit is best described by the table below.

| Inputs | | Points | | | | | | |
|---|---|---|---|---|---|---|---|---|
| V | W | A | B | C | Gate 1 | Gate 2 | Gate 3 | Output |
| 0 | 0 | 0 | 1 | 1 | DISABLED | ENABLED | DISABLED | N |
| 0 | 1 | 0 | 0 | 0 | DISABLED | DISABLED | ENABLED | P |
| 1 | 0 | 1 | 0 | 1 | ENABLED | DISABLED | DISABLED | M |
| 1 | 1 | 1 | 0 | 1 | ENABLED | DISABLED | DISABLED | M |

Note that this is a well-designed circuit because no combination of inputs enables more than one gate.

---

### 5-7.2 Other 3-state Drivers and Tranceivers

The **74365**, **74366**, **74367**, and **74368** are other 3-state drivers. They each have six data inputs and six 3-state outputs. The enable lines are common to many gates. These gates also have enhanced drive capability; they can sink 32 mA instead of the 16 mA for a **74125** or **74126**. **These are essentially the same as the 8095, 8096, 8097, and 8098** drivers manufactured by National Semiconductor Corporation. See the TTL data book for more information.

Octal bus drivers such as the **74LS241** and **74LS244** are also available. They are capable of driving 8 outputs, but they come in a 20-pin package.

Most microprocessor data lines are *bidirectional*; they handle data going

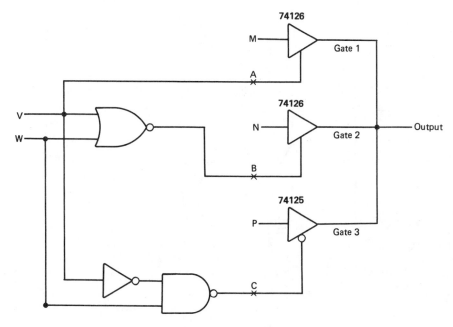

FIGURE 5-15   Circuit for Ex. 5-6.

either to or from the microprocessor. External devices are often connected to a bidirectional line by *tranceivers* such as a **74LS245** that can transfer data in either direction. A typical circuit for one line is shown in Fig. 5-16. The ENABLE line must be brough LOW for the chip to function. If it is LOW, the DIRECTION line determines the direction of data flow by enabling one of the 3-state gates. If DIRECTION is LOW, for example, gate 2 is enabled and data flows from B to A.

## 5-8   STROBED GATES, EXPANDABLE GATES, AND EXPANDERS

Some gates can be strobed or can accept expanded inputs. Careful use of these gates increases the designer's ability to optimize his circuit.

### 5-8.1   Strobed Gates

A strobed gate is simply a gate with a *strobe input*. Typically, if the strobe input is HIGH, the gate will function normally, but if the strobe input is LOW, the output will go to a particular logic level and remain there, regardless of the other inputs. The most common strobed gate is the **7425**

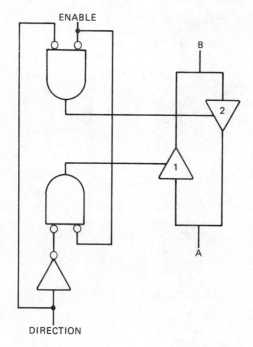

FIGURE 5-16   One line of a bidirectional transceiver.

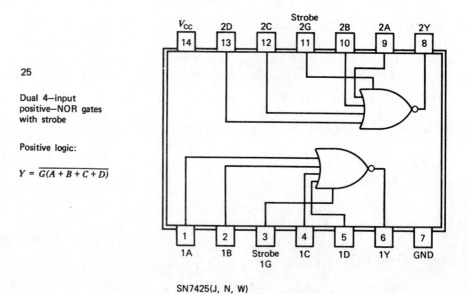

**25**

**Dual 4—input
positive—NOR gates
with strobe**

**Positive logic:**

$$Y = \overline{G(A + B + C + D)}$$

SN7425(J, N, W)

FIGURE 5-17   The **7425** strobed 4-input NOR gate. (*The TTL Data Book for Design Engineers,* 2nd ed., Texas Instruments, Inc. Courtesy of Texas Instruments, copyright 1976.)

dual 4-input NOR gate, shown in Fig. 5-17. If the strobe input is HIGH, the 7425 functions as a 4-input NOR gate. But if the strobe is LOW, its output will be HIGH as output equation indicates.

## 5-8.2 Expandable Gates and Expanders

*Expandable gates* have inputs that allow additional logic to be introduced, thus making them more versatile. A second gate, called an *expander*, is used to provide the additional logic.

The **7423**, shown in Fig. 5-18, is a dual 4-input NOR gate with strobe. Gate 1 is expandable. It is designed to work with a **7460** expander, and the expandable inputs are the X and $\overline{X}$ inputs (pins 1 and 15). If X and $\overline{X}$ are left open, the circuit functions as a strobed 4-input NOR gate.

The schematic diagram and circuit of the **7460** are shown in Fig. 5-19. Note that the X and $\overline{X}$ outputs are simply the open-collector and emitter of a transistor. When connected to a **7423** it functions as an additional phase-splitter transistor and causes the output to be 0 when it is on. The output equation of the expandable gate of the **7423** is

$$Y = \overline{G(A + B + C + D) + X}$$

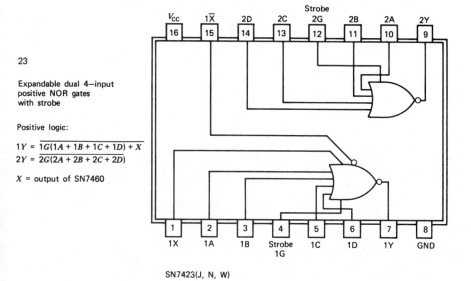

SN7423(J, N, W)

FIGURE 5-18   The **7423** 4-input NOR gate. Both gates have strobes and gate 1 is expandable. (*The TTL Data Book for Design Engineers.* 2nd ed., Texas Instruments, Inc. Courtesy of Texas Instruments, copyright 1976.)

**60**

Dual 4—input expanders
'60

Positive logic:

$X = ABCD$ when connected to $X$ and $\overline{X}$ inputs
of SN7423, SN7450, or SN7453

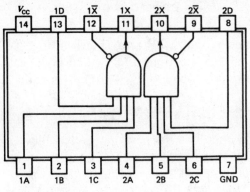

SN7460(J, N)

(a) Circuit symbol

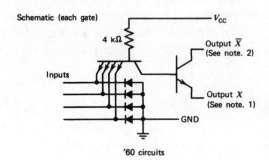

'60 circuits

*Notes.* 1. Connect to $X$ input of '23, '50, or '53 circuit.
2. Connect to $\overline{X}$ input of '23, '50, or '53 circuit.

Resistor value shown is nominal and in ohms.

(b) Schematic

FIGURE 5-19   The **7460** dual 4-input expander. (From *The TTL Data Book for Design Engineers,* 2nd ed., Texas Instruments, Inc. Courtesy of Texas Instruments, copyright 1976.)

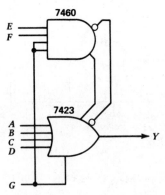

FIGURE 5-20   Implementation of the expression $Y = \overline{G(A + B + C + D + EF)}$

where $X$ is the expandable input expression. If the 7423 is connected to a 7460, this becomes

$$Y = \overline{G(A + B + C + D) + EFHJ}$$

where $E$, $F$, $H$, and $J$ are the 7460 inputs. Note that the strobe applies to the 7423 inputs, but not to the expander inputs.

---

**EXAMPLE 5-7**

Design a circuit to produce the output equation:

$$Y = \overline{G(A + B + C + D + EF)}$$

**SOLUTION**

Perhaps the simplest solution is to use a combination of the 7423 and 7460. This solution is shown in Fig. 5-20. The strobe input to the 7423 must also be used as an input to the expander to produce the term $GEF$.

---

## 5-9 AND-OR-INVERT GATES

SOP expressions can be implemented by AND-OR logic as shown in Sec. 3-3.5. To facilitate this implementation, a series of AND-OR-INVERT gates are available in the 7400 line. These consist of a group of AND gates connected to a NOR gate, all on the same chip.

The AND-OR-INVERT (AOI) gates available in the 7400 series are shown in Fig. 5-21. Note that the word "wide" in each gate refers to the number of inputs to the NOR gate. The 7450 is a dual 2-wide AOI gate with one gate expandable, so that additional logic can be introduced via a 7460 expander. The 7451 is the same as the 7450, except that it lacks the expander capability. The 7453 and 7454 are 4-wide AOI gates with the 7453 having an expandable input. Other AOI gates and some gates that are only AND-ORs (no inversion) exist in other series. A typical problem using AOI gates is shown in Ex. 5-8.

---

**EXAMPLE 5-8**

A *register* is a group of logic outputs, usually associated with each other to form a code representing information (a number, a letter, etc.) An $n$-bit register has $n$ outputs. Given two 4-bit registers, $A$ and $B$, and a select line, design a circuit so that the output is the complement of register $A$ if the select line is LOW, and the complement of register $B$ if the select line is HIGH.

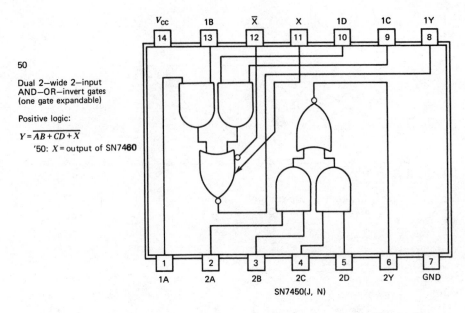

**50**

Dual 2—wide 2—input
AND—OR—invert gates
(one gate expandable)

Positive logic:

$Y = \overline{AB + CD + X}$

    '50: $X$ = output of SN7460

SN7450(J, N)

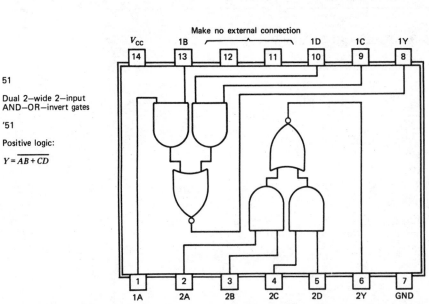

**51**

Dual 2—wide 2—input
AND—OR—invert gates

'51

Positive logic:

$Y = \overline{AB + CD}$

FIGURE 5-21   AOI gates available in the standard **7400** series. (From *The TTL Data Book for Design Engineers*, 2nd ed., Texas Instruments, Inc. Courtesy of Texas Instruments, copyright 1976.)

**53**

Expandable 4–wide
AND–OR invert gates

'53

Positive logic:

$$Y = \overline{AB + CD + EF + GH + X}$$
$$X = \text{output of SN7460}$$

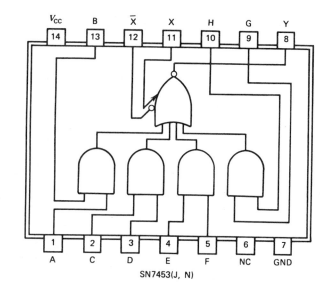

SN7453(J, N)

**54**

4–wide
AND–OR–invert gases

'54

Positive logic:

$$Y = \overline{AB + CD + EF + GH}$$

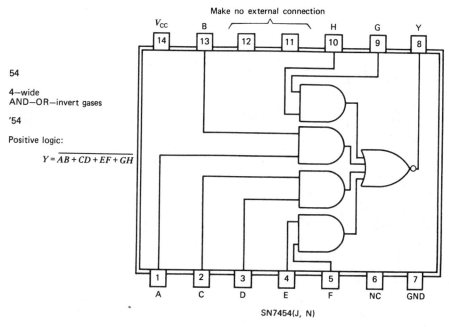

SN7454(J, N)

FIGURE 5-21 (Continued)

## SOLUTION

In either case, the output is the complement of a 4-bit register; therefore, the output must contain 4 bits. This design can be realized by setting up four, 2-wide, 2-input AOI gates and allowing the select line and its complement to make the choice. Consequently, it can be built with two **7451**s, as shown in Fig. 5-22. When the select line is LOW, all the B inputs to the AND gates are disabled and the A inputs

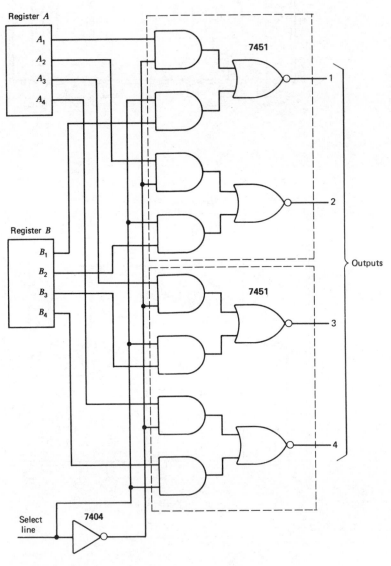

FIGURE 5-22   Circuit for Ex. 5-8.

come through the AOI gates. When the select line is HIGH, only the AND gates connected to the B register are enabled, and the 4-bit output is the complement of the B register.

## 5-10 THE EXCLUSIVE-OR GATE

An EXCLUSIVE-OR (XOR) gate is a 2-input gate whose output is the same as an OR gate except that it produces a 0 output if *both* inputs are 1. The symbol $\oplus$ is used to indicate the XOR operation. The symbol and truth table for the 7486 quad 2-input XOR gate are shown in Fig. 5-23.

XOR gates and circuits built from them have a large variety of important applications, and are discussed further in Chapter 13. To familiarize the reader with their use, however, two examples are presented here.

---

**EXAMPLE 5-9**

Given two 4-bit registers, A and B, design a circuit to determine whether the numbers in the two registers are the same.

**SOLUTION**

The required circuit is an equality detector; it must do a bit-by-bit comparison of the register outputs. If the output of each pair of corresponding bits are equal ($A_1 = B_1$, $A_2 = B_2$, etc.), the numbers in the two registers are equal. This comparison can be made by using XOR gates. If the output of an XOR gate is LOW, its two inputs are equal. Four XOR gates, one to compare each bit, are required and the output of each gate must be LOW for equality. Therefore, if the outputs are fed to a 4-input NOR gate, the final output will be HIGH only if all its inputs are LOW, indicating that the two registers contain the same number. The circuit is shown in Fig. 5-24. A **7425** is used as the 4-input NOR gate. It is shown in its negative-NAND input representation with the strobe clamped. The circuit usually works if the strobe is left unconnected.

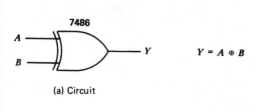

(a) Circuit

$Y = A \oplus B$

| A | B | Y |
|---|---|---|
| 0 | 0 | 0 |
| 0 | 1 | 1 |
| 1 | 0 | 1 |
| 1 | 1 | 0 |

(b) Table truth

FIGURE 5-23 EXCLUSIVE-OR gates.

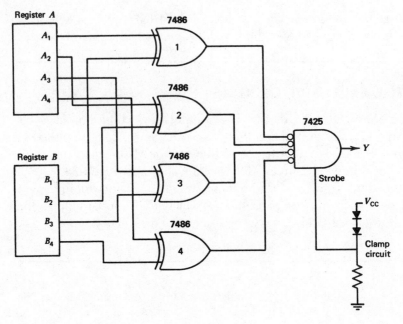

FIGURE 5-24 A 4-bit equality detector. Gate 1 compares bit 1 of both registers; gate 2 compares bit 2; and so on.

## EXAMPLE 5-10

Given a 4-bit register and a select line, design a circuit such that:
  (a) If the select line is HIGH, the circuit output is the same as the register.
  (b) If the select line is LOW, the output is the complement of the register outputs.

## SOLUTION

This circuit can be designed using AOI gates and inverters (see Problem 5-12), but it can be designed more simply by using XOR gates. A little thought reveals that if *one of the inputs to an XOR gate is a 1, the gate inverts the other input.* However, *if one input to an XOR gate is a 0, the output is the same as the other input.* Therefore, the circuit is designed by tying the select line and the register outputs together in four XOR gates as shown in Fig. 5-25. When the select line is HIGH, one input to each XOR gate is 0 and it does not invert. If the select line is LOW, however, one input to each XOR gate is HIGH and the four outputs are the complements of the four bits of the register.

## SUMMARY

In this chapter, the basic SSI gates available in the 7400 series were introduced and their characteristics studied. Some simple circuits using these gates were de-

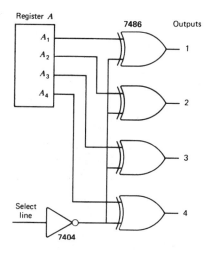

FIGURE 5-25   Design of the complementation circuit of Ex. 5-10.

signed. In addition, open-inputs, wire-ANDing, strobed gates, expanders, and EX-CLUSIVE-ORs were introduced and examples of their use were presented. These basic gates are used as components in the more elaborate circuits to be covered in succeeding chapters.

## GLOSSARY

**Schmitt trigger.** A gate that turns on at a different voltage than the voltage that turns it off.

**SSI.** Small scale integrated circuit.

**Pull-up resistor.** A resistor connected between $V_{CC}$ and a point in the circuit.

**Clamp.** A circuit that ties unused inputs to a voltage in the 1 range, which is less than $V_{CC}$.

**Wire-ANDing.** Tying the outputs of several gates together.

**Open-collector gate.** A gate whose output is a collector with no internal connection to $V_{CC}$.

**Three-state gate.** A gate having either a 1, 0, or high impedance output.

**Strobe.** An input signal that can activate or disable a gate.

**Expandable gate.** A gate with inputs to accept additional logic.

**Expander.** A gate with special outputs to provide additional logic to an expandable gate.

**AOI.** AND-OR-INVERT gate.

**XOR.** EXCLUSIVE-OR-gate.

**Transceiver.** A bidirectional driver-receiver.

## REFERENCES

**Morris** and **Miller,** *Designing With TTL Integrated Circuits*, McGraw-Hill, New York, 1971.

*The TTL Data Book for Design Engineers*, Texas Instruments, Inc., 2nd ed., 1976.

**George K. Kostopoulos,** *Digital Engineering*, Wiley, New York, 1975.

**Dan I. Porat** and **Arpad Barna,** *Introduction to Digital Techniques*, Wiley, New York, 1979.

**Christopher E. Strangio,** *Digital Electronics*, Prentice-Hall, Englewood Cliffs, N.J., 1980.

## PROBLEMS

5-1.   How long are the responses of the NAND gate and Schmitt trigger of Ex. 5-2 to a 2-V, 500-kHZ sine wave?

5-2.   Design a **7413** Schmitt trigger detector to determine if the output of a sine wave is greater than 5.1 V. If the input frequency is 1 MHz, what is the minimum width of the output pulse?

5-3.   A voltage spike that starts at 0 and rises at the rate of 1 V/$\mu$s until it reaches 3 V, after which it falls at the rate of 1 V/$\mu$s, as shown in Fig. P5-3, is applied to a **7413**. Show the output of the **7413** as a function of time.

5-4.   An 8-input NOR gate has five unused inputs. Will it function properly if:

  (a) The unused inputs are tied to used inputs.

  (b) The unused inputs are tied to clamp.

  (c) The unused inputs are tied to pull-up resistors.

  (d) The unused inputs are tied to ground.

5-5.   Why shouldn't an open-collector gate be tied to a totem-pole gate?

5-6.   An incandescent lamp is used to monitor various points in a circuit, as shown in Fig. 5-11. Is the lamp ON or OFF if the gate is a **7406** and the input is connected to:

  (a) A logic 1.

  (b) A logic 0.

  (c) The input is not connected to anything (open).

Repeat this problem if the gate is a **7407**.

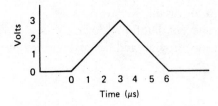

FIGURE P5-3

5-7. If the output of a disabled 3-state gate is connected to the input of a TTL gate, does the TTL gate see a 1 or a 0? Explain.

5-8. Given three 3-state gates, $A$, $B$, and $C$, and two select inputs, $D$ and $E$, design a circuit to function in accordance with the table below:

| Select Inputs | | Gate that Controls the Output |
|---|---|---|
| D | E | |
| 0 | 0 | A |
| 0 | 1 | A |
| 1 | 0 | B |
| 1 | 1 | C |

5-9. For the circuit of Fig. P5-9, find the output for each combination of inputs.

5-10. Use expander and AOI gates to produce the following outputs:

(a) $Y = \overline{AB + CD + EF + GH + JK}$

(b) $Y = \overline{AB + CD + ACEF}$

5-11. Identify the gates and find the output expression for the circuits of:

(a) Fig. P5-11a.

(b) Fig. P5-11b.

What indicates that the gates of Fig. P5-11a are open-collector gates?

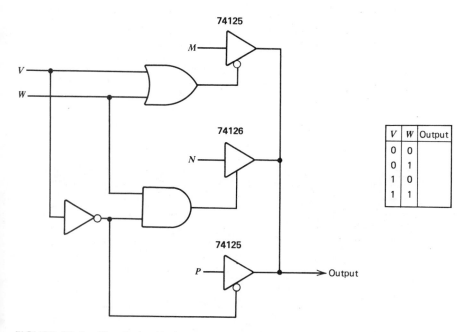

| V | W | Output |
|---|---|---|
| 0 | 0 | |
| 0 | 1 | |
| 1 | 0 | |
| 1 | 1 | |

FIGURE P5-9 Circuit for Prob. 5-9.

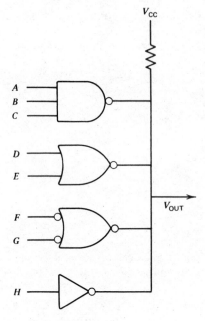

FIGURE P5-11a

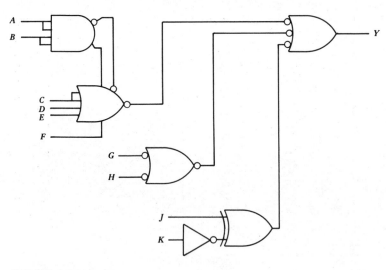

FIGURE P5-11b

5-12.   Use AOI gates to solve Ex. 5-10.

5-13.   Design an EXCLUSIVE-OR gate using:

(a) Only NAND gates.

(b) Only NOR gates.

5-14.   Implement the function:

$$A \oplus B \oplus C$$

(a) Using 7486s.

(b) Express the function in SOP form and implement it without 7486s.

5-15.   The function $AB + \overline{AB}$ can be generated by an XOR gate and an inverter. Show that the function will be generated regardless of whether the inverter is placed on the output of the XOR gate or on one of the inputs.

5-16.   Show that $A \oplus (A \oplus B) = B$.

After attempting the problems the student should return to the questions of Sec. 5-2 and be sure their answers are clear. If the student does not understand them all, he or she should reread the appropriate sections of the chapter to find the answers.

CHAPTER

# 6

# FLIP-FLOPS

## 6-1  INSTRUCTIONAL OBJECTIVES

This chapter explains what flip-flops (FFs) are, how they work, and why they are used in certain circuits. After reading it the student should be able to:

1.  Construct a FF from NAND gates and from NOR gates, and explain how they respond to SET and CLEAR pulses.
2.  Explain how D and J-K FFs react to pulses on their SET, CLEAR, and CLOCK lines, and how these FFs react to pulse trains that are given by timing charts.
3.  Explain how each parameter listed in Sec. 6-14 affects the action of a FF.
4.  Design registers and counters using FFs.
5.  Design circuits where FFs monitor and react to sequences of events.

## 6-2  SELF-EVALUATION QUESTIONS

As the student reads the chapter, he or she should be able to answer the following questions:

1.  What are the stable states of a FF? How are they sensed?
2.  How do D and latch-type FFs react to inputs on their D and CLOCK/ENABLE lines? What is the difference between a D FF and a latch?

3.   What is the difference between master-slave FFs, edge-triggered FFs, and FFs with data lockout?
4.   How do direct SETS and direct CLEARS function?
5.   What are the limitations on the speed of FFs?
6.   How do basic counters and registers work? Why do they employ FFs?
7.   How can FFs be used to handle sequence problems?
8.   What are races and glitches? How can a glitch be useful?

## 6-3   INTRODUCTION

The circuits considered in previous chapters were *combinatorial* circuits whose outputs depended *solely* upon the inputs. The outputs of such circuits do not depend upon the sequence in which those inputs were applied, nor upon the state of the circuit before the inputs were applied. In circuits of any size or complexity, however, the *sequence* of events quickly becomes critical and the logic designer must cope with the additional dimension of time.

Consider, for example, a problem of the automobile manufacturer, as stated to the logic designer. Before allowing a car to start, he wants the driver to be seated and to buckle the seat belt. To accomplish this, the manufacturer installs appropriate sensors in both the seat and belt. At first it seems that a simple AND gate will suffice. If the driver is seated *and* the belt is buckled, the car can be started without the nasty warning buzzer. But can the manufacturer frustrate the wily driver who dislikes seat belts and, therefore, buckles the belt first and then sits on it? Now the manufacturer specifies a circuit design requiring:

1.   The driver is first seated.
2.   The belt is then buckled.
3.   The driver must have sat down before buckling the seat belt.

It is logic problems like these involving sequences of events in time that make logic design more difficult and more challenging.

## 6-4   THE BASIC FLIP-FLOP

To keep track of any sequence of events, a device having the capability of remembering things (memory) is required. The simplest and most widely used *memory cell* is the *bistable multivibrator*, commonly called the *flip-flop* (FF).

The most basic FF is the *SET-RESET* FF shown in Fig. 6-1. This FF produces two outputs. The output labeled $Q$ is also called the SET (or 1) output; the other output, labeled $\bar{Q}$, is called the RESET (or CLEAR or 0) output. When the SET-RESET FF is operating properly, the $Q$ and $\bar{Q}$

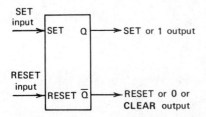

FIGURE 6-1   The basic SET-RESET flip-flop.

outputs are always *complements* of each other. The FF is considered to be in the SET or logic 1 state if the Q output is HIGH and the $\overline{Q}$ output is LOW. Conversely, the FF is considered to be RESET or CLEARED (or to contain a 0) if the Q output is LOW and the $\overline{Q}$ output is HIGH.

There are two basic inputs to a FF called SET and RESET. Each of these inputs has two levels; *active* and *quiescent*. An active level makes things happen; a quiescent input is passive. When an input is active, it forces the FF output to assume its state (i.e., an active SET input causes a FF to SET), but a quiescent level does not affect the output of the FF.

The FF of Fig. 6-1 operates as follows.

1.  An active (1) signal on the SET line causes the FF to SET.
2.  Once SET, it remains in its SET state, even after the SET signal has been removed. It remains SET until an active signal is applied to the RESET line.
3.  This RESET signal clears the FF and it remains RESET until a SET signal is again applied.
4.  For proper operation, the SET and RESET signals should not be applied simultaneously.

A simple FF is often described as a *one-bit memory*. When both inputs are quiescent (neither SET nor RESET is active), the FF "remembers" which input was most recently active. If a SET signal was received last, the FF output will be SET ($Q = 1$, $\overline{Q} = 0$) and if a RESET signal was received last, the FF output will be RESET.

One problem remains. The internal circuitry of most FFs is symmetrical, and when power is applied after the circuit has been turned OFF, there is no way of telling whether the FF will come ON in the SET or RESET state. Many sophisticated systems use a POWER-ON CLEAR signal, generated whenever power is first applied, to clear all critical FFs before operation begins.

## 6-5   NOR GATE FLIP-FLOPS

Perhaps the simplest practical FF is a FF constructed from two **7402** NOR gates, as shown in Fig. 6-2. It illustrates all the important points of Sec. 6-

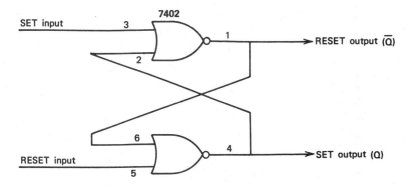

FIGURE 6-2   The NOR gate flip-flop.

4. The active signal levels are HIGH and the quiescent (inactive) signal levels are LOW. Note that because of the inversion property of the NOR gates the Q or SET output is the output of the lower gate, whose input is connected to the RESET line, and the $\overline{Q}$ or RESET output is the output of the upper gate. The FF operates as follows.

1.   If a 1 is applied to the SET input (pin 3) it causes pins 1 and 6 to go LOW, and the $\overline{Q}$ output (pin 1) is a 0.
2.   With the RESET input quiescent (0), both pins 6 and 5 are 0, causing pin 4 (the SET or Q output) to be HIGH. This HIGH signal is also applied to pin 2.
3.   If the SET input becomes quiescent (0), there is still a 1 input to the top NOR gate at pin 2. Therefore, its output remains LOW and the output of the lower NOR gate, which has two 0 inputs, remains HIGH. The SET output is still 1 and the FF has not changed state; it "remembers" that a SET pulse is the most recent pulse.
4.   When the RESET or CLEAR input at pin 5 becomes a 1, pins 4 and 2 go LOW.
5.   The LOW inputs at pins 3 and 2 cause pin 1 to go HIGH. The FF is now in its RESET state (Q = 0 and $\overline{Q}$ = 1).
6.   When the RESET input returns to 0, pin 6 is still a 1 and the FF remains in the CLEAR state until the next SET pulse occurs.

   To clarify sequential circuits, engineers often construct timing charts that show the time relationship of voltages at various points in the circuit. A timing chart for the NOR gate FF is shown in Fig. 6-3. The FF starts in the CLEAR state because the RESET input is initially HIGH. After the RESET signal goes LOW, the FF remains RESET until the leading edge of the SET pulse occurs. The FF then SETS (the SET output goes to 1 and the RESET output goes to 0) and remains SET until the RESET input again goes HIGH.

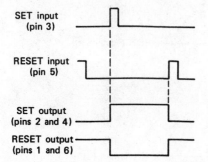

FIGURE 6-3   Timing chart for the NOR gate flip-flop (Fig. 6-2).

## 6-6   NAND GATE FLIP-FLOPS

The most common SET-RESET FF is built with two **7400** NAND gates as shown in Fig. 6-4. For clarity the NAND gates are shown as NORs with negative inputs because *the active level of the input signals is negative.* Negative active signals are often written with a bar above the signal name, as shown in the figure. The FF operates as follows.

1.   Assume the FF is being SET. This means the SET input to pin 1 is LOW, which causes the SET output at pin 3 to go HIGH.
2.   Since the RESET input is quiescently HIGH, both pins 4 and 5 are HIGH, causing pins 6 and 2 (the RESET output) to be LOW.
3.   If the SET input now goes HIGH, the FF remains in its SET state because the level at pin 2 remains LOW and both inputs to the lower NAND gate remain HIGH.
4.   This condition continues until a RESET input causes pin 5 to go LOW, which forces the RESET output HIGH. Now pins 1 and 2 are both HIGH, causing the SET output to go LOW.
5.   The FF remains in this reset condition until the next SET pulse causes pin 1 to go LOW again.

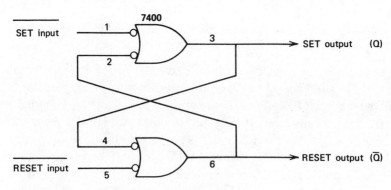

FIGURE 6-4   The NAND gate flip-flop.

The timing chart of Fig. 6-5 was constructed to help explain the NAND gate FF:

1.  It is assumed that the circuit is turned on at $t = 0$.
2.  With both the SET and RESET inputs quiescent (HIGH) we cannot tell in which state the FF came up; therefore, the outputs are represented with a question mark.
3.  At $t = 3$, the SET input goes LOW, setting the FF, as indicated by the fact that the SET output goes HIGH and the RESET output goes LOW.
4.  Despite the disappearance of the SET input at $t = 4$, the FF remains SET until the leading edge of the RESET pulse occurs at $t = 7$.
5.  This resets the FF and it remains RESET until the next SET pulse at $t = 11$.
6.  At $t = 13$ both the SET and RESET become active (LOW) simultaneously. This forces both the Q and $\bar{Q}$ outputs HIGH. Thus the *penalty* paid for having both inputs active is the *loss of* FF *action* because Q and $\bar{Q}$ are no longer complementary.
7.  When the SET input goes HIGH at $t = 14$, the FF assumes the RESET state since the RESET is still active.
8.  For $t$ greater than 15 neither input is active, but the FF remains RESET. It "remembers" that the RESET input was the last active input.

---

## EXAMPLE 6-1

The circuit of Fig. 6-6 uses three **7420** NAND gates to form a 3-state FF. This is a circuit with 3 inputs—$\bar{A}, \bar{B}$, and $\bar{C}$—and 3 outputs—A,B, and C. Only one of the inputs may be LOW (active) at any one time and only the output corresponding

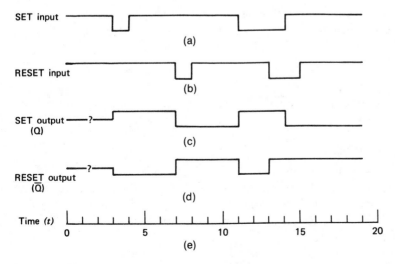

FIGURE 6-5 Timing chart for the **7400** NAND gate flop of Fig. 6-4.

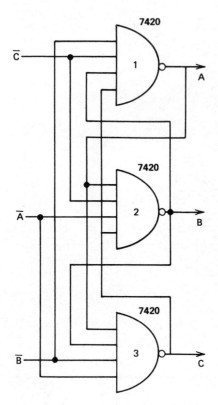

FIGURE 6-6  A **7420** 3-state flip-flop.

to the LOW input should be LOW. This output must remain LOW until the next active input occurs.

This circuit might be used in a digital computer that is required to remember whether the results of the last arithmetic operation was greater than, less than, or equal to 0.

Explain how the circuit operates as:

1. $\overline{A}$ goes LOW.
2. $\overline{A}$ returns to its quiescent HIGH state.
3. $\overline{C}$ goes LOW.

### SOLUTION

1. When $\overline{A}$ goes LOW, it forces outputs B and C HIGH. Since $\overline{B}$ and $\overline{C}$ are HIGH (quiescent), all four inputs to gate 1 are HIGH and output A goes LOW, holding the inputs to gates 2 and 3 LOW.
2. Output A remains LOW after $\overline{A}$ returns to its HIGH state and holds B and C HIGH.
3. When $\overline{C}$ goes LOW, it forces the outputs of gates 1 and 2 HIGH. This causes four HIGH inputs to be present at gate 3 and output C now goes LOW.

## 6-7 D-TYPE FLIP-FLOPS

Circuits designed specifically as FFs by TTL manufacturers fall primarily into two categories: D-type FFs and J-K FFs. The most commonly used D FF is the **7474**, dual D, positive-edge-triggered FF. The **7474** has two identical FFs in a 14-pin package. The circuit, pin layout, and function table are shown in Fig. 6-7.

Note first the direct CLEAR and direct SET inputs that enter the first FF on pins 1 and 4 (Fig. 6-7a). These are called the *asynchronous* inputs because their effect is *independent* of the clock (CK). If only these two inputs are used, the FF will function as a simple SET-RESET FF. The bubbles shown on the asynchronous inputs (pins 1, 4, 10, 13) indicate that the active level is LOW.

The direct SET or PRESET input, pin 4, is generally drawn at the top of the FF, as shown in Fig. 6-7a. A negative pulse on this input immediately SETS the FF regardless of the clock or the D input, as shown on line 1 of Fig. 6-7b. Conversely, the direct CLEAR or RESET input is generally drawn at the bottom of the FF and a negative pulse at this point clears the FF. If the SET and CLEAR inputs are both active, (LOW) simultaneously, both the $Q$ and $\bar{Q}$ outputs go to a 1, as shown on line 3 of Fig. 6-7b, and remain there until either one of the active inputs is removed. This mode of operation is not recommended (and is avoided by all but the most adventuresome engineers). If the SET and CLEAR inputs are not used, they may be left open, or preferably tied to a clamp (see Sec. 5-5.2) or some other HIGH level.

When the direct inputs are inactive, the synchronous inputs may be used. The clock inputs are used to control the FF. On each *positive edge* of the clock, the level on the D input (pin 2 on FF1 and pin 12 on FF2) is set into the FF. Thus, if the D input is HIGH (1) when the clock goes HIGH,

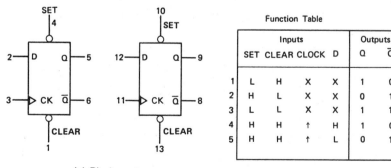

(a) Pin layout

(b) Function table for either FF

FIGURE 6-7 The **7474** dual-D positive-edge-triggered flip flop and its function table. In (b), X = irrelevant.

the Q output goes to 1 and the FF remains SET until the next clock pulse. Conversely, if the D input is LOW a clock pulse causes the Q output to go LOW. The inputs are called *synchronous* because the outputs are *synchronized to the clock* and *only change when it goes positive*. This is also shown on lines 4 and 5 of Fig. 6-7*b*, where the up-arrow symbol ( ↑ ) indicates that changes occur on positive transitions only.

---

### EXAMPLE 6-2

Devise a circuit to toggle a D FF. Toggling means causing the FF to reverse its state on each clock pulse.

### SOLUTION

In order to force the D FF to toggle, a 0 must be clocked in whenever the FF is SET and a 1 must be clocked in whenever the FF is CLEAR. But the Q̄ output is LOW whenever the FF is SET, and HIGH whenever it is CLEAR. Therefore, if the Q̄ output is connected to the D input of the same FF, it will toggle on each clock pulse. The circuit and timing chart solutions are shown in Fig. 6-8.

---

Another D-type FF often used because of its packaging density is the **74174** hex D shown in Fig. 6-9. Six D-type FFs have been placed inside one 16-pin DIP. To achieve such packaging density, the following price was paid.

1. A positive transition ( ↑ ) at pin 9 clocks *all* FFs simultaneously.
2. A negative pulse or level at pin 1 CLEARS *all* six FFs.
3. The Q̄ outputs are not available.

The Q̄ outputs, if required, can be obtained by connecting an inverter to the Q outputs.

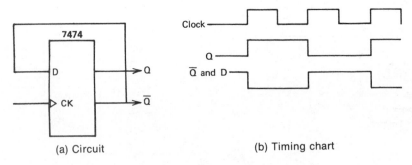

(a) Circuit  (b) Timing chart

FIGURE 6-8 Circuit and timing chart for Ex. 6-2. *Note:* The direct SET and direct CLEAR are not shown and are assumed to be HIGH.

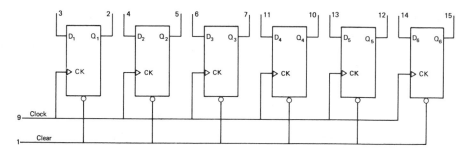

FIGURE 6-9   The **74174** HEX D flip-flop. $V_{CC}$ is on pin 16; GND is on pin 8.

## 6-8   BISTABLE LATCHES

Bistable latches are a variation of the D-type FF. They have a D input and an ENABLE input. If the ENABLE input is HIGH, the output follows the D signal. But if the ENABLE signal goes LOW, the output remains where it was at the last instant the ENABLE was HIGH, and is not affected by any changes on the D input. The action of a bistable latch is illustrated by the timing chart of Fig. 6-16 (see also Ex. 6-6).

The most popular bistable latch is the **7475** QUAD LATCH shown in Fig. 6-10. Note that four FFs are contained in the 16-pin DIP, but there are only two ENABLE gates (labeled G), each connected to a pair of latches, and there are no direct CLEARS or SETS.

## 6-9   J-K MASTER-SLAVE FLIP-FLOPS

J-K FFs are very versatile and widely used. These FFs are so named because they have a J and a K input as well as a clock. Typically, the outputs of a J-K FF change on the *negative-going* edge of the clock, although positive-edge triggered J-K FFs are also available.

J-K FFs conform to the function table of Fig. 6-11, where $Q_N$ is the state of the FF before the clock transition and $Q_{N+1}$ is the state of the FF after the transition. A J-K FF operates as follows.

1.   If both J and K inputs are 0 (LOW), the FF will not change state. The table indicates that $Q_{N+1} = Q_N$.
2.   If J = 1 and K = 0, the FF SETS on the next negative clock transition. In this case the $Q_{N+1}$ output is 1.
3.   Conversely, if K = 1 and J = 0, the FF CLEARS.
4.   If both J and K are 1, the FF toggles, or changes state on each clock transition, as indicated in the function table, which shows that $Q_{N+1} = \overline{Q}_N$.

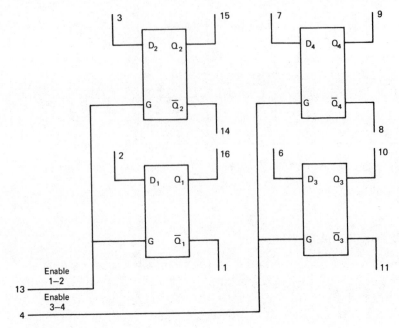

FIGURE 6-10 The **7475** quad latch flip-flop. $V_{CC}$ is on pin 5; GND is on pin 12.

---

## EXAMPLE 6-3

A clock line is a line that alternates between HIGH and LOW levels at a predetermined frequency. Typically, a clock line can be produced by a square wave generator (SWG). Given an input line and a clock line, design a circuit to the following specifications:

1. The output can only change on negative clock transitions.
2. The output must become 1 if the input is 1 and the present state of the output is 0.

| Inputs | | Output |
|---|---|---|
| J | K | $Q_{N+1}$ |
| 0 | 0 | $Q_N$ |
| 1 | 0 | 1 |
| 0 | 1 | 0 |
| 1 | 1 | $\bar{Q}_N$ |

FIGURE 6-11 Function table for a J-K flip-flop.

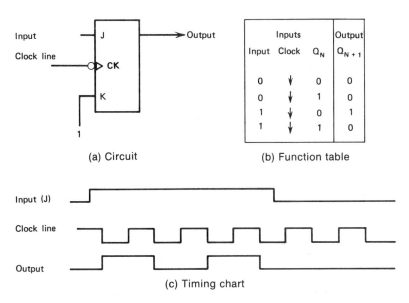

(a) Circuit  (b) Function table

(c) Timing chart

FIGURE 6-12 Solution for Ex. 6-3.

3. If the input is 0, the output must be 0 after the next clock pulse.
4. The output must not remain a 1 for two consecutive clock pulses.

The input-output relationships are also defined by the function table of Fig. 6-12, where the down-arrow ( ↓ ) indicates a negative transition of the clock. A negative transition is also indicated by the bubble shown on the clock input.

### SOLUTION

The solution and timing chart are also shown in Fig. 6-12. The input line is connected to the J input of a J-K FF, while the K input is clamped HIGH. Therefore, if J is 0, the FF CLEARS, while if J is a 1 as it is for the first four negative transitions, the FF toggles.

Most J-K FFs in current use are master-slave FFs. The pin layout of the 74107 dual J-K master-slave FF is shown in Fig. 6-13. To get two FFs into a 14-pin package, the direct SET inputs were sacrificed. The direct CLEAR acts as it does on a 7474; any negative pulse on this line CLEARS the FF regardless of the other inputs.

The master-slave FF has complicated internal circuitry and can be somewhat tricky to use. A simplified equivalent circuit is presented in Fig. 6-14a. This shows four essential parts: the input gates, the master FF, the transition gates, and the slave FF. Both the master and slave FFs are shown as NAND gate FFs that require a negative pulse to SET (or CLEAR) them.

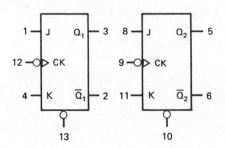

FIGURE 6-13 Pin layout for the **74107** dual J-K master-slave flip-flop. $V_{CC}$ is on pin 14; GND is on pin 7.

The output is taken from the slave FF. Data are gated into the master FF when the clock is HIGH and then transferred to the slave whenever the clock goes LOW.

The master-slave circuit of Fig. 6-14 operates as follows.

1. The input gates are only open when the clock is HIGH. At this time the transition gates are closed by the inverted clock; thus the master FF is disconnected from the slave FF.

2. The J gate is connected to the $\bar{Q}$ output, and the K gate is connected to the Q output. Therefore, if the slave is SET, no pulse can get through the J gate, and if the slave is CLEAR, no pulse can get through the K gate.

3. If the slave FF is SET, the master can only be CLEARED (though it may remain SET if it was initially SET) and if the FF is CLEAR, the master can only be SET.

4. The outputs of the J-K FF can only change when the clock goes LOW. This cuts off the input gates, which prevents the master FF from changing, and opens the transition gates to transfer the master output to the slave FF.

5. The FF changes on the negative edge of the clock because at that time the master output is sent to the slave FF and the master FF cannot change while the clock remains LOW.

---

### EXAMPLE 6-4

Using the master-slave FF of Fig. 6-14a, explain how it behaves as the clock changes:
   (a) If both J and K are 0.
   (b) If both J and K are 1.
Assume initially the clock is LOW and the FF is SET.

### SOLUTION

   (a) 1. With the clock LOW initially, the master FF is connected to the slave. Since the slave is in the 1 state, the master must also be in the 1 state.

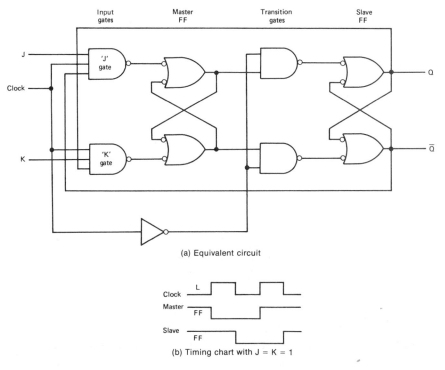

(a) Equivalent circuit

(b) Timing chart with J = K = 1

FIGURE 6-14 The master-slave FF equivalent circuit and timing chart for Ex. 6-4.

2. If the J and K inputs are both 0, they cut off both input gates. No pulse can get through to change the master FF. Therefore, the master FF remains in the 1 state and causes the output to remain a 1. This conforms to the function table.

(b) 1. To understand the situation when J and K are both 1, refer to the timing chart of Fig. 6-14b. Initially, the master and slave FFs are HIGH and the clock is LOW, as specified. When the clock goes HIGH, only the K gate is enabled (the LOW output on $\bar{Q}$ holds the J gate off), and the master FF CLEARS.

2. When the clock next goes LOW, the master FF becomes connected to the slave and CLEARS it. Now when the clock goes HIGH, $\bar{Q}$ is a 1 and Q is a 0; therefore, only the J gate is enabled, setting the master FF.

3. This causes the slave FF to SET when the clock again goes LOW. Therefore, the FF changes state, or toggles, on each negative clock transition, as the function table specifies.

One crucial problem remains. If the FF is SET and the clock is HIGH, a 1 on the K input *at any time* enables the K gate and causes the master FF to RESET. It is not necessary that the K input be a 1 at the negative edge of the clock.

In the function tables for J-K FFs, some manufacturers use the pulse symbol (⊓) instead of the down-arrow ( ↓ ) to indicate that the master FF may change whenever the clock is HIGH.

To summarize, *when the clock is HIGH, a 1 on the input opposite the sense of the FF* (the J input if the FF is CLEAR; the K input if the FF is SET) *at any time causes the FF to toggle at the next negative transition.* In most cases, this problem will not occur, but when it does, it can cause the unwary engineer much grief and frustration. One solution that may help is to use an *asymmetrical clock* that is only HIGH for a short time. But generally the engineer must analyze the specific circuit to determine what is causing the problem, and figure out a way to design it out or design around it.

There are many variations of the basic J-K FF available. The 7473 is a master-slave FF identical to the 74107 except that power and ground are not on pins 14 and 7, as they are in most IC chips. This is often inconvenient for engineers who buy IC panels with power and ground prewired to the corner pins of each socket.

The 7476 is a 16-pin dual master-slave FF. The extra two pins make direct SETS available. Some chips are available in 14-pin packages with common CLOCKS and common CLEARS. This makes direct PRESETS available for both FFs. Once the specific system requirements are determined, the engineer will have to examine the manufacturers' catalogs to determine the best chip for his application, but he should try to choose *popular* chips. Chips that are rarely used tend to be unavailable from distributors and require long delivery times from the manufacturers.

Symmetrical FFs can be "flipped around" if this suits the application. For example, if a FF must have a direct SET but a direct CLEAR is not required, a 74107 suffices. If the J and K inputs are interchanged (put the normal J input into the K pin and the K input into the J pin), and Q and Q̄ are interchanged, the CLEAR input now functions as a direct SET. Refer back to Fig. 6-13; if the J input is applied to pin 1, and the K input is applied to pin 4, and Q is considered as the output at pin 2 instead of pin 3, a negative pulse at pin 13 now acts as a direct SET (see Ex. 6-13).

## 6-10  EDGE-TRIGGERED FLIP-FLOPS

Edge-triggered J-K FFs can be used where it is necessary to eliminate the problem of setting the master FF while the clock is HIGH. They react to the J and K inputs *only* at the *negative edge* of the clock. The 74113 and 74114 are examples of edge-triggered FFs.

In addition, J-K FFs with data lockout are also available. Using data lockout, the data are *stored* in the *master* FF on the *positive* edge of the clock pulse and *transferred* to the *output* on the *negative* edge of the clock pulse. The **74111** is an example of a dual J-K master-slave FF with data lockout.

---

**EXAMPLE 6-5**

Is is possible for the circuit of Ex. 6-3 to fail?

**SOLUTION**

If the circuit is built using a J-K master-slave FF, a positive level on the J input at any time when the clock is HIGH and the output is LOW causes it to fail, as the timing chart of Fig. 6-15 shows. The first and third output pulses represent failures because the J input was LOW when the negative edge of the clock occurred, but a pulse resulted because the input and clock were HIGH at $t = 1$ and $t = 17$ and set the master FF. The circuit will not fail if a J-K negative-edge-triggered FF is used or if the input does not change while the clock is HIGH. There are also other circuits that solve Ex. 6-3 (see Problem 6-9).

---

## 6-11 TIMING CHARTS

Timing charts are helpful in clarifying the operation of the various FFs as well as in showing how the voltages throughout a circuit vary as a function of time.

---

**EXAMPLE 6-6**

Given the input and clock shown on lines 2 and 3 of Fig. 6-16, find the output:

1.   If the input and clock are applied to the D and CLOCK inputs of a 7474.
2.   If the input and clock are applied to the D and ENABLE of a latch-type FF (the 7475, for example).

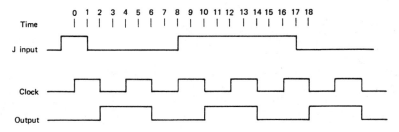

FIGURE 6-15   Timing chart for Ex. 6-5 using J-K master-slave FFs.

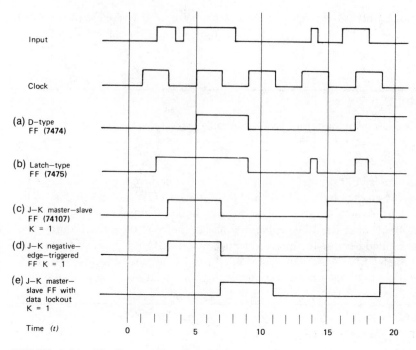

FIGURE 6-16 Flip-flop timing charts for Ex. 6-6.

3. If the input is applied to the J input of a J-K master-slave FF like the **74107**, while the K input is tied HIGH.

4. If the input is applied to the J input of an edge-triggered J-K FF with the K input tied HIGH.

5. If the input is applied to the J input of a J-K master-slave FF with data lockout, such as the **74111**, and the K input is tied HIGH.

## SOLUTION

1. (Figure 6-16$a$) The D-type FF changes state at positive clock edges only. These occur at $t = 1, 5, 9, 13,$ and 17. But since the input (D) is only HIGH at $t = 5$ and 17, two positive pulses will occur at these times.

2. (Figure 6-16$b$) The latch output is the same as the input whenever the clock is HIGH (times between 1 and 3, 5 and 7, 9 and 11, etc.). Therefore, the short pulse at $t = 4$ is missed because the clock is LOW, but the short pulse at $t = 14$ is transferred to the output because the clock is HIGH. Note that the latch retains the last output level whenever the clock goes LOW.

3. (Figure 6-16$c$) The J-K master-slave FF only changes on negative clock pulses, which occur at $t = 3, 7, 11, 15,$ and 19. Since at $t = 3$, both J and K are 1, the FF toggles to the SET state, and since at $t = 7$, both J and K are also 1, the FF toggles back to the CLEAR state. At $t = 11$, J is a 0; therefore, the FF remains CLEAR.

The pulse at $t = 14$ SETS the master FF, causing the output to SET at $t = 15$. It CLEARS at $t = 19$, because K is 1.

4.   (Figure 6-16d) The J-K negative-edge-triggered FF behaves as the master-slave FF, except that is does not respond to pulse at $t = 14$ because the J and K inputs are only effective at the negative edge (at $t = 15$). At $t = 15$, $J = 0$, and $K = 1$; therefore, the FF does not change state and it does not produce the second output pulse.

5.   (Figure 6-16e) The J-K master-slave FF with data lockout examines its inputs at the positive edge of the clock and changes state at the next negative edge. It sees 1s on the J inputs at $t = 5$ and $t = 17$ and toggles on the following negative clock edges.

## 6-12   DIRECT SETS AND DIRECT CLEARS

Direct CLEARS and direct SETS (the asynchronous inputs) override *any* clock inputs. They are internally wired to prevent any clock action from affecting the FF when they are active. For example, a clock cannot SET a 1 into a FF while the direct CLEAR is LOW. The timing charts of Fig. 6-16 were drawn on the assumption that the direct CLEARS and SETS were inactive, allowing the FFs to respond to clock pulses only. A CLEAR pulse at any time, however, would have RESET any of the FFs.

The asynchronous outputs are active only on LOW inputs. This means FFs should respond only to their clock inputs if the direct inputs are left unwired. As a precaution, many engineers wire unused direct inputs to clamp. Usually both the direct and clocked inputs are used to make optimum use of a FF.

### EXAMPLE 6-7

An engineer must CLEAR a **7474** when a pulse line (a line carrying a train of pulses of varying durations) is negative, and clock in the level on the D input when the pulse goes positive. He builds the circuit of Fig. 6-17a, reasoning that a LOW level on the pulse line CLEARS the FF and the following positive edge both removes the direct CLEAR and clocks in the D level. Is his reasoning valid? Explain reasons for your answer.

### SOLUTION

The direct CLEAR will remain active anywhere from 10 to 30 ns after the CLEAR input has gone HIGH. In the circuit of Fig. 6-17a, the clock occurs simultaneously with the removal of the direct CLEAR.

The CLEAR does not have sufficient time to release and still overrides the clock. The **7474** will remain in the CLEAR state regardless of the level on the D input. The circuit does not meet the specifications.

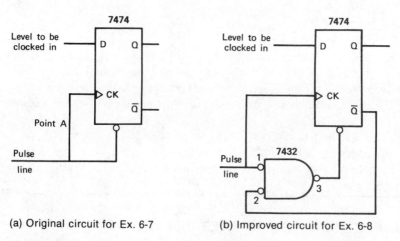

(a) Original circuit for Ex. 6-7            (b) Improved circuit for Ex. 6-8

FIGURE 6-17   Circuits for Exs. 6-7 and 6-8.

One of the most common errors made by designers is their attempt to clock a FF too soon after the removal of a direct input, although it is rarely as obvious as it is in Ex. 6-7. The time required between the removal of a direct input and the successful application of a clock is not specified by the manufacturer, but experience has shown it to be between 10 and 30 ns. Inserting two inverters at point A of Fig. 6-17 may delay the clock long enough to allow the circuit to work, but this also depends upon the transition times of the inverters as well as the speed of the direct CLEAR.

## 6-13   RACE CONDITIONS

A race condition exists if a circuit output depends upon which of two *nearly* simultaneous inputs arrive at a point in the circuit first, or where there may not be enough time between the removal of one input and the arrival of another. Inserting inverters at point A of Fig. 6-17a may create the latter type of race condition.

Circuits that contain race conditions are unreliable. If the same circuit is built several times or mass-produced, the output of each circuit is liable to be different due to the varying speeds of the circuit components. What is worse, the output of the same circuit may be different at different times if this output depends upon the results of a *very* close race, because those results may not always be the same. Consequently, engineers devote considerable time and effort toward eliminating race conditions. A thorough examination of the specific circuit usually reveals a satisfactory way of controlling pulses so that races are eliminated. Only when all else fails would we resort to brute force techniques, such as using cascaded inverters, to delay inputs. Superior logic design often eliminates race conditions as Ex. 6-8 demonstrates.

---

**EXAMPLE 6-8**

Design a circuit to meet the specifications of Ex. 6-7 without race problems.

**SOLUTION**

If the CLEAR input is removed as soon as the FF is RESET, it will no longer override the clock! The solution is shown in Fig. 6-17b. If the pulse line goes LOW while the FF is SET, a negative pulse comes through the **7432** to CLEAR the FF. This releases the direct CLEAR by causing pins 2 and 3 of the **7432** to go HIGH. Since the direct CLEAR is now released at the beginning of the negative input pulse, the clock will be effective when it goes HIGH.

---

The circuit of Fig. 6-17b was set up in the laboratory with the D input held HIGH and the resultant waveforms are shown on the oscilloscope (CRO) traces in Fig. 6-18. The top trace is the clock, the second trace is the FF output, and the bottom trace is the direct CLEAR or the output of the **7432**. It can be seen that the FF output goes HIGH on the positive

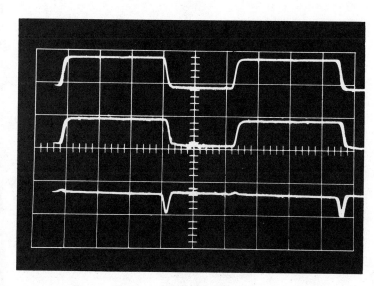

Scales:  Vertical **5 V/cm**

Horizontal **200 ns/cm**

FIGURE 6-18   Oscilloscope traces for the circuit of Fig. 6-17b.

edge of the clock, and when the clock goes negative, the spike on the CLEAR input RESETS the FF, eliminating any and all race difficulties.

## 6-14 FLIP-FLOP PARAMETERS

TTL FFs operate properly if all pulses applied to them are longer than 50 ns. For engineers who must design higher speed circuits, the following parameters are specified by the manufacturer and should be observed.

1.   $f_{max}$. This is the highest frequency at which clock pulses may be applied to a FF and still maintain proper and stable clocking. For a 7474 or a 74107, the minimum value of $f_{max}$ is 15 MHz. The manufacturer's state that typical 7474s and 74107s will toggle at 25- and 20-MHz rates, but they do not guarantee the FFs will toggle faster than 15 MHz.

2.   $t_{setup}$. This is the time a signal must be present on one terminal before an active transition occurs at another terminal. For a 7474, $t_{setup}$ is 20 ns (listed in some catalogs as 20 ↑). This means that the D input must be held constant for at least 20 ns before a positive clock edge to assure a reliable output.

3.   $t_{hold}$. This is the time a signal must remain at a terminal *after* an active transition occurs. For a 7474, this is 5 ns. The signal at the D input should be removed no sooner than 5 ns after the positive edge of the clock. There is zero hold time for most J-K FFs.

4.   **Clock high pulse width.** This is the minimum time a clock must remain in its HIGH state for reliable clocking.

5.   **Clock low pulse width.** This is the minimum time a clock must remain in its LOW state for reliable clocking.

6.   **PRESET or CLEAR LOW.** This is the minimum time a PRESET or CLEAR pulse must be LOW for reliable setting or clearing.

In addition, the propagation times required for the CLOCK, PRESET or CLEAR to affect the output are also specified; this enables the designer to estimate the delay time through a series of gates and FFs.

| Parameter | 74107 | 7474 | 74111 | Unit |
|---|---|---|---|---|
| $f_{max}$ | 15 | 15 | 15 | MHz |
| $t_{SETUP}$ | 0 ↑ | 20 ↑ | 0 ↑ | ns |
| $t_{HOLD}$ | 0 ↓ | 5 ↑ | 30 ↑ | ns |
| CLOCK HIGH | 20 | 30 | 25 | ns |
| CLOCK LOW | 47 | 37 | 25 | ns |
| PRESET or CLEAR LOW | 25 | 30 | 25 | ns |
| CLOCK LOAD | 2 | 2 | 2 | Standard TTL load |
| CLEAR LOAD | 2 | 2 | 3 | Standard TTL load |

FIGURE 6-19   Flip-flop specifications.

The table of Fig. 6-19[1] specifies the above times and loading for the common FFs.

---

**EXAMPLE 6-9**

How fast can a 7474 be clocked?

**SOLUTION**

The maximum clock frequency is 15 MHz. This indicates that positive transitions on the clock cannot come at a rate faster than 15 MHz or they must be separated by a time of $\frac{1}{15}$ MHz or 66.7 ns. A consideration of the clock LOW and clock HIGH times indicates that the pulses must be no closer than 67 ns apart. (The clock must be HIGH for 30 ns and LOW for 37 ns.) If only a symmetric (50 percent duty cycle) clock is used, the clocks can be no closer than 74 ns or 13.5 MHz.

---

FFs may respond properly to shorter pulses than those specified. However, the manufacturer does not guarantee this; therefore, operating with short pulses is risky and should be avoided. On the other hand, the manufacturer does not guarantee that a very short, sharp pulse, often called a *glitch*, will not affect the output. For the 7474, the CLEAR line must be held LOW for at least 30 ns to *guarantee* that the FF will CLEAR, but a 5-ns glitch may also CLEAR the FF. System failures are often traced to glitches on critical inputs.

The manufacturers' specifications on FFs must also be checked for *loading*. This is done by checking the unit load or $I_{IL}$ specifications and remembering that one standard load is $-1.6$ mA. For the 7474, $I_{IL}$ is $-1.6$ mA, or one standard load, for the D and PRESET inputs. But it is $-3.2$ mA, or two standard loads for the CLOCK and CLEAR inputs. Figure 6-19 shows that both the CLEAR and CLOCK inputs of the 7474 or 74107 are each the equivalent of two standard gate loads. Therefore, only five CLEAR or CLOCK inputs can be driven from an ordinary driver. Buffer drivers are often used when the loading becomes very heavy.

## 6-15 USES OF FLIP-FLOPS

Flip-flops (FFs) are the logic designers most versatile tool and are used throughout the industry. Some of the more common and more interesting uses of FFs are now described.

---

[1]This table is extracted from the FF characteristics as listed in the *TTL Data Book*, Texas Instruments, Inc., pp. 6-46 and 6-47.

### 6-15.1  Registers

Registers were defined in Chapter 1 as a repository for a group of bits. Registers are usually composed of a group of FFs; that is, an N-bit register consists of N FFs. Very often no logic or arithmetic operations are performed on the bits in the register; the function of the register is simply to retain (store) the word for a period of time.

An example of a 4-bit register consisting of two 7474 FFs is shown in Fig. 6-20. The register may be cleared, if necessary, by a single pulse on the CLEAR line. Data on the D line are clocked in on the positive edge of each clock, and replace the previous contents of the register.

Where all stages of a register are to be cleared and loaded simultaneously, it is often wiser to build the register with 74174 chips because they contain 6 FFs in a single package. They also save on loading. The CLOCK and CLEAR inputs of a 74174 take only one load each, whereas it requires a drive capability of 12 loads to CLOCK or CLEAR six individual 7474 FFs. (See Sec. 6-14.)

Another popular register is the 74173, whose pinout and function table is shown in Fig. 6-21. This chip is popular because it has 3-state outputs and can be placed directly on a microprocessor bus.

The 74173 contains four D-type FFs with a common clock and a common clear. To enter data, both DATA ENABLE inputs, G1 and G2, must be LOW. Then the function table shows that data are entered on positive clock edges. The OUTPUT CONTROL pins determine whether the chip is enabled. If they are both LOW the Q outputs of the FFs drive the line; otherwise the IC is disabled and the output is high impedance.

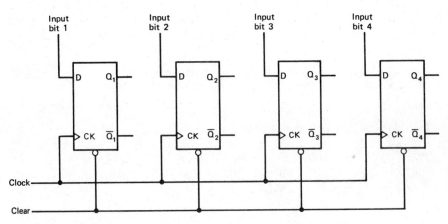

FIGURE 6-20  A 4-bit register composed of **7474** FFs.

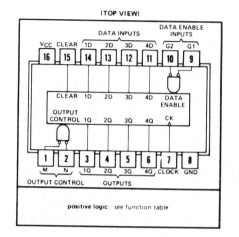

(TOP VIEW)

positive logic: see function table

(a) Pinout

| FUNCTION TABLE | | | | | |
|---|---|---|---|---|---|
| INPUTS | | | | | OUTPUT |
| CLEAR | CLOCK | DATA ENABLE | | DATA | Q |
| | | G1 | G2 | D | |
| H | X | X | X | X | L |
| L | L | X | X | X | $Q_0$ |
| L | ↑ | H | X | X | $Q_0$ |
| L | ↑ | X | H | X | $Q_0$ |
| L | ↑ | L | L | L | L |
| L | ↑ | L | L | H | H |

When either M or N (or both) is (are) high the output is disabled to the high-impedance state; however sequential operation of the flip-flops is not affected.

(b) Function Table

FIGURE 6-21 The **74173**. (From the *TTL Data Book for Design Engineers*, 2nd ed., Texas Instruments, Inc. Courtesy of Texas Instruments, copyright 1976.)

## 6-15.2 Sequence Problems

As stated at the beginning of this chapter, FFs can be used to keep track of sequences of events. The auto seat belt problem is now restated in more formal terms and solved in Ex. 6-10.

---

### EXAMPLE 6-10

Given two input lines, A and B, design a circuit to produce a HIGH output only if both A and B are both HIGH and A becomes HIGH before B becomes HIGH. (For the auto problem, the A line corresponds to the event "the passenger is seated," and the B line corresponds to the event "the seat belt is buckled.")

### SOLUTION

The FF and gate of Fig. 6-22 determine whether the three conditions necessary for proper operation are satisfied:

1. A is HIGH (the driver is in the car).
2. B is HIGH (the seat belt is buckled).
3. A went HIGH before B went HIGH (the driver sat down before he buckled his seat belt).

If both inputs to the AND gate (7408) are HIGH, the circuit is acting properly. A LOW output from the 7408 indicates an alarm condition.

The FF sets only if A goes HIGH (the driver sits down), placing a 1 on the D input and removing the direct CLEAR before B goes HIGH (the belt is buckled).

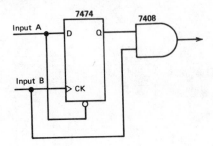

FIGURE 6-22   Circuit for solution of Ex. 6-10.

Now the positive-going edge of the B input sets the FF. This causes two HIGH inputs to appear at the AND gate, indicating proper operation. Note that we have taken the precaution of clearing the FF to cause an alarm condition whenever A goes LOW (the driver is not seated).

Unfortunately, the FF output cannot be directly connected to the alarm because we must still guard against the remaining possibility that:

1. A goes HIGH.
2. B goes HIGH setting the FF.
3. B goes LOW (this will not reset the FF).

This corresponds to the driver sitting down, buckling the seat belt, and then unbuckling it.

By connecting the B input (the belt sensor) to the AND gate, we prevent this possibility. Now the alarm will sound whenever B is LOW (the belt is unbuckled).

## 6-15.3   Counters

Digital counters are often needed to count pulses or events. Counters are usually composed of FFs forming a register, and the pulses to be counted are clocked into the least significant stage of the register.

A 3-bit up-counter consisting of three **74107** FFs is shown in Fig. 6-23. The J and K inputs to all FFs are clamped HIGH and the pulses to be counted are applied to the CLOCK input of the least significant stage. The Q output of each stage is connected to the CLOCK input of the next (more significant) stage.

To build an up-counter, which increases its count by 1 (increments) as each clock pulse appears, the following procedure can be applied.

1. Examine the least significant stage; if it is a 0, change it to a 1 and stop.
2. If it is a 1, change it to a 0, go to the second least significant stage and repeat the foregoing procedure.
3. Continue until a stage that contains a 0 is found. Change this stage to a 1 and stop.

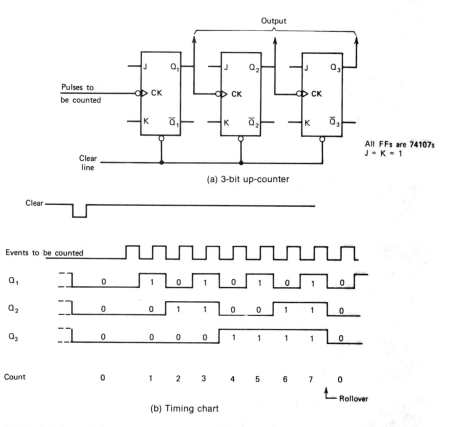

(a) 3-bit up-counter

(b) Timing chart

FIGURE 6-23  A 3-stage up-counter and timing chart.

The circuit and timing chart of Fig. 6-23 implement this procedure in hardware. If any stage changes from a 0 to a 1, the $Q$ output goes from LOW to HIGH. This provides a positive transition that does *not* clock the next stage and, therefore, does not affect the higher order stages. However, if a stage goes from a 1 to a 0, its $Q$ output goes LOW and toggles the following stage.

Refer to the timing chart (Fig. 6-23*b*); the counter output is initially set to all 0s by the clear pulse. The first clock pulse toggles $Q_1$ HIGH, but the LOW to HIGH transition does not affect $Q_2$ or $Q_3$. Now reading the output as $Q_3Q_2Q_1 = 001$, the output is the binary number 1 indicating that one pulse has passed. On the second pulse, $Q_1$ returns to its 0 state. Since $Q_1$ changes from a 1 to a 0, the negative-going edge clocks $Q_2$ into the 1 state. The LOW to HIGH transition on $Q_2$ does not affect $Q_3$, and the output now reads $Q_3Q_2Q_1 = 010$ or the binary number 2.

The counter continues to increment on successive clock pulses until the count reaches 111 (binary 7). On the next clock pulse, the counter "rolls over" to 000. A three-stage counter has a capacity of $2^3$ or 8 counts before rolling over (as shown in Fig. 6-23b). If additional stages are added in the same manner, each additional stage doubles the capacity of the counter. Also, although the clocks are drawn at equal intervals, this is not a requirement; the counter counts irregularly as well as regularly spaced pulses.

---

### EXAMPLE 6-11

A counter with a 32-count capacity is needed. How can it be built?

### SOLUTION

Since $32 = 2^5$, a 5-stage counter is needed. Adding two additional stages to the circuit of Fig. 6-23a solves the problem.

### EXAMPLE 6-12

A down-counter is one that decrements, or reduces its count by 1, as each pulse occurs. Design a 3-stage down-counter.

### SOLUTION

A counter can be decremented by the following procedure:

1. Examine the least significant stage; if it is a 1, change it to a 0 and stop.
2. If it is a 0, change it to a 1, go to the next stage and repeat the procedure.

The hardware implementation for the solution is shown in Fig. 6-24a, where the $\bar{Q}$ outputs are connected to the CLOCK of the succeeding stages. The counter output is still taken from the Q outputs. As the timing chart (Fig. 6-24b) shows, the first pulse causes the output to change from 000 to 111. This is the rollover situation for a down-counter. Each additional pulse causes the output of the counter to be reduced by 1.

### EXAMPLE 6-13

Design a 4-stage binary up-counter with a LOAD input. Whenever the LOAD input is HIGH, a binary 6 is to be loaded into the counter. If the LOAD input is LOW, the counter must be capable of counting. 74107s must be used in this example.

### SOLUTION

Registers are usually loaded using their direct CLEARS and direct SETS. For the 74107, only the direct CLEARS are available, but as the number to be loaded is

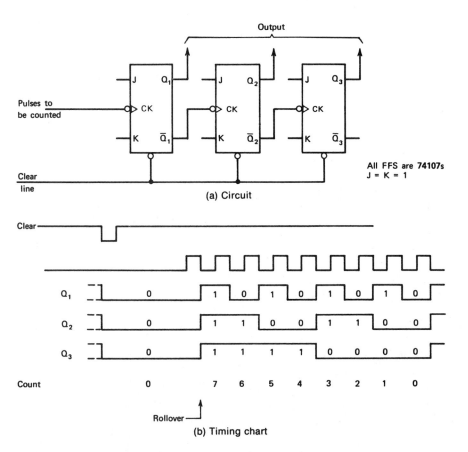

FIGURE 6-24 A 3-stage down-counter and timing chart.

0110, stages 0 and 3 must be CLEARED and stages 1 and 2 must be SET. Stages 1 and 2 of the **74107** must be "flipped around" as described in Sec. 6-9, page 164. The circuit is shown in Fig. 6-25. Because stages 1 and 2 are flipped around, the $\bar{Q}$ outputs of these stages assume the role normally played by the $Q$ outputs of a counter; the output is taken from them, and they clock the following stages on a HIGH to LOW transition. A positive LOAD pulse loads a six into the counter; in the absence of LOAD pulses, the counter increments on each input pulse.

## 6-15.4 Counters Made From D-Type FFs

Counters can also be made from D FFs such as the 7474. To do so, the FFs must be made to toggle by connecting their D inputs to $\bar{Q}$. An up counter can then be built by connecting $\bar{Q}_1$ to $D_2$, $\bar{Q}_2$ to $D_3$, and so on. Connecting the $Q$ outputs to the succeeding D inputs produces a down counter (see Problems 6-20 and 6-21).

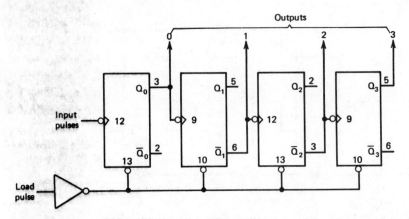

FIGURE 6-25 Solution of Ex. 6-13. *Note* This circuit is built from two **74107** FFs. For all stages J = K = 1.

## 6-16 SYNCHRONIZING FLIP-FLOPS

The operation of many complex digital circuits is controlled by a fixed frequency *clock*, often derived from a crystal oscillator. These are *synchronous circuits* because most or all *state changes are synchronized with the internal clock*. A problem arises when these circuits must also respond to external events, such as signal changes on a switch or other input line, and these events *cannot* be synchronized with the external clock.

There are two ways of handling these events.

1. **Asynchronous.** The external event or transition is used to direct SET or CLEAR a FF when it occurs.

2. **Synchronous.** The external event is used to produce a pulse that is synchronized with the internal clock.

The asynchronous method is slightly simpler to implement in hardware, but dangerous as the author unfortunately learned when he was in industry. In a complex system there are many state changes on each clock edge, and some ICs respond faster than others. If an external event is allowed to affect a system whenever it occurs, a small percentage of the time it will occur near a clock transition, after some ICs have responded, but before others have. This intermingling of transitions often causes erroneous operation. What is worse, these problems are difficult to test for and difficult to foresee because of the random time difference between the events. In the author's case, the problems became apparent only after the equipment was installed at the customer's site, and fixing it meant a 1500-mile airplane ride. Therefore, it is usually wise to prevent transition from occurring at random times.

Transitions can be synchronized with the internal clock by the use of synchronizing FFs, as Ex. 6-14 shows.

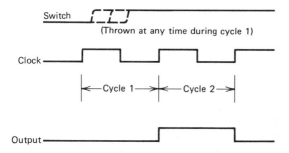

FIGURE 6-26 Synchronizing an output with a clock.

---

**EXAMPLE 6-14**

A debounced switch[2] is connected to a circuit driven by a clock. Typically the clock is much faster than the switch, which is operated by humans, so the switch is up or down for many clock cycles. Design a circuit so that a switch thrown at any time within cycle 1 will produce a single output pulse during cycle 2, as shown in Fig. 6-26. Notice that the output pulse is synchronized with the clock as required.

**SOLUTION**

This type of circuit usually requires two FFs. The first FF is SET asynchronously by the switch. The second FF can then be set by the clock provided that the first FF is set. The circuit is shown in Fig. 6-27.

The switch sets FF1 and FF2 sets on the next positive clock edge. When FF2 sets, it clears FF1 and FF2 then clears on the following positive clock edge.

---

## 6-17 GLITCHES

Glitches are short, sharp spikes that occur on signal lines. If they are un-expected, they can be very troublesome, especially if they cause FFs to CLEAR or SET. On the other hand, *controlled glitches* can be useful in certain circuits.

A controlled glitch is one that has been *deliberately* designed into a circuit to produce certain events. Controlled glitches are designed in the following manner.

1. Event 1 initiates a glitch.
2. The glitch causes event 2.
3. The occurrence of event 2 terminates the glitch.

Here the glitch must be very short and sharp, since the event caused by the glitch terminates it.

---

[2]Switch debouncing is discussed in Sec. 7-8

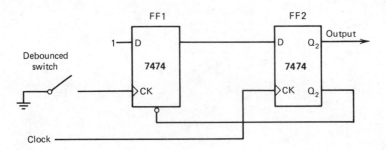

FIGURE 6-27  Circuit for Ex. 6-14.

---

**EXAMPLE 6-15**

Given two input lines, A and B, design a circuit to the following specifications:
(a) The output goes to 1 on every positive transition of the A line.
(b) The output goes to 0 on every positive transition of the B line.

The output must be capable of being SET by another A line transition immediately after being CLEARED, regardless of the level of the B line.

**SOLUTION**

A controlled glitch should occur whenever the B line makes a positive transition.
A circuit to solve this problem is shown in Fig. 6-28. Its operation is as follows.

1. $Q_1$ is SET by every positive transition of line A.
2. If $Q_1$ is SET, FF2 is enabled and SETS on a positive transition of line B.
3. This causes FF1 to CLEAR, which now clears FF2 via its CLEAR input.

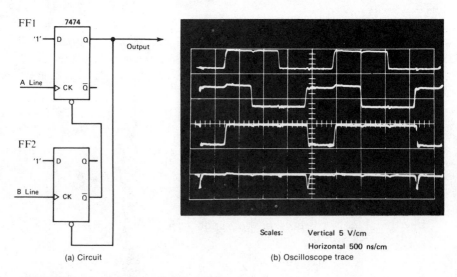

Scales:     Vertical 5 V/cm
            Horizontal 500 ns/cm

(a) Circuit          (b) Oscilloscope trace

FIGURE 6-28  Solution of Ex. 6-15.

4. Thus FF2 is only SET for the time it takes itself to SET, plus the time it takes for FF1 to CLEAR. Since this is about two propagation delay times, FF2 creates a glitch of perhaps 40 ns. Here event 1 is the positive transition of the B line.
5. This causes the glitch, which is the setting of FF2.
6. When FF2 is SET, it causes event 2, the clearing of FF1, and when FF1 CLEARS, it CLEARS FF2 terminating the glitch.

---

This circuit (Fig. 6-28$a$) was set up in the laboratory, and the important CRO waveforms are shown in Fig. 6-28$b$. The top trace is the A line and the second trace is the B line. The output is shown on the third trace and the glitch is seen on the bottom trace. Clearly, the positive edge of A causes the output to rise, and the positive edge of B causes the glitch, which causes the output to fall, as noted in the specification of Ex. 6-14.

## SUMMARY

The basic FFs available in TTL were introduced in this chapter, and some examples of circuits using these FFs were presented. Other FFs exist (consult the manufacturers' catalogs for a complete list of available FFs) and may be useful in special cases, but the vast majority of circuits are designed using D and J-K FFs. NAND and NOR gate FFs are often used where only SET-RESET FFs are required, but D and J-K FFs are generally preferred; they are more versatile because they can be CLEARED or SET directly and information can also be clocked into them.

There is no formula we can give to enable one to design a circuit that will satisfy a given specification. To be successful, the engineer must first be very familiar with the characteristics of the available FFs and gates; then a review of the design examples presented in this chapter will (hopefully) lead to ideas necessary for the design.

## GLOSSARY

**SET.** The state of a FF when $Q = 1$ and $\bar{Q} = 0$.
**RESET.** The state of a FF when $Q = 0$ and $\bar{Q} = 1$.
**Active signals.** Signal levels that cause a FF to change state.
**Quiescent level.** The signal level that does not cause a FF to change state.
**Toggling.** Causing a FF to change or reverse its state.
**Clock.** A continuous square wave of a constant frequency.
**Glitch.** A short, sharp pulse on a signal line.
**Up-counter.** A counter that increments on every pulse.
**Down-counter.** A counter that decrements on every pulse.

## REFERENCES

**J. Millman** and **H. Taub**, *Pulse and Digital Circuits*, McGraw-Hill, New York, 1965.

L. **Nashelsky,** *Introduction to Digital Computer Technology,* Wiley, New York, 1977.

**Morris** and **Miller,** *Designing with TTL Integrated Circuits,* McGraw-Hill, New York, 1971.

*The TTL Data Book for Design Engineers,* Texas Instruments, Inc., 1976.

**Dan I. Porat** and **Arpad Barna,** *Introduction To Digital Techniques,* Wiley, New York, 1979.

## PROBLEMS

6-1.  How would the circuit of Fig. 6-4 operate if the SET input (pin 1) of the NAND gate FF became disconnected?

6-2.  How would the circuit of Fig. 6-2 operate if the SET input (pin 3) of the NOR gate FF became disconnected?

6-3.  What levels would the outputs of the NOR gate FF assume if both the SET and RESET inputs are HIGH?

6-4.  Design a 3-state FF using **7425** ICs.

6-5.  Design a 3-state FF using 2-input open collector ICs. The design should use **7403** chips. The input should come in through **7405** gates. *Note:* This circuit works very well with DTL or other logic that can tolerate wire ANDing.

6-6.  Design a latch FF using only two input logic gates (**7400s, 7402s, 7408,** etc.) and inverters.

6-7.  Using the circuit of Fig. 6-14, explain how the FF works if $J = 0$ and $K = 1$. Assume initially the FF is SET and the clock is LOW.

6-8.  Redraw the circuit of Fig. 6-14 to show how you would add a CLEAR line to the master-slave FF.

6-9.  Solve the problem of Ex. 6-3 using only a **7474** and a **7408** chip.

6-10.  Refer to Fig. 6-16; the input and clock are both HIGH for $t$ between 17 and 18. Explain why the master FF of the **74107** does not SET at this time and why the output goes LOW at $t = 19$.

6-11.  For Fig. P6-11, sketch the output if the input is:

   (a) The D input to a **7474.**

   (b) The input to a latch.

   (c) The J input to a master-slave FF **74107** and the K input is tied HIGH.

   (d) The J input to a negative edge-triggered FF and the K input is tied HIGH.

   (e) The J input to a J-K master-slave FF with data lockout and the K input is tied HIGH.

6-12.  Explain the operation of the circuit of Fig. P6-12 as it is clocked. Draw a timing chart for $Q_1$ and $Q_2$ as the clocks proceed.

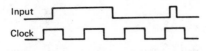

FIGURE P6-11

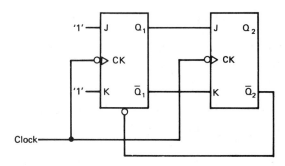

FIGURE P6-12

6-13. On a logic laboratory training board a variable is provided from a switch input as shown in Fig. P6-13. A **74107** and a source of clock pulses are also available. To obtain the variable and its complement the rest of the circuit of Fig. P6-13 was constructed. Explain how it works and which outputs contain the variable and its complement.

6-14. Considering HIGH and LOW clock widths only, find the maximum frequency at which you can toggle a **74107**. Use the TTL Data Book or Fig. 6-19.

6-15. Explain the symbol 30 ↑ for the hold time of a **74111**.

6-16. Draw a 6-stage register made up of **74174**s. What must you add if the $\overline{Q}$ outputs are also required?

6-17. Design a 16-bit register:
  (a) Using **7474**s.
  (b) Using **74174**s.
Assume the clocks and clears are originally driven by **7400** NAND gates.

6-18. Design a circuit to produce a 1 output if $ABC = 1$ and only if $A$ precedes $B$, and $B$ precedes $C$.

6-19. Using two FFs, $Q_1$ and $Q_2$, which form a 2-stage counter whose outputs are shown in Fig. P6-19, design a circuit whose output is HIGH for counts 1 and 2 and LOW for counts 0 and 3.

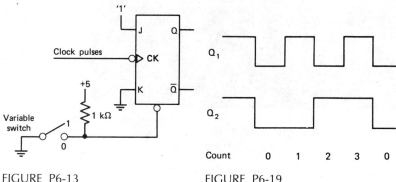

FIGURE P6-13                    FIGURE P6-19

(a) Use a third FF. Assume the $\bar{Q}$ outputs are also available.

(b) Use another gate but no additional FFs.

*Note:* The solution of Problem 6-19(b) was used to generate the waveforms shown in Fig. P6-20. Observe and explain the notches in the fourth trace.

6-20.   Design a 3-stage up-counter using 7474s.

6-21.   Design a 3-stage down-counter using 7474s.

6-22.   Design a 3-stage up-down counter. An up-down line is provided as well as a clock line. If the up-down line is HIGH, the counter is to increment on each clock pulse, and if the up-down line is LOW, the counter is to decrement.

6-23.   Design a 4-bit register with the following inputs:

(a) LOAD 1—on the leading edge of the LOAD 1 pulse four bits of input data are loaded into the FF.

(b) CLEAR—whenever the CLEAR is LOW the register is cleared.

(c) LOAD 2—whenever the LOAD 2 input is LOW the register must contain a binary 13.

Design it on the assumption that the CLEAR and LOAD 2 inputs cannot occur simultaneously.

6-24.   Repeat Problem 6-23 with the additional specification that the CLEAR and LOAD 2 pulse may occur simultaneously, and when they do, the LOAD 2 pulse predominates.

6-25.   Design a 5-stage counter that can have a binary 13 loaded into it:

(a) Using 7474s.

(b) Repeat using 74107s.

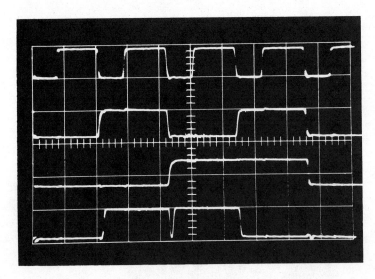

Scales:   Vertical 5 V/cm

Horizontal 500 ns/cm

FIGURE P6-20   Waveforms for Problems 6-19. Scales: Vertical 5 V/cm; horizontal 500 ns/cm.

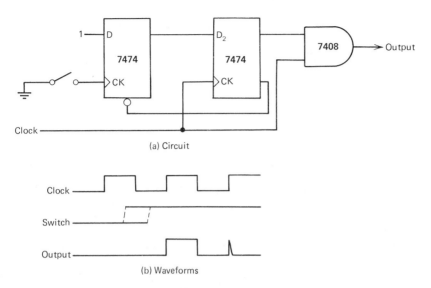

(a) Circuit

(b) Waveforms

FIGURE P6-28   (a) Circuit. (b) Waveforms.

6-26.   Draw a timing chart for the circuit of Fig. 6-25.

6-27.   Given a constant clock and an input line, design a circuit so that if there is a LOW on the input line at any time during cycle 1, the output will be HIGH throughout cycle 2. Note that the output can be HIGH for many consecutive cycles if the input is constantly LOW. This circuit can be considered a *glitch-latch* since it transforms a glitch into a pulse whose duration is one clock cycle.

6-28.   The circuit of Fig. P6-28a was designed to produce a single pulse on the cycle following the switch throw only when the clock is HIGH. The output waveform is shown in Fig. P6-28b.

(a) Explain why the trailing glitch occurred.

(b) Redesign the circuit to eliminate the glitch.

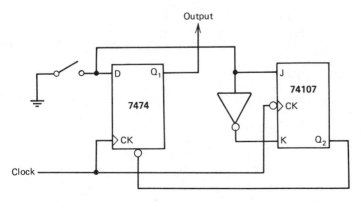

FIGURE P6-30

6-29.   Repeat Problem 6-27 if the output is to be HIGH only if the input was LOW during the previous cycle and the clock is HIGH.

6-30.   A student designed the circuit of Fig. P6-30 to solve the single pulse problem. Like the student, the circuit worked half the time. Explain what happened by drawing timing charts to show how the circuit operates:

   (a) If the switch is thrown when the clock is LOW.
   (b) If the switch is thrown when the clock is HIGH.

After attempting the problems, return to Sec. 6-2 and be sure you can answer the questions. If any of them still seem difficult, review the appropriate sections of the chapter to find the answers.

CHAPTER

# 7

# ONE-SHOTS

## 7-1 INSTRUCTIONAL OBJECTIVES

This chapter introduces the student to the uses of *one-shots* (monostable multivibrators) in digital circuits. It also covers the construction of clock generators (oscillators) and methods of switch debouncing. After reading it, the student should be able to:

1. Use a one-shot to produce a pulse of predetermined duration.
2. Design triggering circuits for one-shots.
3. Use one-shots to control the timing or sequence of events of a circuit.
4. Design one-shot, **555,** and Schmitt trigger oscillators to run at a pre-determined frequency.
5. Debounce a switch.
6. Design a one-shot discriminator.

## 7-2 SELF-EVALUATION QUESTIONS

Watch for the answers to the following questions as you read the chapter. They should help you understand the material presented.

1. What is the function of the resistor and capacitor connected to a one-shot?
2. Why is the maximum duty cycle a limitation on the maximum frequency of triggers to a one-shot?

3.   Explain the conditions necessary to trigger a **74121**.
4.   What are the differences between a *retriggerable* one-shot and an ordinary one-shot?
5.   Explain how to design one-shot timing chains.
6.   What is switch bounce? What problems does it cause? How can switch bounce problems be eliminated?

## 7-3   INTRODUCTION TO ONE-SHOTS

The purpose of a one-shot (also called a monostable multivibrator) is to produce an output pulse having a duration or pulse length determined by the designer. One-shots are used to set timing and control the sequence of events in a digital system.

A one-shot is essentially a FF with only one stable state. Like a FF, it has two outputs, $Q$ and $\bar{Q}$. Normally it is in the CLEAR state ($Q = 0$ and $\bar{Q} = 1$). When a trigger (a pulse edge for 7400 ICs) is applied, the one-shot turns on or flips to the 1 state. It remains there for a time period determined by a resistor and capacitor, connected to it for this purpose, and then switches back to the 0 state. The 1 state of a one-shot is called *quasi-stable* because it must return to the 0 state when the output pulse time expires. *If a one-shot is continually in the 1 state, it indicates either the IC or the circuit is faulty.*

A trigger initiates the one-shot action. If additional triggers are applied to a one-shot when it is in the 1 state, they are ineffective unless the one-shot is *retriggerable*. Nonretriggerable one-shots are also limited by their *duty cycles*. They need time to recover after being fired. Therefore, trigger pulses should not be applied so often as to cause the ON time of the one-shot to exceed the duty cycle specified by the data sheet. *Duty cycle* is defined by Eq. (7-1) as the ratio of the ON time to the total time (ON time plus OFF time) of a repetitive wave form, as shown in Fig. 7-1.

$$\text{Percent duty cycle} = \frac{T_{ON}}{T_{ON} + T_{OFF}} \times 100 \qquad (7\text{-}1)$$

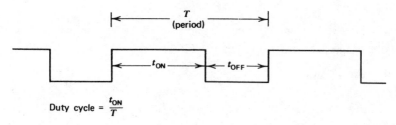

FIGURE 7-1   Duty cycle of a repetitive "square" wave.

Whenever the maximum duty cycle of a one-shot is exceeded, its output pulse will *jitter*. This means the width of each pulse will not be constant.

---

**EXAMPLE 7-1**

A one-shot is designed to produce a 100 $\mu$s pulse and is specified to have a maximum duty cycle of 75 percent. Calculate the minimum time that should be allowed between triggers.

**SOLUTION**

Since the maximum duty cycle is 75 percent, the minimum time between triggers is

$$T_{OFF} + T_{ON} = \frac{T_{ON} \times 100}{\text{percent duty cycle}} = \frac{100 \times 100}{75} = 133.3 \ \mu s$$

Example 7-1 shows the one-shot has 33.3 $\mu$s between pulses to recover. More frequent triggers could cause the output pulse to jitter; its duration would not be 100 $\mu$s consistently.

---

## 7-4  THE 74121 ONE-SHOT

The most popular and commonly used one-shot in the TTL series is the 74121,[1] whose schematic and function table are shown in Fig. 7-2. The NAND gate (shown as a NOR gate with negative inputs) and the Schmitt trigger AND gate shown in Fig. 7-2a are part of the one-shot, and are *not* external gates. In order to trigger the one-shot, which causes it to produce an output pulse, there must be a *rising pulse edge* at point Z (Fig. 7-2a) that is inside the **74121** and *not* available to the engineer. The **74121** can be triggered in one of two ways:

1.  One or both of the A inputs is LOW and the B input goes HIGH.
2.  The B input is HIGH and one of the A inputs goes LOW while the other A input remains HIGH or both A inputs go LOW simultaneously.

### 7-4.1  Triggering the 74121 One-Shot

The triggering of the **74121** is described in detail in the function table, Fig. 7-2b. The outputs are either the *quiescent* state ($Q = 0$, $\overline{Q} = 1$), indicating the one-shot has *not* been triggered, or pulses indicating that the inputs have succeeded in triggering the one-shot. A line-by-line examination of the function table reveals the following.

1.  Line 1 indicates that if input A1 is continuously LOW and input B is continuously HIGH, there is no trigger because both inputs to the internal AND gate

---

[1]Designers can now buy a **74221** that is essentially two **74121**s in a single IC.

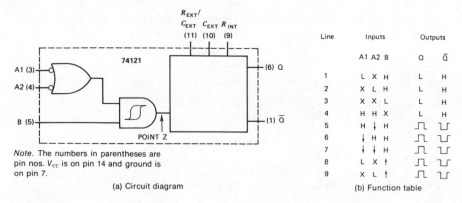

FIGURE 7-2 The **74121** one-shot circuit diagram and function table.

are HIGH. Therefore, its output is continuously HIGH, and no pulse edge occurs at point Z. Consequently, the one-shot does not trigger. This condition exists regardless of the level or transitions of input A2. Since the A2 input does not affect the output, an X, which means irrelevant, is placed in the function table for it.

2. Line 2 is essentially the same as line 1. If A2 is continuously LOW and B is continuously HIGH, the one-shot does not trigger regardless of what happens at input A1.

3. Line 3 shows that the one-shot cannot trigger if the B input is LOW, regardless of the state of the A inputs. If the B input to the AND gate is LOW, its output must be LOW, and point Z cannot receive the positive edge necessary to trigger the one-shot.

4. Line 4 shows that the one-shot cannot fire if both A inputs are HIGH. This places a LOW input on the top leg of the internal AND gate and prevents the one-shot from firing.

5. Lines 5, 6, and 7 show how to trigger the one-shot successfully using the A inputs. First the B input must be HIGH. Then, if one of the A inputs is HIGH and the other goes LOW ( ↓ ) (lines 5 and 6), or if both A inputs go LOW simultaneously (line 7), the one-shot triggers.

6. Lines 8 and 9 show how to trigger the one-shot using the B input. If either or both of the A inputs are LOW, a positive-going edge ( ↑ ) on the B input triggers the one-shot.

## 7-4.2 Timing

The timing of the one-shot, or the duration of its output pulse, is determined by the timing resistor and capacitor[2] connected to it. A timing capacitor is

[2]Most commercially available IC mounting panels make provision for mounting some discrete components. These external components consist of pullup resistors and decoupling capacitors, as well as timing resistors and capacitors.

generally required and is connected between pins 10 and 11 of the **74121**. For long pulses, large value capacitors are required. If electrolytic capacitors are used, the positive terminal must be connected to the $C_{EXT}$ input (pin 10). For the timing resistor, the designer has two choices:

1. The internal timing resistor (nominally 2 k$\Omega$) can be utilized by connecting pin 9 to $V_{CC}$.
2. An external timing resistor can be used by connecting it between $V_{CC}$ and pin 11. The timing resistor must be between 1.4 and 40 k$\Omega$, and pin 9 *must be left open*. (It may not be clamped or tied HIGH.)

The duration of the output pulse is given by

$$t_W = 0.7 C_T R_T \tag{7-2}$$

where $C_T$ and $R_T$ are the values of timing capacitor and resistor, respectively. Manufacturers also supply curves giving the out-put pulse width as a function of the timing resistors and capacitors, but for the **74121**, Eq. (7-2) is simpler and easier to use.

Because of variations between ICs and the tolerances on commercial resistors and capacitors, measured pulse durations may differ by as much as 20 percent from the values calculated by the above equations. We recommend that the circuit be breadboarded by using the values obtained from the equations (or the closest standard values of resistors and capacitors). Then, using the CRO, apply whatever tweaking (adjusting) is necessary to precisely time the pulses.

The *minimum* output pulse width is obtained using no external capacitor (there is always some stray capacity between pins 10 and 11) and the internal resistor. Typically, this results in an output pulse width of 30 to 35 ns. The maximum allowable value for an external capacitor is 1000 $\mu$F (see the manufacturer's specifications). The longest pulse width results from using both the maximum external capacitor and resistor:

$$t_W(MAX) = 0.7 \times (1000 \times 10^{-6}) \times (40 \times 10^3) \tag{7-2}$$
$$= \textbf{28 seconds}$$

By the proper choice of external resistors and capacitors, a pulse width anywhere between 30 ns and 28 seconds can be obtained. Longer pulse widths can be generated by setting up an oscillator followed by a divide-by-N circuit (see Problem 7-17).

---

**EXAMPLE 7-2**

Use a **74121** to produce an output pulse of 1 $\mu$s. Find the required external capacitance using:
(a) The internal timing resistor.
(b) An external timing resistor of 10 k$\Omega$.

**SOLUTION**

(a) The value of the internal timing resistor is 2 k$\Omega$. Therefore,

$$T = 0.7 \, R_1 C \qquad (7\text{-}2)$$

$$C = \frac{T}{0.7R} = \frac{10^{-6}}{0.7 \times (2 \times 10^3)} = 714 \text{ pF}$$

The circuit is shown in Fig. 7-3a.
(b) Using an external resistor of 10 k$\Omega$, we have

$$C = \frac{T}{0.7R} = \frac{10^{-6}}{7000} = 143 \text{ pF}$$

The connections for this circuit are shown in Fig. 7-3b.

---

## 7-4.3 Duty Cycle Limitations

The recovery time required for one-shots imposes a duty cycle limitation on them as explained in Sec. 7-3. The duty cycle depends upon the value of the timing resistor. For the **74121** the maximum allowable duty cycle is 67 percent for a 2-k$\Omega$ timing resistor, but rises to 90 percent if an external timing resistor of 40 k$\Omega$ is used. The one-shot of Ex. 7-2a has a 2-k$\Omega$ resistor. Therefore, its triggers should be more than 1.5 $\mu$s apart. This allows at least 0.5 $\mu$s for recovery time and provides a maximum duty cycle of 67 percent.

If the maximum duty cycle is exceeded, the one-shot still triggers, but its pulse duration is no longer stable. The variation of pulse length is called *jitter*. Generally jitter is undesirable because one-shot output pulses are usually required to have a specific duration.

## 7-4.4 Schmitt Trigger Input

The B input to the **74121** is a Schmitt trigger input (note the hysteresis symbol on the gate in Fig. 7-2a) that responds to very slowly changing

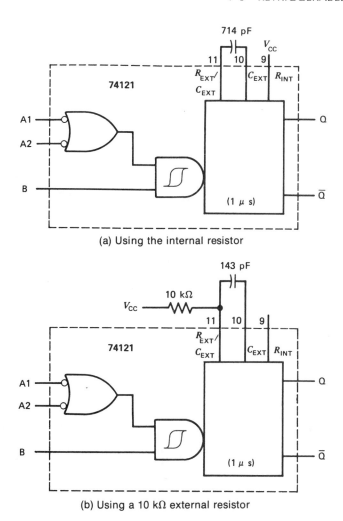

(a) Using the internal resistor

(b) Using a 10 kΩ external resistor

FIGURE 7-3 **74121**s designed to produce a 1-$\mu$s pulse.

inputs. The A inputs are normal TTL inputs and should change faster than 1 V/$\mu$s, but the B inputs can respond to pulse edges as slow as 1 V/s. Consequently, if slow waveforms are required to trigger a one-shot, they are applied to the B input.

## 7-5 RETRIGGERABLE ONE-SHOTS

A *retriggerable* one-shot responds to a trigger when it is ON as well as when it is quiescent. If a trigger occurs when the one-shot is ON, it resets the timing. The one-shot does not turn OFF until one pulse duration after the last trigger. Retriggerable one-shots have two other advantages over a **74121**:

1.   Their duty cycle is unlimited. (They can be ON continuously if triggered by a pulse train of the proper frequency.)
2.   They have a CLEAR input. The presence of a LOW signal on the CLEAR input immediately *resets* the one-shot.

---

### EXAMPLE 7-3

A retriggerable one-shot is timed to produce a 10-$\mu$s pulse. Find its output if triggers occur at times of 1, 7, 20, and 23 $\mu$s:

   (a) If no CLEAR pulses are applied.
   (b) If CLEAR pulses are applied at times of 12, 22, and 32 $\mu$s. Assume the trigger and CLEAR pulses are of very short duration.

### SOLUTION

The solution is shown in Fig. 7-4.

   (a) Without CLEAR pulses the one-shot is ON from $t = 1$ until $t = 17$ $\mu$s (10 $\mu$s after the trigger at $t = 7$) and from $t = 20$ to $t = 33$ $\mu$s as shown in Fig. 7-4a.

   (b) The output is SET by the triggers as in part A, but it is RESET whenever a CLEAR pulse occurs, as Fig. 7-4b shows.

---

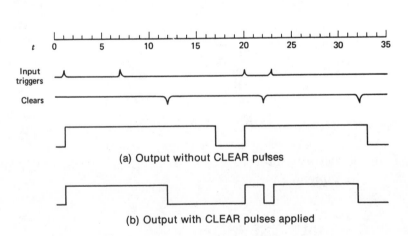

**(a) Output without CLEAR pulses**

**(b) Output with CLEAR pulses applied**

FIGURE 7-4   Response of a 10-$\mu$s retriggerable one-shot to a series of triggers and clears.

## 7-5.1  The 74122

The **74122** is a single retriggerable one-shot in a 14-pin DIP package. Its circuit, function table and timing chart are shown in Fig. 7-5. The internal timing resistor is 10 k$\Omega$ and can be used by connecting pin 9 to $V_{cc}$. The timing capacitor is connected between pins 11 and 13 (the two $C_{EXT}$ connections). An external resistor, if used, is connected between pin 13 ($R_{EXT}$/

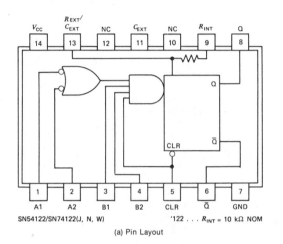

SN54122/SN74122(J, N, W)     '122 . . . $R_{INT}$ = 10 k$\Omega$ NOM

(a) Pin Layout

122          FUNCTION TABLE

| CLEAR | A1 | A2 | B1 | B2 | Q | Q̄ |
|---|---|---|---|---|---|---|
| L | X | X | X | X | L | H |
| X | H | H | X | X | L | H |
| X | X | X | L | X | L | H |
| X | X | X | X | L | L | H |
| X | L | X | H | H | L | H |
| H | L | X | ↓ | H | ⊓ | ⊔ |
| H | L | X | H | ↓ | ⊓ | ⊔ |
| H | X | L | H | H | L | H |
| H | X | L | ↑ | H | ⊓ | ⊔ |
| H | X | L | H | ↑ | ⊓ | ⊔ |
| H | H | ↓ | H | H | ⊓ | ⊔ |
| H | ↓ | ↓ | H | H | ⊓ | ⊔ |
| H | ↑ | H | H | H | ⊓ | ⊔ |
| ↑ | L | X | H | H | ⊓ | ⊔ |
| ↑ | X | L | H | H | ⊓ | ⊔ |

(b) Function table

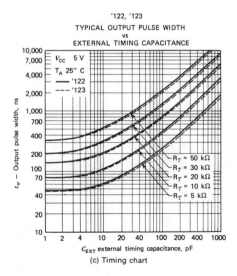

'122, '123
TYPICAL OUTPUT PULSE WIDTH
vs
EXTERNAL TIMING CAPACITANCE

$V_{CC}$  5 V
$T_A$ 25" C
—— '122
- - - '123

$R_T$ = 50 k$\Omega$
$R_T$ = 30 k$\Omega$
$R_T$ = 20 k$\Omega$
$R_T$ = 10 k$\Omega$
$R_T$ = 5 k$\Omega$

$t_w$ — Output pulse width, ns

$C_{EXT}$ external timing capacitance, pF
(c) Timing chart

FIGURE 7-5   The **74122** retriggerable one-shot. (From *The TTL Data Book for Design Engineers*, 2nd ed., Texas Instruments, Inc. Courtesy of Texas Instruments, copyright 1976.)

$C_{EXT}$) and $V_{CC}$, with pin 9 left open. The external timing resistor must be between 5 and 50 k$\Omega$.

The function table (Fig. 7-5$b$) gives the response to 15 possible input conditions. The salient points on the function table are as follows:

1. If the CLEAR line is LOW, the output must be RESET.
2. A positive-going edge at the output of the internal 4-input AND gate is required to trigger the one-shot.
3. Any input conditions that cause the output of the internal 4-input AND gate to be constant will not trigger the one-shot.

The essential difference between the B inputs and the CLEAR input is that the B input can go LOW (while the one-shot is fired) *without ending the output pulse.*

An examination of the function table shows, however, that while a LOW on the CLEAR input causes the one-shot to CLEAR, it will retrigger when the CLEAR goes HIGH again unless prevented by a level on the A or B inputs (see Problem 7-19).

To select the time for the output pulse of a **74122**, do the following:

1. If $C_{EXT} < 1000$ pF, use the chart of Fig. 7-5$c$. (This is valid for pulse widths less than 2 to 10 $\mu$s, depending on $R_T$.)
2. If $C_{EXT} \geqslant 1000$ pF ($t_w \geqslant 2$ to 10 $\mu$s), use Eq. (7-3).

$$ t_W = K_D\, R_T C_{EXT} \left( 1 + \frac{0.7}{R_T} \right) \tag{7-3} $$

where $t_W$ is the pulse width in ns, $R_T$ is the timing resistance in k$\Omega$s, $C_{EXT}$ is the external timing capacitance in pF, and $K_D = 0.32$ for a **74122**.

3. When the CLEAR input or electrolytic capacitors are used, the circuit of Fig. 7-6 should be used to prevent reverse voltage across $C_{EXT}$. This changes $K_D$ to 0.28 for the **74122**.

Electrolytic capacitors should be connected with their positive terminal to $R_{EXT}/C_{EXT}$. They will generally be used when the pulse time is greater than 3 ms.

---

### EXAMPLE 7-4

A system obtains its basic power from the 120V, 60-Hz commercial power lines. Design a circuit to monitor these lines. If the commercial ac power ever misses a cycle, a buzzer is to sound continuously until a reset pushbutton is manually depressed.

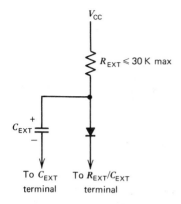

FIGURE 7-6 Timing component connection when CLEAR or electrolytic capacitors are used.

## SOLUTION

The design is shown in Fig. 7-7 and proceeds as follows.

1. The circuit to reduce the 120 V power to TTL levels and speeds has already been found in Ex. 5-2 and shown in Fig. 5-3. The output of the Schmitt trigger is a 60-Hz square wave, which means it produces pulse edges every 16.7 ms.
2. If these pulse edges are fed to a **74122** retriggerable one-shot, whose time is set slightly longer than 16.7 ms, say 20 ms, the one-shot never is reset (its 20-ms output pulse never expires) unless a cycle is missed.
3. The circuit and equation of Fig. 7-6 should be used because the CLEAR input is used and because the time is greater than 3 ms, so electrolytic capacitors will probably be used. A 10-kΩ external resistor was also chosen.

The capacitor required is

$$t_W = 0.28\, R_T C_{EXT} \left( 1 + \frac{0.7}{R_T} \right) \tag{7-3}$$

$t_W$ is in nanoseconds      20 ms = $20 \times 10^6$ ns
$R_T$ is in kΩ      $R_T = 10$ kΩ
$C_{EXT}$ is in pF

Substituting in Eq. (7-3) yields

$$C_{EXT} = \frac{t_W}{0.28\, R_T \left( 1 + \dfrac{0.7}{R_T} \right)} = \frac{20 \times 10^6}{(0.28) \times (10) \times (1.07)}$$

$$= 6.67 \times 10^6 \text{ pF} = 6.67 \ \mu F$$

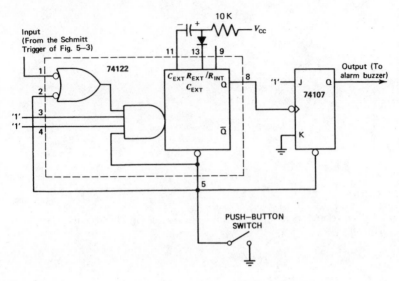

FIGURE 7-7 Circuit for commercial power dropout detection.

To use standard capacitors and allow a small margin for error, a 7 $\mu$F (5 $\mu$F in parallel with 2 $\mu$F) capacitor is chosen.

4. When the **74122** output goes LOW it sets the **74107**. This drives the alarm buzzer. If its voltage and current requirements are high, the alarm buzzer may be coupled to the **74107** through a buffer/driver. If an ac cycle is missed the **74122** output goes LOW setting the **74107**.

5. When the pushbutton is depressed it clears the FF and also puts a low voltage on inputs $A_2$ and $B_1$ of the **74122**.

6. When the pushbutton is released, it triggers the **74122** and the circuit resumes normal operation. Note that it will sound the alarm again in 20 ms if the ac power is *not* working properly.

## 7-5.2 The 74123

The **74123** is a *dual* retriggerable one-shot in a 16-pin DIP package. The circuit and function table are shown in Fig. 7-8. The **74123** is a popular IC because it contains *two* one-shots in a single chip. To achieve this packing density, the internal logic has been simplified and there is *no* internal timing resistor. An external timing resistor between 5 and 50 k$\Omega$ must be connected between $V_{CC}$ and the $R_{EXT}/C_{EXT}$ pin on the IC.

The function table (Fig. 7-8*b*) indicates that the **74123** triggers in the same manner as a **74121** or **74122**. A positive-going edge at the output of the internal AND gate is required to trigger the one-shot.

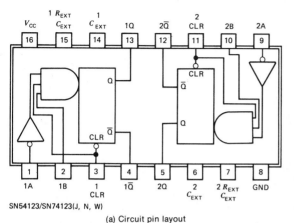

FUNCTION TABLE

| INPUTS | | | OUTPUTS | |
|--------|---|---|---------|---|
| CLEAR | A | B | Q | Q̄ |
| L | X | X | L | H |
| X | H | X | L | H |
| X | X | L | L | H |
| H | L | ↑ | ⊓ | ⊔ |
| H | ↓ | H | ⊓ | ⊔ |
| ↑ | L | H | ⊓ | ⊔ |

SN54123/SN74123(J, N, W)

(a) Circuit pin layout

(b) Function table

FIGURE 7-8   The **74123** dual retriggerable one-shot. (From *The TTL Data Book for Design Engineers,* 2nd ed., Texas Instruments, Inc. Courtesy of Texas Instruments, copyright 1976.)

The timing chart of Fig. 7-5c can be used to set the pulse width of a 74123 if $C_{EXT} < 1000$ pF. For $C_{EXT} > 1000$ pF, the pulse width is

$$t_W = 0.28 \, R_T C_{EXT} \left( 1 + \frac{0.7}{R_T} \right) \tag{7-4}$$

where, again, $t_W$ is the pulse width in nanoseconds, $R_T$ is the timing resistor in kΩ, and $C_{EXT}$ is the timing capacitor in pF. If the diode (Fig. 7-6) is required, the constant in (7-4) becomes 0.25.

## 7-6  INTEGRATED CIRCUIT OSCILLATORS

Oscillators are needed in digital circuits to generate clock pulses (clocks) required to control the timing of the circuit. As previously mentioned, highly stable crystal-controlled oscillators are available in DIP packages and produce TTL compatible outputs. If the tolerance on the required frequency is not extremely critical, however, serviceable oscillators can be built from a variety of TTL circuits.

### 7-6.1  The 74121 Oscillator

Two **74121** one-shots can be coupled together to produce an oscillator, as Ex. 7-5 shows. When a pulse on one of the one-shots expires, the resulting negative-going edge triggers the other one-shot. If this is allowed to continue indefinitely, the circuit functions as an oscillator or clock generator.

## EXAMPLE 7-5

Design an oscillator, using two **74121**s, to produce output pulses of 1 $\mu$s ON, 2 $\mu$s OFF, and so forth. Include a START switch and a SINGLE-CYCLE/CONTINUOUS switch. When the latter switch is in the SINGLE-CYCLE position, and the START switch is flipped ON, each one-shot should pulse *once*. But if the switch is in CONTINUOUS mode, a continuous train of pulses should result when the START switch is turned ON.

## SOLUTION

The circuit is shown in Fig. 7-9. It is designed as follows:

1.  The timing of the two one-shots is set for 1 $\mu$s and 2 $\mu$s, respectively. For convenience, their 2-k$\Omega$ internal resistors are used, and the capacitors (using Eq. 7-2) are 714 and 1428 pF, respectively.
2.  The START switch is connected to the B input of the first one-shot. For simplicity, we assume it is free of bounces (see Sec. 7-8). When the START switch is turned ON, it fires the first one-shot for a period of 1 $\mu$s. When the first one-shot turns OFF, the negative edge at $Q_1$ triggers the second one-shot, which fires for a time of 2 $\mu$s.
3.  When the pulse output of the second one-shot expires, its negative-going edge retriggers the first one-shot. Thus the circuit continues to oscillate and produce

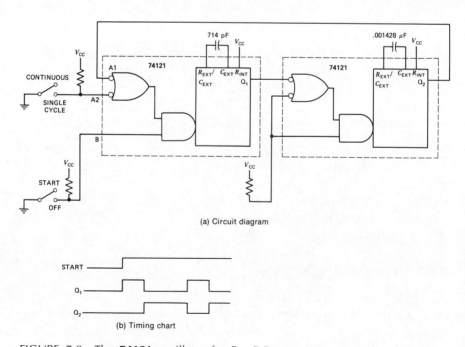

(a) Circuit diagram

(b) Timing chart

FIGURE 7-9 The **74121** oscillator for Ex. 7-5.

these clock pulses as long as the START switch is HIGH. The output waveforms are shown in Fig. 7-9b.

4.   The SINGLE-CYCLE/CONTINUOUS switch is connected to the $A_2$ input of the first one-shot. If the switch is open, $A_2$ is HIGH and the circuit functions as described above. If the switch is closed, however, the LOW level at $A_2$ forces the output of the internal negative-OR gate HIGH, regardless of what happens at $A_1$. When the START switch is opened, each one-shot fires in turn. But the negative edge at $Q_2$ cannot retrigger the first one-shot. Consequently, each one-shot fires only once. This is SINGLE-CYCLE operation.

## EXAMPLE 7-6

Design a circuit to produce a continuous clock that is HIGH for 10 $\mu$s and LOW for 1$\mu$s. Use 74121s.

## SOLUTION

At first glance it appears that a **74121** with a 1-$\mu$s output pulse width and a **74121** with a 10-$\mu$s output pulse width are needed. Unfortunately, were this circuit built, the latter one-shot would have a duty cycle of 10/11 = 91 percent. This is beyond the allowable specifications (Sec. 7-4.3) even if an external resistor is used. The solution, therefore, is to use *two* 5-$\mu$s one-shots and a 1-$\mu$s one-shot, which trigger each other in turn, as shown in Fig. 7-10a. The output can be taken from $Q_3$ as the timing chart of Fig. 7-10b shows. For clarity, the timing resistors and capacitors are not shown. Instead the length of each pulse is shown in parentheses within the one-shot.

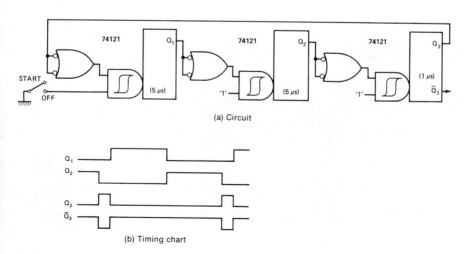

(a) Circuit

(b) Timing chart

FIGURE 7-10   Circuit and timing chart for Ex. 7-6.

## 7-6.2 The 74123 Oscillator

Oscillators can also be built using **74123** retriggerable one-shots. These oscillators have three advantages over a **74121**:

1. Since there are two one-shots in a package, the oscillator can be built with one chip.

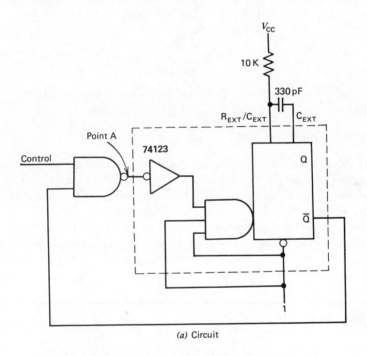

*(a)* Circuit

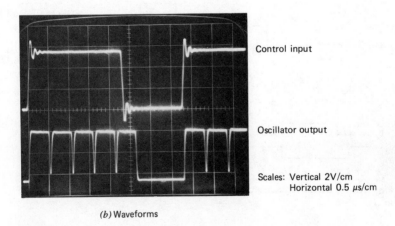

*(b)* Waveforms

FIGURE 7-11 The single one-shot oscillator.

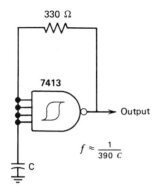

$$f \approx \frac{1}{390 \ c}$$

FIGURE 7-12  A Schmitt-trigger oscillator.

2.   Since the **74123** is retriggerable, there are no duty cycle problems.
3.   The oscillations can be stopped on command by driving the CLEAR input LOW.

There are also drawbacks: external resistors must be used, and there is less on chip logic available for gating in such things as single-cycle/continuous operation.

If the CLEAR is used to stop oscillations, both halves of the **74123** will SET when the CLEAR line goes HIGH. They will remain SET until the section with the shorter time delay goes down. This retriggers the section with the longer time delay and normal operation resumes.

An oscillator can also be made from a single **74123** by tying its Q output to its A input. When the pulse expires, Q goes LOW, retriggering the one-shot. The outputs of this oscillator are *glitches*, but the time between the glitches is controlled by the RC of the **74123**.

A variation of this circuit that provides for external control is shown in Fig. 7-11a.[3] When the CONTROL input is LOW, point A is always HIGH and the oscillator is quiescent. Oscillations start when the CONTROL input goes HIGH. By using the chart (Fig. 7-5) the time of this oscillator was set for 0.5 $\mu$s and the oscillators output is shown in Fig. 7-11b. The glitches last for about 40 ns. This oscillator can also be controlled by using its CLEAR input.

## 7-6.3   Schmitt-Trigger Oscillators

A very simple oscillator can be made from a Schmitt trigger as shown in Fig. 7-12. The external resistor must be kept small or the circuit will fail

[3]A circuit like this has been successfully used to drive the high frequency clock and shift register in a CRT driver kit.

to oscillate (see Problem 7-14). A 330-$\Omega$ resistor is a good choice for this oscillator. Schmitt trigger oscillators can be designed to produce outputs from 0.1 Hz to 10 MHz.

The oscillator of Fig. 7-12 operates as follows.

1. The **7413** inverts. If its output is HIGH the capacitor charges. When the capacitor voltage reaches $V_{T+}$ (the positive-going threshold voltage of 1.7 V), the **7413** output goes LOW.

2. Now the capacitor discharges toward $V_{OL}$ (0.2 V, typical). When it reaches $V_{T-}$ (0.9 V) the Schmitt-trigger output goes HIGH again and the capacitor starts to recharge again.

3. The frequency of oscillation depends on the resistor and capacitor. For a 330-$\Omega$ resistor:

$$f \approx \frac{1}{390C} \tag{7-5}$$

---

**EXAMPLE7-7**

Design a 100-kHz Schmitt-trigger oscillator.

**SOLUTION**

From (7-5) we obtain

$$C = \frac{1}{390f} = \frac{1}{390 \times 10^5} = 0.0256\mu F$$

The circuit should be breadboarded as designed above and its output observed on a CRO. Any tweaking necessary to produce a precise pulse width is done by changing values of C.

---

## 7-6.4  The 555 Timer

The **555** timer is a very popular and commonly used oscillator.[+] It comes in an 8-pin DIP package (about half the size of an ordinary IC) and can be set up either as a one-shot or an astable circuit (clock).

A simplified equivalent circuit for the **555** set up as a one-shot as shown in Fig. 7-13. Internally, the **555** has five basic parts:

1. The lower comparator.
2. The upper comparator.

---

[+]Available as a 555 from National Semi-Conductor, Signetics and Fairchild or as an MC 1455 from Motorola.

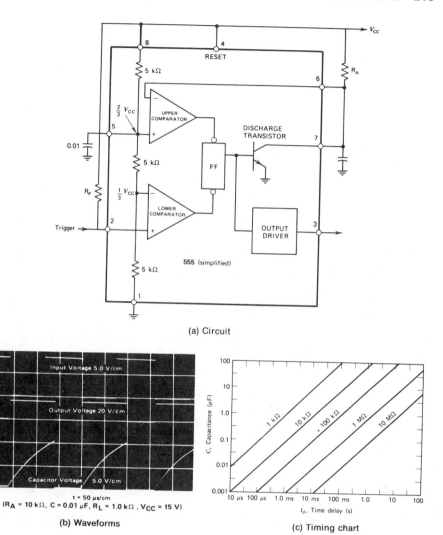

(a) Circuit

**(b) Waveforms**

(c) Timing chart

FIGURE 7-13 The **555** timer connected as a one-shot. (Taken from Motorola MC 1455/1555 data sheets and Motorola Semiconductor Data Library, Vol. 7, pp. 2-25. Courtesy of Motorola Integrated Circuits Division.)

3. The internal FF.
4. The discharge transistor.
5. The output driver.

The lower comparator compares the voltage on its input (pin 2) with $\frac{1}{3}V_{CC}$. $V_{CC}$ can be any voltage from 4.5 to 18 V. If the voltage on pin 2 is less than $\frac{1}{3}V_{CC}$, the lower comparator produces a LOW output voltage, which CLEARS the FF.

The upper comparator compares the voltage on its input (pin 6) with $\frac{2}{3}V_{CC}$. If its input is greater than $\frac{2}{3}V_{CC}$, it SETS the FF. Pin 5 of the 555 can be used to control or change the threadhold voltage of the upper comparator. Normally this feature is not used and pin 5 is connected to ground through a 0.01 $\mu$F capacitor.

If the internal FF is SET, it saturates the discharge transistor and also causes the output driver to produce a 1 output. Conversely, if the FF is CLEAR, the discharge transistor is cutoff and the output driver produces a 0 output.

For monostable (one-shot) operation, an external resistor and capacitor are connected as shown in Fig. 7-13. In the quiescent state the FF is SET, and the discharge transistor is ON. It effectively shorts the capacitor to ground. The voltage at pin 6 is therefore $V_{CE(sat)} \approx 0.2$ V. Pin 2 is used as the trigger input. Normally the voltage at pin 2 is HIGH because of the resistor $R_p$. A trigger causes one-shot operation as follows.

1. An input trigger drives pin 2 to ground.
2. This fires the lower comparator, resetting the FF, which causes the discharge transistor to turn OFF.
3. The timing capacitor now charges towards $V_{CC}$, through $R_A$ and raises the voltages on pin 6.
4. When the voltage on pin 6 reaches $\frac{2}{3}V_{CC}$, the upper comparator fires.
5. This SETS the FF, which turns on the discharge transistor and shorts out the capacitor. The waveforms are shown in Fig. 7-13b.
6. The time of the pulse is given by the timing chart of Fig. 7-13c, or:

$$t_W \approx 1.1\, R_A C \tag{7-6}$$

7. If pin 4 (RESET) is connected to ground, it turns ON the discharge transistor. This immediately discharges the capacitor and prevents it from charging. If it is not used, it is connected to $V_{CC}$. For clarity, the reset logic is not shown in the simplified schematic.
8. The trigger pulse must be shorter than the output pulse, and additional triggers, while the output is LOW, do not affect the pulse width.

---

## EXAMPLE 7-8

Design a 10-ms one-shot using a 555.

## SOLUTION

We arbitrarily choose a 1-$\mu$F capacitor. From the chart or Eq. (7-6), find

$$R_A = 9.1\ k\Omega$$

---

The 555 can also be used as an astable circuit or clock generator, by connecting it as shown in Fig. 7-14a. Here, both comparator inputs (pins 2 and 6) are connected to the capacitor. The circuit operates as follows:

1.   Assume the discharge transistor is off. Then the capacitor charges towards $V_{CC}$ through $R_A$ and $R_B$.

2.   When the capacitor voltage reaches $\frac{2}{3}V_{CC}$ the upper comparator SETS the FF. This turns on the discharge transistors and the capacitor discharges toward ground through $R_B$.

3.   When the capacitor voltage becomes as low as $\frac{1}{3}V_{CC}$, the lower comparator CLEARS the FF. The discharge transistor turns OFF and the capacitor starts to charge again. This process continues indefinitely, resulting in an oscillator or clock generator.

4.   The output frequency can be obtained from the timing chart (Fig. 7-14c) or

$$f = \frac{1.44}{(R_A + 2R_B)C} \tag{7-7}$$

5.   The output and capacitor voltage waveforms for an 11-kHz oscillator are shown in Fig. 7-14b. Note that $R_A + 2R_B \approx 13\ \text{k}\Omega$, and a frequency of 11 kHz results using either Eq. (7-7) or the chart.

---

**EXAMPLE 7-9**

Design a 100-kHz oscillator using a 555.

**SOLUTION**

Using Eq. (7-7) we find

$$f(R_A + 2R_B)C = 1.44$$

If we choose $C = 10^{-9}$ (0.001 $\mu$F), then $R_A + 2R_B = 14,400\ \Omega$. This is a good choice as it keeps $R_A$ and $R_B$ in the k$\Omega$ range. There are many ways to select $R_A$ and $R_B$. One choice, which makes them approximately equal, is to set $R_A = 5\text{k}\Omega$ and $R_B = 4.7\ \text{k}\Omega$.

To check, we use the timing chart with $R_A + 2R_B = 14,4000\ \Omega$ and $C = .001$ $\mu$F. This combination does yield a frequency of approximately 100 kHz.

---

## 7-7   TIMING GENERATION PROBLEMS

Engineers are very frequently required to design circuits that produce a sequence of pulses to control the operation of a larger circuit or system. Generally these pulses must conform to a timing chart. There are several

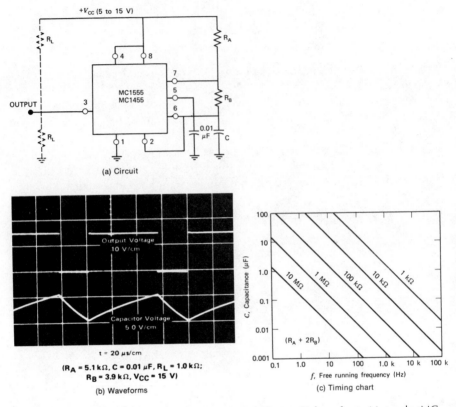

FIGURE 7-14 The **555** connected as an oscillator. (Taken from Motorola MC 1455/1555 data sheets and Motorola Semiconductor Data Library, Vol. 7, pp. 2–25. Courtesy of Motorola Integrated Circuits Division.)

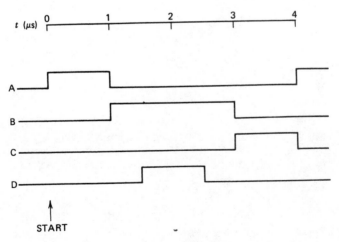

FIGURE 7-15 Timing chart for Ex. 7-10.

ways to design such a timing circuit. Designs using one-shots are considered below, and designs using shift registers are considered in Chapter 9.

## 7-7.1 The Design of Timing Circuits Using One-Shots

Circuits to generate pulse sequences can be built from one-shots by having the *negative-going* edge (generated when the one-shot pulse expires) trigger the next succeeding one-shot. Simple examples of this were presented in Exs. 7-5 and 7.6. SINGLE-CYCLE or CONTINUOUS modes of operation can be easily incorporated into this design, as shown below.

---

### EXAMPLE 7-10

Design a circuit using 74121s to generate a series of pulses conforming to the timing chart given in Fig. 7-15. Include a START switch and a SINGLE-CYCLE/CONTINUOUS mode switch.

### SOLUTION

This circuit requires five one-shots as follows.

1. A one-shot to generate pulse A.
2. A one-shot to generate pulse B (triggered by the expiration of pulse A).
3. A one-shot to generate pulse C (triggered by the expiration of pulse B).
4. A one-shot to generate pulse D.
5. A one-shot to generate the time required between the expiration of pulse A and the start of pulse D. This one-shot fires at the end of pulse A, and when it expires, it triggers pulse D.

The final circuit is shown in Fig. 7-16. The START switch (assumed debounced) is connected to the B input of the A one-shot. The output of the C one-shot is fed back to the A one-shot, so that in CONTINUOUS mode the expiration of the C one-shot triggers the A one-shot and restarts the cycle. The switch on the A2 input of the A one-shot controls the mode of operation.

---

## 7-8 SWITCH BOUNCE

Switches are often used to start or control a circuit's action, but whenever a switch is designed into a circuit, we must be very careful to avoid problems that could be caused by switch "bounce." The action of a switch is shown in Fig. 7-17a. Most toggle switches in current use are the BREAK-BEFORE-MAKE type. This means that, in the process of switching, the switch blade first leaves one contact (BREAK), touches neither contact for a few milliseconds during its traverse, and then mates with the other contact (MAKE).

Unfortunately, both the MAKE and BREAK processes consist of a series

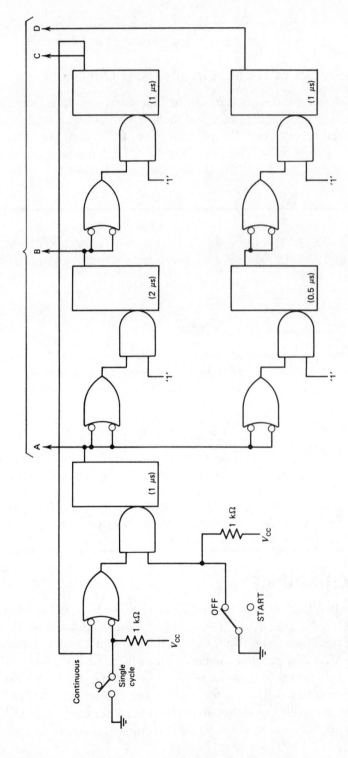

FIGURE 7-16 A one-shot timing chain for Ex. 7-10. Notes: 1. All one shots are **74121**s 2. Inputs marked '1' should be clamped or pulled up to $V_{CC}$. 3. Timing resistors and capacitors not shown for clarity.

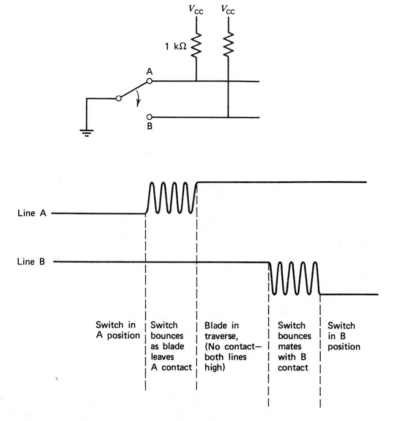

FIGURE 7-17   The action of a toggle switch as it is thrown from A to B.

of connections and disconnects called *bounces*. The action of a switch as it is toggled is shown in Fig. 7-17*b* and the bounces are plainly seen. Some switches bounce for a duration of up to 50 ms.[5]

Many digital circuits require a single transition from a switch. The oscillator of Fig. 7-9, for example, when operated in SINGLE-CYCLE mode produces pulses that last for only 3 $\mu$s. If, however, the START switch bounced for several milliseconds, it would generate many START pulses and cause many cycles to occur before the switch stopped bouncing. To convert the bouncy output of a switch to a clean pulse, switch debouncing circuits must be used. Three simple debouncing circuits are described below.

---

[5]Some switch manufacturers include a maximum bounce time as part of their specifications.

### 7-8.1  7474 Debouncing

A noisy switch can be debounced by connecting its outputs to the direct SET and CLEAR of a 7474 FF as shown in Fig. 7-18*a*. When the switch is in the UP position, the LOW input at point A keeps the FF in the SET position. The bounces which occur when the switch leaves the *upper* position do not affect the FF. When the switch starts to engage the *lower* contact, the first bounce clears the FF. The other bounces merely serve as additional CLEAR pulses and have no further effect on the FF. Therefore, the FF produces a single output change, as required, when the switch is thrown from UP to DOWN.

The situation is similar when the switch is thrown to the UP position. The FF does not set until the first negative bounce occurs when the blade starts to engage the upper contact. The following bounces merely serve as additional SET pulses. The FF of Fig. 7-18*a* is a successful debouncing circuit because only a single transition occurs for each throw of the switch.

### 7-8.2  7400 Debouncing

NAND gate FFs built from 7400s are also often used to debounce switches. The circuit for a NAND gate debouncer is shown in Fig. 7-18*b* and its

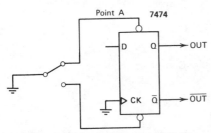

(a) Flip-flop debouncing with a **7474** FF

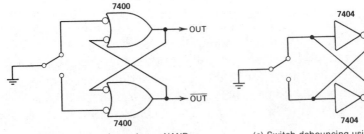

(b) Switch debouncing using a NAND gate FF

(c) Switch debouncing using inverters

FIGURE 7-18   Three methods of debouncing switches.

operation is the same as the **7474** debouncer. The **7400** FF is also SET (or CLEARED) by the first negative bounce after the switch has been thrown, and further bounces have no effect on it.

### 7-8.3   7404 Debouncing

A clever debouncing circuit can be built using two **7404** inverters, as shown in Fig. 7-18c. When the switch is in the UP position, the output of the upper inverter is HIGH and the output of the lower inverter, which is shorted to ground, is LOW. When the switch enters the break portion of its downward traverse, the 0 output of the lower inverter keeps the upper output HIGH, until the first negative bounce when the switch blade touches the lower contact. This negative spike causes the output of the lower inverter to go HIGH and the output of the upper inverter to go LOW. The LOW output of the upper inverter holds the lower inverter HIGH during the bounces and a clean pulse is produced.

---

**EXAMPLE 7-11**

In the circuit of Fig. 7-16 (see Ex. 7-10), which switches should be debounced?

**SOLUTION**

Certainly the START switch should be debounced. The first pulse is only 1 $\mu$s long, but the first one-shot would be erratically triggered for several milliseconds if the START switch were not debounced.

Generally mode control switches like the SINGLE-CYCLE/CONTINUOUS switch do not need debouncing. They are set before the START switch is thrown. Since setting the two switches are human actions, whose time durations are long compared to switch bounce times, it is safe to assume the mode switch stops bouncing long before a person can throw the START switch.

---

## 7-9   DEBOUNCING CLASS A SWITCHES

Class A or single-pole, single-throw switches, have only two contacts (they are either open or closed) and cannot be debounced by the methods of Sec. 7-8.

The simplest method of debouncing class A switches is to connect them to the D input of a FF and clock it with a constant clock whose period is longer than the bounce time of the switch. This is shown in Fig. 7-19.

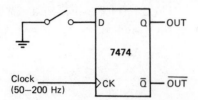

FIGURE 7-19 Debouncing a class A switch using a FF.

## 7-9.1 Schmitt-Trigger Debouncing

If the proper frequency clock is unavailable, class A switches can generally be debounced by connecting them to an RC network and then a Schmitt trigger, as shown in Fig. 7-20. The time constant of the RC network should be longer than the switch bounce time so that the capacitor charges relatively slowly.

The circuit of Fig. 7-20 operates as follows.

1. When the switch is opened, the bounces discharge the capacitor and it does not have time to charge to $V_{T+}$ until the bounces stop completely. Only then does the Schmitt-trigger output go LOW.

2. When the switch closes the first bounce discharges the capacitor, causing the Schmitt-trigger output to go HIGH. Since the capacitor does not have sufficient time to charge up to $V_{T+}$ between bounces, the Schmitt-trigger output remains HIGH.

## 7-9.2 One-Shot Debouncing

One-shots can also be used to debounce switches as shown in Fig. 7-21. The first one-shot is set for a long pulse, compared to the bounce time of the switch (8.8 ms in Fig. 7-21). When the switch stops bouncing and

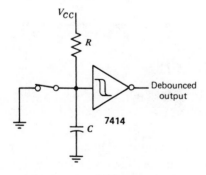

FIGURE 7-20 Debouncing a single-pole, single-throw switch.

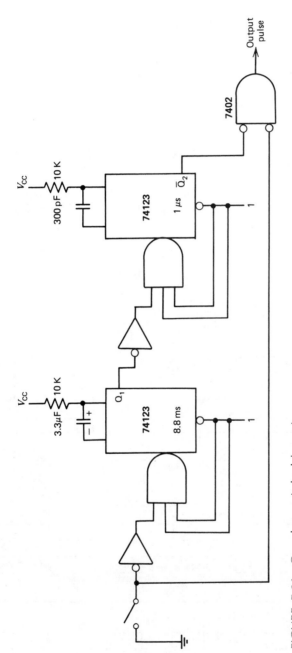

FIGURE 7-21 One-shot switch debouncing.

retriggering the first one-shot, it times out and triggers the second one-shot (for 1 $\mu$s in Fig. 7-21). The function of the NOR gate is to produce an output pulse when the switch is thrown from HIGH to LOW (because the switch line will be LOW when the second one-shot fires) but not when the switch is thrown from LOW to HIGH. This circuit produces a single output pulse for each switch throw. The length of that pulse is variable and determined by the second one-shot.

## 7-10 THE ONE-SHOT DISCRIMINATOR

A single pair of one-shots (a single 74123) can be used as an FM *discriminator* that determines which one of two signal frequencies is being received. This

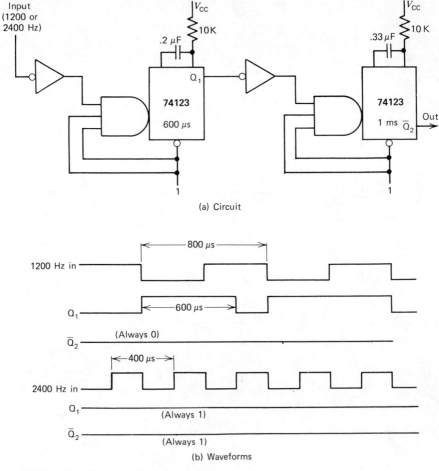

(a) Circuit

(b) Waveforms

FIGURE 7-22 The one-shot discriminator.

can be used in Frequency Shift Keying (FSK, see Sec. 18-4) circuits. The time of the first one-shot should be set between the periods of the two input frequencies, and the first one-shot should trigger the second one-shot. If the lower frequency (longer pulse width) signal is applied to the first one-shot it will constantly time out and provide a series of triggers to the second one-shot, which will always be SET. If the higher frequency is being received, the first one-shot will be constantly SET; the second one-shot will not see any trigger pulses and will remain CLEAR. This discriminator is illustrated in Ex. 7-12 (also see Problem 7-18).

---

**EXAMPLE 7-12**

The "Kansas City Standard" used by some engineers for transmitting digital data to tape cassettes uses a 2400 Hz signal for a 1 and a 1200 Hz signal for a 0. Design a one-shot discriminator to produce a 1 output when the input is 2400 Hz and a 0 output when the input is 1200 Hz.

**SOLUTION**

The solution is shown in Fig. 7-22. The periods of the inputs waves are approximately 400 $\mu$s (for a 2400 Hz input) and 800 $\mu$s (for a 1200 Hz input). The time of the first one-shot is set between them or at approximately 600 $\mu$s. If the 1200-Hz signal is coming in, the first one-shot resets every 800 $\mu$s, as shown in the waveforms of Fig. 7-22b. The second one-shot is set for a pulse width of 1 ms and it is always triggered. If the input frequency is 2400 Hz, the first one-shot is always SET; the second one-shot gets no trigger pulses and is always CLEAR. Thus, $\bar{Q}_2$ will be a 1 if the input frequency is 2400 Hz and a 0 if the input frequency is 1200 Hz.

---

## SUMMARY

In this chapter the design of one-shots to generate pulses of predetermined lengths and the design of TTL oscillators of any desired frequency were studied. Switch debouncing circuits were also covered. These circuits are widely used to generate and control the timing or sequence of events in a digital system. Consequently, the engineer should know how to design them.

## GLOSSARY

**Astable circuit.** A square wave oscillator or clock generator.

**Debouncing circuit.** A circuit designed to produce a clean output in response to a switch closure.

**Discriminator.** A circuit whose output depends on the frequency of the input wave.

**Duty cycle.** The ratio of ON time to total pulse period for a repetitive square wave.

**Jitter.** One-shots that produce output pulses of different lengths are said to jitter.

**Monostable circuit.** See one-shot.

**One-shot.** A circuit that produces an output pulse for a fixed period of time in response to a trigger, and then returns to its quiescent state.

**Recovery time.** The time required by a one-shot to recover (to recharge its internal capacitor) after having been fired.

**Retriggerable one-shot.** A one-shot that can be restarted by triggers when it is in its 1 state.

**Switch bounce.** Fluctuations in output levels when a switch MAKES or BREAKS with a contact.

**Trigger.** A short pulse or edge that initiates a one-shot pulse.

**Tweaking.** A small change made in resistance or capacitance to time a circuit precisely.

## REFERENCES

*The TTL Data Book for Design Engineers*, Texas Instruments, Inc., 2nd ed., 1976.

Thomas R. Blakeslee, *Digital Design With Standard MSI & LSI*, Wiley, New York, 1975.

**Millman** and **Taub**, *Pulse, Digital, and Switching Waveforms*, Chapter 11, McGraw-Hill, New York, 1965.

**Herbert Taub** and **Donald Schilling**, *Digital Integrated Electronics*, McGraw Hill, New York, 1977.

**Christopher E. Strangio**, *Digital Electronics*, Prentice-Hall, Englewood Cliffs, N. J., 1980.

## PROBLEMS

7-1.   A traffic light is red for 25 seconds and green for 45 seconds. Calculate its duty cycle.

7-2.   Design a **74121** to produce a 5-ms pulse using:
(a). The internal resistor.
(b). A 40-kΩ external resistor.
What is the minimum time that should be allowed between pulses in each case?

7-3.   A **74121** has a timing capacitor of 1 $\mu$F. The timing resistor consists of a 5-kΩ fixed resistor in series with a 20-kΩ potentiometer. Calculate:
(a). The minimum pulse width of the output.
(b). The maximum pulse width of the output.
(c). Why is it unwise to use a potentiometer without the fixed resistor?

7-4.   Design a one-shot to produce pulses of $0.5$ $\mu$s using:
   (a). A **74121**.
   (b). A **74122**.
   (c). A **74123**.
   In each case above calculate the required timing resistors and capacitors and draw the circuit showing how they are connected to the one-shot.

7-5.   Repeat Problem 7-4 if the required pulse is $0.5$ ms.

7-6.   A one-shot has a timing resistor of 10 k$\Omega$ and a timing capacitor of $0.5$ $\mu$F. Calculate its output pulse width if the one-shot is:
   (a). A **74121**.
   (b). A **74122**.
   (c). A **74123**.

7-7.   Add an ENABLE-DISABLE toggle switch to the circuit of Fig. 7-7. If the switch is in the DISABLE position the circuit should never alarm, but if the switch is in the ENABLE position the circuit should function normally.

7-8.   Given two 100-kHz clocks, A and B, and two output points, C and D. If *both* clocks are operational, A must be connected to C and B to D. If either clock fails, the working clock must drive both points. Design a circuit to detect a clock failure and produce the required outputs. (A clock failure means the clock stops making transitions.) (*Hint:* Retriggerable one-shots are excellent devices for this type of problem.)

7-9.   Design a clock that is ON for 19 ms and OFF for 1 ms using:
   (a). **74121**s.
   (b). **74123**s.

7-10.   Design a one-shot to produce a pulse width of 5 ms using:
   (a). A **74121**.
   (b). A **74122**.
   (c). A **74123**.
   (d). A **555**.

7-11.   Design an oscillator to produce a 50-kHz square wave using:
   (a). Two **74121**s.
   (b). One **74123**.
   (c). A Schmitt trigger.
   (d). A **555**.

7-12.   Explain with diagrams, how you would build a circuit to produce the waveform of Fig. P7-12. Include a SINGLE-CYCLE/CONTINUOUS option. De-bounce switches where necessary. Show specifically where you would take the output to obtain the continuous waveform shown.
   (a). Use two **74121**s.
   (b). Use a single **74123**.

FIGURE P7-12

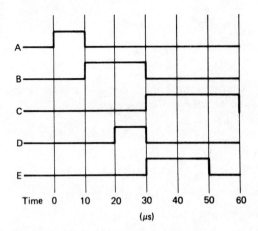

FIGURE P7-15

7-13.   A **555** timer has $R_A = R_B = 100$ k$\Omega$ and $C = 1$ $\mu$F. Find its frequency of oscillation from the curves (Fig. 7-14) and verify it by the equation.

7-14.   An attempt was made to build a Schmitt-trigger oscillator using a 1-k$\Omega$ resistor. It would not oscillate, but it was found that the Schmitt-trigger output voltage was low continuously and its input voltage was 1 V. Using these clues, explain what happened and why the Schmitt trigger would not oscillate.

7-15.   Design a circuit to generate the timing pulses shown in Fig. P7-15.

7-16.   A one-shot is to be triggered from a switch. Is it necessary to debounce the switch if the pulse time is:

   (a). 5 $\mu$s?

   (b). 500 $\mu$s?

   (c). 500 ms?

   Explain the reasons for your answer in each case.

7-17.   Design a 10-minute clock by building a 1-Hz oscillator and using IC counters to divide the frequency down.

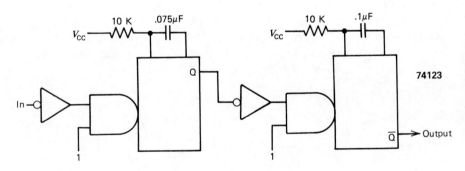

FIGURE P7-18

7-18. The circuit of Fig. P7-18 is a discriminator. The output is different if a 4 kHz or a 5 kHz wave is applied.

(a). Find the "on time" of each one-shot.

(b). Make a timing chart for each input frequency.

(c). What is the output in each case?

7-19. Design a 200-kHz oscillator made up of two **74123**s with the following stipulations:

(a). When an external CLEAR line goes LOW both sections of the oscillator shall CLEAR immediately.

(b). When the CLEAR line goes HIGH only the first section of the oscillator shall trigger.

After attempting the problems, return to Sec. 7-2 and review the self-evaluation questions. If you cannot answer certain questions, return to the appropriate sections of the chapter to find the answers.

CHAPTER

# 8

# COUNTERS

## 8-1 INSTRUCTIONAL OBJECTIVES

After reading this chapter, the student should be able to:

1. Design a counter to count to N, and recycle, where N is any given number.
2. Design a circuit to divide down a clock frequency to a desired submultiple.
3. Design a counter to count in an irregular or shortened sequence.
4. Explain the principle underlying the design of a synchronous counter.
5. Design counters using the 7490 series of TTL counters.

## 8-2 SELF-EVALUATION QUESTIONS

Watch for the answers to these questions as you read this chapter. They should help you understand the material presented. When you have finished the chapter, return to this section and be sure you can answer all of the questions.

1. Why are crystal oscillators used? Why must their frequency be as high as the highest frequency clock required in the circuit?
2. How is a specific count decoded from an N-bit counter?
3. What is the difference between synchronous and ripple counters? What are the advantages and disadvantages of each?

4. What are the steps necessary in the design of an irregular counter?

5. How are the CARRY and BORROW outputs of a **74193** used to construct a cascaded up-counter or down-counter?

6. How does the LOAD input function on a **74193**?

## 8-3 INTRODUCTION

Counters are special purpose digital circuits, composed of FFs or ICs designed specifically as counters, whose function is to count any number of events required by the system specifications. A count-by-N circuit, which is capable of counting up to a specified number N, will require at least K FFs, where $2^K \geq N$. We have already considered a count-by-8 or 3-bit counter in Sec. 6-15.3. In this chapter various methods of designing counters are examined.

## 8-4 DIVIDE-BY-N CIRCUITS

A divide-by-N circuit can count to a given number, N. It is continuously clocked, and is allowed to recycle, or rollover to 0, after reaching the Nth count. If the output is taken from the most significant stage of the counter, it provides a pulse that is N times as long as the original clock pulse. The frequency of this output is equal to the frequency of the original clock divided by N: hence the name divide-by-N. The circuit of Fig. 6-23 is a 3-bit counter, or a divide-by-8 circuit, composed of three FFs. The frequency of the $Q_3$ output is one-eighth the frequency of the clock.

Complex systems that require several clocks of various frequencies are designed in accordance with these principles. If the timing for the system is critical, it is usually provided by a *crystal controlled oscillator*, which is highly stable and jitter-free. The frequency of the crystal is chosen to be as high or higher than the highest frequency clock required by the circuit. The various clocks needed throughout the circuit are then obtained by counting down the clock pulses of the crystal oscillator. Crystal oscillators in DIP packages, which operate from a 5-V supply and have TTL compatible outputs, are commercially available. For inexpensive oscillators, typical crystal frequencies range from 4 to 20 MHz.

---

**EXAMPLE 8-1**

The basic timing for a system comes from a 1-MHz crystal controlled oscillator. At a point in the circuit a 500-Hz clock is required. What is the capacity of the counter and how many FF stages will be required to build it?

**SOLUTION**

To reduce 1 MHz to 500 Hz a divide-by-2000 counter is required (1 MHz ÷ 500 = 2000). Since 11 is the smallest number such that $2^{11} \geq 2000$, at least 11 stages will be required for a divide-by-2000 counter. It is now specified and can be designed by using the principles developed later in this chapter.

## 8-5 RIPPLE COUNTERS

Ripple counters are the simplest counters to build; consequently, they are very commonly used. The circuit of Fig. 6-23 is a 3-bit ripple counter. It is called a ripple counter because the clock ripples through the counter stage by stage. Whether stage two changes depends on the output of stage one, and whether the stage three changes depends on the output of stage two. (This limits the speed of the counter or the frequency at which the counts to be counted may occur, and is the main disadvantage of a ripple counter.) For a K-bit counter, there are some counts for which all K stages change. There must be sufficient time for the clock to ripple through the K stages, plus any additional gating and decoding time, before the next clock pulse can occur.

**EXAMPLE 8-2**

How fast can an 11-stage ripple counter built of **74107**s be clocked? Allow an additional 60 ns for decoding or gate delays.

**SOLUTION**

We must first determine the propagation delay for a single **74107**. This can be found from the manufacturer's specifications. From page 6-47 of the TTL *Data Book* the worst case propagation delay from the clock to the $Q$ or $\bar{Q}$ output is seen to be 40 ns. To produce a reliable output the circuit will require

$$11 \text{ stage delays} + 1 \text{ gate delay} = \text{total delay}$$
$$11 \text{ stages} \times 40\text{ns/stage} + 60 \text{ ns} = 500 \text{ ns}$$

Thus the events to be counted may not be closer than 500 ns apart and the clock frequency is limited to 2 MHz.

### 8-5.1 Decoding Specific Counts

Digital systems are often required to generate a pulse when a counter reaches a specified number. A *decoder* is a circuit that can produce this pulse. The simplest way to design a decoder is to AND together the $Q$ outputs of all stages that are a 1 and the $\bar{Q}$ outputs of all stages that are a 0 for the required count. An example of a 4-stage counter decoding a count of 6 is shown in Fig. 8-1. For the moment ignore the dashed line. Since 6 is binary 0110, the $\bar{Q}$ outputs of stages 0 and 3 and the $Q$ outputs of stages 1 and 2 are ANDed together. The output of the NAND gate is low only when the counter contains a binary 6.

Counters are most often decoded at counts of 0 or when they reach their maximum count (all 1s). When the count is 0, the $\bar{Q}$ outputs are all HIGH. Therefore, to decode a count of 0, the $\bar{Q}$ outputs of each stage are connected to a NAND gate decoder, while to decode the maximum count the $Q$ outputs of each stage are connected to a NAND gate decoder.

Counters that are connected to decoders often produce glitches at the outputs of the decoders. This occurs because the clock edge causes the FFs to change state, and not all FFs change at precisely the same time. Therefore, some false states may exist for a few nanoseconds before the transition is complete.

The circuit of Fig. 8-1 was connected in the laboratory and the results are shown in Fig. 8-2. The top CRO trace is the clock input. The most significant stage of the counter is shown on trace 2 (it is a 1 for counts 8 through 15); the second most significant stage is shown in trace 3. Trace 4 shows the decoder output. It is LOW at a count of 6, but a glitch can be seen clearly when the count changes from 7 to 8. Whether these unwanted glitches cause a real problem depends on the rest of the circuit. If the glitch is applied to the CLOCK, PRESET, or CLEAR of a FF, it could cause the FF to change state, which might be very troublesome. The glitch of Fig. 8-2 was tested and found strong enough to direct clear a FF. The causes and effects of glitches are considered further in Chapter 10.

Counters can be made glitchfree by tying the clock to the input of the decoder as shown by the dashed line of Fig. 8.1. If the dashed line is connected, the negative edge of the clock cuts off the decoder before the

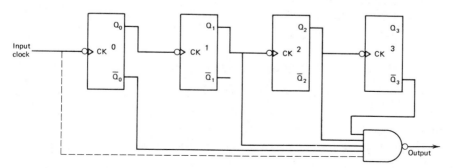

FIGURE 8-1 A 4-bit ripple counter with a decoder for a count of 6.

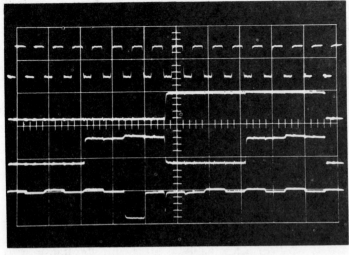

**Scales:**   **Vertical  5 V/cm**

**Horizontal   1 μs/cm**

FIGURE 8-2   CRO traces for a 4-bit counter and the output of a decoder for a count of 6. Vertical scale 5 V/cm; horizontal scale 1 μs/cm.

glitches occur. Now the decoder cannot function until the clock goes HIGH, but if this delay in the output is acceptable, a glitchfree output is obtained.

## 8-6  SYNCHRONOUS COUNTERS

Synchronous counters are often used to reduce the time delay and glitch problems associated with ripple counters. A synchronous counter has one common clock that is connected to all FFs; consequently, all FFs change within one propagation delay time regardless of the number of stages. Also, with all stages changing together, glitches are greatly reduced because of the shorter transition time interval. Glitches are not completely eliminated because one stage may respond to the same clock faster than another stage.

### 8-6.1  K-Bit Synchronous Counters

As in ripple counters, the simplest K-bit synchronous counter counts to N, where $N = 2^K$. A synchronous counter is designed in accordance with the following principle:

*The Mth stage toggles only on clock pulses that occur when all less significant stages are one.*

Consider, for example, counting numbers in binary. The least significant bit changes on each count. The second least significant bit counts 00110011 . . . . It changes on counts that occur after the least significant bit is a 1. The third least significant stage counts four 0s, four 1s, four 0s, and so on. It changes at counts of 4, 8, 12, . . . or on the pulses *following* counts of 3, 7, 11, . . . . But 3, 7, 11, . . . are exactly those counts where the two least significant bits are both 1. This conforms to the principle stated above.

---

**EXAMPLE 8-3**

Design a divide-by-16 synchronous counter using **74107s**.

**SOLUTION**

The counter was designed as follows.

1.  Since there are 16 states and $16 = 2^4$, at least 4 FFs are required.
2.  Because this is a synchronous counter the clocks of all stages must be connected together. The circuit and timing chart for the 4-stage synchronous counter is shown in Fig. 8-3. To develop this circuit, first the four FFs were drawn, then the common clock was sketched in.
3.  The first stage of the counter toggles on each clock pulse, so $J_1$ and $K_1$ are tied high.
4.  According to the synchronous principle, the second stage toggles only when the first stage is a 1. This is accomplished by tying $Q_1$ to the J and K inputs of stage 2.
5.  Stage 3 toggles only when $Q_1$ and $Q_2$ are both HIGH. $Q_1$ and $Q_2$ are ANDed together at point A, which is wired to $J_3$ and $K_3$.
6.  FF4 cannot toggle unless $Q_1$, $Q_2$, and $Q_3$ are all HIGH. The second AND gate is used to produce this signal at point B ($J_4$ and $K_4$).

---

## 8-6.2  Timing Synchronous Counters

In a synchronous counter[1] all FFs toggle simultaneously but sufficient time must be allowed for the gates to establish the proper J and K levels on all FFs before the next pulse can occur.

The maximum propagation delay time for the synchronous counter of Fig. 8-3 may be calculated by adding the two delays caused by the **7408s** to the propagation time for the FF, plus the decoding time. The worst case propagation delay time for a **7408** is 27 ns (from the manufacturer's

---

[1]This section contains advanced material and may be omitted without loss of continuity.

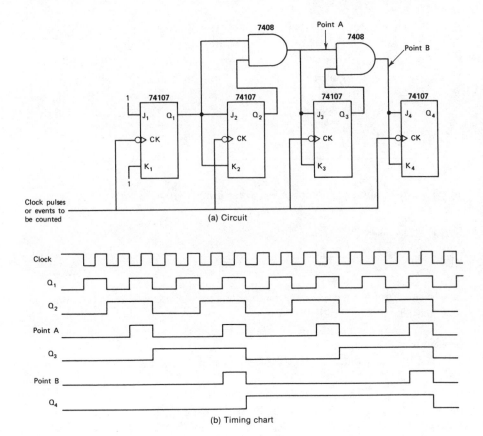

FIGURE 8-3   A 4-bit synchronous counter and timing chart.

specifications), and if the decoding time is again assumed to be 60 ns the propagation delay for the counter is:

One FF propagation delay + two 7408 delays + decoder delay
= total delay 40 ns + 54 ns + 60 ns = 154 ns

This compares to a worst case delay of 220 ns for a 4-stage ripple counter.

To increase the number of stages of the synchronous counter, we have two choices. The first choice is to connect another 7408 to point B of Fig. 8-3a to AND $Q_4$ with $Q_1$, $Q_2$, and $Q_3$. By adding an AND gate for each additional stage the amount of gates required remains reasonable. However, this adds 27 ns per stage to the overall circuit propagation delay. This approach saves hardware at the expense of delay time.

The second approach is to tie all the less significant stages to a larger NAND gate, invert the output, and tie it to the J and K inputs of the next

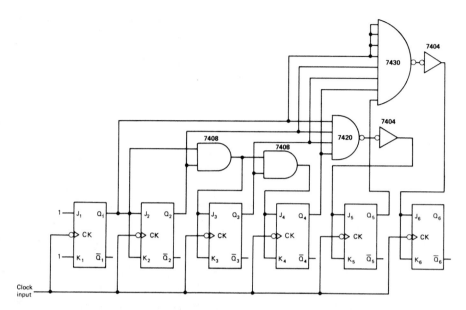

FIGURE 8-4  A 6-stage synchronous counter. *Note:* All FFs are **74107**s.

stage. This is shown in Fig. 8-4 where all inputs have a maximum of two gate delays.

For the counter of Fig. 8-4 the maximum delay is

$$\text{One FF propagation delay} + \text{two gate delays} + \text{decoder delay}$$
$$= \text{total delay}$$
$$40 \text{ ns} + 54 \text{ ns} + 60 = 154 \text{ ns}[2]$$

This delay is now independent of the number of stages. Using a single 7430 per stage, up to 3 additional stages can be accommodated. If more than 9 stages are required the outputs of the NAND gates can be fed to NOR gates, to produce the required logic without introducing more than two gate delays. As the number of stages increases, the number of gates required becomes formidable. This approach saves time at the expense of hardware.

For slow speed circuits, where speed is no problem, ripple counters are usually satisfactory. For higher speed circuits, we must accept the additional gating and complexity of synchronous counters, and for very high speed circuits, the designer is forced to Schottky or ECL. Choosing the optimum

[2]NAND gates and inverters are slightly faster than AND gates. If they were used instead of the 7408s, the worst case delay would drop to 144 ns.

compromise between speed, cost, and complexity is one of the decisions engineers often make.

## 8-7 THE 3s COUNTER

If a certain clock frequency is needed in a circuit, it is necessary to divide the master oscillator by the number required to obtain the desired clock frequency. Therefore, we must be able to design circuits that count to, or divide by, any given number.

Divide-by-N circuits where $N = 2^K$ have already been discussed. When N is not equal to $2^K$ the problem is more complex. The smallest such number is 3. Fortunately, a 3s counter or divide-by-3 circuit is easy to build and requires no additional gating. The circuit and timing chart of a 3s counter are shown in Fig. 8-5. By connecting $\overline{Q}_2$ to $J_1$, FF1 cannot toggle when $Q_2 = 1$, which causes the circuit to divide by 3. Note that the $K_1$ input could be connected either to $J_1$ or clamped HIGH. This circuit also gives us the capability of building a divided-N counter if N can be expressed as $3 \times 2^K$.

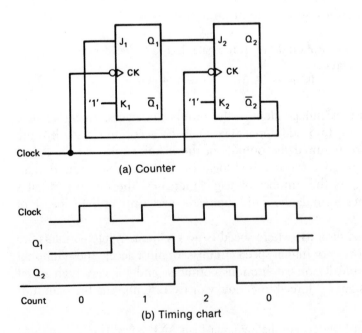

(a) Counter

(b) Timing chart

FIGURE 8-5   A 3s counter and its timing chart.

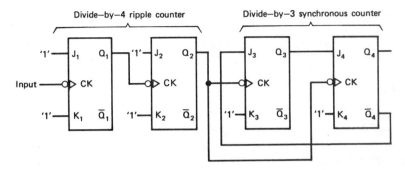

FIGURE 8-6  A divide-by-12 counter.

## EXAMPLE 8-4

Design a divide-by-12 counter using the minimum number of FFs and no external gates.

### SOLUTION

The solution proceeds as follows.

1.  Since 4 is the smallest number such that $2^4 > 12$, four FFs are required.
2.  The fact that $12 = 3 \times 2^2$ suggests that a 2-stage counter and a 3s counter solve the problem.
3.  If the count is to go from 0 to 11 the 3s counter must be placed at the most significant stages.
4.  The circuit is shown in Fig. 8-6 where a divide-by-4 counter precedes a divide-by-3. The circuit is a hybrid because the count ripples through the divide-by-4 counter but the output of the divide-by-4 acts as a synchronous input to the divide-by-3 stages.

## 8-8  IRREGULAR AND TRUNCATED COUNT SEQUENCES[3]

Synchronous circuits requiring irregular count sequences, or truncated counters (a truncated counter is a K stage counter that counts 0, 1, . . . , n, *where n is not equal to* $2^K$), can be designed using Karnaugh maps. Before attacking the problem of designing counters for an arbitrary count or sequence, however, let us analyze a counter that has an irregular count sequence. An irregular count sequence is a sequence that does not proceed 0, 1, 2, . . . , n, but allows skips and jumps in the counting sequence.

[3]This section contains advanced material and can be omitted without loss of continuity.

## EXAMPLE 8-5

Determine the count sequence of the counter of Fig. 8-7a. Assume $Q_3$ is the most significant digit and the count starts at 0.

## SOLUTION

Each condition of the FFs is called a *state*. Each state is given a number that corresponds to the count, read as the binary number $Q_3Q_2Q_1$. To determine the progress of the counter as each clock pulse occurs, a *state checkout table* is drawn as shown in Fig. 8-7b. This is a systematic way to attack the problem. The step-by-step procedure is as follows.

1.   The problem specifies that the count starts at 0; therefore, 0 becomes the first entry in the present state column of the table. The corresponding values of $Q_3Q_2Q_1$ are written as 000.

2.   Next the J and K values that depend upon the present values of $Q_1$, $Q_2$, and

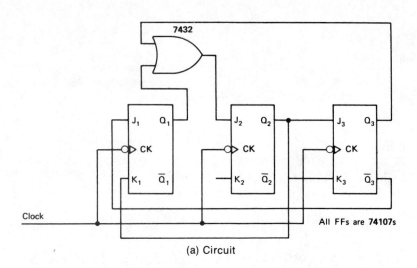

(a) Circuit

| Present State | $Q_3 Q_2 Q_1$ | $J_3 K_3$ | $J_2 K_2$ | $J_1 K_1$ | Next State $Q_3 Q_2 Q_1$ | Numerical Next State |
|---|---|---|---|---|---|---|
| 0 | 0 0 0 | 0 0 | 0 1 | 1 0 | 0 0 1 | 1 |
| 1 | 0 0 1 | 0 0 | 1 1 | 1 0 | 0 1 1 | 3 |
| 3 | 0 1 1 | 1 1 | 1 1 | 1 1 | 1 0 0 | 4 |
| 4 | 1 0 0 | 0 0 | 1 1 | 0 0 | 1 1 0 | 6 |
| 6 | 1 1 0 | 1 1 | 1 1 | 0 1 | 0 0 0 | 0 |

(b) State checkout table

FIGURE 8-7   Circuit and state checkout table for Ex. 8-5.

$Q_3$ are determined from the circuit diagram and written on the top line of the table. In this counter we see that $J_3$, $K_3$, and $K_1$ are connected to $Q_2$. Therefore, they are all 0 because $Q_2$ is initially 0. $J_1$ is connected to $\overline{Q}_3$. Since $Q_3$ is initially 0, $\overline{Q}_3$ and $J_1$ are 1. $J_2$ is 0 because both inputs to the OR gate are 0.

3.    Once all the Js and Ks are listed, the next state can be determined. Since $J_3$ and $K_3$ are both 0, FF3 will not change state. It is currently a 0 and will remain a 0. With $J_2$ a 0 and $K_2$ a 1, FF2 will also remain a 0. FF1 sets, however, because $J_1 = 1$ and $K_1 = 0$. The state after the next clock pulse is 001 or binary 1. This completes the first line of the checkout state table.

4.    The second line is started by listing the *next state on the first line as the present state on the second line*. This state is 1 and the Js and Ks for this state are calculated. They are the same as on the first line except for $J_2$, which is a 1 because the $Q_1$ input to the OR gate is now 1. This causes FF2 to toggle to a 1 after the next clock pulse, and the counter goes to state 3.

5.    Now the present state of the third line is written as a binary 3 ($Q_3Q_2Q_1 = 011$). If we use these values of Qs, all the Js and Ks become 1s and all the FFs toggle. Therefore, $Q_3$ changes to a 1 and $Q_1$ and $Q_2$ toggle to 0s, making the next state 4.

6.    The table is continued in this manner until a *next state is reached that is the same as a previous present state*. In this problem when 6 is the present state, the next state is 0, so the counter has rolled around and the count sequence has been found to be 0, 1, 3, 4, 6, 0, . . . . This is a divide-by-5 counter with an irregular sequence.

7.    There can never be more than eight steps in this count sequence because the counter contains only three FFs. If a sequence of 9 to 16 steps were needed, an additional FF would be required.

## 8-8.1    Design of a Truncated Counter

Before a counter for an irregular or truncated count sequence can be designed, the basic transition table for the type of FF to be used must be constructed. The basic transition table for the J-K FF is shown in Fig. 8-8, where $Q(t)$ is the state of a FF before the clock pulse and $Q(t + 1)$ is the desired state of the FF after the transition. The states of the J and K

| $Q(t)$ | $Q(t + 1)$ | J | K |
|--------|-----------|---|---|
| 0 | 0 | 0 | $d$ |
| 0 | 1 | 1 | $d$ |
| 1 | 0 | $d$ | 1 |
| 1 | 1 | $d$ | 0 |

FIGURE 8-8    The basic transition table for J-K FFs.

inputs required to cause the transition are also given in the table. For example, if the present state of a FF is a 0, and the next state is to be a 0, the table shows that J must be a 0 but K is a don't care. If J is a 0, the FF will remain cleared regardless of the K input. Once the desired states of a counter are specified the counter is designed as follows.

1.  A state table for the counter is constructed.
2.  The J and K inputs required for each FF are found and Karnaugh maps for the realization of each input are constructed.
3.  The circuit to be built is drawn and should be checked out on paper using a state checkout table.

This procedure is best illustrated by examples.

---

### EXAMPLE 8-6

Design a divide-by-5 counter using 74107s and check it using a state checkout table.

### SOLUTION

The solution proceeds as follows.

1.  Since there are to be five separate states, three FFs are needed because three is the smallest number, N, such that $2^N \geq 5$. There are many ways to choose a sequence of five states. We have already seen an irregular count sequence in Ex. 8-5. Here we shall choose the truncated count sequence, 0, 1, 2, 3, 4, 0, . . . .
2.  A state table must be drawn up as shown in Fig. 8-9a. The three FFs are designated as $Q_3$, $Q_2$, and $Q_1$, where $Q_1$ is the least significant bit. All possible states of the three FFs are listed in column 1, and their corresponding Q outputs are listed in columns 2, 3, and 4 just as they would be written in a truth table.
3.  The desired next state, which is the state that must follow the present state, is listed in column 5, and the corresponding FF outputs are then listed in columns 6, 7, and 8. For example, if the present state of the counter is 0 ($Q_3Q_2Q_1 = 000$), the next state must be 1 ($Q_3Q_2Q_1 = 001$) as specified by the given count sequence, or if the present state is 4, the next state must be 0.
4.  Since the count sequence goes from 0 through 4, the counter should never be in states 5, 6, or 7. The next state for these present states is arbitrary. To simplify the design, the next states are listed as don't cares.
5.  Columns 9 through 14, containing the required J and K conditions to cause each transition, can now be filled in. Here, use is made of the J-K FF transition table, Fig. 8-8. When the counter goes from 0 to 1, $Q_2$ and $Q_3$ do not change, but $Q_1$ changes from 0 to 1. Therefore, $J_3$ and $K_3$, and $J_2$ and $K_2$, both have values of 0 and d, respectively, since the J-K transition table indicates these are the values J and K must have for a FF in the 0 state to remain in the 0 state. But $Q_1$ SETS, so that $J_1$ must be a 1 while $K_1$ can be either 1 or 0.

| 1 | 2 3 4 | 5 | 6 7 8 | 9 10 | 11 12 | 13 14 |
|---|---|---|---|---|---|---|
| Present State | $Q_3$ $Q_2$ $Q_1$ | Next State | $Q_3$ $Q_2$ $Q_1$ | $J_3$ $K_3$ | $J_2$ $K_2$ | $J_1$ $K_1$ |
| 0 | 0 0 0 | 1 | 0 0 1 | 0 $d$ | 0 $d$ | 1 $d$ |
| 1 | 0 0 1 | 2 | 0 1 0 | 0 $d$ | 1 $d$ | $d$ 1 |
| 2 | 0 1 0 | 3 | 0 1 1 | 0 $d$ | $d$ 0 | 1 $d$ |
| 3 | 0 1 1 | 4 | 1 0 0 | 1 $d$ | $d$ 1 | $d$ 1 |
| 4 | 1 0 0 | 0 | 0 0 0 | $d$ 1 | 0 $d$ | 0 $d$ |
| 5 | 1 0 1 | ? | $d$ $d$ $d$ | $d$ $d$ | $d$ $d$ | $d$ $d$ |
| 6 | 1 1 0 | ? | $d$ $d$ $d$ | $d$ $d$ | $d$ $d$ | $d$ $d$ |
| 7 | 1 1 1 | ? | $d$ $d$ $d$ | $d$ $d$ | $d$ $d$ | $d$ $d$ |

(a) State table

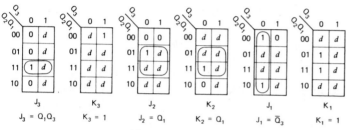

(b) Karnaugh maps

$J_3 = Q_1 Q_3$    $K_3 = 1$    $J_2 = Q_1$    $K_2 = Q_1$    $J_1 = \overline{Q}_3$    $K_1 = 1$

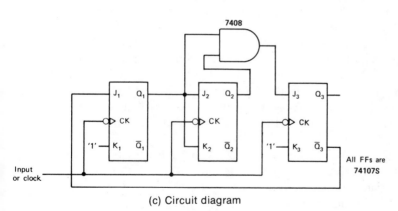

(c) Circuit diagram

| Line No. | Present State | $Q_3$ $Q_2$ $Q_1$ | $J_3$ $K_3$ | $J_2$ $K_2$ | $J_1$ $K_1$ | Next State $Q_3$ $Q_2$ $Q_1$ | Numerical Next State |
|---|---|---|---|---|---|---|---|
| 1 | 0 | 0 0 0 | 0 1 | 0 0 | 1 1 | 0 0 1 | 1 |
| 2 | 1 | 0 0 1 | 0 1 | 1 1 | 1 1 | 0 1 0 | 2 |
| 3 | 2 | 0 1 0 | 0 1 | 0 0 | 1 1 | 0 1 1 | 3 |
| 4 | 3 | 0 1 1 | 1 1 | 1 1 | 1 1 | 1 0 0 | 4 |
| 5 | 4 | 1 0 0 | 0 1 | 0 0 | 0 1 | 0 0 0 | 0 |

(d) State checkout table

FIGURE 8-9   Design of a divide-by-5 counter.

6.    To go from state 1 to state 2 (the second line in the table), $Q_1$ must CLEAR, $Q_2$ must SET, and $Q_3$ must remain CLEARED. The J and K inputs required to cause these transitions are listed in columns 9 through 14 of line 2. The rest of the J and K inputs were also obtained by going from the present state to the next state in accordance with the J-K transition table. Where the next state is a don't care, all the J and K inputs are written as don't cares.

7.    At this point, six Karnaugh maps must be drawn (see Fig. 8-9b), one for each J and K input. The Q variables in the Karnaugh maps correspond to the present state table. Since $Q_3$ is the most significant variable, it is written across the top of the map, and the $Q_2Q_1$ variables are written down the side in order of significance. For example, $J_3$ is a 0 for present states 0, 1, and 2, a 1 for state 3, and a don't care otherwise. These values are placed in the $J_3$ Karnaugh map and the minimal simplification results in the equation: $J_3 = Q_1Q_2$.

8.    Fortunately, many don't cares simplify the Karnaugh maps and the circuitry. The minimal expression for each J and K input is obtained and the six equations are written below each Karnaugh map. Where the Karnaugh map contains no 0s ($K_3$ and $K_1$ in this example), the J or K input should be permanently HIGH or 1.

9.    The circuit is drawn as shown in Fig. 8-9c. It is started by drawing the three FFs and the clock. Then the J and K connections are made in accordance with the six equations obtained from the Karnaugh maps, and the circuit is complete. For this counter only $J_3$ required a gate. All other Js and Ks were either connected to 1s or to another Q or $\overline{Q}$ output.

10.    After the circuit is drawn it should be checked out on paper by constructing the state checkout table as shown in Fig. 8-9d. Since 0 is in the count sequence, it is convenient to start by assuming a present state of 0. (Any other number in the count sequence would also work.) Therefore, $Q_1$, $Q_2$, and $Q_3$ are all 0. Looking at the circuit diagram (Fig. 8-9c) we see that $J_2$, $K_2$, and $J_3$ are also 0 but that $J_1$, $K_1$, and $K_3$ are all 1s. The J and K values are now written into the state checkout table.

11.    The Q outputs for the next state can now be determined. Since $J_1$ and $K_1$ are both 1, FF1 will toggle to the 1 state. Since $J_2$ and $K_2$ are both 0, FF2 will not change state, and FF3 will also remain a 0 because $J_3$ is a 0 and $K_3$ is a 1. The next state Q outputs are found to be 001 so that the next state is 1.

12.    The second line of the checkout table is written using the next state from the line above. Thus the present state on the second line is 1, the next state result from line one. The J and K inputs for this state are found and a new next state is determined. This procedure continues until the circuit returns to its original state. The count found using the table should agree with the specification of the problem.

---

## 8-8.2   Design of Counters Having Irregular Count Sequences

The methods of Sec. 8-8.1 may also be used to design counters that must have highly irregular count sequences, as Ex. 8-7 illustrates.

## EXAMPLE 8-7

Design a counter to produce the following sequence of outputs: 0, 2, 4, 3, 6, 7, 0, . . . .

## SOLUTION

The solution proceeds as follows.

1. Since in this sequence there are six distinct counts, three FFs are required.
2. As before, the outputs of these FFs are labeled $Q_3$, $Q_2$, and $Q_1$ and the state table is constructed as shown in Fig. 8-10a.
3. All present state possibilities are listed in order, but note the irregular next state listings caused by the specifications.
4. The corresponding next state values of $Q_3$, $Q_2$, and $Q_1$ are then entered in the table and the Js and Ks calculated. Since states 1 and 5 are not in the count sequence, the Js and Ks for these states are don't cares.
5. The Karnaugh maps are drawn as shown in Fig. 8-10b, and the minimal equations obtained. Here $K_3$ requires an OR gate with $Q_1$ and $\bar{Q}_2$ as inputs, and $K_2$ appears to require two AND gates and an OR gate. For $K_2$, however, a little inspiration simplifies the circuit:

$$K_2 = \bar{Q}_1\bar{Q}_3 + Q_1Q_3 = \overline{(Q_1 \oplus Q_3)} = Q_1 \oplus \bar{Q}_3$$

Hence $K_2$ is the output of a 7486 gate whose inputs are $Q_1$ and $\bar{Q}_3$.
6. The circuit is then drawn as shown in Fig. 8-10c.
7. Finally the state checkout table is carefully constructed as shown in Fig. 8-10d. It verifies that the count sequence conforms to the specifications of the problem.

Synchronous counters can also be designed using D-type FFs by realizing that the Q outputs for the next state are the same as the present D inputs. Usually a circuit with D FFs requires more gates than the same circuit built with J-K FFs because there are more don't cares in the J-K transition table and the don't cares allow the designer more options and versatility. Consequently, this type of circuit is rarely designed with D FFs.[+]

## 8-9 IC COUNTERS

Because of the general need for counters, many are already packaged in TTL ICs. The three simplest counters available are the 7490, 7492, and 7493. They are a decade counter, a divide-by-12 and a divide-by-16 (binary) counter, respectively. Unfortunately, these counters have power on pin 5

[+]For a specific example, see the first edition of *Practical Digital Design Using ICs*.

| Present State | $Q_3\,Q_2\,Q_1$ | Next State | $Q_3\,Q_2\,Q_1$ | $J_3\ K_3$ | $J_2\ K_2$ | $J_1\ K_1$ |
|---|---|---|---|---|---|---|
| 0 | 0 0 0 | 2 | 0 1 0 | 0 $d$ | 1 $d$ | 0 $d$ |
| 1 | 0 0 1 | ? | $d$ $d$ $d$ | $d$ $d$ | $d$ $d$ | $d$ $d$ |
| 2 | 0 1 0 | 4 | 1 0 0 | 1 $d$ | $d$ 1 | 0 $d$ |
| 3 | 0 1 1 | 6 | 1 1 0 | 1 $d$ | $d$ 0 | $d$ 1 |
| 4 | 1 0 0 | 3 | 0 1 1 | $d$ 1 | 1 $d$ | 1 $d$ |
| 5 | 1 0 1 | ? | $d$ $d$ $d$ | $d$ $d$ | $d$ $d$ | $d$ $d$ |
| 6 | 1 1 0 | 7 | 1 1 1 | $d$ 0 | $d$ 0 | 1 $d$ |
| 7 | 1 1 1 | 0 | 0 0 0 | $d$ 1 | $d$ 1 | $d$ 1 |

(a) State table

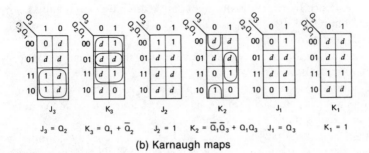

$$J_3 = Q_2 \qquad K_3 = Q_1 + \overline{Q}_2 \qquad J_2 = 1 \qquad K_2 = \overline{Q}_1\overline{Q}_3 + Q_1Q_3 \qquad J_1 = Q_3 \qquad K_1 = 1$$

(b) Karnaugh maps

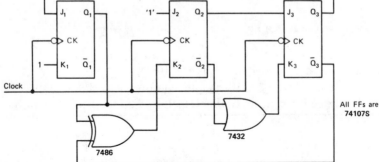

(c) Circuit diagram

| Present State | $Q_3\,Q_2\,Q_1$ | $J_3\ K_3$ | $J_2\ K_2$ | $J_1\ K_1$ | Next State $Q_3\,Q_2\,Q_1$ | Numerical Next State |
|---|---|---|---|---|---|---|
| 0 | 0 0 0 | 0 1 | 1 1 | 0 1 | 0 1 0 | 2 |
| 2 | 0 1 0 | 1 0 | 1 1 | 0 1 | 1 0 0 | 4 |
| 4 | 1 0 0 | 0 1 | 1 0 | 1 1 | 0 1 1 | 3 |
| 3 | 0 1 1 | 1 1 | 1 0 | 0 1 | 1 1 0 | 6 |
| 6 | 1 1 0 | 1 0 | 1 0 | 1 1 | 1 1 1 | 7 |
| 7 | 1 1 1 | 1 1 | 1 1 | 1 1 | 0 0 0 | 0 |

(d) State checkout table

FIGURE 8-10   Design of the counter of Ex. 8-7.

and ground on pin 10,[5] so care must be taken when wiring them into a circuit. Also, to some extent they are all ripple counters, and may produce glitches on decoder outputs.

## 8-9.1    The 7490 Decade Counter

The 7490 is a decade counter that actually consists of a single J-K FF and a divide-by-5 circuit. The chip pin configuration, functional block diagram, and count sequence are shown in Fig. 8-11. In the 7490, FF A functions as a divide-by-2 counter, and the other three FFs function as a divide-by-5 counter. The design of the divide-by-5 circuit is essentially the same as shown in Fig. 8-9. If the input signal is connected to input A, and the $Q_A$ output is connected to input B, a decade counter results and the output is given by the BCD count table of Fig. 8-11c.

The $R_{0(1)}$ and $R_{0(2)}$ inputs form a direct clear. The Q outputs of the 7490 are all LOW whenever both $R_{0(1)}$ and $R_{0(2)}$ are HIGH, as shown by the RESET/COUNT function table of Fig. 8-11c. The counter can also be direct SET to a count of 9 if both the $R_{g(1)}$ and $R_{g(2)}$ inputs are HIGH, which is a useful feature when performing decimal arithmetic. If the counter is to count normally, at least one of the $R_0$ inputs and one of the $R_g$ inputs must be held LOW.

---

**EXAMPLE 8-8**

Design a divide-by-25 counter using 7490s.

**SOLUTION**

Two divide-by-5 circuits in two 7490s can be cascaded, and the A FFs left unconnected, to build a divide-by-25 counter. The circuit is shown in Fig. 8-12. Note that the count sequence does not progress from 0 to 24. Note also that one of the $R_0$ and $R_g$ inputs on each IC are tied to ground.

---

## 8-9.2    The 7492 Divide-by-12 Counter

The 7492 is a divide-by-12 circuit that consists of a J-K FF followed by a divide-by-6 counter. The pinout, block diagram, and count sequence are shown in Fig. 8-13. FF A is simply a divide-by-2 circuit; FFs B and C form a divide-by-3 counter, exactly as shown in Fig. 8-5. FF D is another divide-by-2 circuit. A divide-by-12 counter is made by putting the signal into input

---

[5] Manufacturers now produce a 74290 and a 74293 that are identical to the 7490 and 7493 except for the pinout. Power and ground are on pins 14 and 7.

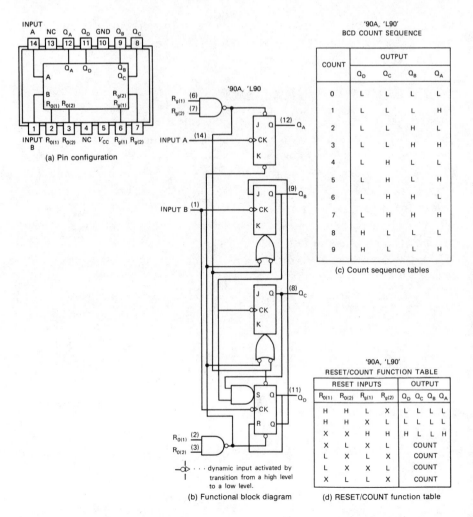

FIGURE 8-11 The **7490** decade counter. (From the *TTL Data Book for Design Engineers,* 2nd ed., Texas Instruments, Inc. Courtesy of Texas Instruments, copyright 1976.)

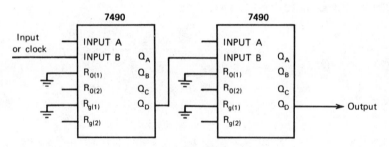

FIGURE 8-12 A divide-by-25 counter using **7490**s.

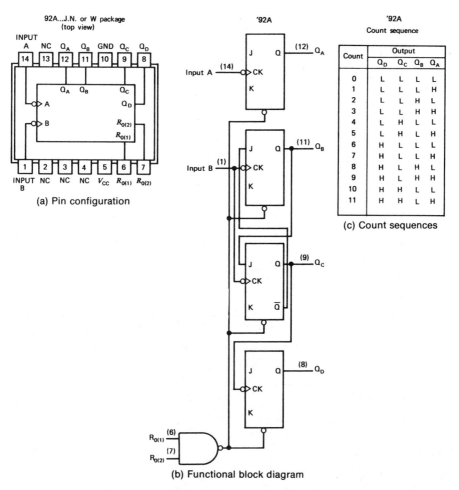

(a) Pin configuration

(b) Functional block diagram

(c) Count sequences

FIGURE 8-13 The **7492** divide-by-12 counter. (From the *TTL Data Book for Design Engineers*. Texas instruments, Inc. Courtesy of Texas Instruments, copyright 1976.)

A and connecting output $Q_A$ to input B. As in the **7490**, $R_{0(1)}$ and $R_{0(2)}$ are ANDed together to form a direct clear. At least one of these inputs must be LOW in order for the IC to count.

An examination of the count sequence table for the **7492** connected as a divide-by-12 counter (Fig. 8-13c) reveals that it does not count in a 0 to 11 sequence. This is because the 3s counter ($Q_B$ and $Q_C$) is before the final J-K FF ($Q_D$). Actually it counts the binary equivalent of 0 through 5 followed by 8 through 13. However, the $Q_D$ output is symmetric when a continuous pulse train is applied, which is an advantage in some circuits.

---

### EXAMPLE 8-9

The basic oscillator for a system runs at 1.2 MHz. Design a circuit that will produce a symmetric 2000-Hz square wave from the basic system clock.

### SOLUTION

Since 1.2 MHz ÷ 2000 = 600, a divide-by-600 counter is needed. It can be built by using two divide-by-10 counters followed by a divide-by-6 counter, which suggests two 7490s and the divide-by-6 portion of a 7492. The circuit is shown in Fig. 8-14. There are also other ways to design this circuit (see Problem 8-19).

---

## 8-9.3   The 7493 4-Bit Binary Counter

The 7493 is a simple 4-bit binary counter, consisting of a single FF, $Q_A$, followed by three cascaded FFs that form a divide-by-8 counter. The pin layout, block diagram, and count sequence are shown in Fig. 8-15. If $Q_A$ is connected to input B, the circuit forms a divide-by-16 ripple counter. It counts in a straight-forward binary sequence, as the count sequence table (Fig. 8-15c) indicates. Again the $R_{0(1)}$ and $R_{0(2)}$ inputs are ANDed together to form a direct CLEAR for all four stages of the counter.

---

### EXAMPLE 8-10

Design a divide-by-128 counter.

### SOLUTION

This counter could be built using seven stages from four 74107s. Similar counters were designed in Chapter 6. Less ICs are needed, however, if 7493s are used. This approach is shown in Fig. 8-16, where the first 7493 is used as a 4-stage counter and cascaded to the 3-stage counter of the second 7493.

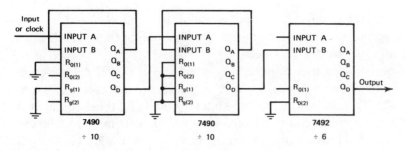

FIGURE 8-14   A divide-by-600 counter.

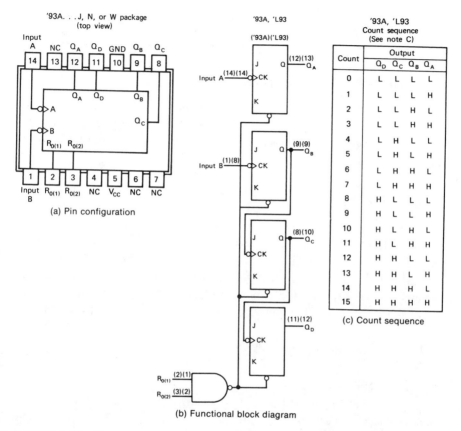

FIGURE 8-15 The **7493** 4-stage binary counter. (From *the TTL Data Book for Design Engineers,*Texas Instruments, Inc. Courtesy of Texas Instruments, copyright 1976.)

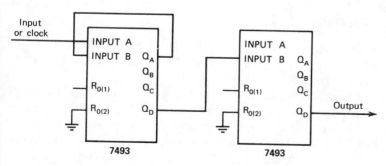

FIGURE 8-16 A divide-by-128 counter using **7493**s.

An alternate solution is to connect both 7493s as 4-stage counters and take the output from $Q_C$ of the second 7493.

---

## 8-10 UP-DOWN COUNTERS

*Up-down counters* are counters that can either count up or count down depending on the mode of the input. Two of the most popular 4-bit up-down counters available are the **74192**, which is a decade counter, and the **74193**, which is a binary counter.

### 8-10.1 The 74193

The **74193** is a synchronous, 4-bit, up-down counter with direct CLEAR and direct LOAD capabilities. This IC contains many useful features, which will be described in detail. The block diagram of the **74193** is shown in Fig. 8-17 with the pin numbers for each input and output signal in parenthesis. The T on each FF stands for toggle and the bubble indicates the FF toggles on the negative-going edge of a pulse. This is equivalent to a J-K FF where J = K = 1.

There are six outputs available from a **74193**:

1.  The CARRY output.
2.  The BORROW output.
3.  The four FF outputs—$Q_A$, $Q_B$, $Q_C$, and $Q_D$—which retain the count. $Q_D$ is the most significant bit of the count.

The inputs to the **74193** consist of:

1.  A CLEAR line.
2.  A LOAD input.
3.  Four data inputs (A,B,C,D).
4.  A COUNT-UP input.
5.  A COUNT-DOWN input.

The function of each input and output is explained below and is best understood by referring to Fig. 8-17.

1.  The CLEAR input is an asynchronous input that causes the count to be 0 (resets all the FFs) whenever it is HIGH. The CLEAR input should be wired to ground if it is not to be used.
2.  The LOAD input is a negative-active signal that will load data inputs A through D into the corresponding FFs whenever it is LOW. The CLEAR input dominates the LOAD. If they are both active simultaneously, the counter will CLEAR.
3.  When the CLEAR and LOAD signals are inactive, the counter will increment on each positive edge of the COUNT-UP line provided the COUNT-DOWN line is HIGH.

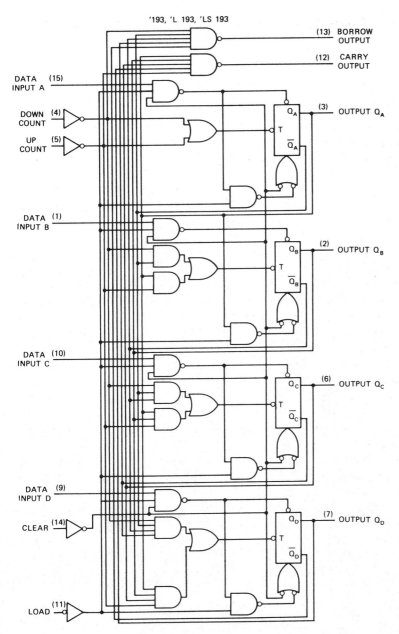

FIGURE 8-17 Block diagram of the **74193.**(From the *TTL Data Book for Design Engineers,* 2nd ed., Texas Instruments, Inc. Courtesy of Texas Instruments, copyright 1976.)

4.  The counter will decrement on each positive edge of the COUNT-DOWN input if the COUNT-UP line is HIGH. If either count input is held LOW the other count input is inhibited.

5.  The CARRY output goes LOW only if the counter reaches 15 (all Q outputs are 1) and the COUNT-UP input is LOW.

6.  The BORROW output goes LOW only if the counter is at 0 and the COUNT-DOWN line goes LOW.

The CARRY and BORROW outputs are used when cascading several **74193**s to accommodate a larger count than a single IC can handle. To cascade stages of an UP-COUNTER, the CARRY output of the least significant **74193** must be connected to the COUNT-UP input of the next **74193,** as shown in Fig. 8-18a. If, for example, the count in IC1 is 15, the counter operates as follows.

1.  When the COUNT-UP line goes LOW, the CARRY-OUT line, which is connected to the COUNT-UP line of IC2, also goes LOW.

2.  The next positive transition of the COUNT-UP line causes IC1 to roll over from all 1s to all 0s.

3.  This, in turn, causes the CARRY-OUT line to go HIGH, applying a COUNT-UP pulse to IC2.

4.  Therefore, after the count pulse, IC1 is 0 and IC2 contains a count of 1. The output of the counter is now:

$$\underbrace{0000}_{\text{IC3}} \quad \underbrace{0001}_{\text{IC2}} \quad \underbrace{0000}_{\text{IC1}}$$

This is binary 16.

5.  IC2 does not increment again until IC1 rolls over once again.

IC3 receives a COUNT-UP pulse only when both IC1 and IC2 have outputs which are all 1s. Then the counter functions as follows.

1.  When the input goes LOW, the CARRY output of IC1 goes low.

2.  IC2 now has a low input on its COUNT-UP line and, since it contains all 1s, it produces a low level at its CARRY output. This is connected to the COUNT-UP line of IC3.

3.  When the input again goes HIGH, IC1 and IC2 both receive a COUNT-UP pulse, which rolls them over to all 0s.

4.  IC3 also receives a COUNT-UP pulse. Therefore, IC3 increments once for every 256 input pulses.

When it is counting down, a **74193** rolls over from all 0s to all 1s if it receives a DOWN-COUNT pulse when its count is 0. DOWN-COUNTERS may be cascaded by connecting the BORROW-OUT line to the COUNT-

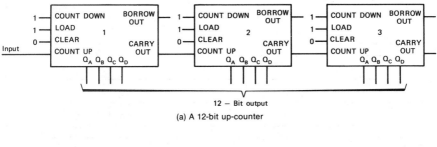

(a) A 12-bit up-counter

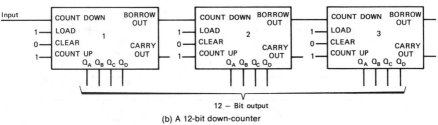

(b) A 12-bit down-counter

FIGURE 8-18   12-bit counters made by cascading **74193**s.

DOWN line of the next IC, as shown in Fig. 8-18b. Assume, for example, that the count in the DOWN-COUNTER is 48, so that IC1 contains a 0 and IC2 contains a 3. This makes the count:

$$\underbrace{0000}_{\text{IC3}} \quad \underbrace{0011}_{\text{IC2}} \quad \underbrace{0000}_{\text{IC1}}$$

Where the four least significant bits are in IC1, the next four bits are in IC2, and the most significant bits are in IC3. The circuit now operates as follows.

1.   When the DOWN-COUNT line goes LOW, the BORROW output of IC1 and the DOWN-COUNT input to IC2 also go LOW.
2.   When the DOWN-COUNT line goes HIGH again, IC1 rolls over to all 1s.
3.   The BORROW output of IC1 now goes HIGH, causing IC2 to receive a DOWN-COUNT pulse and decrement to 2.
4.   The count now reads:

$$\underbrace{0000}_{\text{IC3}} \quad \underbrace{0010}_{\text{IC2}} \quad \underbrace{1111}_{\text{IC1}}$$

This is the binary equivalent of 47.

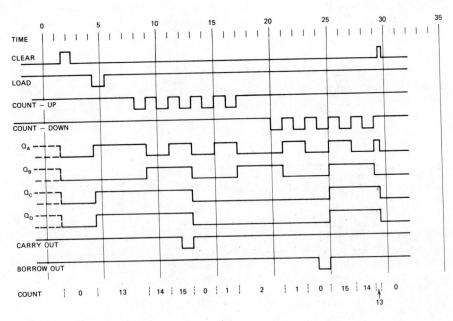

FIGURE 8-19  Timing chart for a **74193**.

If all stages of a DOWN-COUNTER contain 0s, the next downcount will roll the entire counter over to all 1s, but a BORROW-OUT pulse will appear at the output of the most significant IC to flag the event.[6]

The timing chart of Fig. 8-19 has been prepared to illustrate all the features of the **74193**.

1.  The count is undetermined until $t = 1.5$, when the leading edge of the CLEAR pulse resets all the Q outputs to 0.

2.  At $t = 4.5$ a LOAD pulse occurs. The inputs (not shown) are assumed to be 13 (input A = 1, B = 0, C = 1, D = 1) and are transferred to the outputs, making the count 13. The A, B, C, and D inputs have no effect unless the LOAD input is LOW.

3.  The LOAD pulse is followed by a series of COUNT-UP pulses. The positive transitions of these pulses occur at $t = 9, 11, 13, 15,$ and $17$ and increment the counter to 14, 15, 0 (rollover), 1, and 2.

4.  For $t$ between 12 and 13 the count is 15 and the COUNT-UP line is LOW, so that a CARRY-OUT pulse is generated.

[6]In 2s complement arithmetic (see Sec. 14-5), an array of all 1s represents the number $-1$. Therefore, if the counter is at 0 and it receives a DOWN-COUNT pulse, it will roll over to all 1s or the number $-1$. This is very convenient for many applications using 2s complement arithmetic.

5.  A series of COUNT-DOWN pulses follows, with positive transitions at $t =$ 21, 23, 25, 27, and 29. This causes the count to decrement from 2 to 1, 0, 15 (rollover), 14, and then 13.

6.  A BORROW-OUT pulse is generated when t is between 24 and 25, where the count is 0 and the COUNT-DOWN level is LOW.

7.  A positive pulse or spike occurs on the CLEAR line at $t = 29.5$, and RESETS all the outputs.

## 8-10.2   The 74192 Decade Counter

The 74192 is a decade counter, which is identical to the 74193 in all other respects. All input and output signals are the same, and they are placed on the same pins. Since the 74192 is a decade counter, it rolls over from 9 to 0 when counting up and from 0 to 9 when counting down. CARRY-OUT and BORROW-OUT signals are generated when the count is 9 or 0 and the proper count line goes LOW. The 74192 is a useful counter for circuits that work on a decimal base.

## 8-11   DIVIDE-BY-N CIRCUITS USING COUNTERS

Divide-by-N circuits, where $N = 2^K$, can be constructed using the methods of Sec. 8-8, but for N large the procedure becomes cumbersome and un-wieldy. Fortunately, it is not difficult to design a divide-by-N circuit using the counters described in Sec. 8-10.

### 8-11.1   Counter Using 7490s, 7492s, and 7493s

Divide-by-N counters can be constructed from 7490s, 7492s, or 7493s by using their direct CLEARS and a minimum amount of gating. The general procedure is to decode the number N and use it to clear the counter. This type of counter generates a glitch that clears itself and the count is N only for the duration of the glitch, after which it is 0. The output is usually taken from the most significant stage of the counter that changes, and is usually not symmetric. Nevertheless, it produces one output pulse for every N input pulses and satisfies the specifications. If N is an even number, a symmetric output can always be obtained by building an N/2 counter and following it by a divide-by-2 (J-K) FF.

---

**EXAMPLE  8-11**

(a) Design a divide-by-87 counter using 7490s. What will the output look like?

(b) If the input is a constant clock of 200 kHz, calculate the time the output is 1 and the time the output is 0 for each output cycle.

## SOLUTION

(a) Since the count is less than 100 the counter can be built by cascading two 7490s. The circuit is shown in Fig. 8-20a. The most significant IC of the counter, which is connected to $R_{0(2)}$, goes HIGH at a count of 80. The AND gates decode a count of 7 from the least significant stage and are connected to $R_{0(1)}$. At a count of 87 both $R_{0(1)}$ and $R_{0(2)}$ are HIGH and the counter resets.

The waveforms are shown in the CRO traces of Fig. 8-20b. The top trace is $Q_{D2}$, which is also used to trigger the oscilloscope. In circuits of this type the oscilloscope must be synchronized by the lowest frequency generated, otherwise the oscilloscope won't trigger repetitively (see Sec. 8-11.2).

The bottom trace is $Q_{A1}$ and the glitch is clearly seen. When $Q_{A1}$ is LOW, the count is even because $Q_{A1}$ is the least significant bit (LSB) of the count. The waveforms show that when $Q_{D2}$ goes HIGH, $Q_{A1}$ goes LOW; this is a count of 80.

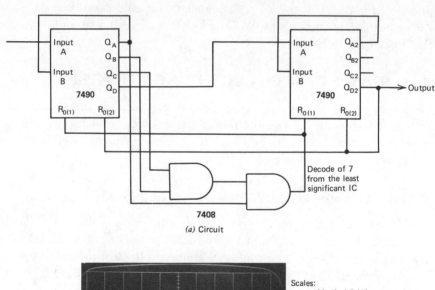

(a) Circuit

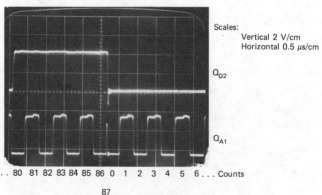

(b) Waveforms

FIGURE 8-20  A divide-by 87 counter.

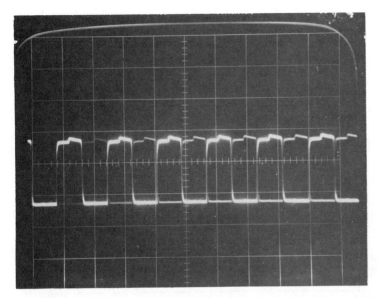

FIGURE 8-21 Oscilloscope shadows caused by triggering on an almost repetitive waveform.

The count then progresses to a count of 87, which is the glitch that resets the counter to 0.

If the input frequency were decreased, all of the pulses in Fig. 8-20b would get longer except the glitch. Thus the glitch would occupy a smaller portion of the trace and would be harder to see.

(b) The output should be taken at the point that changes least often, $Q_{D2}$ in this case. If the input is 200 kHz, then each input point is 5 $\mu$s (the input to Fig. 7-20b is much faster to bring out the glitch.) $Q_{D2}$ is up for 7 counts (80–86) plus the glitch time, and down for 80 counts (0–79) minus the glitch time. If the glitch time is ignored, $Q_{D2}$ is up for 35 $\mu$s and down for 400 $\mu$s. Thus a 5-$\mu$s input produces a 435-$\mu$s output and the circuit divides the input frequency by 87.

## 8-11.2   Oscilloscope Shadows[7]

If the waveforms of complex circuits are to be observed on an oscilloscope (CRO), the CRO should always be triggered on the stage that changes least often ($Q_{D2}$ in Fig. 8-20). Figure 8-21 shows what happens if we attempt to observe $Q_{A1}$ while also triggering from it; a common mistake. Note that the trace is clear on the left side, but shadows start to come in and get progressively deeper as we move to the right.

[7]This section contains advanced material and may be omitted on first reading.

The reason for the shadows is that the CRO is being triggered by an *almost repetitive waveform*. $Q_{A1}$ is an identical pulse 86 times, but a glitch the $87^{th}$ time. If the glitch is within the trace, it inverts the waveform and causes the shadow. As we move farther to the right, there are more pulses following the trigger, and a higher probability that the glitch is within them; hence the deepening shadows.

## 8-11.3   Divide-by-N Counters Using 74193s and 74192s

The simplest way to design divide-by-N counters using 74192s and 74193s is to decode the desired count and connect the output of the decoder to the CLEAR input of the IC. When the count reaches N the decoder output clears the counter, and it starts counting again from 0. The IC is cleared by a glitch. This is similar to designing a counter made up of 7490s (Sec. 8-11.1).

---

**EXAMPLE 8-12**

Design a divide-by-147 counter using 74192s or 74193s. Glitch clears are acceptable.

**SOLUTION**

Since 147 is a number requiring three decimal digits, three 74192s would be required, but 147 = 10010011, which is an 8-bit binary number and can be contained in two 74193s; therefore, the 74193 is selected for this design. The circuit is shown in Fig. 8-22. The binary equivalent of 147 is decoded by the 7420 and applied to the CLEAR inputs of the 74193s. The 7404 inverts the decoded pulse to make the CLEAR pulse positive.

---

Down counters provide a more elegant solution to the problem of Ex. 8-12, a solution that requires no external gates. The number N can be

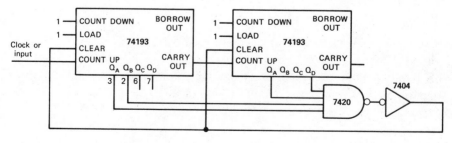

FIGURE 8-22   A divide-by-147 counter using **74193s** as an up counter.

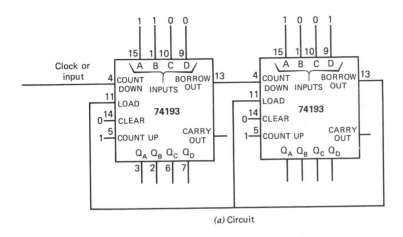

(a) Circuit

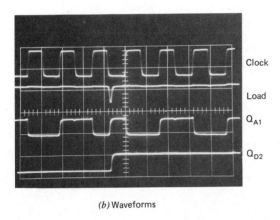

(b) Waveforms

FIGURE 8-23 A divide-by-147 counter using **74193**s as a down counter.

placed on the inputs to the **74193**s as shown in Fig. 8-23a, and the BOR-ROW-OUT signal of the most significant **74193** connected to the LOAD input of both **74193**s. The input pulses are applied to the COUNT-DOWN line. When the count reaches 0 and the clock goes LOW, the BORROW-OUT pulse from the most significant **74193** loads the number N back into the counter. Figure 8-23a is a divide-by-147 counter using this circuit.

The waveforms are shown in Fig. 8-23b. The top trace is the input.[8] The second trace is the BORROW-OUT of the most significant stage. It is a glitch that only occurs when the count is 0 and the input goes LOW. It loads the number 147 into the counter. The third trace is $Q_{A1}$, the least

[8] Delayed triggering was used to obtain the CRO traces.

significant bit of the count. When the counter rolls over $Q_A$ is 0 for half an input pulse (when the count is 0, but before the input goes LOW), and 1 for half a count (when 147 is loaded into the counter). Otherwise $Q_A$ changes only on positive edges of the input clock. The bottom trace is $Q_{D2}$, the output and MSB of the counter. It is 0 when the count is LOW, but goes HIGH when 147 is loaded into the counter. Note that the count changes on the positive edge of the input but the glitch occurs on the negative edge because the BORROW-OUT goes LOW at this time. During a complete output cycle there are 146 full counts (counts 146–1), and the counts 0 and 147 each last for half of the input pulse. There are no glitches on $Q_A$ in this circuit; the only glitch is on the BORROW-OUT line.

---

### EXAMPLE 8-13

A 200-kHz, 50-percent duty cycle square wave is applied to the input of Fig. 8-23a. The output is taken from $Q_{D2}$ during each cycle. How long is the output HIGH and how long is the output LOW?

### SOLUTION

$Q_{D2}$ goes HIGH when the **74193**s are loaded by the glitch on BORROW-OUT-2. It stays HIGH until the count decrements to 127. Thus it is HIGH for 19.5 counts (half a count at 147 plus counts 146 through 128), or $19.5 \times 5\ \mu s = 97.5\ \mu s$. It is low for 127 counts plus half the count at 0 or for 637.5 $\mu s$. The final output takes $97.5 + 637.5$, or 735 $\mu s$, so the 5-$\mu s$ input wave has been lengthened (and the frequency divided) by 147.

### SUMMARY

In this chapter we considered the design of counter to count to any arbitrary number N. First ripple and synchronous counters were discussed; then, the design of counters for truncated and irregular sequences was explained. In the latter part of the chapter the use of existing IC counters was discussed. Simple methods were developed for making these IC counters count to any desired number.

### GLOSSARY

**Ripple counter.** A counter built so that the stages change sequentially. A particular stage will change only if the previous stage changes.

**Synchronous counter.** A counter built so that all stages change simultaneously in response to a clock pulse.

**Decoder.** A circuit or gate that detects when a counter reaches a specific count.

**Irregular count sequence.** A count sequence that does not progress 0, 1, 2, . . . , but contains skips and jumps.

**Truncated count sequence.** A count sequence of a K stage counter that progresses 0, 1, 2, . . . $n$, where $n < 2^K$.

**Binary counter.** A K stage counter that recycles after $2^K$ counts.

**Decade counter.** A counter that recycles after 10 pulses.

**UP-DOWN counter.** A counter that will count either up or down, depending upon its inputs.

**LOAD.** An input that causes a counter to be directly set.

**CARRY-OUT.** A pulse indicating that a counter is about to overflow.

**BORROW-OUT.** A pulse indicating that a counter is about to underflow.

## REFERENCES

**Dan I. Porat** and **Arpad Barna**, *Introduction to Digital Techniques*, Wiley, New York, 1979.

**Christopher E. Strangio**, *Digital Electronics*, Prentice-Hall, Englewood Cliffs, N. J., 1980.

**Ronald J. Tocci**, *Digital Systems*, Prentice-Hall, Englewood Cliffs, N. J., 1980.

## PROBLEMS

8-1.    Design a circuit to decode a count of 23 from 5-stage ripple counter.

8-2.    Design a counter to decode a count of 0 from a 16-stage ripple counter using ICs that have been previously discussed. Find the worst case delay through your counter and decoder.

8-3.    The circuit of Fig. P8-3 is the third stage of a synchronous counter using 7474s. Explain how it works.

8-4.    Design a 4-stage synchronous counter using **7474s.**

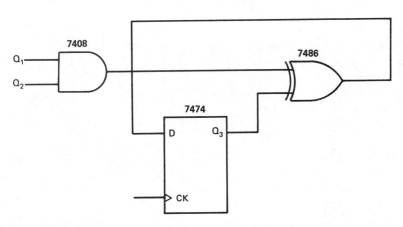

FIGURE P8-3

8-5.    Design a 5-stage synchronous counter using 74107s under the following assumptions:

(a) Minimum time delay is the most important consideration.

(b) Minimum hardware is the most important consideration.

8-6.    Calculate how many chips of each type you would need to build a 16-stage synchronous counter. Assume the clock comes from the output of a single **7400** and take loading into account. Use the minimum hardware configuration.

8-7.    Design an 11-stage synchronous counter with a maximum of two gate delays.

8-8.    Design the inputs to the 33rd and 32nd stages of a 33-stage synchronous counter that has no more than two gate delays. (*Hint*; Use four **7430**s and a **7425** per stage.)

8-9.    Design a synchronous divide-by-12 counter.

8-10.   Design a divide-by-48 counter that counts from 0 to 47.

8-11.   Design a divide-by-18 counter and construct your count table.

8-12.   (a) Design a divide-by-27 counter by cascading three divide-by-3 counters and construct your count table.

(b) Design a divide-by-27 counter using five FFs and a decoder.

8-13.   (a) Draw the timing chart and find the count sequence for the counter of Fig. 8-6.

(b) Repeat this problem if the 3s counter precedes the divide-by-4 counter.

8-14.   Design a divide-by-7 counter:

(a) Use 74107s.

(b) Use 7474s.

8-15.   Design a counter to count the following sequence:

$$0, 1, 4, 5, 6, 0, \ldots$$

(a) Use 74107s.

(b) Use 7474s.

8-16.   Determine the count sequence of the circuit of Fig. P8-16. Start at 000.

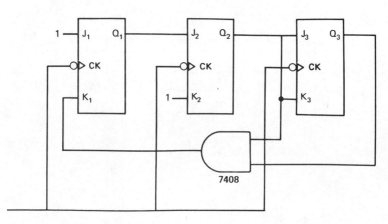

FIGURE P8-16

8-17.   Draw a timing chart for the four outputs of a 7490 as the input is clocked. Notice the $Q_D$ output is not symmetrical. How can the $Q_D$ output be made symmetrical?

8-18.   Given a 1-MHz clock, design a circuit to produce a 25-kHz clock.

8-19.   Example 8-9 could have been solved by using a divide-by-12 followed by a divide-by-50, or a divide-by-6 followed by a divide-by-100. Draw these circuits. Which have symmetrical outputs?

8-20.   Given a 1.2-MHz clock, design a circuit to produce a 100-Hz clock.

8-21.   Given a 1-MHz clock, design a circuit to produce a 12.5-kHz clock. Use no more than two chips.

8-22.   Design a divide-by-2048 counter.

8-23.   Design a circuit to count from 5 to 15 and then repeat. Use only one chip.

8-24.   Build a divide-by-75 counter using:
   (a) 7490s
   (b) 7492s
   (c) 7493s
In each case show specifically where you take the outputs from and describe the outputs.

8-25.   What is the count of the circuit of Fig. P8-25 if the FFs are:
   (a) 7490s
   (b) 7492s
   (c) 7493s

8-26.   Design a divide-by-138 counter using two 7490s and a 74107. Produce a symmetrical output.

8-27.   Design a divide-by-205 counter:
   (a) Using 74193s as UP-COUNTERS.
   (b) Using 74193s as DOWN-COUNTERS.

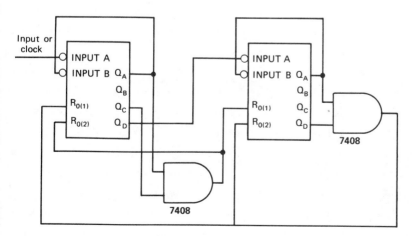

FIGURE P8-25

8-28. Design a divide-by-166 counter. Show where you take your outputs.

(a) Use two **7490**s and a J-K FF.

(b) Use two **74193**s in the countdown mode.

(c) If the input is a 100-kHz symmetric square wave, how long is your output high and how long is your output low in each case?

8-29. An input line is supposed to produce a microsecond pulse every 9 milliseconds. Occasionally, it misses a pulse, but it never misses two in a row. Design a circuit to detect when it has missed 73 pulses and sound an alarm. The alarm is to sound until a reset toggle switch is flipped to the off position. When the toggle switch goes to the on position the circuit restarts. (*Hint*: Retriggerable one-shots do this job nicely.)

8-30. Given: An input line and a pushbutton switch.

(a) Build a 100-kHz clock.

(b) Input data is to be sampled on the positive edge of the clock. The third consecutive 1 and subsequent 1s are to be counted. When the count reaches 213, a lamp is to light and the clock must stop. Depressing the pushbutton clears everything. When the pushbutton is again released, the circuit restarts. Ignore switch bounce problems.

8-31. You are an oscilloscope manufacturer and want to include a delay feature on your oscilloscope. You have three BCD thumbwheel switches. Design the circuit to delay the scope by the number of $\mu$s set into your thumbwheel switches. Include a delay/no delay option. (*Hint*: **74192**s might be good ICs to use.)

After attempting the problems the student should return to the questions of Sec. 8-2 and be sure all can be answered. If the student does not understand them all, he or she should reread the appropriate sections of the chapter to find the answers.

## 9-1 INSTRUCTIONAL OBJECTIVES

This chapter introduces the student to the design and application of shift registers. After reading it, he or she should be able to:

1. Design shift registers for both left and right shifting.
2. Include such features as parallel loading and control of serial inputs in shift registers.
3. Utilize currently existing TTL shift registers.
4. Design timing chains using shift registers.
5. Design parallel-to-serial and serial-to-parallel converters using shift registers.
6. Utilize MOS shift registers.

## 9-2 SELF-EVALUATION QUESTIONS

Watch for the answers to the following questions as you read this chapter. They should help you to understand the material presented.

1. When a register shifts, which bits are shifted out of the register? Which stages are vacated?
2. How can multiplication and division of binary numbers by 2 be accomplished using shift registers? Is this analogous to a procedure using decimal arithmetic?
3. What is circular shifting?
4. Explain why a **74164** is used to perform a serial-to-parallel conversion and a **74165** or **74166** to perform a parallel-to-serial conversion.

5. In a shift register timing chain, why must only one pulse be in the chain at any one time?

6. What is the major functional difference between an MOS shift register and a TTL shift register?

7. Explain the difference between static and dynamic shift registers.

## 9-3 THE BASIC SHIFT REGISTER

A *shift register* consists of a group of FFs connected so that *each FF transfers its bit of information to the next most significant FF of the register when a clock occurs.* The action of a shift register is illustrated in Fig. 9-1, where each bit shifts one place to the right after each clock. In Fig. 9-1 it is assumed, for clarity, that bits shifting out of the most significant stage (the rightmost stage) are lost, and that 0s shift into the least significant stage (the leftmost stage).

### 9-3.1 Shift Registers Built from D-Type FFs

Shift registers can be built from D-type FFs, as shown in Fig. 9-2*a*, by connecting the Q output of each stage to the D input of the next succeeding stage. Since the D input to each stage is the same as the Q output of the previous stage, data will be shifted one bit to the right on the positive edge of each clock pulse.

|  | Least Segment Bit (LSB) | | | | | | | Most Significant Bit (MSB) |
|---|---|---|---|---|---|---|---|---|
| Initial Data | 1 | 1 | 0 | 0 | 1 | 1 | 0 | 1 |
| Clock pulse 1 | 0 | 1 | 1 | 0 | 0 | 1 | 1 | 0 |
| Clock pulse 2 | 0 | 0 | 1 | 1 | 0 | 0 | 1 | 1 |
| Clock pulse 3 | 0 | 0 | 0 | 1 | 1 | 0 | 0 | 1 |
| Clock pulse 4 | 0 | 0 | 0 | 0 | 1 | 1 | 0 | 0 |
| Clock pulse 5 | 0 | 0 | 0 | 0 | 0 | 1 | 1 | 0 |
| Clock pulse 6 | 0 | 0 | 0 | 0 | 0 | 0 | 1 | 1 |
| Clock pulse 7 | 0 | 0 | 0 | 0 | 0 | 0 | 0 | 1 |
| Clock pulse 8 | 0 | 0 | 0 | 0 | 0 | 0 | 0 | 0 |

FIGURE 9-1 The action of a shift register.

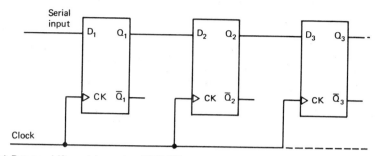

(a) D-type shift register using **7474** FFs

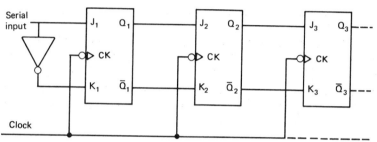

(b) Shift register built from J-K FFs

FIGURE 9-2   Shift registers built from D and JK FFs.

## 9-3.2   Shift Registers Built from J-K FFs

Shift registers can also be built from J-K FFs by connecting the $Q$ and $\bar{Q}$ outputs of each stage to the J and K inputs of the succeeding stage, as Fig. 9-2b shows. The information in the register shifts one stage to the right on each negative transition of the clock. Since the data are constant for an entire clock period, master-slave-type problems are not encountered in a J-K shift register.

Note that shift registers are generally shown as shifting to the right,[1] as in Fig. 9-2. Therefore, if the *contents* of a shift register are to be considered as a *binary number* the *least significant bit* (LSB) must be on the *left*. Consequently, in this chapter, numbers are represented with the *LSB* on the *left* and the *most significant bit* (MSB) on the *right*, to correspond to the *normal* direction of shifting in a shift register. Unfortunately, this is a *reversal* of the usual way of representing numbers, but it is clearer when the numbers are contained in shift registers.

[1]Also see Figs. 9-3, 9-7, and 9-9.

---

## EXAMPLE 9-1

A 10-bit shift register contains the number 1011011101. What does it hold after two shift clocks if 0s are shifted into the vacated positions?

## SOLUTION

The clock pulses shift the word two places to the right and shift 0s into the two least significant places. The resulting output is therefore **0010110111**. Note that the two most significant (rightmost) bits have been shifted out of the register and lost.

## 9-4 LEFT-RIGHT SHIFT REGISTERS

A LEFT-RIGHT shift register can shift information either left or right. The direction of shifting is controlled by a toggle switch or a control FF. Computers require some form of a LEFT-RIGHT shift register to execute shift instructions.

---

## EXAMPLE 9-2

Design a LEFT-RIGHT shift register using 7474 D-type FFs. The mode of shifting is to be controlled by a toggle switch.

## SOLUTION

1.  The toggle switch positions are labeled LEFT and RIGHT to indicate the direction of shifting. If the toggle switch is in the RIGHT position, the D input to a stage must be connected to the Q output of the next stage to the left.
2.  If the toggle switch is in the LEFT position, the D input to a stage must be connected to the next stage to the right.
3.  An AND-OR-INVERT (AOI) gate can be used to implement this logic (see Sec. 5-9). Each contact of the toggle switch is connected to one of the AND gates so that only one AND gate is enabled at any time.
4.  The AOI gate could be followed by an inverter to compensate for the AOI inversion. A more elegant solution, however, is to move the inversion to the front of the AOI gate by using the $\bar{Q}$ output of the FF instead of the Q output. This eliminates an inverter at each stage.
5.  Figure 9-3 shows three stages of a LEFT-RIGHT shift register with the mode switch and interconnections required to make it function properly.

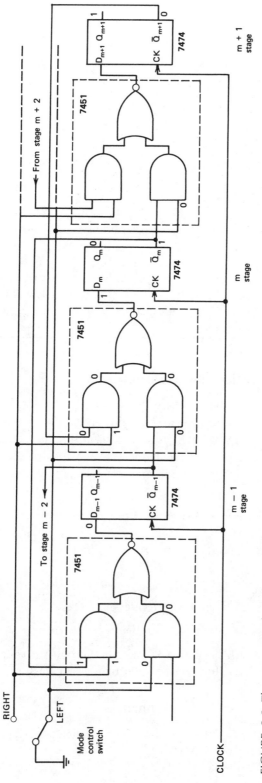

FIGURE 9-3 Three stages of a LEFT-RIGHT shift register.

## EXAMPLE 9-3

By placing the correct 1s and 0s on Fig. 9-3, show that the shift register functions properly if:

1. It is in SHIFT-LEFT mode.
2. $Q_{m+1} = 1$
3. $Q_m = 0$

## SOLUTION

The proper 1s and 0s are already placed on Fig. 9-3. With the mode control switch in the LEFT position, all the lower AND gates in the AOI ICs are connected to ground (0). The mode inputs to the upper AND gates are open and act as a 1. Since $Q_{m+1} = 1$, the 0 at $\overline{Q}_{m+1}$ is connected to the AOI gate of stage m, which causes the input, $D_m$, to be HIGH. Similarly, since $Q_m = 0$, the 1 on $\overline{Q}_m$ is fed to the leftmost AOI gate and causes $D_{m-1}$ to be 0. Therefore, at the next clock pulse, the D inputs are entered into the FFs; $Q_m$ becomes 1, and $Q_{m-1}$ becomes 0. The 1 and 0 in $Q_{m+1}$ and $Q_m$ are transferred to $Q_m$ and $Q_{m-1}$ and a LEFT SHIFT occurs as specified.

## 9-4.1 Multiplication and Division by 2

A binary number can be multipled by 2 simply by placing it in a shift register and shifting it one stage to the right (providing a 1 is not shifted out of the most significant stage of the shift register). Similarly, a binary number can be divided by 2 by shifting it one stage to the left. In division any remainder is lost unless additional circuitry is used to preserve it.

## EXAMPLE 9-4

The number 25 is in an 8-stage shift register. Multiply the number by 8.

## SOLUTION

Initially the number 25 in an 8-stage register looks like:

$$10011000$$

Note that the leftmost bit is the LSB. Multiplication by 8 is equivalent to three multiplications by 2, or three shifts to the right. After three shifts, the register contains

$$00010011$$

This is $128 + 64 + 8 = 200$, or $8 \times 25$. This example worked properly because 1s were not shifted out of the MSB.

## EXAMPLE 9-5

Given four 10-bit registers, each containing a binary number, describe how one could obtain the average of the four numbers.

### SOLUTION

Since the average is the sum of the four numbers divided by 4, an adder circuit is needed to sum the numbers. The output of the adder could be transferred into a shift register. Two left shifts would then divide the output by four and leave the average in the register. Unfortunately, any remainder would be lost (i.e., the fractional part of the average would be discarded. See Problems 9-4 and 9-5).

## 9-5  SERIAL INPUTS AND PARALLEL LOADING OF SHIFT REGISTERS

When shifting right, the least significant stage of the shift register is vacated. The designer has the option of entering any data into this stage. The input to the least significant stage is generally called the *serial input*. In Fig. 9-2a, for example, the serial input is connected to $D_1$, the D input of the first stage.

There are four common ways of using the serial input to a shift register.

1.  **Pulling in 0s.** This places 0s in all the vacated stages of a shift register. It is accomplished by tying the serial input ($D_1$) to ground.
2.  **Pulling in 1s.** This places 1s in all the vacated slots of a shift register and is accomplished by tying $D_1$ HIGH.
3.  **Circular shifting.** This connects the output of the most significant stage to $D_1$. Consequently, the MSB before the shift becomes the LSB after the shift. Circular shifting can also be accomplished in SHIFT-LEFT mode by connecting the output of the LSB to the input of the MSB. Most computers use these circuits to execute circular shift instructions. In microprocessors circular shift instructions are called ROTATES.
4.  **Serial data input.** Serial input from an external source can be shifted into the register by tying the output of the external source to $D_1$. Care must be taken to synchronize the external data with the shift clocks.

## EXAMPLE 9-6

Use a 4-position rotary switch to implement the 4 shift register options listed above.

### SOLUTION

The solution is shown in Fig. 9-4. The rotary switch selects one gate of a 4-wide AOI gate (**7454**) and the proper data are connected to the other input. The output

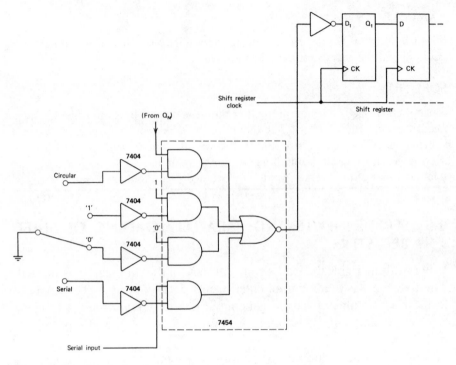

FIGURE 9-4 Circuit for Ex. 9-6.

is then inverted and fed to the D input of the least significant stage of the shift register. The 7404 inverters cause the inputs to all AND gates except the selected gate of the AOI circuit to be LOW.

## 9-5.1   Microprocessor Shifting

Shifting in microprocessors ($\mu$Ps) generally makes use of the CARRY FF. This is a FF within the $\mu$P chip whose prime function is to retain the state of the carry after an arithmetic operation. In SHIFT or ROTATE instructions it is often used both to provide a place to retain the bit shifted out and to supply the bit being shifted in.

The operation of a microprocessor ROTATE instruction is shown in Fig. 9-5. Note that in a microprocessor $B_0$ is the LSB and $B_7$ the MSB. In a ROTATE LEFT instruction the MSB of the 8-bit (1-byte) $\mu$P word is shifted into the CARRY FF, while it supplies the input to the LSB. This is really a 9-bit ROTATE using the 8 bits of the register plus the CARRY FF. RIGHT ROTATES are accomplished similarly by reversing the arrows as shown in Fig. 9-5b. Since the contents of the CARRY FF can be retained and the CARRY FF manipulated in other ways, the $\mu$P user can readily control shift operations using the CARRY FF.

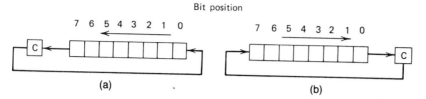

Bit position

FIGURE 9-5 The **6800** Rotate instructions. (a) Rotate Left (ROL), (b) Rotate Right (ROR). (From Greenfield and Wray. Using Microprocessors and Microcomputers: The **6800** Family. Copyright John Wiley & Sons, Inc. 1981. Reprinted by permission of John Wiley & Sons, Inc.)

## 9-5.2 Parallel Loading

Many shift registers must be capable of accepting data from an external source as well as shifting bits. These shift registers operate in SHIFT mode or LOAD mode. In SHIFT mode they operate normally, but in LOAD mode the external data are jammed into the register and replace any previous data.

---

**EXAMPLE 9-7**

Design a shift register controlled by a switch. When the switch output is LOW the register shifts RIGHT, but when the switch output is HIGH, external data are loaded into the register.

**SOLUTION**

The first two stages of the shift register are shown in Fig. 9-6. When the switch is down, or in the SHIFT position, the output of each of the **7400** gates is HIGH, and these outputs do not affect the direct SETS and CLEARS of the FFs. If the switch is in the LOAD position, however, the external data bits are jammed into the shift register FFs through the NAND gates. If data bit 1 is a 1, for example, the direct set to stage 1 is LOW, setting it; while if data bit 1 is a 0, the lower NAND gate has two HIGH inputs, and its LOW output clears FF1.

---

## 9-6 PARALLEL LOAD AND PARALLEL OUTPUT SHIFT REGISTERS

Many shift registers are available in TTL ICs. The important features of these shift registers are as follows.

1. **The number of bits in the shift register.** If longer shift registers are required, they can be built up by cascading ICs.

2. **Parallel load.** The ability to parallel load data simultaneously into the shift register.

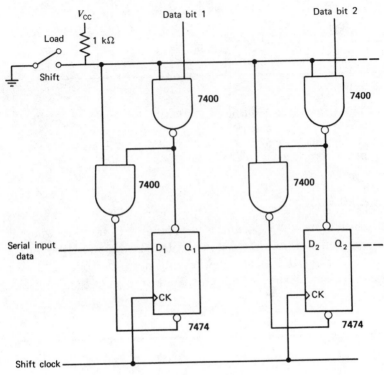

FIGURE 9-6  A shift register with shift-load capability.

3.  **Serial input.** The ability to serially shift data into the shift register.
4.  **Parallel output.** The ability to obtain all output bits of a shift register simultaneously in parallel.
5.  **Serial output.** The ability to obtain serial output from a shift register.
6.  **Shift frequency.** The maximum permissable frequency of the shift clock.
7.  **LEFT-RIGHT shift register.** The ability of the shift register to shift data in either direction.

To acquaint the student with shift registers, the 74164, 74165, and 74166 shift registers are described in detail.

## 9-6.1  The 74164 Serial-In Parallel-Output Shift Register

The 74164 is an 8-bit shift register in a 14-pin package with all 8 outputs available. The pin configuration and block diagram are shown in Fig. 9-7. A parallel-load is not available on this IC and the data are loaded serially via an input AND gate. Serial inputs A and B must both be HIGH to load

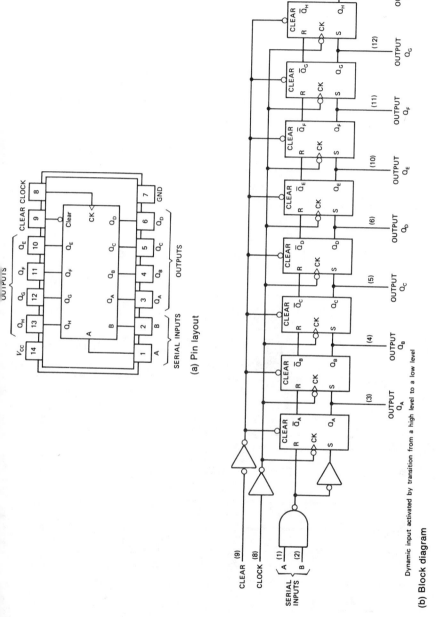

(a) Pin layout

Dynamic input activated by transition from a high level to a low level

(b) Block diagram

FIGURE 9-7 The **74164** shift register. (From *The TTL Data Book for Design Engineers*, 2nd. Ed., Texas Instruments, Inc. Courtesy of Texas Instruments, copyright 1976.)

a 1 into the shift register. The **74164** also has an asynchronous CLEAR, which clears all stages of the shift register whenever it goes LOW.

The operation of the **74164** is illustrated by the timing chart of Fig. 9-8.

1. The clear pulses at $t = 1$ and $t = 24$ clear all outputs.
2. The positive clock transitions, which clock in the input data, occur at $t = 1$, 3, 5,.. . . The only transition times when inputs A and B are both HIGH are at $t = 9$, 11, and 15. $Q_A$ goes HIGH for one clock period following each of these times.
3. Since the $Q_A$ output is shifted to $Q_B$ one clock period later, $Q_B$ is HIGH at $t = 11$, 13, and 17.
4. The data continue to shift right through the shift register on succeeding clock pulses.
5. Note that the output pulses on $Q_E$, $Q_F$, $Q_G$, and $Q_H$ are truncated by the clear pulse at $t = 24$.

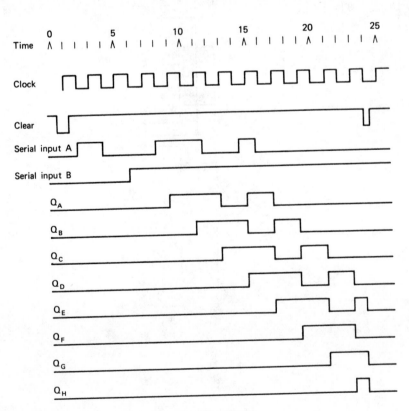

FIGURE 9-8   A timing chart for the **74164**.

## 9-6.2 The 74165 and 74166 8-bit Parallel-Load Shift Registers

The 74165 and 74166 are two very similar 8-bit parallel-in, serial-out shift registers. The 74165 is first described in detail, then the differences between the 74165 and the 74166 are explained.

The pin layout and block diagram of the 74165 are shown in Fig. 9-9. From it we see that there are 8 parallel inputs and a SHIFT/LOAD input (pin 1). When the SHIFT/LOAD input is LOW (LOAD mode), the clocks are inhibited and the input data are jammed into the 8 stages of the shift register as it was in Ex. 9-7. When the SHIFT/LOAD input is HIGH, loading is inhibited, and data are shifted one bit to the right on each positive transition of the clock provided the CLOCK INHIBIT input is LOW. An examination of Fig. 9-9*b* reveals that the CLOCK input is ineffective if the CLOCK INHIBIT input is HIGH. Note also that the CLOCK and CLOCK INHIBIT inputs can be interchanged. The CLOCK INHIBIT line should be changed from LOW to HIGH only while the clock line is HIGH. A positive transition of the CLOCK INHIBIT line when the clock line is LOW causes an extra clock to occur. A serial input line is also provided on pin 10. The serial data are shifted into the A FF on each clock pulse.

It was impossible to make the parallel outputs available and package the 74165 in a 16-pin DIP. Therefore, only the outputs of the most significant stage are available ($Q_H$ and $\overline{Q}_H$).

To study the operation of the 74165 in detail, consult the timing chart of Fig. 9-10. For clarity the output of each *internal* FF is shown, but only the $Q_H$ output is *actually available* at an external connection. The 74165 operates as follows.

1. Positive edges or clocks occur at $t = 1, 3, 5, \ldots$.
2. The first pulse on the serial input is clocked into $Q_A$ at $t = 1$.
3. This pulse shifts to the right and appears at the output, $Q_H$, seven clock pulses later ($t = 15$).
4. The short pulse at $t = 8$ does not enter the shift register because no clock occurs while it is HIGH.
5. The pulse on the serial input from $t = 16.5$ to $t = 19.5$ starts a double-width pulse down the shift register, because clocks occur at $t = 17$ and $t = 19$ while it is HIGH. This double-width pulse, however, is truncated at $t = 20$.
6. From $t = 20$ to $t = 22$, the shift register is in the LOAD mode. The data loaded in are assumed to be 11011001 (input H to input A, respectively).
7. At $t = 22$ the shift register returns to shift mode. The next shift clock occurs at $t = 23$, and the word starts to form at the output.
8. From $t = 25.5$ to $t = 28.5$, the CLOCK INHIBIT input goes high. Note that it makes a LOW-to-HIGH transition when the clock is HIGH. The negative

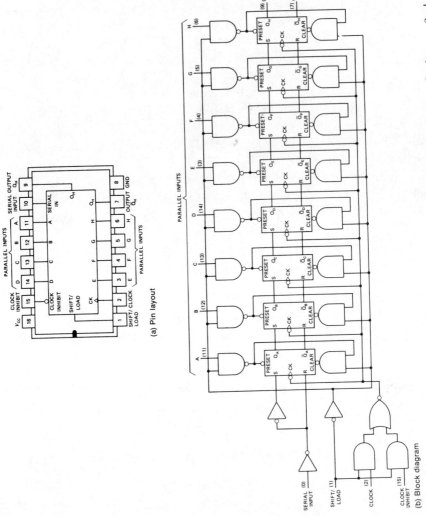

FIGURE 9-9 The **74165** shift register. (From the *TTL Data Book In Design Engineers*, 2nd ed., Texas Instruments, Inc. Courtesy of Texas Instruments, copyright 1976.)

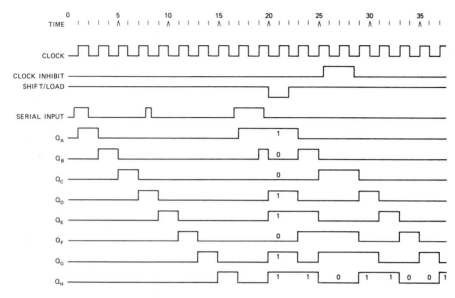

FIGURE 9-10   A timing chart for the **74165**.

transition of the CLOCK INHIBIT input can be made at any time. This pulse inhibits the clock at $t = 27$, and the shift register stages cannot change at this time.

9.   At $t = 29$ shift clocks resume and the data loaded in are shifted out at $Q_H$, one bit at a time.

10.   The data, which appear serially at $Q_H$, are the parallel-loaded data, 11011001. The first 0, however, is twice as long as the other pulses because of the inhibited clock at $t = 27$. This shows that one must be careful when using the CLOCK INHIBIT line.

The **74166** is an 8-bit PARALLEL-LOAD, SERIAL-OUTPUT shift register. The differences between the **74165** and the **74166** are as follows.

1.   The **74166** is *synchronously* loaded. If the SHIFT/LOAD input is LOW, it only loads on the positive edge of the clock. Referring to Fig. 9-10, if a **74166** were used it would have loaded at $t = 21$, instead of $t = 20$. The pulse at this time would be only two time units long ($t = 21$ to $t = 23$) for the asynchronously loaded **74165**. Therefore, the time of the output pulses of the **74166** would be more uniform.

2.   The **74166** does not have $\overline{Q}_H$ available.

3.   The **74166** has a DIRECT CLEAR, which asynchronously clears all stages.

### 9-6.3   The 74198 LEFT-RIGHT Shift Register

The **74198** is an 8-bit shift register in a 24-pin package. The extra pins allow many desirable features to be added. The **74198** has left or right shifting, parallel outputs, and parallel load.

The pinout and function table of the **74198** are shown in Fig. 9-11. The function table shows that the operation of the shift register is controlled by the mode inputs, S1 and S0. If they are both 1s, the IC is in LOAD mode, and data on the input pins will be loaded into the register on the positive edge ( ↑ ) of the clock. Like the **74166**, the **74198** is synchronously loaded.

If S1 and S0 are 0 and 1, respectively, the register is in SHIFT RIGHT mode. Data will be shifted from A towards H on each clock pulse and the level on the SHIFT RIGHT serial input will be entered into the $Q_A$ FF. If S1 and S0 are 1 and 0 respectively, the **74198** will shift left on each clock

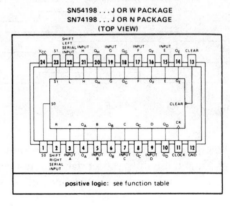

(a) Pinout

'198
FUNCTION TABLE

| CLEAR | MODE | | CLOCK | SERIAL | | PARALLEL | OUTPUTS | | | | |
|---|---|---|---|---|---|---|---|---|---|---|---|
| | S₁ | S₀ | | LEFT | RIGHT | A...H | $Q_A$ | $Q_B$ | ... | $Q_G$ | $Q_H$ |
| L | X | X | X | X | X | X | L | L | | L | L |
| H | X | X | L | X | X | X | $Q_{A0}$ | $Q_{B0}$ | | $Q_{G0}$ | $Q_{H0}$ |
| H | H | H | ↑ | X | X | a...h | a | b | | g | h |
| H | L | H | ↑ | X | H | X | H | $Q_{An}$ | | $Q_{Fn}$ | $Q_{Gn}$ |
| H | L | H | ↑ | X | L | X | L | $Q_{An}$ | | $Q_{Fn}$ | $Q_{Gn}$ |
| H | H | L | ↑ | H | X | X | $Q_{Bn}$ | $Q_{Cn}$ | | $Q_{Hn}$ | H |
| H | H | L | ↑ | L | X | X | $Q_{Bn}$ | $Q_{Cn}$ | | $Q_{Hn}$ | L |
| H | L | L | X | X | X | X | $Q_{A0}$ | $Q_{B0}$ | | $Q_{G0}$ | $Q_{H0}$ |

H = high level (steady state), L = low level (steady state)
X = irrelevant (any input, including transitions)
↑ = transition from low to high level
a . . . h = the level of steady-state input at inputs A thru H, respectively.
$Q_{A0}$, $Q_{B0}$, $Q_{G0}$, $Q_{H0}$ = the level of $Q_A$, $Q_B$, $Q_G$, or $Q_H$, respectively, before the indicated steady-state input conditions were establ
$Q_{An}$, $Q_{Bn}$, etc. = the level of $Q_A$, $Q_B$, etc., respectively, before the most-recent ↑ transition of the clock.

(b) Function table

FIGURE 9-11 The **74198** Left-Right shift register. (From the *TTL Data Book for Design Engineers*, 2nd ed., Texas Instruments, Inc. Courtesy of Texas Instruments, copyright 1976.)

pulse. If S1 and S0 are both 0s, nothing happens; this is a "no operation" mode.

The **7494** and **7495** are 5-bit shift registers in 14-pin packages with many of the features of the **74198.** Many other shift registers are also being manufactured and the reader should consult the manufacturer's literature to stay current.

## 9-7  APPLICATIONS

There are many applications of shift registers. Three examples are considered in this section.

### 9-7.1  Circulating a Single Pulse Through a Shift Register

A single negative pulse can be circulated through a shift register by using the circuit of Fig. 9-12. The occurrence of this pulse can be used to control timing circuits (see Problem 9-11) or sequences of events (see Ex. 11-1). For repetitive operation the shift register must be driven by a constant frequency clock.

If any of the A through G inputs of Fig. 9-12 are LOW, the SERIAL INPUT is HIGH and 1s are shifted into the **74164.** When the A through G inputs are all HIGH, the SERIAL INPUT to the **74164** goes LOW. The next clock pulse clocks the LOW input to output A, which now causes the

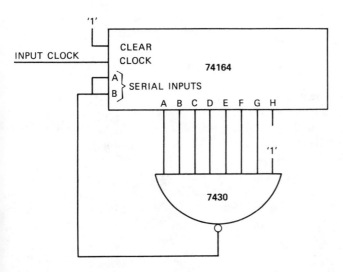

FIGURE 9-12 Circulating a single negative pulse through a shift register.

output of the **7430** to be 1. At any one time only one output is LOW,[2] and the LOW output progresses down the line with each clock pulse. The input clock frequency should be set at the basic timing for the circuit.

## 9-7.2   Timing Circuits

Sequences of timing pulses can be generated and controlled by the circuit of Fig. 9-12, but a more elegant solution that requires less gates is given in Ex. 9-8. The basic idea is to shift 1s into the shift register until the timing sequence is complete. The pulses then cause the shift register to be cleared. EXCLUSIVE-OR gates are used to select the proper timing pulses.

---

**EXAMPLE 9-8**

A circuit is required to produce a series of four pulses repetitively every 70 $\mu$s. The pulses are to be HIGH during the following times:

$$
\begin{array}{ll}
A & 0\text{--}20 \ \mu s \\
B & 15\text{--}45 \ \mu s \\
C & 35\text{--}60 \ \mu s \\
D & 55\text{--}70 \ \mu s
\end{array}
$$

Design the circuit.

**SOLUTION**

The solution is shown in Fig. 9-13. It was designed as follows.

1.   An examination of the pulses shows that the shortest time required is 5 $\mu$s. Therefore, the shift register must be driven by a 200-kHz clock to shift in 1s at 5-$\mu$s intervals. The SERIAL INPUTS are tied HIGH so that 1s are constantly shifted into the register until it clears.

2.   Because the total cycle time is 70 $\mu$s, 14 outputs are needed and two **74164** shift registers must be used.

3.   Assuming the shift register is cleared at $t = 0$, $Q_{A1}$ goes HIGH at 5 $\mu$s, $Q_{B1}$ goes HIGH at 10 $\mu$s, and so on. $Q_{F2}$ goes HIGH at 70 $\mu$s. This is inverted and CLEARS both shift registers, thus causing a 70$\mu$s cycle. Note that the output at $Q_{F2}$ is a glitch; it is HIGH only until the shift registers clear, which clears $Q_{F2}$ and removes the CLEAR pulse.

4.   The A output can be obtained simply by inverting $Q_{D1}$. $Q_{D1}$ goes LOW when the shift register is cleared and goes HIGH at 20 $\mu$s.

5.   The D output can be obtained directly from $Q_{C2}$. It goes HIGH at 55 $\mu$s and LOW at 70 $\mu$s, when the shift registers CLEAR.

---

[2]This may not be true for the first seven clock pulses following power turn on because of residual LOWS that may appear in the shift register at that time.

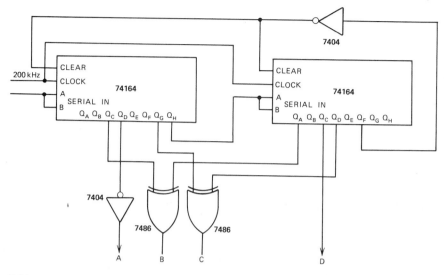

FIGURE 9-13  Timing circuit for Ex. 9-8.

6.  The B output is obtained by using an EXCLUSIVE-OR gate with inputs $Q_{C1}$ and $Q_{A2}$. From $t = 0$ to 15 $\mu$s, both inputs and the output are 0. At 15 $\mu$s $Q_{C1}$ goes HIGH. At 45 $\mu$s $Q_{A2}$ goes HIGH; so both inputs are HIGH and the output of the EXCLUSIVE-OR goes LOW.

7.  The C output is obtained similarly by tying the EXCLUSIVE-OR inputs to $Q_{G1}$ (which goes HIGH at 35 $\mu$s) and $Q_{D2}$ (which goes HIGH at 60 $\mu$s).

## 9-7.3  Serial-to-Parallel and Parallel-to-Serial Conversion

It is often necessary to convert serial data to parallel or vice-versa. Computers, for example, transmit and receive data in parallel form, but many of the devices they communicate with, such as teletypes, require the data in serial form. Therefore, a parallel-to-serial conversion must be made when going from a computer to a teletype, and a serial-to-parallel conversion must be made when going from a teletype to a computer. Shift registers are generally used to make these conversions.

## EXAMPLE 9-9

A 12-bit computer (a PDP 8 perhaps) presents 12 data bits in parallel. Assume that when the 12 bits are available a negative-going pulse, 2 $\mu$s long, appears on a

COMPUTER DATA READY line, which is controlled by the computer. Design a circuit to do the following when the COMPUTER DATA READY pulse expires:

1. Raise a signal called DATA AVAILABLE.
2. Place the bits on a single output line called the DATA line. The bits are to appear serially and each bit must remain for 1 $\mu$s.
3. After the 12 bits have appeared, the DATA AVAILABLE line must go LOW and 0s must appear on the DATA line. The DATA line must remain 0 until the next DATA AVAILABLE pulse.

## SOLUTION

When the specifications get this complex, it is best to draw a timing chart as shown in Fig. 9-14a. If we refer to the timing chart, we see that the following solution steps are required for the designer.

1. A parallel-to-serial conversion is required. This suggests 74165s.
2. Since 12 bits are involved in the conversion, two 74165s are needed. They can be cascaded to form a 16-bit shift register by tying $Q_H$ of the first shift register to the SERIAL INPUT of the second shift register.
3. Because 0s are required on the DATA line after the 12 data bits have passed, the four unused inputs and the unused SERIAL INPUT to the least significant shift register are connected to ground.
4. The COMPUTER DATA READY signal can be connected to the SHIFT/ LOAD input and used to load the data.
5. To shift out the data, as required, the shift register must be clocked by a 1-$\mu$s clock, which starts at the trailing edge of the COMPUTER DATA READY pulse. Here we can use a 1-$\mu$s one-shot oscillator (see Sec. 7-6). Connecting COMPUTER DATA READY to the B input of the first oscillator stage shuts the oscillator off while the pulse is LOW and synchronizes the clocks with it.
6. The 74165 is clocked on a positive edge, so the shift register clocks are tied to the $\overline{Q}_2$ output of the one-shot oscillator. The first positive-going edge at this point occurs exactly 1 $\mu$s after the expiration of COMPUTER DATA READY.
7. To generate DATA AVAILABLE, 12 clocks must be counted. Therefore a 7492 is used. The DATA AVAILABLE FF is SET by COMPUTER DATA READY, but it is masked from the output by the AND gate until COMPUTER DATA READY expires. The DATA AVAILABLE FF is cleared by the 7492 after 12 counts. Note that the $Q_D$ output of the 7492 must be inverted to produce a positive edge at this time. Note also that the 7492 is cleared by COMPUTER DATA READY (inverted and applied to $R_{0(1)}$ and $R_{0(2)}$) so that it always starts at a count of 0.
8. The shift register continues to shift and the 7492 to count after the 12 data pulses expire. However, since the shift register is continually shifting out 0s and the DATA AVAILABLE FF remains clear, the output is unaffected by this activity.
9. The next COMPUTER DATA READY pulse stops the clock and clears the 7492. The circuit is ready to restart when the pulse expires.

10. The final circuit, designed with these considerations in mind, is shown in Fig. 9-14b.

The circuit for Ex. 9-9 is a typical, but simple, illustration of digital design. No formulas exist to solve these types of design problems. The solution depended on thought, imagination, familiarity with ICs, and the use of circuits considered here and in previous chapters.

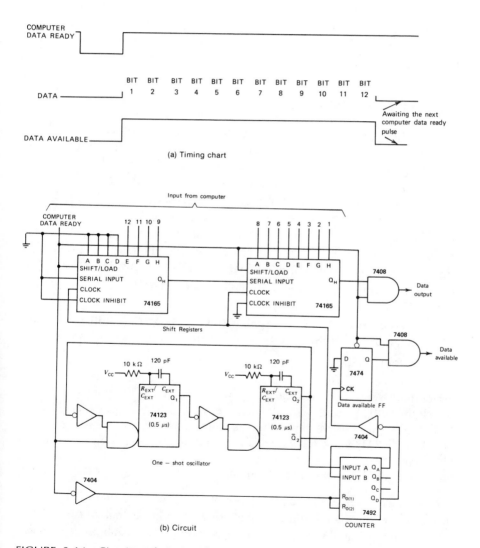

(a) Timing chart

(b) Circuit

FIGURE 9-14 Circuit and timing diagram for Ex. 9-9. Capacitor values obtained from Fig. 7-5c.

## 9-8 MOS SHIFT REGISTERS

MOS shift registers are quite popular because of their ability to store many bits of information in a small package. They are often used as serial memories for this reason. Because of the large number of bits, such features as parallel outputs or parallel loading are not feasible.

There are two types of MOS shift registers; *static* and *dynamic*. Static shift registers hold their data in FFs, and can maintain the data indefinitely. Dynamic shift registers, which usually contain more bits than static shift registers, store their information on internal capacitors between the shift register stages. Because information tends to leak off the capacitors as time passes, dynamic shift registers must be periodically *clocked* or *refreshed*. Consequently, they have a *minimum clock frequency*.

### 9-8.1 Dynamic Shift Registers

Two examples of typical dynamic shift registers are the **2524** and **2525**, manufactured by Signetics, Inc. The **2524** is a 512-bit shift register and the **2525** is a 1024-bit shift register; otherwise they are identical. Information pertaining to these chips is given in Fig. 9-15. From the pin configuation (Fig. 9-15a) we see that the shift registers are packaged in an 8-pin DIP, so they are about half the size of an ordinary IC. Figure 9-15a also shows that the **2524/25** requires both phases of a 2-phase clock ($\phi_1$ and $\phi_2$). A 2-phase clock is shown in Fig. 9-16. The clocks are alternately active (LOW) and there is a definite quiescent period when both clocks are HIGH. Specific data on the timing of the clocks should be obtained from the manufacturer's specifications.

The basic operation of the shift registers can be determined from the block diagram and truth table (Fig. 9-15b and 9-15c). It is apparent that:

1. If the WRITE input is a 0, data are recirculated. The output of the shift register is fed back to its input through the upper leg of the input AND-OR gate. The lower leg of the AND-OR gate is disabled by the 0 on the WRITE input.
2. If the WRITE input is a 1, the upper leg of the AND-OR gate is disabled and new data (the data on the input line) are written into the shift register.
3. If the READ input is a 1, the current shift register output bit is available at the output.
4. If the READ input is a 0, the output is also a 0.

The maximum clock frequency is obtained from Fig. 9-15d. Note that the clock requires an amplitude of about 15 V. Clocks that go from $+5$ to $-10$ V are typical. At 15 V the maximum clock rate is about 4.5 MHz.

The minimum clock frequency required to maintain the data is given in Fig. 9-15e. There is wide discrepancy between the *guaranteed* and *typical*

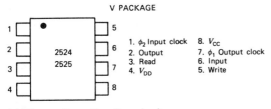

1. $\phi_2$ Input clock  8. $V_{CC}$
2. Output  7. $\phi_1$ Output clock
3. Read  6. Input
4. $V_{DD}$  5. Write

(a) Pin configuration (top view)

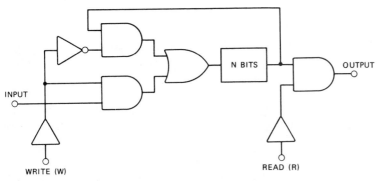

NOTE =
N = 512 or 1024 '0' = OV, '1' = +5V.

(b) Block diagram

| WRITE | READ | FUNCTION |
|-------|------|----------|
| 0 | 0 | Recirculate, Output is '0' |
| 0 | 1 | Recirculate, Output is Data |
| 1 | 0 | Write Mode, Output is '0' |
| 1 | 1 | Read Mode Output is Data |

(c) Truth table

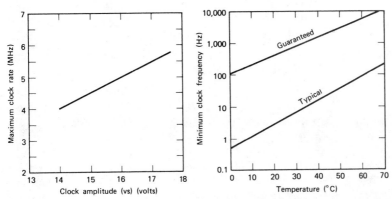

(d) Maximum clock rate versus clock amplitude

(e) Minimum operating clock frequency versus temperature

FIGURE 9-15  The Signetics **2524/25** shift registers. (Permission to reprint granted by Signetics Corp., a subsidiary of U.S. Philips Corp., 811 E. Arques Avenue, P.O. Box 409, Sunnydale, Calif. 94086.)

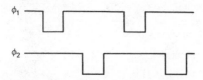

FIGURE 9-16 A 2-phase clock.

curves. At 30°C the minimum clock frequency must be 900 Hz to guarantee the integrity of the data, but typical devices can run successfully at 9 Hz.

## 9-8.2 Static Shift Registers

Static shift registers are very attractive because they can store data indefinitely; continual clocking is not required. Two examples of static shift registers are the **2521** and **2522**, shown in Fig. 9-17.

The **2521** has 128-bit shift registers within it, and the **2522** contains 132-bit shift registers. The **2522** was designed specifically for use with computer driven line printers that are capable of printing 132 characters on each line.

Both the **2521** and **2522** are dual shift registers, which means there are 2 shift registers within each package. Therefore, two bits are written in parallel, and two bits come out in parallel on each clock pulse. Figure 9-17a shows that the entire device is packaged in an 8-pin DIP. The truth table and block diagram (Figs. 9-17b and 9-17c) indicate that if the RE-CIRCULATE input is 0, the data in each of the shift registers are recirculated, whereas if the RECIRCULATE input is 1, the data on the $IN_1$ and $IN_2$ lines are written into the shift registers. The block diagram also shows that, internally, the shift registers require a 3-phase clock. Fortunately, there is a clock generator built into the shift register that accepts the external clock ($\phi_{IN}$) and generates the 3-phase clock ($\phi_1, \phi_2,$ and $\phi_3$); therefore, the user need not concern himself with this problem.

One application of the **2522** is to drive line printers, as shown in Fig. 9-18. The line printer illustrated requires 10 input bits for each character to be printed and prints 132 characters on each line. Its operation, in conjunction with a computer, might proceed as follows.

1. Since 10 bits in parallel are required, five **2522**s are necessary.
2. The computer presents one entire line of data to the line printer, then waits for the printer to print it before presenting the next line.
3. To accomplish this, the computer sets aside 132 characters in its memory as a printer buffer.
4. When the 132-character printer buffer is full, and the line printer is idle, the computer must transfer its information to the line printer. The computer does this as rapidly as possible, so it can go on to other tasks while the printer is printing.

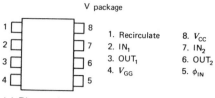

V package

1. Recirculate    8. $V_{CC}$
2. $IN_1$    7. $IN_2$
3. $OUT_1$    6. $OUT_2$
4. $V_{GG}$    5. $\phi_{IN}$

(a) Pin configuration (top view)

| Recirculate | Input | Function |
|---|---|---|
| 0 | 0 | Recirculate |
| 0 | 1 | Recirculate |
| 1 | 0 | "0" is Written |
| 1 | 1 | "1" is Written |

NOTE: "0" = OV; "1" = +5 V.

(b) Truth table

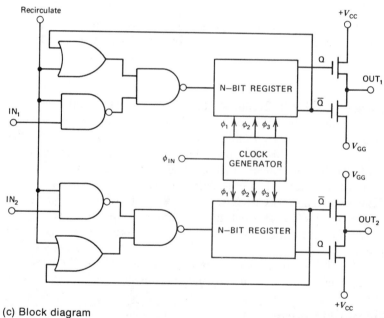

(c) Block diagram

FIGURE 9-17 The **2521/22**.(Permission to reprint granted by Signetics Corp., a subsidiary of U.S. Philips Corp., 811 E. Arques Avenue, P. O. Box 409, Sunnydale, Calif. 94086.)

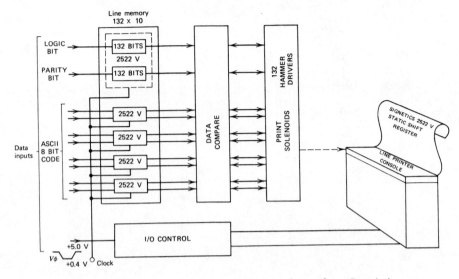

FIGURE 9-18  Block diagram of a line printer and interface.(Permission to reprint granted by Signetics Corp., a subsidiary of U.S. Philips Corp., 811 E. Arques Avenue, P. O. Box 409, Sunnydale, Calif. 94086.)

5.  The computer sets the **2522**s into WRITE mode and transfers the data into them as fast as possible. The **2522** has a typical maximum clock frequency of 2 MHz (see the manufacturer's specifications), but the computer input-output transfer rate is generally slower than this and limits the rate of data transfers. In any case, this rate is far faster than the rate at which the line printer can print.

6.  The line to be printed is now in the shift registers.

7.  The **2522**s are placed in RECIRCULATE mode and the clock is slowed down to be compatible with the rate at which the line printer can print the data.

8.  As the **2522** is clocked, the data are fed into the line printer, a character at a time, and printed.

9.  When the line has been printed, the printer is again idle and the computer can present the characters to be printed on the next line.

### 9-8.3  Bubble Memories

Bubble memories are a recently developed, highly promising memory technology. Basically, a bubble memory is a large scale circular shift register. The bits are retained on small magnetized domains called *bubbles*, rather than FFs or capacitors, and these bubbles are constantly shifted throughout a magnetic media. Being magnetic, bubbles will retain their information even when power is removed.

Bubble memories consist of the bubble package plus various ICs to shift the data, read or write them, and amplify them for output. The bubbles are very small and can be shifted rapidly, allowing bubble memories to hold a large amount of information. Because of the constant shifting they are not *random access* memories (see Chapter 15), but they are expected to compete with magnetic disks and tapes, and they have the advantage of having no moving parts.

## SUMMARY

In this chapter we introduced shift registers. The design of shift registers with special features such as LEFT-RIGHT and circular shifting was covered. Bipolar shift registers and their use as parallel-to-serial and serial-to-parallel converters was considered. Finally, the characteristics of MOS shift registers were explained so that they can be incorporated into designs, where applicable.

## GLOSSARY

**Shift register.** A register of FFs that shift data in response to a clock.

**LEFT-RIGHT shift register.** A shift register capable of shifting data in either direction.

**Circular shifting.** Shifting in which the MSB is shifted into the LSB position or vice versa.

**Parallel loading.** Simultaneously loading all bits of a shift register in parallel.

**Serial input.** The input to the least significant stage of a shift register.

**Static shift register.** A shift register with no limitation on the minimum clock frequency.

**Dynamic shift register.** A shift register whose clock must run above a certain frequency in order to preserve the data.

## REFERENCES

Geroge K. Kostopoulos, *Digital Engineering*, Wiley, New York, 1975.

Morris and Miller, *Designing with TTL Integrated Circuits*, McGraw-Hill, New York, 1971.

*The TTL Data Book for Design Engineers*, Texas Instruments, Inc., 2nd ed., 1976.

*MOS Silicon Gate Technology—2500 Series*. Signetics Corporation, Sunnydale, California, 1971.

Dan I. Porat and Arpad Barna, *Introduction to Digital Techniques*, Wiley, New York, 1979.

## PROBLEMS

9-1.   If the content of a 12-bit shift register is

100101101101

what does it contain:
  (a). After three right shifts?
  (b). After four left shifts?

9-2.   Design a LEFT-RIGHT shift register using J-K FFs.

9-3.   A 12-bit shift register contains the number 65.
  (a). How would you multiply it by 16?
  (b). How would you divide it by 16?

9-4.   Find the average of the numbers 5, 7, 7, and 8 using the method of Ex. 9-5. Show what goes into the adder, what goes into the shift register, and what comes out of the shift register.

9-5.   Make a simple improvement on the method of Ex. 9-5 to round the answer (i.e., if true average is 6 or 6.25, the circuit output should be 6, but if the true average is 6.50 or 6.75, the circuit output should be 7).

9-6.   Design a shift register that performs circular left or right shifts.

9-7.   Design the Nth stage of a shift register that responds to SHIFT-LEFT, SHIFT-RIGHT, and asynchronous LOAD commands. Assume SHIFT LEFT/RIGHT, SHIFT/LOAD, $Q_{N-1}$, $Q_{N+1}$, and the input data bit are your inputs.

9-8.   Design a synchronous SHIFT/LOAD shift register. If the register is in LOAD mode, data must be entered only on a positive clock transition.

9-9.   Redesign the cricuit of Fig. 9-4 to eliminate two inverters.

9-10.   Design a shift register controlled timing circuit to produce the pulses of Fig. P7-15.

9-11.   When power is applied to the circuit of Fig. 9-12, outputs A, C, and F come up LOW. Draw a timing chart showing how each of the eight outputs behaves as clocks are applied to the shift register.

9-12.   Enlarge the timing chart of Fig. 9-14a by adding the outputs of the one-shots and the counter.

9-13.   A device presents 16 bits on a serial input line along with a DATA READY pulse. Each bit lasts for 10 $\mu$s. The DATA READY pulse goes positive at the start of the first bit and remains HIGH until the end of the 16th bit. Design a circuit to accept the 16 serial bits and present them to a computer in parallel. The circuit must present a positive INPUT READY signal to the computer when it has the data, and the computer will reply with a negative-going ACKNOWLEDGE pulse when it has accepted the data. The ACKNOWLEDGE pulse must terminate INPUT READY.

9-14.   Given 13 inputs and a negative-going synchronizing strobe that occurs occasionally and randomly. The data on the 13 input lines are valid only when the synchronizing strobe is low. Design a circuit to:
  (a) Produce 0s as an output when the synchronizing strobe is low.

(b) Put out the 13 data bits in a 500-$\mu$s serial bit stream when the synchronizing pulse goes high.

(c) Put out 1s after the data are finished.

9-15.  Given a square wave clock, design a circuit to produce a two-phase clock:

(a) Use FFs.

(b) Use a **74164**.

9-16.  Write 1s and 0s on the diagram of Fig. 9-15 to show how it operates if:

(a) 1 is being written.

(b) 0 is being recirculated.

9-17.  The following series of pulses are to be produced cyclically every 70 $\mu$s.

*Time in $\mu$s*

A.   0–15

B.   15–45

C.   30–65

D.   5–25

(a) Use one-shot timing. Identify the one-shots you use and give the values of the timing resistors and capacitors you use.

(b) Use shift register timing. It is not necessary to include a single cycle option.

9-18.  For the circuit of Fig. P9-18, sketch the outputs as a function of time. They are repetitive. Start with the **74164** in the clear state.

9-19.  Design timing circuits to produce the pulses of Fig. P9-19.

(a) Use one-shot timing. Show your resistors, capacitors, and their connections to the one-shots.

(b) Use shift register timing.

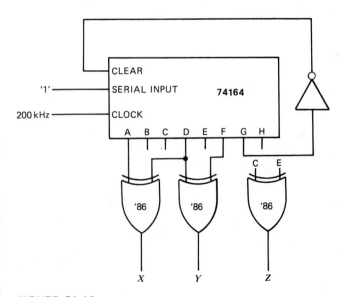

FIGURE  P9-18

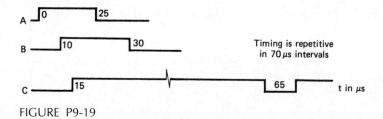

FIGURE P9-19

9-20. Given: Four lights in a row. Design a circuit to do the following:

| $T$ (sec). | $L_1$ | $L_2$ | $L_3$ | $L_4$ |
|---|---|---|---|---|
| 0–1 | OFF | OFF | OFF | OFF |
| 1–2 | ON | OFF | OFF | OFF |
| 2–3 | ON | ON | OFF | OFF |
| 3–4 | ON | ON | ON | OFF |
| 4–5 | ON | ON | ON | ON |
| 5–6 | OFF | OFF | OFF | OFF |
| 0–1 | OFF | OFF | OFF | OFF |
| REPEAT | | | | |

Assume your clock input is a 60-Hz square wave.

After attempting the problems, the student should return to Sec. 9-2. If the student still cannot answer some of the self-evaluation questions, he or she should review the appropriate sections of the chapter to find the answers.

# 10

# CONSTRUCTION AND DEBUGGING OF IC CIRCUITS

## 10-1 INSTRUCTIONAL OBJECTIVES

In this chapter we depart from the main thrust of this book, the study of IC devices and their application, to consider methods of building and debugging IC circuits. After reading the chapter the student should be able to:

1. Write a specification for a circuit.
2. Draw up logicals, wire lists, and module charts.
3. Buzz-out (trace) a circuit.
4. Design and run an acceptance test.
5. Locate faulty ICs and faulty circuit wiring.
6. Start up the timing chain of a circuit.
7. Begin to debug a circuit.

## 10-2 SELF-EVALUATION QUESTIONS

Watch for the answers to the following questions as you read the chapter. They should help you understand the material presented. When you have finished the chapter, return to this section and be sure you can answer all of the questions.

1. What are the advantages of wire-wrapping construction over printed circuit boards?

2.   What are the advantages of printed circuit boards?

3.   What criteria are used to determine the best method of building a circuit?

4.   What are specifications? Why are they important?

5.   Why should wire runs spanning more than one logical be named?

6.   What are the typical causes of errors in a combinatorial circuit?

7.   Why is it more difficult to find errors in sequential circuits than in combinatorial circuits?

8.   Are synchronous or asynchronous circuits generally easier to debug? Explain why.

## 10-3   WIRE WRAPPING

There are two major methods of building digital circuits. The first is to place the ICs in sockets and wire wrap the interconnecting wires. The alternate method is to solder the ICs into a printed circuit board where the interconnecting lines are already printed in copper on the board. Wire wrapping is more flexible and versatile than printed circuits and is considered first.

For use in wire wrapping, the ICs are plugged into a socket or a panel. *Sockets* are generally small, most contain only 14 or 16 contacts, and are designed to accommodate a single IC in a dual-in-line (DIP) package. *Panels*

FIGURE 10-1   An IC panel. (Photo courtesy of Cambridge Thermionic Corp.)

are generally larger, contain several rows of contacts, and can accommodate *many* ICs. A wire-wrap panel with several ICs in it is shown in Fig. 10-1.

In both sockets and panels, the contacts that engage the IC pins are connected to wire-wrap posts. These posts are usually square, 0.025 in. (0.635 mm) on a side, and long enough to accept three wraps of wire. The wire used is thin, solid wire (No. 30 AWG is the most popular) and is wrapped around the post. The sharp, rectangular edges of the post cut into the wire slightly to make a solid electrical contact. The ideal wrap consists of about six turns of bare wire plus a turn of insulated wire. A wire-wrap socket with three separate wraps of wire on one pin is shown in Fig. 10-2. The wire is generally wrapped by an electrically (or pneumatically) driven tool called a *wire-wrap gun* (Fig. 10-3): A small plastic hand-operated wire-wrap tool is also available. About 1.5 in. (3.8 cm) of insultation is stripped from the solid wire and it is inserted into the smaller hole in the gun or tool. The larger hole in the center of the gun is then placed over the wire-wrap post and the gun spins the wire onto the post to make the connection. The other end of the wire is then brought to the next post in the wire run and the process of stripping and wrapping is repeated to connect the wire.

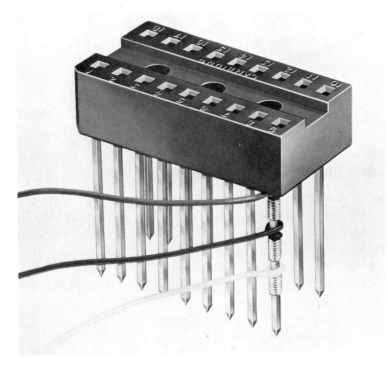

FIGURE 10-2  An IC socket with three wraps of wire. (Photo courtesy of Cambridge Thermionic Corp.)

*(a)*

FIGURE 10-3a  An electric wire wrap gun. (Photo courtesy of Cambridge Thermionic Corp.)

## 10-3.1   Semi-Automatic Wire Wrapping

For commercial production or large quantity runs of electronic equipment, *semi-automatic* wire-wrapping machines are often used. The panel to be wrapped is placed on the machine and the wire-wrap gun automatically positions itself for the first wrap as shown in Fig. 10-4. The operator then selects the proper wire from a group of prestripped and presized wires and inserts it into the gun. The gun wraps the wire and then moves to the next position to be wrapped.

The motion of the gun is numerically controlled by punched cards or a paper tape prepared from a *wire list* (see Sec. 10-5.5). The device being built must be thoroughly tested and the wire list correct before the punched card deck or paper tape is made, otherwise identical errors will appear in each piece of the production run.

FIGURE 10-3b  A hand wire wrap tool. (Photo courtesy of Cambridge Thermionic Corp.)

## 10-4  PRINTED CIRCUITS

The design and construction of a printed circuit (PC) board can start only after a circuit has been thoroughly developed, tested and debugged. A copper-clad board is used and the circuit is built as follows:

1.  Holes are drilled into the board to accommodate the IC pins and the leads of any discrete components in the circuit.
2.  The copper is then etched off the board by an acid solution, leaving only the pads and copper paths that serve as interconnecting wires.
3.  The components are then soldered to the PC board. Usually the entire board is either *wave soldered* or *flow soldered* so that all components are connected simultaneously.
4.  If a system is large enough to require several PC boards, fingers (contact strips) are printed on the edge of each board. The boards are inserted into card connectors, which are then wired together to interconnect the boards. Figure 10-5 shows a general purpose printed circuit board with edge contact fingers to plug into a connector.

FIGURE 10-4   Semi-automatic wire wrapping. (Photo courtesy of Cambridge Thermionic Corp.)

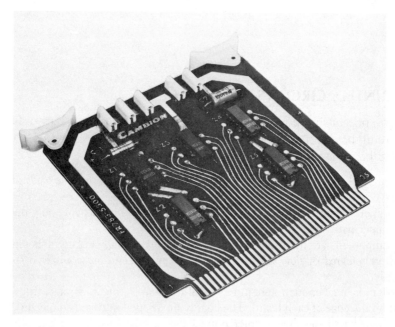

FIGURE 10-5   A printed circuit board. (Photo courtesy of Cambridge Thermionic Corp.)

## 10-4.1   Comparison of Printed Circuits and Wire Wrapping

Almost all preproduction testing, or prototyping, is done using wire-wrap connections. The advantages of wire wrapping are:

1.   Wire wrapping requires a much shorter lead time. Once the wire list is complete, wire wrapping can start immediately. The production of a PC board, which has been described briefly, is a much more laborious procedure. It requires sophisticated drafting to make a mask, followed by photographing, etching, component insertion, and soldering. Only rarely is a PC board ready in less than two weeks after the circuit has been finalized.

2.   It is much easier to correct errors or make changes in a wire-wrap circuit. A faulty wire can be removed simply and quickly using a small *unwrap* tool. A wire change on a PC board generally requires cutting the print with a knife or razor blade to break a connection, and soldering a wire bridge between existing printed wires to form new connections.

3.   Once the circuit is being tested, it is easier to replace faulty ICs because ICs in a wire-wrap circuit are inserted in sockets, whereas ICs are usually soldered into PC boards.

PC boards are best used when large production quantities are involved. These are the advantages of PC boards:

1.   **Wiring.** PC boards do not require any wiring, thus eliminating *human* wiring error in a production run.

2.   **Cost.** In large quantity, after the board has been designed, additional boards can be built quickly and economically.

3.   **Production speed.** It is much faster to mount components on a PC board and solder them automatically by machine than to build the equivalent wire-wrap circuit.

4.   **Repeatability of errors.** If an error is made in a PC board, its repair is painful, but at least it occurs at the *same* place on every board. When several identical circuits are built using wire wrap, the wiring errors occur in different places for each circuit. This increases the time, cost, and aggravation of debugging wire-wrap circuits.

The foregoing clearly indicates that the wire-wrap method is better for circuits that have not been fully tested and debugged, or for circuits that are to be produced in small quantities. In prototype development, it is almost mandatory. If a large number of identical circuits are to be produced, however, the PC board advantages predominate. Portable radios, television sets, calculators, and other *mass-produced* items use PC boards.

## 10-5   CONSTRUCTION OF WIRE-WRAP CIRCUITS

There are five steps in the construction of a wire-wrap circuit described in detail in this section. Each step must be carefully documented with appropriate paperwork. It is almost impossible to debug a prototype circuit *without complete* and *precise documentation.*

### 10-5.1 Specifications

The *specifications* describe how a circuit is to operate. They are written before work is started and should state exactly how the final circuit will perform. Engineering contracts are generally based on the specifications, which are written by the customer. The vendor, usually an engineering firm, estimates the cost of the circuit from the specifications and bids on the job. It often happens that during the design, building, and testing of a circuit, situations arise that were not anticipated in the original specifications. Often disputes arise between the customer and the vendor as to why the circuit doesn't work and who will bear the cost of repairing or redesigning it. In extreme cases, threats of legal action are heard. It is extremely important, therefore, that *specifications are written very carefully,* and cover as many situations and contingencies as can be envisioned.

### 10-5.2 Logicals

After the specifications are finished, the engineer designs the circuit by drawing *logic diagrams* (called *logicals*) as we have done in this book. Figure 9-14, for example, is a logical. Generally the engineer roughly sketches the logicals. The drafting department then redraws them professionally. Since a complex circuit may require many pages of logicals, careful labeling is extremely important. To increase the clarity and usefulness of the logicals, the following suggestions are made.

1.  Label every FF with a *descriptive name* that relates to its *function.* For example, in Fig. 9-14, the FF was called the DATA AVAILABLE FF, because it controlled the DATA AVAILABLE signal.
2.  Each *stage* or *output* of a shift register or counter should also be labeled. If, for example, a **74164** shift register with outputs A through H is used to control the timing of a circuit, then TIMING SHIFT REGISTER-B is a reasonable way to describe the B output of the **74164**. Counters should be similarly labeled. If a four-stage counter is built of FFs, WORD COUNTER 3-Q is a way to describe the Q output of the third stage of that counter.
3.  When a single wire run appears on several pages of logicals, it should *leave* the righthand side of the drawing that contains its *source* and enter all drawings that contain *loads* on the left side. At the point it leaves the source drawing, all the logical page numbers where it appears should be in parentheses. When it enters another logical where it is feeding loads only, the source drawing should be referenced in parentheses. It is also useful to name each wire run that appears on several logicals. Sometimes the source location is used for the name of a run. For example, K17-6 means the source is pin 6 of the IC at location K17.

All the above points are illustrated in Fig. 10-6, which shows parts of sheets 4 and 5 of a drum interface. Two signals, PATTERN COUNTER-D and GATED PATTERN COUNTER, leave sheet 4 on the right. The numbers in parentheses indicate that PATTERN COUNTER-D goes to sheets 2, 5, and 7, and GATED PATTERN COUNTER goes to sheets 5 and 6.

The left side of sheet 5 where the signals enter is also shown. Note that on sheet 5, they only reference sheet 4, their source sheet. Note also that these logicals are complete because they contain the pin numbers, module locations, and IC numbers for each gate.

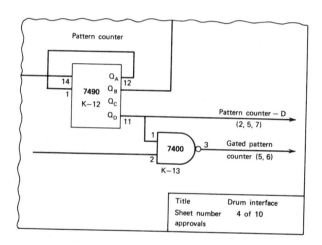

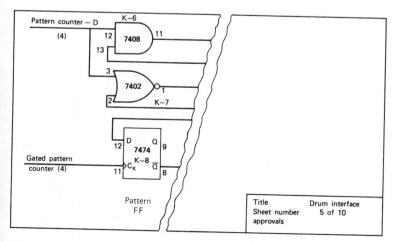

FIGURE 10-6  Part of a set of Drum Interface logicals, complete with pin numbers and module locations.

### 10-5.3   Module Placement and Pin Numbering

In the process of drawing the logicals, all gates should be labeled with their type number (i.e., all 2-input NAND gates are labeled **7400**, all 3-input NAND gates are labeled **7410**, etc.). Before the logicals are finished, each gate must be assigned a physical location and correct pin numbers.

A *module chart* is used to assign physical locations to the various ICs. The module chart should conform as nearly as possible to the panel on which the ICs are to be mounted. For example, if the ICs in a system are to be plugged into 24 socket wire-wrap panels, a module chart, shown in Fig. 10-7, could be used. Each rectangular box represents one IC socket. We recommend subdividing the box into a number of slots. The IC type number is placed in the top slot and an *additional slot is allocated for each gate in the IC*. The drawing or page number of the logical where that gate is used is then placed in the slot.

FIGURE 10-7   A blank module chart for a 24-socket IC panel.

Figure 10-8 is an example of such proper use of a module chart. Assume a 2-input NAND gate appears on page 1 of the logicals. A **7400** is required, so it is placed in the top slot of an available socket on the module chart (location 1 in Fig. 10-8). The first NAND gate uses pins 1, 2, and 3. These pin numbers are written on the logicals and the number 1 (to indicate sheet 1 of the logicals) is placed in the first of the slots reserved for the 4 gates of the **7400**. When a second NAND gate is encountered on sheet 1, pins 4, 5, and 6 are assigned and a 1 is placed in the second slot on the module chart. On page 2 a third NAND gate is encountered. A 2 is placed in the third slot of the module chart, and pins 9, 10, and 8 are assigned to the NAND gate. The fourth slot remains vacant, indicating there is a NAND gate available in case one has been overlooked, or a NAND gate is needed to correct an error. If additional gates are needed after all the slots are filled, additional ICs must be used. Figure 10-8 also shows that there are only two slots assigned to the **7474** in location 3, because a **7474** contains only two FFs. The **7404** in location 7, however, has six slots assigned to it, one for each inverter in the IC.

## 10-5.4   Wire Lists

A wire list is simply a list of the routing or path of each wire in a system. A typical wire list form for a circuit built at R.I.T. is shown in Fig. 10-9. A wire list should contain an entry for each wire in the system. The first column in the list (name) is used only when the wire is designated by a name on the logical. Otherwise it is left blank and the name of the run is taken from the source.

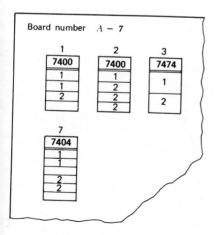

FIGURE 10-8   A portion of a module chart.

| Name | Source | To | To | To |
|------|--------|------|------|------|
| $X_1$ | C1 — 1 | A5 — 1 | A1 — 1 | |
| $Y_1$ | C6 — 1 | A5 — 2 | A1 — 2 | |
| $X_2$ | C1 — 2 | A6 — 1 | A1 — 9 | |
| $Y_2$ | C6 — 2 | A6 — 2 | A1 — 10 | |
| $X_3$ | C1 — 2 | B1 — 12 | A2 — 1 | |
| $Y_3$ | C6 — 3 | B1 — 13 | A2 — 2 | |
| $X_4$ | C1 — 4 | B1 — 9 | A2 — 9 | |
| $Y_4$ | C6 — 4 | B1 — 10 | A2 — 10 | |
| $X_5$ | C1 — 5 | B1 — 4 | A3 — 1 | |
| $Y_5$ | C6 — 5 | B1 — 5 | A3 — 2 | |
| Ad enable | C1 — 6 | A4 — 9 | A4 — 4 | A4 — 1 |
| | | A3 — 12 | A3 — 9 | A3 — 4 |
| Carry 1 | A1 — 3 | A1 — 4 | A5 — 4 | |
| Carry 2 | B2 — 3 | A1 — 12 | A5 — 9 | |
| Carry 3 | B2 — 6 | A2 — 4 | A5 — 12 | |
| Carry 4 | B2 — 8 | A2 — 12 | B1 — 1 | |
| Carry 5 | B2 — 11 | A3 — 5 | | |
| | A6 — 3 | A1 — 5 | A5 — 5 | |
| | B1 — 11 | A1 — 13 | A5 — 10 | |
| | B1 — 8 | A2 — 5 | A5 — 13 | |
| | B1 — 6 | A2 — 13 | B1 — 2 | |
| | A2 — 11 | B2 — 12 | | |
| | A2 — 6 | B2 — 9 | | |
| | A1 — 11 | B2 — 4 | | |
| | A1 — 6 | B2 — 1 | | |
| | A1 — 8 | B2 — 2 | | |
| | A2 — 3 | B2 — 5 | | |
| | A2 — 8 | B2 — 10 | | |
| | A3 — 3 | B2 — 13 | | |
| Sum 1 | A5 — 3 | A4 — 10 | | |
| Sum 2 | A5 — 6 | A4 — 5 | | |
| Sum 3 | A5 — 8 | A4 — 2 | | |
| Sum 4 | A5 — 11 | A3 — 13 | | |
| Sum 5 | B1 — 3 | A3 — 10 | | |
| Bsum 1 | A4 — 8 | C1 — 9 | | |

FIGURE 10-9  A typical wire list.

The entry in the *source* column is the module location and pin number of the gate that is the source of the wire run. If a run has multiple sources (3-state or open collector gates are examples), the additional sources could be identified by an asterisk. The entries in the TO columns are the various loads or destinations of the wire run. If a run contains more than three destinations (such as the AD ENABLE run in Fig. 10-9), the TO columns in the succeeding line are filled, with the SOURCE column left blank. No entry in the SOURCE column indicates that the source of the run can be found on the line above.

## 10-5.5  Wiring and Buzz-Out

After the logicals, module charts, and wire list are completed, the circuit is ready for wire wrapping. The simplest procedure is to follow the wire list. When wiring from the list, a third wire should *never* be wrapped on a post. If a wire is listed for a post that already has two wires on it, this is surely an indication of an error, probably the same point listed on two different runs, and the wire list should be checked against the logicals before proceeding.

After completing the wiring, the circuit should be "buzzed out" *before* inserting the ICs and applying power. Buzz-out means checking the continuity of each run in the wire list. In industry this check is often made using a buzzer, or a signal injector and tracer but it can be made with an ohmmeter. As part of the buzz-out, one should always check to be sure *power and ground have* **not** *been wired together.* If this error is made, the results, when power is applied, may be spectacular!

## 10-5.6  Acceptance Test

The *acceptance test* is a series of tests performed on the completed system to determine whether it has met the specifications. Usually if a system passes the acceptance test, it is delivered to the customer. The vendor is then entitled to payment in full, as provided in the contract. Therefore, the customer should document the acceptance test so it is thorough and comprehensive, to be sure that the unit satisfied the requirements. Sometimes partial acceptance tests are made at fixed times after the award of contract so the customer can determine the vendor's progress. These tests, and the penalty for failure, should also be well-documented in the specifications (Sec. 10-5.1).

## 10-5.7  Manuals

For any complex circuit, the vendor must usually write a manual as part of the contract. The manual usually contains a written description of the system, instructions on how to operate it, and a detailed explanation of the systems operation. Most manuals also contain logicals, module charts, wire lists, mechanical drawings, maintenance procedures, and aids for trouble-shooting.

The design engineers are usually responsible for the manuals. Although they may get some help from technical writers, the engineers are the people who really understand how the system operates. Consequently, they must generate at least the first draft of the manual. Engineers are normally used to write specifications, proposals, and instruction sheets, as well as manuals, and *an engineer who writes well is a very valuable employee.*

## 10-6 ERROR DETECTION IN COMBINATORIAL CIRCUITS

We estimate that a circuit with five ICs has no better than a 50 percent chance of working when power is first applied. As the complexity increases, the probability of the circuit working on the first try decreases rapidly, and it is wildly optimistic to expect a complex circuit to work when first turned on.

The process of locating and correcting the faults in a circuit is colloquially called *"debugging."* The engineers who designed the circuit are often called upon to fix or debug it because they best know how it should work. The principal causes of error are faulty components, faulty wiring, and faulty design. Often all three are present, especially in a new design.

The ability to debug a complex circuit is one of the criteria which separates the competent from the mediocre engineer. Routine debugging is often left to technicians, but when a *new system is being tested the engineer is indis-pensible because design errors are as prevalent as any other problem.*

To debug successfully, *one must start by knowing exactly what the circuit is supposed to do.* The circuit should then be tested by placing it in the mode where it *makes errors most frequently,* continually if possible. Now, comparing the way the circuit actually works to the way it was designed to work should reveal the source of the problems. In complex circuits there are often many errors of various types working together to befuddle the engineer. As *soon as any part of a circuit is found to be faulty, it should be repaired.* An entire logic system usually will not work in response to a single fix (in fact, it may appear worse than before), but only by killing the bugs one at a time can one hope for eventual success.

### 10-6.1 Faulty ICs

In combinatorial circuits, errors usually occur for particular values of the input variables. The circuit should be set up using these input variables and then investigated step by step until a gate is found that is not operating as its function table specifies. If a faulty IC is the cause of a problem, its output is generally constant regardless of the inputs. Faulty ICs are usually relatively easy to find, compared to wiring and design errors.

---

**EXAMPLE 10-1**

A circuit is constructed to produce the function

$$f(X,Y,Z) = \Sigma(0,4,5,6)$$

which simplifies, using Karnaugh maps, to:

$$X\bar{Z} + X\bar{Y} + \bar{Y}\bar{Z}$$

The circuit is built as shown in Fig. 10-10 and produces a high output in response to a 1 input (X = 0, Y = 0, Z = 1). Locate the problem.

**SOLUTION**

These values of XYZ are set into the circuit and the logic levels throughout the circuit are checked. The results are also shown in Fig. 10-10. An examination of the circuit reveals that gate 3 is not operating as it should. The 0 input to the top leg of the NAND gate should produce a 1 output.

The problem illustrated in Ex. 10-1 could be caused by a faulty IC and once the IC is located, it should be replaced. As often as not, however, the problem will persist. These other possible sources of error should then be checked:

1. The IC is inserted backwards.
2. Power and ground are not wired properly to the IC socket.
3. Faulty wiring (see Sec. 10-6.2 below).

---

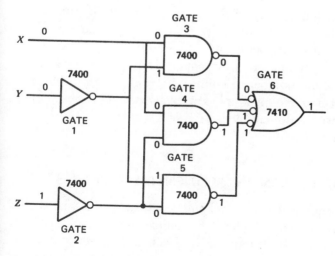

FIGURE 10-10   Circuit for Ex. 10-1.

## 10-6.2 Faulty Wiring

There are several possible ways in which faulty wiring can cause erroneous operation of a circuit.

1. **Floating inputs.** Floating inputs are gate inputs that are not connected to an output of another gate or a switch. Typically they act like a logic 1, but may appear as a logic 0 or a "maybe" (a voltage between 0.8 and 2.0 V, which is neither a 1 or a 0, is sometimes called a maybe) on a CRO. If the top leg of gate 3 is floating, it would cause the error shown. If the CRO indicates a LOW where a floating input is suspected, connect the point to $V_{CC}$ through a 1-k$\Omega$ resistor. If the point is actually floating, it goes to $V_{CC}$, but if the point is genuinely LOW (because of a gate or switch output), it remains LOW despite the pull-up resistor.

2. **Open wires.** Sometimes wires break or open inside the insulation where a break is not visible. Such breaks act as a floating input. An open wire in the X run could cause the problem of Ex. 10-1. Open wires can be detected by noting a different voltage level at each end of the same wire.

3. **Parallel sources.** The outputs of two or more TTL ICs should *never* be tied together, except for open-collector or 3-state ICs. Even when this design rule is followed, a misrouted wire can result in tying two outputs together or tying an output to ground. This can cause an IC to appear faulty and is one of the more difficult errors to detect. A parallel output can cause the output under investigation to be LOW or "maybe" when it should be HIGH. This could be the cause of the error in Fig. 10-10. If the wire between gate 3 and gate 6 were connected to another output, which was LOW, the effect would be as shown. If parallel sources are suspected as the cause or error, and all else fails, unwrap the output. It should go HIGH. Then by reconnecting the wires to the output one at a time and observing which wire causes the output to go LOW, one can begin to trace the faulty wire run and find the other source.

The buzz-out procedure (Sec. 10-5.5) checks each point in a run to see if it is connected. The run could be faulty, however, if it is connected to other points in addition to the points on the wire list. One way to check for this is to count the number of *wire terminations*. This should equal $2N - 2$, where N is the number of points on the list. If a run has five points on the wire list (a source and four loads), each pin should be examined and the number of wire wraps counted. The total should be 8. Any other total indicates a faulty wire run.

## 10-7   ERROR DETECTION IN SEQUENTIAL CIRCUITS

Error detection in sequential circuits (circuits containing FFs, one shots, shift registers, etc.) is far more complex than finding errors in combinatorial circuits. Of course, sequential circuits are susceptible to all the problems of combinatorial circuits, but a circuit that fails because of a faulty FF or a missing wire presents a relatively easy debugging problem. The more difficult and more interesting problems occur when the circuit does different things at different times. This intermittent operation or inconsistent behavior is usually because of FFs being SET or CLEARED at unexpected times, and the problem of determining the cause of these CLEARS or SETS is often quite difficult.

As with combinatorial circuits, the engineer who observes the following principles has the best chance of successfully debugging a sequential circuit.

1. *Thoroughly understand the circuit and know what it should be doing at all times.*
2. *Set the circuit up to operate so that errors occur as frequently as possible.*
3. *Expect multiple errors.*
4. *Compare what the circuit is doing to what it should be doing and correct each specific discrepancy as it is discovered.*

The first principle—thoroughly understanding the circuit—is achieved by studying the specifications and the logicals. The second step must be accomplished by setting the circuit up so that *it does the same thing consistently*, preferrably the *wrong* thing, as often as possible. This erroneous behavior gives the engineer an excellent starting point for debugging.

## 10-7.1  Light and Switch Panels

Light and switch panels help debug complex circuits. These panels, which consist simply of a group of lights and switches, are similar to computer front panels and help the engineer determine how the circuit is functioning. The switches allow him or her to set in various combinations of input variables and the lights monitor the system response. At the start of a project it may seem like extra work to build these panels, but they always pay for themselves by saving debugging time and aggravation.

## 10-7.2  Start-up Procedures

The operation of a sequential circuit is generally controlled by the basic timing of the circuit, and this should be debugged first. If the circuit is controlled by an oscillator there is usually little problem in getting the circuit going, but if it is controlled by a one-shot timing chain (see Ex. 7-10), the one-shots may fail to start. If the timing chain is initiated by a switch, it may be difficult to determine where the problem is. Fortunately, most CROs have lights on them that flash whenever the CRO is triggered. By attaching the CRO probe to the output of the first one-shot and causing the output signal to trigger the CRO, we can determine whether the first one-shot is firing, or if there is something wrong with it. If it *is* firing, the problem is further down the line. By proceeding through the timing chain, one circuit at a time, the problems can be found and eventually the timing circuit will start.

Once the timing chain is operative, repetitive circuit behavior is best observed on the CRO and the engineer stands a much better chance of debugging a circuit. In Chapter 7 many circuits had both SINGLE/CYCLE and CONTINUOUS modes of operation. It is wise to incorporate the CONTINUOUS mode in circuits that are not designed to work continuously or repetitively. In practice (a computer memory would be an example) the CONTINUOUS mode is used while debugging and is well worth the extra effort required to incorporate it.

### 10-7.3  Glitch Hunting

Once the basic timing has been debugged and the circuit placed in repetitive operation, most troubleshooting problems are solved fairly easily. Only the hardiest bugs will survive. Unfortunately, these require the shrewdest engineers to exterminate them.

At this stage of debugging, *most problems are caused by FFs being SET or CLEARED at the wrong time because of design errors and/or glitches.* Design errors generally allow pulses to occur at the wrong time. Glitches are often caused by race conditions and are shorter pulses that are much more difficult to see on a CRO. Unfortunately, by the time the error is apparent, its causes have generally disappeared. Therefore, if the circuit can be run in a loop and FF inputs observed before the FF erroneously SETS, we may succeed in observing the error and tracing its cause. When building a computer, for example, it is wisest to debug the jump instruction first. Then the computer can be set up to execute an instruction or series of instructions, jump back and do it again. This sets up a repetitive loop and the action of the instructions can be observed.

Some errors may occur only when the circuit is in a certain mode of operation, or when a certain FF is SET or CLEAR, and these conditions may occur relatively rarely. The frequency of occurrence of these conditions can be increased by the use of temporary jumpers to ground, which can artificially hold FFs CLEARED or SET gate outputs to desired levels. Although the entire circuit will not operate properly with these jumpers in place, they can often help to isolate and detect an error.

Glitches are caused by race conditions (Sec. 6-13), transient circuit outputs (such as the decoder of Sec. 8-5.1), or electrical noise or crosstalk. Careful attention must be paid to the design of the SET, CLEAR, and CLOCK inputs of FFs to prevent spurious SETS. In worst cases, a glitch may occur very infrequently and last for only a few nanoseconds. It is almost impossible to see on a CRO and its cause must be inferred from the action of the circuit. A digital circuit that fails one time in a million is virtually useless! If a computer executes 200,000 instructions in a second and errs once in every million instructions, we can expect it to run for only 5 seconds before chaos sets in! Finding the cause of these invisible and infrequent glitches is perhaps the most difficult and challenging of debugging problems.

When working with a computer, keep in mind most errors can also be caused by the program. Determining whether the source of an error is in hardware or software requires a grasp of programming. This is another problem in advanced debugging that is beyond the scope of this book.

## 10-7.4   Turn-on Procedures

Every system should incorporate a MASTER CLEAR or POWER ON RESET circuit. It is the function of this pulse to CLEAR or SET all the critical FFs so that a system may start properly. This pulse should also be capable of being generated electronically by a switch or circuitry so that it can be used as a last resort to clear a circuit that "hangs up," or locks into a particular state and will not change. Hangup conditions are another source of error that must be avoided.

## 10-7.5   Synchronous and Asynchronous Behavior

In *synchronous* circuits, all events are controlled by a clock and errors or glitches normally occur near the clock edges. In *asynchronous* circuits, events are often sequential; one event starts when the preceding event finishes. Because of the greater variation in time, glitches and races can occur more frequently in asynchronous circuits.

Sometimes even synchronous circuits must communicate with other circuits that send it pulses asynchronously. It is often wise to synchronize these pulses by gating them with the circuit clock, rather than allowing them to enter whenever they are generated. This guards against the possibility of a pulse entering at precisely the wrong time and causing an unforseen race or glitch. The use of synchronizing FFs (see Sec. 6-16) is recommended for troublesome circuits.

## 10-7.6   Logic Probes

Logic probes are a new innovation in digital testing. A typical logic probe is powered from $V_{CC}$ of the system under test. It contains a metal tip that is touched to the point under test. There are usually two indicator lights that indicate a 1 or a 0. If neither indicator glows, the point under test is floating (unconnected). The ability to recognize a floating input is an advantage of logic probes.

Most logic probes also have the ability to detect single pulses. A short single pulse causes a light on the probe to glow for perhaps 200ms, so that its occurrence is clearly visible. Some probes contain a light that stays on after being triggered by a pulse until a switch is depressed. Logic probes can detect the occurrence of short pulses where they should not normally appear.

The circuit for a logic probe is shown in Fig. 10-11. A 0 on the input lights the top Light-Emitting Diode (LED) and a 1 lights the lower LED. If a short pulse or glitch appears on the input, it triggers the **74121,** which lights the pulse LED for 100 ms, so the user can see it and is aware of its existence.

## 10-8   LOGIC ANALYZERS[1]

As digital systems become larger and more complex, the problems of de-bugging them have increased. Modern digital systems consist of many con-stantly changing lines. A typical microprocessor ($\mu$P), for example, has 8 data lines, 16 address lines, and several control lines, all of which change on every $\mu$P clock cycle. To diagnose and fix a fault in such a system requires sophisticated test instruments such as a *logic analyzer* to supplement or replace the oscilloscope.

A typical logic analyzer is the **7D01,** manufactured by Tektronix, Inc., shown in Fig. 10-12. Logic analyzers consist of five basic parts:

1. Multiple parallel inputs
2. A memory
3. A sample clock
4. A trigger
5. A display

### 10-8.1   Logic Analyzer Inputs

The multiple inputs to a logic analyzer are brought in via *pods*. A pod is shown in Fig. 10-13, and is a group of input probes. The **7D01** uses two pods to bring in 16 data lines plus clock and probe qualifier information. The probes, which are connected to the lines under test, are high impedance and low capacitance (1 M$\Omega$ in parallel with 5 pf for the 7D01), so that high speed integrity is maintained and the circuit under test is not disturbed.

---

[1]The reader should be familiar with memories to fully understand logic analyzers. Those students who have no experience with memories may want to defer reading this section until they have read Chapter 15 (Memories).

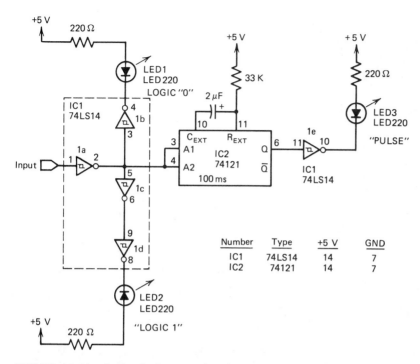

FIGURE 10-11 A simple logic probe that uses two integrated circuits. When a logic-0 signal voltage is applied to the input, the "logic 0" LED will light. When a logic-1 signal voltage is applied to the input, the "logic 1" LED indicator will light. If the input oscillates between the 0 and 1 states, the "Pulse" LED indicator will also light. (Courtesy of Steve Ciarcia and *Byte* Magazine.)

## 10-8.2 The Memory

Logic analyzers require a high speed memory. When sampling the data, inputs are written into memory on each positive or negative edge of the sampling clocks. The dimensions of the memory can be changed depending on the number of data lines coming in. The **7DO1** has a 4K memory that can be partitioned as 256 words by 16 bits, 512 words by 8 bits, or 1K words by 4 bits.

## 10-8.3 The Sample Clock

When in sample mode data are constantly being clocked into the memory, with the oldest data being overwritten by the newest data. The logic analyzer can generate the sample clock rate. In the **7DO1** sample rates run from 10 ns to 5 ms. The sample rate should be selected to catch all data changes.

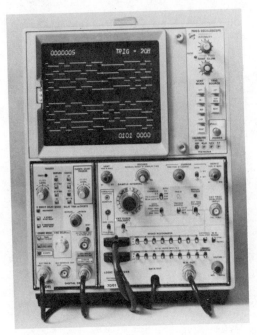

FIGURE 10-12    The Tektronix 7D01 Logic Analyzer. (Photo courtesy of Tektronix, Inc.)

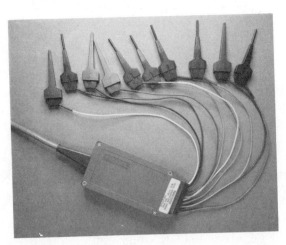

FIGURE 10-13    A logic analyzer pod. (Photo courtesy of Ed Pickett.)

Thus a system that changes slowly will generally be sampled at a low rate, while a high speed system will often be sampled at rates from 100 ns to 1 $\mu$s.

In the 7DO1 the sample clock rate is selected by a rotary switch on the front panel. The switch has an EXTERNAL position, which is often used. In EXTERNAL the clock rate is determined by the clock of the system under test. This *synchronizes* the logic analyzer with the test system, and the logic analyzer records changes as they occur in the test system. On the 7DO1 the clock line can come on the low order pod.

## 10-8.4   The Trigger

Logic analyzers are set up to accept a trigger pulse. When the trigger occurs the analyzer fills the memory and then stops sampling and freezes the memory so it can be displayed and the progress of the system up to and shortly after the trigger can be observed.

The trigger can enter the logic analyzer in one of three ways:

- **External**—the logic can be triggered by an external pulse.
- **Channel 0**—a pulse on channel 0 of the input pod can cause a trigger.
- **Word Recognizer**—the word recognizer is a group of switches on the front panel. There is one switch for each input. These are three position switches for 1, 0, and don't care. *The logic analyzer triggers only when the word recognizer switches match the data inputs* to the logic analyzer. Setting a switch in the don't care position effectively removes that switch from the comparison.

There are three modes of triggering; center triggering, pre-triggering, and post-triggering. If the logic analyzer contains a 256-word memory, for example, and center triggering is used, the logic analyzer will retain the 128 words before and after the trigger. The user who is primarily interested in data after the trigger can choose the pre-trigger mode. The analyzer will then display only 16 words before the trigger and 240 words after the trigger. The post trigger mode is similar except that it displays 240 words before the trigger and 16 words after.

## 10-8.5   Displays

After the trigger, sampling stops so the user can examine the data in the memory. One of the ways of displaying these data is by a timing diagram, as shown in Figure 10-12.

The timing diagram shows the data on all the input lines as a function of time and allows the user to check the progress of the system.

The timing display gives the *illusion* of being a multiple trace oscilloscope display, but it is really *discrete samples* connected together by straight lines.

A *binary state table* is a second mode of display. The 1s and 0s on the inputs at each clock sample can be displayed, as shown in Fig. 10-14. Hex is an alternate form of this display and is preferred by many users. A third type of display, called mapping, also exists. In this mode the screen is divided into a 64 by 64 grid, giving 65,536 data points. The progress of the data on this grid can be followed. Stray data points appearing in any area where they do not normally occur can quickly be found in the mapping mode.

---

**EXAMPLE 10-2**

Due to a fault, a microprocessor ($\mu$P) is going to a location around 2100. How can a logic analyzer help solve this problem?

**SOLUTION**

Even though the erroneous address is not fully known, the logic analyzer can help. The inputs can be put on the $\mu$P's address bus. The word recognizer can be set to $(21XX)_{16}$, where XX means that eight least significant switches are set to don't cares, and sampling started using the $\mu$P's clock. When the $\mu$P reaches the first address starting with 21, wherever it is, the logic analyzer will trigger and display not only the first address, but the addresses that preceded the jump to the 2100 area. This

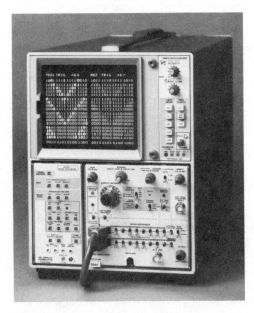

FIGURE 10-14 The **7001** displaying in binary mode. (Photo courtesy of Tektronix, Inc.)

should allow the user to examine these locations to discover what caused the erroneous jump.

## 10-8.6    Other Logic Analyzer Features

There are many other features available in logic analyzers. Space limitations preclude a full explanation of these features but we can list them briefly:

- **Personality modules**—these tailor a logic analyzer to a specific $\mu$P and display the data as that particular $\mu$P's instructions. This is called a disassembler.
- **Probe qualification**—some logic analyzers contain an additional input called a *probe qualifier*. This is not a data input (it cannot be displayed), but is used to affect triggering. When used, the word recognizer will not trigger unless all data and the probe qualifier match the corresponding switches.
- **Clock qualification**—some logic analyzers contain a *clock qualifier*. The logic analyzer will not accept a sample clock unless the clock qualifier is also correct.
- **Reference memory**—this is an auxiliary memory. The data in the main memory can be transferred into the auxiliary memory and then compared with the data entered into another section of the main memory, or data entered into the main memory at a later time. This is valuable for tracking down intermittents and for comparing the operation of the system at two different times. The 7D01 can highlight differences between the main memory and the reference memories.
- **Glitch latch**—this is an additional circuit used to capture and display any glitches that may not be detected by an ordinary analyzer if the glitch occurs between the sample clocks.

Logic analyzers are rapidly evolving, and becoming more complex. Most of the newer ones are programmable. We regret that we could only touch on the highlights in this section, but a thorough description would be highly dependent on the particular logic analyzer being used, and would require a small book of its own.

### SUMMARY

In this chapter, advantages and disadvantages of wire wrapping and PC board construction were explored. A method of documenting wire wrap construction was presented. Most companies set their own drafting, wire list, and module chart standards, but usually they are similar to those presented here.

The chapter concludes with two sections on debugging. Frankly, debugging is as much art as science and depends on the insight and ingenuity of the engineer. No book can hope to present a specific method for curing the myriad of problems that arise in practice. Here we presented some basic methods used to attack faulty circuits. Modern, sophisticated test instruments such as logic analyzers were introduced. Using these as starting points, we hope the reader can proceed until his or her particular circuit is functioning properly.

## GLOSSARY

**Prototype.** A preproduction model of a system built for testing and debugging.

**Wire list.** A list of the routing of each wire in a system.

**PC board.** Printed circuit board.

**Debugging.** Correcting the faults in a circuit or system.

**Maybe (Colloquial).** A logic level between 0 and 1.

**Buzz-out.** A check of the wiring of a circuit.

**Logical.** A drawing showing the ICs in a system and their interconnections.

**Module chart.** A chart showing the locations of the ICs in a system.

**Floating input.** An input that is not tied to a source.

**Hang-up condition.** A condition where the state of a system will not change.

## REFERENCES

Thomas R. Blakeslee, *Digital Design With Standard MSI and LSI*, Wiley, New York, 1975.

Steve Ciarcia, "Build A Low-Cost Logic Analyzer," BYTE *Magazine*, April 1981.

# 11

# MULTIPLEXERS AND DEMULTIPLEXERS

## 11-1  INSTRUCTIONAL OBJECTIVES

This chapter introduces multiplexers and demultiplexers, and presents some of their applications. After reading it, the student should be able to:

1.  Digitally multiplex an $n$-line input to a 1-line output.
2.  Understand and utilize 7400 series multiplexers.
3.  Design 1-line-to-$n$-line demultiplexers and utilize 7400 series demultiplexers.
4.  Design and utilize decoders.
5.  Use multiplexers to generate combinatorial functions of several variables.
6.  Design a keyboard decoder.

## 11-2  SELF-EVALUATION QUESTIONS

Watch for the answers to the following questions as you read the chapter. They should help you understand the material presented.

1.  Define *multiplexing* and *demultiplexing*.
2.  How many select lines are required by an $n$-line-to-1-line multiplexer?
3.  Is an unselected output of a TTL demultiplexer 0 or 1? Explain.
4.  What is the level of the outputs of an unselected demultiplexer (a demultiplexer that is inactive because of a HIGH strobe input)?

5. What is the difference between a demultiplexer and a decoder? How can a demultiplexer be made to function as a decoder?

6. Explain why an *n*-1 input multiplexer is used to generate a function of *n* variables.

7. On a keyboard decoder, why is scanning essential?

## 11-3 MULTIPLEXERS

*Multiplexing is the funneling of information from one of several input lines to a single output line.* The inputs to a multiplexer are the several input lines, and the SELECT inputs determine which one of the input lines is connected to the output.

The basic multiplexer scheme is shown in Fig. 11-1. Only one switch is closed at any one time, and connects the selected input to the output. There are many instances where one of several possible inputs to a circuit must be selected (an accumulator in a computer is an example), and multiplexers are an ideal solution to the problem.

Multiplexing is often used when information from many sources must be transmitted over long distances, and it is less expensive to multiplex the data onto a single wire for transmission. There are two ways to multiplex information from many sources onto a single line: Frequency Division Multiplexing (FDM) and Time Division Multiplexing (TDM). In both cases, the bandwidth of the transmitting line must be larger than the bandwidth (or frequency components) of the information to be transmitted.

In FDM each signal is *modulated* up to a specific band and transmitted. Thus, if 3 kHz is the highest frequency of each signal (telephone conversations are an example), one signal may be sent from 3–6 kHz, another from 7–10 kHz, and so on until the bandwidth of the transmission media is exhausted.

In TDM each signal gets a *specific time slot* on a transmission line. TDM is used more often with digital circuits. Example 11-1 shows the design of

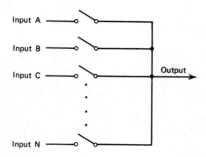

FIGURE 11-1 The basic multiplexer.

a multiplexer using gates and shift registers; it is also an example of TDM because each input is on the line for 1 $\mu s$ during each 8-$\mu s$ cycle.

## EXAMPLE 11-1

Given eight input lines, A–H, design a circuit to connect each input line to the output line sequentially for 1 $\mu s$ (i.e., input A is connected to the output for the first 1 $\mu s$, input B is connected to the output at 2 $\mu s$, etc.). The device is cycled continuously so that input A is connected to the output after input H.

## SOLUTION

One solution is shown in Fig. 11-2. The 1-$\mu s$ oscillator and the shift register control the timing and selection for the circuit (see Sec. 9-7.1). Only one shift register output is LOW at any one time. Because of the 1-MHz clock, the LOW output advances once each microsecond.

The shift register output is ORed with the corresponding input by the **7432** gates. All unselected lines produce a HIGH input to the **7430**. If the selected input is HIGH, all inputs to the **7430** are HIGH, and it produces a LOW output that is inverted to make the final output HIGH. If the selected input is LOW, however, its **7432** gate produces a LOW output that causes the **7430** output to be HIGH and the final output to be LOW. *Thus the final output is the same as the input on the selected line.*

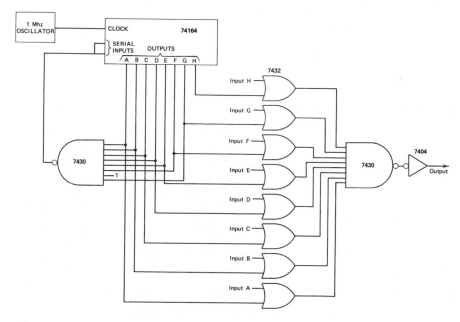

FIGURE 11-2  An 8-line-to-1-line multiplexer.

## 11-3.1 The 74153 4-line to 1-line Multiplexer

Fortunately, a variety of IC multiplexers and demultiplexers exist, so there is little need to construct multiplexers from SSI gates. The functional block diagram and function table for the **74153** are shown in Fig. 11-3. This is a dual 4-line-to-1-line multiplexer and the two SELECT inputs simultaneously control both sections of the multiplexer. If the A and B SELECT inputs are 0 and 1, for example, the data bits at 1C2 and 2C2 are transmitted to the 1Y and 2Y outputs. The data at the other inputs are irrelevant until the SELECT inputs change. The **74153** also has an ENABLE or STROBE line for each section. If the STROBE input to a section is HIGH, the output of that section is LOW.

---

### EXAMPLE 11-2

Given three 4-bit registers, A, B, and C, design a circuit whose output is either the contents of registers A, or B, or C, or the binary equivalent of 9.

### SOLUTION

One of four possible inputs may appear on the output. This suggests a 4-line-to-1-line multiplexer. To choose 1 of 4 inputs, 2 SELECT bits are required. The

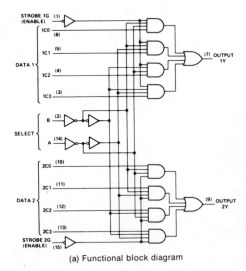

| SELECT INPUTS | | DATA INPUTS | | | | STROBE | OUTPUT |
|---|---|---|---|---|---|---|---|
| B | A | C0 | C1 | C2 | C3 | G | Y |
| X | X | X | X | X | X | H | L |
| L | L | L | X | X | X | L | L |
| L | L | H | X | X | X | L | H |
| L | H | X | L | X | X | L | L |
| L | H | X | H | X | X | L | H |
| H | L | X | X | L | X | L | L |
| H | L | X | X | H | X | L | H |
| H | H | X | X | X | L | L | L |
| H | H | X | X | X | H | L | H |

Select inputs A and B are common to both sections.
H = high level, L = low level, X = irrelevant

(a) Functional block diagram     (b) Function table

FIGURE 11-3   The **74153** multiplexer. (From the *TTL Data Book for Design Engineers,* 2nd ed., Texas Instruments, Inc. Courtesy of Texas Instruments, copyright 1976.)

selection table below was made by arbitrarily assigning an output to each combination of SELECT bits.

| SELECT Bits | | Output |
|---|---|---|
| B | A | |
| 0 | 0 | Register A |
| 0 | 1 | Register B |
| 1 | 0 | Register C |
| 1 | 1 | 9 |

Since a 4-bit output is required, this circuit can be built using two **74153**s, as shown in Fig. 11-4. The selection table requires that Register A is connected to the C0 inputs, Register B to the C1 inputs, Register C to the C2 inputs, and the number 9 is wired to the C3 inputs. The output is one of the four inputs depending on the SELECT line levels as indicated by the selection table.

## 11-3.2 The 74157

The **74157** is a quadruple 2-line-to-1-line multiplexer. Its functional block diagram is shown in Fig. 11-5. A HIGH input on the STROBE line disables the IC by forcing all outputs LOW. With the STROBE LOW, the level on the SELECT line determines whether the A or B inputs are connected to the outputs. The price paid for the larger number of output bits per IC (4) is the reduced number of input choices (2). The **74157** is ideal for choosing one of two long data registers.

**EXAMPLE 11-3**

Two 12-bit registers are to be multiplexed together. Design the circuit.

**SOLUTION**

Since each **74157** accommodates 4 bits, three **74157**s are needed. The first 12-bit register is tied to the A inputs of the three multiplexers, and the second register is tied to the B inputs. A single SELECT line is tied to all the **74157**s. If it is HIGH, the second register is selected and if it is LOW, the first register is selected.

The **74LS257** and **74LS258** are newer 2-line to 1-line multiplexers that function very much like the **74157**. These multiplexers have 3-state outputs, however, and the strobe line is replaced by an OUTPUT CONTROL signal that must be LOW for the multiplexer to drive the bus. Otherwise, the

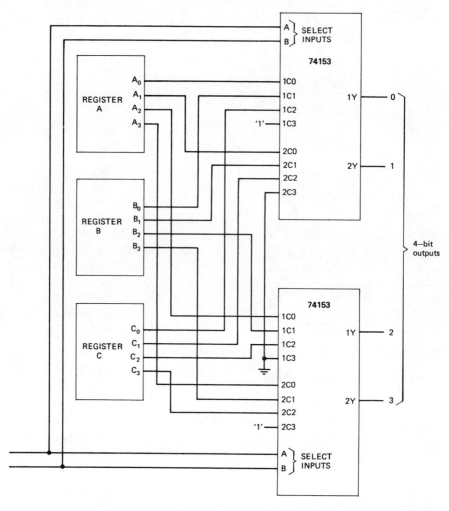

FIGURE 11-4 Circuit for Ex. 11-2.

multiplexer outputs are high impedance. These ICs are ideal for driving multiplexed inputs onto a 3-state bus and are used in the computer of Sec. 16-9 for that purpose. The difference between the '257 and the '258 is that the '258 inverts its outputs.

### 11-3.3 Larger Multiplexers

The 74150, 74151, and 74152 are multiplexers that can accommodate 8 or 16 inputs. The functional diagram of the three multiplexers is shown in Fig. 11-6.

'157, 'L157

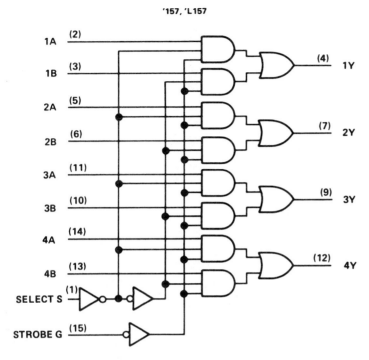

FIGURE 11-5 Functional Block diagram of the **74157**. (From the *TTL Data Book for Design Engineers,* 2nd ed., Texas Instruments, Inc. Courtesy of Texas Instruments, copyright 1976.)

The **74150** is a 16-line-to-1-line multiplexer. Since 16 inputs and 4 SELECT lines are required, the IC comes in a 24-pin DIP. Unlike the **74153** or **74157**, the **74150** *inverts* the selected input. There is also a STROBE on the **74150** that forces the output HIGH whenever it goes HIGH.

The **74151** and **74152** are both 8-line-to-1-line multiplexers. The **74151** comes in a 16-pin package and provides outputs for inverted and noninverted data and a STROBE INPUT. When the STROBE is HIGH, the W output is HIGH and the Y output is LOW regardless of the inputs. The **74152** comes in a 14-pin package. It has no provision for a STROBE and provides inverted output data only.

---

**EXAMPLE 11-4**

Design a 64-line-to-1-line multiplexer. Show how it would select the 37th line.

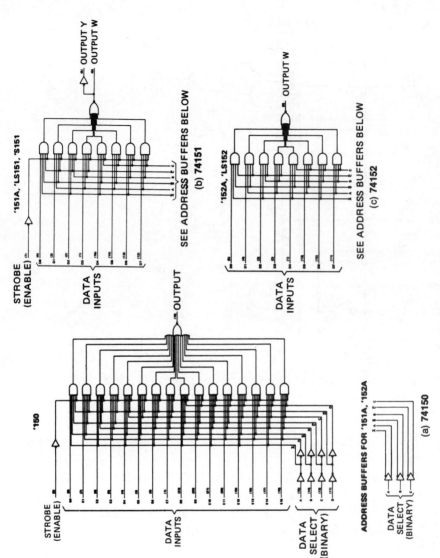

FIGURE 11-6 Functional block diagrams of the **74150**, **74151**, and **74152** multiplexer. (From the *TTL Data Book for Design Engineers*, 2nd ed., Texas Instruments, Inc. Courtesy of Texas Instruments, copyright 1976.)

## SOLUTION

One solution is shown in Fig. 11-7. Since 1 of 64 lines is being chosen, 6 lines are required to make the selection. In Fig. 11-7, the 64 input lines are applied to four **74150**s, and 4 SELECT lines are applied to all the **74150**s. The outputs of the **74150**s are then connected to one section of a **74153**. The two remaining SELECT lines are connected to the **74153** and decide which of its 4 inputs is funneled through to the output. The final output is inverted to compensate for the inversion of the **74150**.

To select the 37th line, the SELECT lines must be binary 37, 100101. The 4 LSBs are applied to each **74150** and select input 5. The 2 MSBs (10) are connected

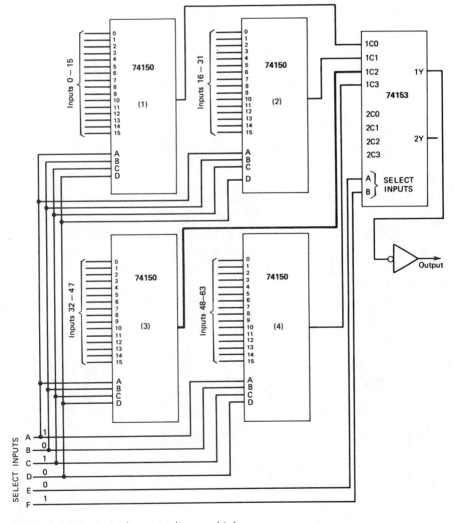

FIGURE 11-7   A 64-line to-1-line multiplexer.

to the SELECT lines on the **74153** and select the third **74150**. Thus input 5 of the third **74150** is the line connected to the output. The selected data path is shown as a dark line on Fig. 11-7.

---

## 11-4 DEMULTIPLEXERS

Demultiplexers reverse the multiplexing operation; they take a single input line and fan it out to one of many output lines. A demultiplexer, therefore, connects the input to the output of the selected line. All other outputs are deselected. For IC demultiplexers, *deselected outputs always assume a HIGH level*.

The demultiplexing principle is illustrated in Fig. 11-8, where it is assumed that only one switch is closed at any one time. Demultiplexers have 1 input and *n* output lines. They also require enough SELECT inputs to specify which one of the *n* output lines is connected to the input line. Like multiplexers, demultiplexers can be designed using SSI gates, but it is generally preferable to use IC demultiplexers.

### 11-4.1 The 74155

The **74155** is a dual 1-line-to-4-line demultiplexer. Its functional block diagram and function table are given in Fig. 11-9. For each section, the data input (labeled 1C or 2C) is connected to one of the four outputs (Y0, Y1, Y2, or Y3), as determined by the two SELECT lines.

The active output level for the demultiplexer is LOW. Consequently, all unselected outputs present a HIGH (inactive) output. If the STROBE input to either section is HIGH, it turns that section off by forcing all outputs HIGH (inactive). The STROBE must be LOW or grounded for the demultiplexer to function.

As seen in the function table (Fig. 11-9b), the selected output of the top section of the **74155** is the complement of the input data on line 1C.

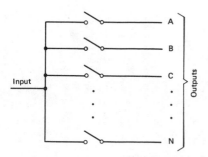

FIGURE 11-8 The basic demultiplexer.

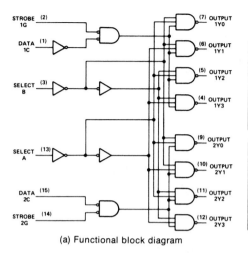

(a) Functional block diagram

FUNCTION TABLES
2-LINE-TO-4-LINE DECODER
OR 1-LINE-TO-4-LINE DEMULTIPLEXER

| INPUTS | | | | OUTPUTS | | | |
|---|---|---|---|---|---|---|---|
| SELECT | | STROBE | DATA | | | | |
| B | A | 1G | 1C | 1Y0 | 1Y1 | 1Y2 | 1Y3 |
| X | X | H | X | H | H | H | H |
| L | L | L | H | L | H | H | H |
| L | H | L | H | H | L | H | H |
| H | L | L | H | H | H | L | H |
| H | H | L | H | H | H | H | L |
| X | X | X | L | H | H | H | H |

| INPUTS | | | | OUTPUTS | | | |
|---|---|---|---|---|---|---|---|
| SELECT | | STROBE | DATA | | | | |
| B | A | 2G | 2C | 2Y0 | 2Y1 | 2Y2 | 2Y3 |
| X | X | H | X | H | H | H | H |
| L | L | L | L | L | H | H | H |
| L | H | L | L | H | L | H | H |
| H | L | L | L | H | H | L | H |
| H | H | L | L | H | H | H | L |
| X | X | X | H | H | H | H | H |

(b) Function tables

FIGURE 11-9 The **74155** demultiplexer. (From the *TTL Data Book for Design Engineers,* 2nd ed., Texas Instruments, Inc. Courtesy of Texas Instruments, copyright 1976.)

Consequently, if the input at 1C is LOW, all outputs are HIGH, regardless of the SELECT and STROBE lines. In the lower section of the multiplexer, the selected 2Y output is the same as input 2C. *The lower section does not invert.* The two sections were designed to be different in order to facilitate decoding (see Sec. 11-5.1).

---

**EXAMPLE 11-5**

(a) Design a 1-line-to-8-line demultiplexer using a **74155.** If the input is HIGH, the output on the selected line must be LOW.

(b) Show how to place a LOW level at output 2Y2.

**SOLUTION**

(a) For an 8-line demultiplexer, 3 SELECT bits are required. The circuit shown in Fig. 11-10 uses a single **74155.** Two SELECT inputs are applied directly to the 74155s. The third SELECT bit is used to drive two STROBES so that one section of the **74155** is always disabled. The data input to section 2 is inverted to compensate for the lack of inversion of the second section.

(b) To cause output 2Y2 to be LOW:

1. SELECT C must be HIGH enabling section 2 and disabling section 1 of the **74155.**

2. The data input line must be HIGH. This places a LOW level on data input 2C. (Section 2 does not invert the data.)

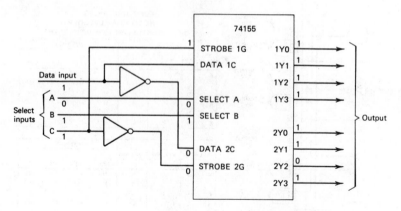

FIGURE 11-10 A 1-line-to-8-line inverting demultiplexer.

3. The A and B SELECT inputs must select 1Y2 and 2Y2 as the outputs. SELECT B must be HIGH and SELECT A LOW. Only output 2Y2 goes LOW because section 1 is disabled. The 1s and 0s for the required output are also shown on Fig. 11-10.

## 11-4.2 The 74154

The **74154** is a 1-line-to-16-line demultiplexer in a 24 pin DIP whose functional block diagram and function table are shown in Fig. 11-11. The four SELECT inputs choose 1 of the 16 outputs. The selected output goes LOW only if inputs G1 and G2 are both LOW, so that one of the G inputs can act as a STROBE and the other can function as the input. If either one of the G inputs is HIGH, all 16 outputs of the **74154** will be HIGH.

**EXAMPLE 11-6**

Design a 1-line-to-32-line demultiplexer.

**SOLUTION**

A solution using two **74154s** is shown in Fig. 11-12. Five SELECT lines are required. Four of them are applied directly to the **74154s**. The fifth SELECT line is applied to the G1 input of one **74154** and its complement is applied to the G1 input of the other. Only the **74154** with the LOW G1 input is enabled. The two G2 inputs are tied together to become the demultiplexer input. If this input is LOW, the selected output is LOW.

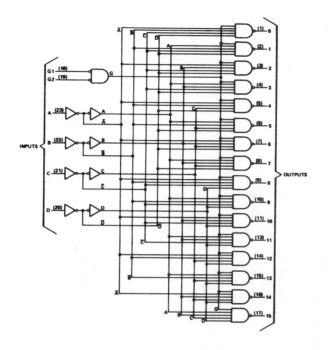

FUNCTION TABLE

| INPUTS | | | | | | OUTPUTS | | | | | | | | | | | | | | | |
|---|---|---|---|---|---|---|---|---|---|---|---|---|---|---|---|---|---|---|---|---|---|
| G1 | G2 | D | C | B | A | 0 | 1 | 2 | 3 | 4 | 5 | 6 | 7 | 8 | 9 | 10 | 11 | 12 | 13 | 14 | 15 |
| L | L | L | L | L | L | L | H | H | H | H | H | H | H | H | H | H | H | H | H | H | H |
| L | L | L | L | L | H | H | L | H | H | H | H | H | H | H | H | H | H | H | H | H | H |
| L | L | L | L | H | L | H | H | L | H | H | H | H | H | H | H | H | H | H | H | H | H |
| L | L | L | L | H | H | H | H | H | L | H | H | H | H | H | H | H | H | H | H | H | H |
| L | L | L | H | L | L | H | H | H | H | L | H | H | H | H | H | H | H | H | H | H | H |
| L | L | L | H | L | H | H | H | H | H | H | L | H | H | H | H | H | H | H | H | H | H |
| L | L | L | H | H | L | H | H | H | H | H | H | L | H | H | H | H | H | H | H | H | H |
| L | L | L | H | H | H | H | H | H | H | H | H | H | L | H | H | H | H | H | H | H | H |
| L | L | H | L | L | L | H | H | H | H | H | H | H | H | L | H | H | H | H | H | H | H |
| L | L | H | L | L | H | H | H | H | H | H | H | H | H | H | L | H | H | H | H | H | H |
| L | L | H | L | H | L | H | H | H | H | H | H | H | H | H | H | L | H | H | H | H | H |
| L | L | H | L | H | H | H | H | H | H | H | H | H | H | H | H | H | L | H | H | H | H |
| L | L | H | H | L | L | H | H | H | H | H | H | H | H | H | H | H | H | L | H | H | H |
| L | L | H | H | L | H | H | H | H | H | H | H | H | H | H | H | H | H | H | L | H | H |
| L | L | H | H | H | L | H | H | H | H | H | H | H | H | H | H | H | H | H | H | L | H |
| L | L | H | H | H | H | H | H | H | H | H | H | H | H | H | H | H | H | H | H | H | L |
| L | H | X | X | X | X | H | H | H | H | H | H | H | H | H | H | H | H | H | H | H | H |
| H | L | X | X | X | X | H | H | H | H | H | H | H | H | H | H | H | H | H | H | H | H |
| H | H | X | X | X | X | H | H | H | H | H | H | H | H | H | H | H | H | H | H | H | H |

H = high level, L = low level, X = irrelevant

FIGURE 11-11   The **74154** 1-line-to-16-line demultiplexer. (From the *TTL Data Book for Design Engineers,* 2nd ed., Texas Instruments, Inc. Courtesy of Texas Instruments, copyright 1976.)

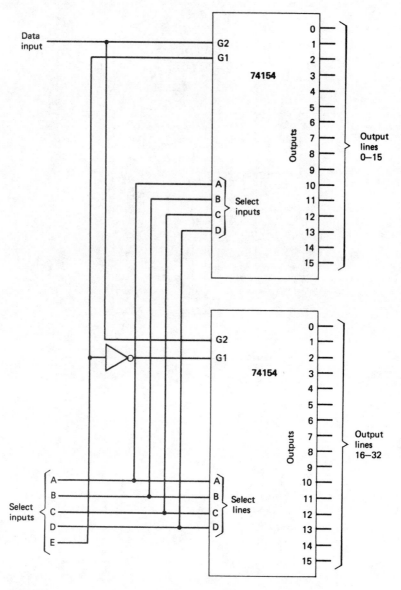

FIGURE 11-12   A 1-line-to-32-line demultiplexer.

## 11-5   PRACTICAL APPLICATIONS

In large digital systems, multiplexers and demultiplexers are used in a variety of ways. Three examples of their use are presented in this section. An example of their use in data transmission systems is presented in Sec. 18-5.2.

## 11-5.1  Decoders

Demultiplexers can be used as digital decoders. A digital decoder has $2^n$ outputs and accepts $n$ inputs. Only the output that corresponds to the binary number on the input lines is activated.

The 74154 (Fig. 11-11) can be used as a 4-line-to-16-line decoder simply by tying both G inputs LOW. Then the output corresponding to the SELECT input is LOW and all other outputs are HIGH.

---

**EXAMPLE 11-7**

Design a 5-line-to-32-line decoder.

**SOLUTION**

The decoder is built by tying the two G2 inputs of the 5-line-to-32-line demultiplexer (Fig. 11-12) to ground. Four of the five required SELECT lines are tied to the 74154s (they select one output on each chip) and the fifth SELECT input is tied to the G1 inputs (to disable one of the 74154s). Consequently, for any combination of the 5 SELECT inputs, one and *only one* output line is decoded. Its output is LOW. The other 31 outputs are all HIGH.

---

A 3-line-to-8-line decoder can be built from a **74155** by tying the data 1C and data 2C lines together. The three required SELECT lines consist of SELECT A, SELECT B, and the common data lines. Because section 1 inverts its data and section 2 does not, one of the outputs is always LOW, provided the STROBES are LOW. A function table for the **74155** used as a 3-line-to-8-line decoder is shown in Fig. 11-13.

Demultiplexers can be used as address selectors for core memories (see Chapter 15). In a 4096 word memory, for example, the 12 required address bits are usually broken into two groups of 6 each. Decoding 1 line out of 64 from the 6 address bits is a typical problem.

Decoders for BCD inputs are also available. The **7442** accepts 4 SELECT inputs in BCD form and places a 0 on the corresponding output line. If the number on the SELECT inputs is invalid (it is greater than 9), all outputs remain HIGH.

## 11-5.2  Multiplexer Logic

In Chapters 2 and 3, combinatorial logic functions of several variables were generated using Boolean algebra or Karnaugh mapping techniques. A combinatorial function can also be generated by a multiplexer:

1.  Write the required function in SOP (sum-of-products) form.

**FUNCTION TABLE**
**3-LINE-TO-8-LINE DECODER**
**OR 1-LINE-TO-8-LINE DEMULTIPLEXER**

| INPUTS | | | | OUTPUTS | | | | | | | |
|---|---|---|---|---|---|---|---|---|---|---|---|
| SELECT | | | STROBE OR DATA | (0) | (1) | (2) | (3) | (4) | (5) | (6) | (7) |
| C† | B | A | G‡ | 2Y0 | 2Y1 | 2Y2 | 2Y3 | 1Y0 | 1Y1 | 1Y2 | 1Y3 |
| X | X | X | H | H | H | H | H | H | H | H | H |
| L | L | L | L | L | H | H | H | H | H | H | H |
| L | L | H | L | H | L | H | H | H | H | H | H |
| L | H | L | L | H | H | L | H | H | H | H | H |
| L | H | H | L | H | H | H | L | H | H | H | H |
| H | L | L | L | H | H | H | H | L | H | H | H |
| H | L | H | L | H | H | H | H | H | L | H | H |
| H | H | L | L | H | H | H | H | H | H | L | H |
| H | H | H | L | H | H | H | H | H | H | H | L |

†C = inputs 1C and 2C connected together
‡G = inputs 1G and 2G connected together
H = high level, L = low level, X = irrelevant

FIGURE 11-13 Function table for a **74155** used as a 3-line-to-8-line decoder. (From the *TTL Data Book for Design Engineers,* 2nd ed., Texas Instruments, Inc. Courtesy of Texas Instruments, copyright 1976.)

2. If $n$ input variables are specified, choose a multiplexer with $n - 1$ SELECT lines.

3. The input lines to the multiplexer numerically correspond to the $n - 1$ MSBs of specified function.

4. The LSB of the specified function determines the signal on each input line. This signal is obtained by examining the SOP form of the specification.

The procedure is given in Ex. 11-8, below.

---

**EXAMPLE 11-8**

If $f(W,X,Y,Z) = \Sigma(0,2,3,6,9,10,14,15)$, realize $f$ using a multiplexer.

**SOLUTION**

1. In SOP form:

$$f(W,X,Y,Z) = \bar{W}\bar{X}\bar{Y}\bar{Z} + \bar{W}\bar{X}Y\bar{Z} + \bar{W}\bar{X}YZ + \bar{W}XY\bar{Z} + W\bar{X}\bar{Y}Z \quad (11\text{-}1)$$
$$+ W\bar{X}Y\bar{Z} + WXY\bar{Z} + WXYZ$$

2. The given function depends upon four variables. Therefore, the required multiplexer must have 3 SELECT lines (step 2); so a **74152**, 8-line-to-1-line multiplexer is chosen.

3. The W, X, and Y lines are connected to the SELECT inputs of the multiplexer.

4. Each multiplexer input line is examined to determine what signal should be applied to it. The table of Fig. 11-14a was constructed to make this determination.

| W X Y | Input Number | Terms Covered | Input |
|-------|--------------|---------------|-------|
| 0 0 0 | 0 | 0, 1 | $\overline{Z}$ |
| 0 0 1 | 1 | 2, 3 | $Z + \overline{Z} = 1$ |
| 0 1 0 | 2 | 4, 5 | 0 |
| 0 1 1 | 3 | 6, 7 | $\overline{Z}$ |
| 1 0 0 | 4 | 8, 9 | $Z$ |
| 1 0 1 | 5 | 10, 11 | $\overline{Z}$ |
| 1 1 0 | 6 | 12, 13 | 0 |
| 1 1 1 | 7 | 14, 15 | 1 |

(a) Table

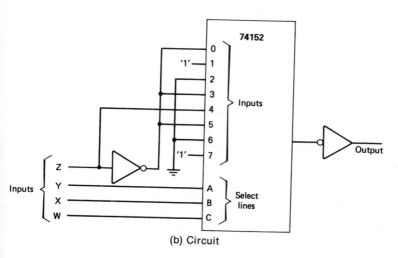

(b) Circuit

FIGURE 11-14  Table and circuit for Ex. 11-8.

5.  The first column of this table lists all the W, X, and Y possibilities. Since these variables are connected to the SELECT lines, they select the input line corresponding to their binary value, as shown in column 2.

6.  Column 3 of the table lists the terms of the specified function covered by each input. The top line, for example, has W = X = Y = 0. This covers SOP terms 0 and 1.

7.  The top line represents $\overline{W}\overline{X}\overline{Y}$. The term $\overline{W}\overline{X}\overline{Y}\overline{Z}$ is part of the SOP form of the specifications, but $\overline{W}\overline{X}\overline{Y}Z$ is not. If $\overline{Z}$ is connected to input line 0 of the multiplexer, it produces a 1 output when W, X, and Y are 0 (selecting input 0) and Z is 0, so that $\overline{Z}$, the input to line 0, is 1.

8.  The second line of the table ($\overline{W}\overline{X}Y$) selects line 1 of the multiplexer and covers terms 2 and 3 of the specification. These terms are both 1, so if input 1 is tied HIGH, a 1 output results whenever W = 0, X = 0, and Y = 1, as required.

9. The third line (input line 2) references SOP terms 4 and 5. Since they are both 0 in this example, input 2 is tied LOW. A 0 output occurs whenever $W = 0$, $X = 1$, and $Y = 0$.

10. The fourth line is $\overline{W}XY$. Since $\overline{W}XY\overline{Z}$ is part of the function, $\overline{Z}$ is tied to input 3.

11. The fifth line is $W\overline{X}\,\overline{Y}$. Here only $W\overline{X}\,\overline{Y}Z$ is part of the function so $Z$ is tied to line 4.

12. The table is continued in this manner and specifies the input to each line of the multiplexer.

13. The final circuit is shown in Fig. 11-14b. The **74152** is followed by an inverter to compensate for the inversion within the multiplexer.

## 11-5.3  Keyboard Decoding

Keyboards are often used to give commands and control the operation of electronic circuits. The keys on a hand calculator, for example, are divided into two groups; the numeric inputs (the digits 0–9) and the command inputs (add, subtract, etc.). Sophisticated calculators have many commands. The HP 21 calculator has 30 keys on its keyboard.

Even larger keyboards exist to provide alphabetic, as well as numeric and command inputs. The teletype is the most common example of an alphanumeric keyboard. A Model 33 teletype has 54 keys (including the space bar) on it, and is often used to communicate with a computer.

The keys on a keyboard are a group of pushbutton switches as shown in Fig. 11-15. They have two normally open contacts that close when the key is depressed. *Keyboard decoding* means determining which one of the keys on a keyboard is depressed, expressing this information digitally, and passing it on to the computer or other device.

Keyboard decoding can be accomplished by using multiplexers and demultiplexers working together. A decoding scheme for 64 switches is shown

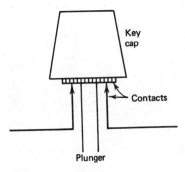

FIGURE 11-15  Diagram of a key switch.

in Figure 11-16. The **74155** is operated as a 3-line-to-8-line decoder and only its selected output is LOW. Any 1 of the 8 decoder output lines can be connected to any 1 of the 8 multiplexer input lines by depressing the switch at the intersection of the lines.

The **74152** multiplexer produces a HIGH output only if its selected input is LOW. A HIGH output indicates a switch is closed, and the combination of the 3 SELECT inputs on the **74152** and the 3 SELECT inputs on the **74155** form a 6-bit code to determine which switch has been depressed.

## EXAMPLE 11-9

The **74152** output goes HIGH when its SELECT inputs are 110 and the SELECT input on the **74155** is 101. Which switch has been depressed?

## SOLUTION

The current path is shown in dark in Fig. 11-16. Output 1Y1 of the **74155** is LOW because of the state of its SELECT lines. This LOW level is fed through the

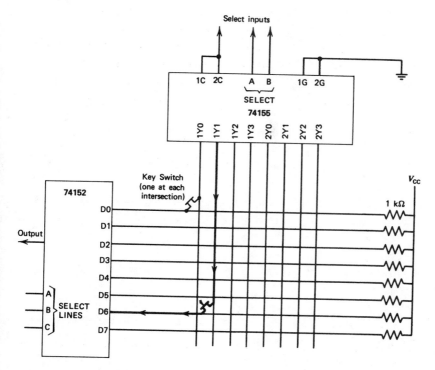

FIGURE 11-16  A method of keyboard decoding.

depressed switch to input 6 of the **74152** and causes its output to go HIGH. The SELECT line bits indicate the switch between lines 1Y1 and D6 has been depressed.

If, for example, the given 6-bit code, 101110, is to represent the letter X, a key cap with the letter X printed or engraved on it can be placed over the key switch at the intersection of lines 1Y1 and D6. When an operator presses the X key it is detected by a HIGH output when the SELECT lines read 101110.

Practical keyboard decoding systems are subject to the following constraints.

1.  The keyboard must be *constantly scanned* to determine if any keys have been depressed.
2.  Whenever any key is depressed, the scanning must stop and a code indicating which key was depressed must be available. A KEYSTROBE signal should also be generated to inform the computer or other receiving device that a key is depressed.
3.  Key switches *bounce* and the circuit must effectively debounce them.
4.  Fast typists tend to press a second key before they release the first key. This is called 2-*key rollover* and should be allowed.

The circuit of Fig. 11-17 resolves all these difficulties for the 64-key circuit of Fig. 11-16. The time of the **74123** one-shot is set to be somewhat longer than the bounce time of the keyswitches. As long as no keys are depressed, the CLOCKING FF remains SET and there is a steady stream of pulses out of the **7408**. These pulses increment the counter and cause the keyboard to be scanned continually. They also clock the retriggerable one-shot so that $\overline{\text{KEYSTROBE}}$ is always HIGH, indicating no character is available.

When a key is depressed, it first bounces, then it makes a solid contact. During the bounce time, the count will be interrupted when the scanner finds the depressed key, but scanning will resume when the switch bounces open. $\overline{\text{KEYSTROBE}}$ will remain HIGH because the one-shot's time is too long. When the key is finally closed firmly, the count will stop at the code for that key, the one-shot will stop getting trigger pulses, and $\overline{\text{KEYSTROBE}}$ will go LOW. If a second key is depressed during this time (2-key rollover), when the first key is released the circuit will scan until it stops at the code for the second key. $\overline{\text{KEYSTROBE}}$ will be HIGH for the one-shot time plus the scanning time and then will go LOW again, indicating a second key has been depressed.

---

**EXAMPLE 11-10**

For the circuit of Fig. 11-17, assume that the clock is 1 $\mu$s, the switches bounce for 1 ms, and the one-shot time is 2 ms. If a switch is depressed for 10 ms, how long is $\overline{\text{KEYSTROBE}}$ LOW?

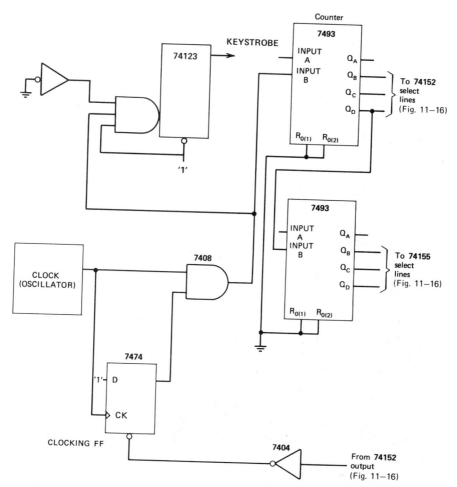

FIGURE 11-17 Scanning logic for a keyboard decoder.

## SOLUTION

During the 1-ms bounce time, $\overline{\text{KEYSTROBE}}$ is HIGH due to the intermittent nature of the interruptions. Note that every time the scan resumes it returns to the selected key every 64 $\mu$s, a relatively short time. After bouncing stops, it takes 2 ms for the one-shot triggers to expire, so $\overline{\text{KEYSTROBE}}$ will be LOW for 7 ms.

## 11-5.4  IC Keyboard Decoders

The circuit of Fig. 11-17 forms the basis for *prepackaged IC keyboard encoders*. An example is the **MM5740** 90-key keyboard encoder manufactured by National Semiconductor Inc. Most keyboard encoders are designed to work with standard typewriter keyboards. They have a 9 × 10 matrix for scanning the keys and provide inputs for SHIFT and CONTROL functions. Thus their outputs are the codes for both upper- and lower-case letters, numbers, punctuation, and a variety of special functions.

## SUMMARY

In this chapter the concept of multiplexing and demultiplexing was explained and TTL multiplexers and demultiplexers were introduced. Methods of building larger multiplexers from the basic multiplexers were explored.

The use of a demultiplexer as a decoder and a multiplexer as a function generator were discussed. Finally, a keyboard decoder, which uses both multiplexers and demultiplexers working together, was explained.

## GLOSSARY

**Multiplexer.** A circuit that connects one of several input lines to a single output line.

**Demultiplexer.** A circuit that connects a single input line to one of several output lines.

**STROBE.** An input that disables multiplexers or demultiplexers when it is HIGH.

**Decoder.** A circuit that decodes its inputs as a binary or BCD number and activates the corresponding output line.

**Keyswitch.** A switch or button on a keyboard.

**Keyboard.** A group of keys used for entering information, like a typewriter or teletype.

## REFERENCES

*The TTL Data Book for Design Engineers*, Texas Instruments, Inc., 1976.

**Thomas R. Blakeslee,** *Digital Design With Standard MSI and LSI*, Wiley, New York, 1975.

**Dan I. Porat** and **Arpad Barna,** *Introduction to Digital Techniques*, Wiley, New York, 1979.

**Ronald J. Tocci,** *Digital Systems*, Prentice-Hall, Englewood Cliffs, N.J., 1980.

## PROBLEMS

11-1.   Design a 32-line-to-line multiplexer using only 14- and 16-pin chips.

11-2.   Design a 24-line-to-1-line multiplexer.

11-3.   For Problems 1 and 2, respectively, show the 1s and 0s if line 21 is selected.

11-4.    Design a 16-line-to-1-line sequential multiplexer that places each input, in turn, on the output line for 10 $\mu$s. Use a **74150** in the circuit.

11-5.    Place one of four 8-bit registers on a set of 8 output lines using:
  (a) **74153**s
  (b) **74157**s

11-6.    For Problem 5, which circuit requires the smallest number of ICs?

11-7.    Design a 1-line-to-8-line demultiplexer using SSI gates.

11-8.    Design a 1-line-to-8-line demultiplexer for sequential demultiplexing by using a **74164** controlled timing sequence similar to that of Fig. 11-2.

11-9.    Design a 1-line-to-64-line demultiplexer.

11-10.   Design a 64-line decoder using:
  (a) **74154**s and SSI gates.
  (b) **74155**s only.

11-11.   Design a decoder with a 100-line output, assuming the inputs are given in two BCD decades.

11-12.   Given $f(w,x,y,z) = \Sigma(0,3,5,6,7,10,12,13)$. Implement the function using a single IC.

11-13.   Given $f(v,w,x,y,z) = \Sigma(0,2,3,6,7,8,11,14,17,18,20,25,26,31)$. Implement the function using multiplexers.

11-14.   (a) Design a full adder using only a single **74153**.
  (b) Design a 3-bit adder using the circuit of part $a$.

11-15.   Design a keyboard decoder for a 128-key keyboard. Show your scanning logic.

11-16.   In Problem 11-15, if the code for the letter B is 1000010, show the location of the B key on your board.

11-17.   Assume a keyboard has the specifications given in Ex. 11-10. At $t = 0$, key 1 is depressed for 10 ms. At $t = 5$ ms a second key, 20 scan stops away from key 1, is depressed for 10 ms. Draw a timing chart for $\overline{\text{KEYSTROBE}}$ during this period.

After attempting the problems, return to Sec. 11-2 and be sure you can answer the self-evaluation questions. If any of them are still difficult, review the appropriate sections of the chapter to find the answers.

# 12

# BINARY CODED DECIMAL

## 12-1 INSTRUCTIONAL OBJECTIVES

This chapter introduces the student to the uses of the Binary Coded Decimal (BCD) representation of numbers. After reading it the student should be able to:

1. Express a decimal number in BCD.
2. Convert from binary-to-BCD or BCD-to-binary using algorithms.
3. Construct circuits to convert from binary-to-BCD or BCD-to-binary.
4. Display a number using light-emitting diodes (LEDs) and 7-segment displays.
5. Use TTL decoder/drivers for 7-segment displays.
6. Multiplex 7-segment displays.

## 12-2 SELF-EVALUATION QUESTIONS

Watch for the answers to the following questions as you read the chapter. They should help you to understand the material presented.

1. Why is 1001 the largest 4-bit number that can appear in a BCD decade?
2. Why are BCD-to-binary conversions necessary? When are they typically used?

3.   Why are binary-to-BCD conversions necessary? When are they typically used?

4.   Explain the addition algorithm for BCD-to-binary conversion and the subtraction algorithm for binary-to-BCD conversion.

5.   Explain the shift-and-add and shift-and-subtract algorithms. When is each used?

6.   Why are some input combinations for the **74184** IC not listed in its function table?

7.   Why are current limiting resistors necessary when driving LEDs from TTL gates?

8.   What is remote blanking? How can it be designed into a display consisting of **7447**s and 7-segment displays?

9.   Why are transistors necessary for driving multiplexed displays?

## 12-3   EXPRESSING NUMBERS IN BINARY CODED DECIMAL

The BCD code for representing numbers was briefly introduced in Sec. 3-6. To review, the BCD code uses four binary bits called a decade to represent a single decimal digit from 0 to 9. Since numbers greater than 9 are *not* used, the 4-bit representation of the numbers from 10 to 15 should *never* appear in a BCD output. The BCD code conversion table originally shown in Fig. 3-22 is reproduced below for convenience.

When a number consisting of several decimal digits is to be represented in BCD form, each digit is represented by its own group of 4 bits. Therefore, there are four times as many bits in the representation as there are decimal digits in the original number.

| Binary — Coded Decimal Representation | Decimal Digit Equivalent |
|---|---|
| 0  0  0  0 | 0 |
| 0  0  0  1 | 1 |
| 0  0  1  0 | 2 |
| 0  0  1  1 | 3 |
| 0  1  0  0 | 4 |
| 0  1  0  1 | 5 |
| 0  1  1  0 | 6 |
| 0  1  1  1 | 7 |
| 1  0  0  0 | 8 |
| 1  0  0  1 | 9 |

FIGURE 3-22   The BCD code conversion table.

## EXAMPLE 12-1

Express the number 6309 in BCD form.

## SOLUTION

From the code conversion table we find that

$$6 = 0110$$
$$3 = 0011$$
$$0 = 0000$$
$$9 = 1001$$

The number 6309 is expressed by stringing these bits together:

$$(6309)_{10} = \underline{0110}\,\underline{0011}\,\underline{0000}\,\underline{1001}$$

$$\qquad\qquad\quad 6 \quad\ 3 \quad\ 0 \quad\ 9$$

Numbers given in BCD form can be converted into decimal numbers simply by dividing them into 4-bit decades, starting at the least significant bit, and assigning the correct decimal digit to each decade.

## EXAMPLE 12-2

Find the decimal equivalent of the BCD number:

$$000101011100001110100$$

## SOLUTION

The given number is divided into groups of 4 bits each and the decimal digit for each decade identified:

$$\underline{0001}\,\underline{0101}\,\underline{1000}\,\underline{0111}\,\underline{0100}$$

$$\quad 1 \quad\ \ 5 \quad\ \ 8 \quad\ \ 7 \quad\ \ 4$$

The decimal equivalent of the given BCD number is **15,874.**

## 12-3.1 Advantages and Disadvantages of BCD

Because most people can only understand and manipulate decimal numbers, any interaction between human beings and calculators, computers, or other devices must be in decimal form. On a calculator, for example, there are 10 digit keys (for the digits 0–9) and numbers are entered by pressing these keys. Internally the calculator receives a BCD code for each digit entered.

Input data to a computer is often entered via a *teletype*. The teletype keyboard contains keys for alphabetic, punctuation, and control characters, as well as keys for the 10 decimal digits. When a digit key is depressed, an 8-bit code is generated. The first 4 bits identify the character as a digit, and the last 4 bits express the particular digit in BCD format.

Unfortunately, it is awkward to perform arithmetic operations on numbers in BCD code, although it can be done (and some calculators and micro-processors operate this way). Computers, however, convert BCD numbers into binary numbers[1] before performing any arithmetic operations on them. Positive integer numbers are converted to binary as soon as they are entered into a computer, and they are stored within the computer in binary form.

After a computer has finished its calculations, the result is in binary. Two steps are required to produce an output that is understandable:

1. The binary number must be converted to BCD.
2. The BCD number must then be used to drive an output device that can be read. Typical output devices are 7-segment displays and printers.

These considerations make it clear that there is a need for methods to convert from BCD to binary numbers (when data are entered into a com-puter), and from binary to BCD (when data are presented by a computer).

## 12-4  CONVERSION USING ALGORITHMS

There are two ways to convert between binary and BCD numbers: *algorithms* and special purpose ICs. The use of algorithms, which involves a manip-ulation of the bits in one form to obtain the other, is described in this section. The hardware implementation of a converter is considered in Sec. 12-5.

### 12-4.1  Conversion from BCD to Binary Using the Add Algorithm

One way to convert from BCD to binary is to assign a *weight* to all the BCD bits, and then *add* the weights for all bits that are 1 in the BCD representation of the number. The weights are assigned in accordance with the value represented by each BCD bit; that is, 1, 2, 4, 8, 10, 20, 40, 80, 100, 200, . . . . A table of the decimal numbers and binary weights cor-responding to each BCD bit is given in Fig. 12-1.

---

[1] Actually most computers use 2s complement arithmetic (see Sec. 14-5), which is a variant of binary arithmetic.

| Bit Position | Decimal Number | Binary Number |
|---|---|---|
| 1 | 1 | 1 |
| 2 | 2 | 1 0 |
| 3 | 4 | 1 0 0 |
| 4 | 8 | 1 0 0 0 |
| 5 | 1 0 | 1 0 1 0 |
| 6 | 2 0 | 1 0 1 0 0 |
| 7 | 4 0 | 1 0 1 0 0 0 |
| 8 | 8 0 | 1 0 1 0 0 0 0 |
| 9 | 1 0 0 | 1 1 0 0 1 0 0 |
| 10 | 2 0 0 | 1 1 0 0 1 0 0 0 |
| 11 | 4 0 0 | 1 1 0 0 1 0 0 0 0 |
| 12 | 8 0 0 | 1 1 0 0 1 0 0 0 0 0 |
| 13 | 1 0 0 0 | 1 1 1 1 1 0 1 0 0 0 |
| 14 | 2 0 0 0 | 1 1 1 1 1 0 1 0 0 0 0 |

FIGURE 12-1 Table of weights for BCD-to-binary conversion.

## EXAMPLE 12-3

Convert the number 169 to binary using the add algorithm.

## SOLUTION

The number 169 is expressed in BCD form as:

$$\underbrace{0001}_{1}\underbrace{0110}_{6}\underbrace{1001}_{9}$$

where the LSB is on the right. There are 1s in positions 1, 4, 6, 7, and 9. The weights corresponding to these numbers are (from Fig. 12-1):

$$
\begin{array}{r}
0001 \\
1000 \\
10100 \\
101000 \\
1100100
\end{array}
$$

The simple addition of these binary numbers yields **10101001**, which is the binary equivalent of decimal 169.

Computers, which generally have binary adders, often use the add algorithm to convert from BCD to binary.

## 12-4.2 Conversion from BCD to Binary Using the Shift Algorithm

It is possible to convert from BCD to binary by shifting and subtracting. The algorithm for this method is:

1. Express the number in BCD.
2. Divide the BCD number into *decades* of 4 bits each, starting at the LSB.
3. Shift the number one place to the right while maintaining the decade boundaries in their original position. The LSB of the BCD number is shifted out of the cells and becomes the LSB of the equivalent binary number.
4. At this time no decade should contain the numbers 5, 6, 7 or 13, 14, 15. Examine the number in each decade. If it is 8 or greater, subtract 3 from it.
5. Return to step 3. Repeat the procedure until all the bits have been shifted out of the decades. The bits shifted out are the binary equivalent of the BCD number.

---

**EXAMPLE 12-4**

Find the binary equivalent of 759 using the shift algorithm.

**SOLUTION**

The solution is shown in Fig. 12-2.

1. Line one—the number 759 is written in BCD and the decade boundaries are drawn.
2. Shift one place to the right.
3. After the shift, decade 2 contains a 10 (1010) and decade 1 contains a 12. The contents of these decades are replaced by the numbers 7 and 9 on line three.
4. Line four—shift.
5. After the shift, decades 2 and 1 contain 11 and 12, respectively. On line five they are replaced with 8 and 9.
6. Line six—shift.
7. Line seven—after the shift, decade 2 contains a 12 and decade 1 contains a 4. In this case, 3 is subtracted from decade 2 only, because decade 1 contains a number less than 8. The numbers on line seven in decades 1 and 2 are 9 and 4, respectively.
8. The shifting procedure continues, in accordance with the algorithm, until the entire number is shifted to the right of decade 1 on line sixteen. This number, 1011110111, is the binary equivalent of 759.
9. The rightmost column in Fig. 12-2 indicates the operation being performed between the lines. Where the operation is subtraction, the numbers in parentheses indicate the decades where the subtraction is performed.
10. When no decade contains an 8 or greater number, the subtraction operation is omitted (line ten is the first example), and another shift follows immediately.

| Line Number | | 7 | | | | 5 | | | | 9 | | | | | | | | | | | | Operation |
|---|---|---|---|---|---|---|---|---|---|---|---|---|---|---|---|---|---|---|---|---|---|---|
| 1 | 0 | 1 | 1 | 1 | 0 | 1 | 0 | 1 | 1 | 0 | 0 | 1 | | | | | | | | | | Shift |
| 2 | | 0 | 1 | 1 | 1 | 0 | 1 | 0 | 1 | 1 | 0 | 0 | 1 | | | | | | | | | Sub 3 (2, 1) |
| 3 | | | 1 | 1 | 0 | 1 | 1 | 1 | 1 | 0 | 0 | 1 | 1 | | | | | | | | | Shift |
| 4 | | | | 1 | 1 | 0 | 1 | 1 | 1 | 1 | 0 | 0 | 1 | 1 | | | | | | | | Sub 3 (2, 1) |
| 5 | | | | 1 | 1 | 0 | 0 | 0 | 1 | 0 | 0 | 1 | 1 | 1 | | | | | | | | Shift |
| 6 | | | | | 1 | 1 | 0 | 0 | 0 | 1 | 0 | 0 | 1 | 1 | 1 | | | | | | | Sub 3 (2) |
| 7 | | | | | 1 | 0 | 0 | 1 | 0 | 1 | 0 | 0 | 1 | 1 | 1 | | | | | | | Shift |
| 8 | | | | | | 1 | 0 | 0 | 1 | 0 | 1 | 0 | 0 | 1 | 1 | 1 | | | | | | Sub 3 (1) |
| 9 | | | | | | 1 | 0 | 0 | 0 | 1 | 1 | 1 | 0 | 1 | 1 | 1 | | | | | | Shift |
| 10 | | | | | | | 1 | 0 | 0 | 0 | 1 | 1 | 1 | 0 | 1 | 1 | 1 | | | | | Shift |
| 11 | | | | | | | | 1 | 0 | 0 | 0 | 1 | 1 | 1 | 0 | 1 | 1 | 1 | | | | Shift |
| 12 | | | | | | | | 1 | 0 | 0 | 0 | 1 | 1 | 1 | 0 | 1 | 1 | 1 | | | | Sub 3 (1) |
| 13 | | | | | | | | 0 | 1 | 0 | 1 | 1 | 1 | 1 | 0 | 1 | 1 | 1 | | | | Shift |
| 14 | | | | | | | | | 1 | 0 | 1 | 1 | 1 | 1 | 0 | 1 | 1 | 1 | | | | Shift |
| 15 | | | | | | | | | | 1 | 0 | 1 | 1 | 1 | 1 | 0 | 1 | 1 | 1 | | | Shift |
| 16 | | | | | | | | | | | 1 | 0 | 1 | 1 | 1 | 1 | 0 | 1 | 1 | 1 | | |

BCD decade No.  3  2  1

FIGURE 12-2  Conversion of 759 to binary using the shift algorithm.

## 12-4.3  Binary to BCD Conversion by Subtraction

It is as necessary to convert from binary to BCD as to convert from BCD to binary. One way to do this is subtraction. This is the reverse of the addition algorithm presented in Sec. 12-4.1. The subtraction algorithm is as follows.

1. Using the binary number column in Fig. 12-1, find the *largest* number in the column that is *less than or equal* to the given binary number.
2. Subtract this number from the given number. Using the result, return to step 1 and repeat the procedure until the remainder is 0.
3. Write a 1 in each bit position corresponding to the binary numbers where a subtraction is performed. Write 0s in all other bit positions. The result is the BCD equivalent of the given binary number.

---

### EXAMPLE 12-5

Find the BCD equivalent of the binary number 1010100111, using the subtraction algorithm.

## SOLUTION

The steps of the solution are shown in Fig. 12-3.

1.   The given number is written on line one.
2.   The largest number in the binary number column of Fig. 12-1, which can be subtracted from 1010100111 and still yield a positive result is 110010000, or the binary equivalent of 400.
3.   The result of this subtraction is listed on line three. Binary 200 is subtracted from this result and the second remainder is shown on line five.
4.   The next two binary numbers from Fig. 12-1, 100 and 80, are investigated. They both prove to be larger than the remainder on line five and no subtraction is performed. These numbers are set aside in Fig. 12-3.
5.   The subtractions continue until the given binary number is reduced to 0 on line 19.
6.   In the process the following numbers were subtracted: 400, 200, 40, 20, 10, 8, and 1. These decimal numbers correspond to bit positions 1, 4, 5, 6, 7, 10, and 11 as shown in Fig. 12-1. The BCD equivalent of the binary number can now be written by placing 1s in these bit positions:

| Bit position | 12 | 11 | 10 | 9 | 8 | 7 | 6 | 5 | 4 | 3 | 2 | 1 |
|---|---|---|---|---|---|---|---|---|---|---|---|---|
| BCD number | 0 | 1 | 1 | 0 | 0 | 1 | 1 | 1 | 1 | 0 | 0 | 1 |

$$\underbrace{0\ \ 1\ \ 1\ \ 0}_{6}\ \ \underbrace{0\ \ 1\ \ 1\ \ 1}_{7}\ \ \underbrace{1\ \ 0\ \ 0\ \ 1}_{9}$$

7.   The decimal number **679** is equivalent to $(1010100111)_2$.

---

| Line | | | | | | | | | | | | | | | | | | |
|---|---|---|---|---|---|---|---|---|---|---|---|---|---|---|---|---|---|---|
| 1 | 1 | 0 | 1 | 0 | 1 | 0 | 0 | 1 | 1 | 1 | | | | | | | Binary Number |
| 2 | | 1 | 1 | 0 | 0 | 1 | 0 | 0 | 0 | 0 | | | | | | | − 400 |
| 3 | | 1 | 0 | 0 | 0 | 1 | 0 | 1 | 1 | 1 | | | | | | | Remainder |
| 4 | | | 1 | 1 | 0 | 0 | 1 | 0 | 0 | 0 | | | | | | | − 200 |
| 5 | | | 1 | 0 | 0 | 1 | 1 | 1 | 1 | | | | | | | | Remainder |
| 6 | | | | | | | | 1 | 1 | 0 | 0 | 1 | 0 | 0 | | | (100) |
| 7 | | | | | | | | 1 | 0 | 1 | 0 | 0 | 0 | 0 | | | (80) |
| 8 | | | | 1 | 0 | 1 | 0 | 0 | 0 | | | | | | | | − 40 |
| 9 | | | | 1 | 0 | 0 | 1 | 1 | 1 | | | | | | | | Remainder |
| 1 0 | | | | | 1 | 0 | 1 | 0 | 0 | | | | | | | | − 20 |
| 1 1 | | | | | 1 | 0 | 0 | 1 | 1 | | | | | | | | Remainder |
| 1 2 | | | | | | 1 | 0 | 1 | 0 | | | | | | | | − 10 |
| 1 3 | | | | | | 1 | 0 | 0 | 1 | | | | | | | | Remainder |
| 1 4 | | | | | | 1 | 0 | 0 | 0 | | | | | | | | − 8 |
| 1 5 | | | | | | | | 1 | | | | | | | | | Remainder |
| 1 6 | | | | | | | | | | | 1 | 0 | 0 | | | | (− 4) |
| 1 7 | | | | | | | | | | | | 1 | 0 | | | | (− 2) |
| 1 8 | | | | | | | 1 | | | | | | | | | | −1 |
| 1 9 | | | | | | | 0 | | | | | | | | | | |

FIGURE 12-3  Subtraction method for converting binary to BCD.

## 12-4.4    Binary to BCD Conversion by the Shift-and-Add Algorithm

A shift-and-add algorithm exists for converting a given binary number into an equivalent BCD number. The procedure is essentially the reverse of the shift-and-subtract algorithm of Sec. 12-4.2.

1.   Divide the BCD area into decades of 4 bits each.
2.   Write the given binary number to the right of the least significant decade. At this point, all BCD decades are vacant and assumed to contain 0s.
3.   Shift the binary number into the BCD decades by shifting left one bit at a time.
4.   Before each shift, examine all the BCD decades. If any of them contain a number *greater than or equal to* 5, add 3 to the contents of *all* such decades.
5.   After each shift, the BCD number is the equivalent of the binary number shifted in up to that point.

---

### EXAMPLE 12-6

Find the BCD equivalent of the binary number 101010111, using the shift-and-add algorithm.

### SOLUTION

The solution is illustrated in Fig. 12-4.

1.   On line 1, the given binary number is written to the right of the vacant BCD decades.
2.   Lines 2, 3, and 4 are the results of successive left shifts.

```
Line Number                                                    Operation

  1                                |1 0 1 0 1 0 1 1 1         Shift
  2                              1 |0 1 0 1 0 1 1 1           Shift
  3                          1 0   |1 0 1 0 1 1 1             Shift
  4                      1 0 1     |0 1 0 1 1 1               Add
  5                  1 0 0 0       |0 1 0 1 1 1               Shift
  6                1|0 0 0 0       |1 0 1 1 1                 Shift
  7            1 0 |0 0 0 1        |0 1 1 1                   Shift
  8        1 0 0   |0 0 1 0        |1 1 1                     Shift
  9        1 0 0 0 |0 1 0 1        |1 1                       Add
 10        1 0 1 1 |1 0 0 0        |1 1                       Shift
 11    1   |0 1 1 1|0 0 0 1        |1                         Add
 12    1   |1 0 1 0|0 0 0 1        |1                         Shift
 13  1 1   |0 1 0 0|0 0 1 1                                   

        3       4        3         BCD  result
```

FIGURE 12-4   Binary-to-BCD conversion by the shift and add algorithm.

3.   On line 4, the least significant decade contains a 5. This is replaced by an 8 on line five.

4.   Line 6 is obtained by shifting line 5. Note that up to this point, the 4 most significant binary bits (1010) have been shifted into the decades. The decades now read 10, or ten, which is equivalent to the binary number 1010.

5.   On lines 6, 7, and 8, no decade contains a number equal to or greater than 5. Therefore, only shift operations are performed.

6.   On line 9, the least significant decade contains a 5 and the next decade contains an 8. On line 10, these are replaced by 8 and 11, respectively.

7.   The process terminates when all the bits are shifted into the BCD decades. This occurs on line thirteen, and **343** can be read as the BCD number. This is the equivalent of the given binary number.

## 12-5  CONVERSION USING ICs

The algorithms considered in the previous section can be implemented in software. Computers often use a program for binary-to-BCD or BCD-to-binary conversions that perform the operations required by the algorithm.

Another way to perform these conversions is to use ICs designed specifically for this purpose, which is the *hardware* approach. It is more costly than doing the conversions in software, but it is much faster. If the device that must make the conversion is not capable of being programmed, hardware may be the only feasible approach to the problem.

### 12-5.1  BCD-to-Binary Conversion Using the 74184

The 74184 IC is a *read only memory* (ROM; see Sec. 15-9) designed specifically to convert BCD to binary. It has 5 inputs and 5 of its 8 outputs ($Y_1$ through $Y_5$) are used for conversion. The 3 remaining outputs ($Y_6$, $Y_7$, and $Y_8$) take the 9s or 10s complement of a BCD number, which is used when performing BCD arithmetic. The 74184 also has an ENABLE input, G, which must be LOW when the chip is functioning. If input G is HIGH, all 8 outputs of the 74184 will be HIGH.

The pin layout, a diagram of the 74184 used as a 6-bit converter, and the function table are shown in Fig. 12-5. The operation of the 74184 must be determined from the function table, Fig. 12-5c, or from Fig. 12-6, where the 5 inputs are written as a binary number and the corresponding outputs given. The 74184 operates on a variant of the shift and subtract algorithm. Notice that all possible combinations on the 5 inputs are *not* listed in the function table, because the input is assumed to be in the BCD form. For example, the input LLHHL is not listed because the C and B inputs of the 74184 are seen in Fig. 12-5b to be connected to the two most significant bits in a BCD decade. These two bits can never both be 1.

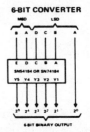

(b) Six-bit converter

**J OR N DUAL-IN-LINE OR
W FLAT PACKAGE (TOP VIEW)**

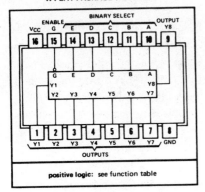

**positive logic:** see function table

(a) Pin layout

| BCD WORDS | INPUTS (See Note A) | | | | | | OUTPUTS (See Note B) | | | | |
|---|---|---|---|---|---|---|---|---|---|---|---|
| | E | D | C | B | A | G | Y5 | Y4 | Y3 | Y2 | Y1 |
| 0-1 | L | L | L | L | L | L | L | L | L | L | L |
| 2-3 | L | L | L | L | H | L | L | L | L | L | H |
| 4-5 | L | L | L | H | L | L | L | L | L | H | L |
| 6-7 | L | L | L | H | H | L | L | L | L | H | H |
| 8-9 | L | L | H | L | L | L | L | L | H | L | L |
| 10-11 | L | H | L | L | L | L | L | L | H | L | H |
| 12-13 | L | H | L | L | H | L | L | L | H | H | L |
| 14-15 | L | H | L | H | L | L | L | L | H | H | H |
| 16-17 | L | H | L | H | H | L | L | H | L | L | L |
| 18-19 | L | H | H | L | L | L | L | H | L | L | H |
| 20-21 | H | L | L | L | L | L | L | H | L | H | L |
| 22-23 | H | L | L | L | H | L | L | H | L | H | H |
| 24-25 | H | L | L | H | L | L | L | H | H | L | L |
| 26-27 | H | L | L | H | H | L | L | H | H | L | H |
| 28-29 | H | L | H | L | L | L | L | H | H | H | L |
| 30-31 | H | H | L | L | L | L | L | H | H | H | H |
| 32-33 | H | H | L | L | H | L | H | L | L | L | L |
| 34-35 | H | H | L | H | L | L | H | L | L | L | H |
| 36-37 | H | H | L | H | H | L | H | L | L | H | L |
| 38-39 | H | H | H | L | L | L | H | L | L | H | H |
| ANY | X | X | X | X | X | H | H | H | H | H | H |

H = high level, L = low level, X = irrelevant

NOTES: A. Input conditions other than those shown produce highs at outputs Y1 through Y5.

B. Outputs Y6, Y7, and Y8 are not used for BCD-to-binary conversion.

(c) Function table BCD to binary converter

FIGURE 12-5  The **74184** BCD-to-binary converter. (From the *TTL Data Book for Design Engineers,* 2nd ed., Texas Instruments, Inc. Courtesy of Texas Instruments, copyright 1976.)

| Input | Input — Output Relationship |
|---|---|
| 0 — 4 | Output = Input |
| 8 — 12 | Output = Input − 3 |
| 16 — 20 | Output = Input − 6 |
| 24 — 28 | Output = Input − 9 |

FIGURE 12-6  Input-output relationship of the **74184.**

## EXAMPLE 12-7

(a) Show how the **74184** converts 36 to a binary number by placing 1s and 0s on the diagram of Fig. 12-5*b*.

(b) Explain why BCD words 36 and 37 are listed on the same line in the function table of Fig. 12-5*c*.

## SOLUTION

(a) Figure 12-5*b* is redrawn with the bits numbered in Fig. 12-7. The 6 bits for the BCD representation of 36 are $\underbrace{110}_{3}\underbrace{110}_{6}$. Because the LSB of a BCD number and the LSB of its binary equivalent are always equal, the LSB falls through. The other 5 bits (11011) are applied to the E to A inputs of the **74184**, and the input HHLHH is located on the 36–37 line of the function table. The corresponding outputs are HLLHL and, following this output by the 0 on the LSB line, we obtain 100100, or 36 in binary.

(b) The difference between the BCD or binary representation of 36 and 37 is only in the LSB, which is not applied to the **74184**. Whether the BCD number is 36 or 37, the identical bits, 11011, are applied to the **74184** and the output is 10010. If this output is followed by an LSB of 0, the number is 36: whereas if it is followed by an LSB of 1, the number is 37. Since the inputs and, consequently, the outputs of the **74184** are the same in either case, they are both listed on the same line of the function table.

Long BCD numbers can be converted to binary by using several 74184s connected in the configuration shown in Fig. 12-8. In this manner a complete conversion of BCD numbers to binary can be achieved in hardware.

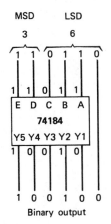

FIGURE 12-7   Conversion of 36 from BCD to binary using a **74184.**

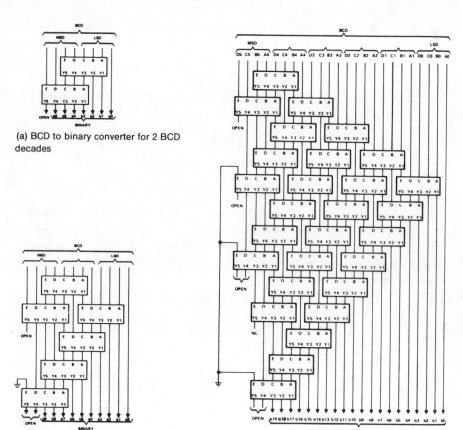

(a) BCD to binary converter for 2 BCD decades

(b) BCD to binary converter for 3 BCD decades

(c) BCD to binary converter for 6 BCD decades

FIGURE 12-8 Circuits for BCD-to-binary conversion. (From the *TTL Data Book for Design Engineers,* 2nd ed., Texas Instruments, Inc. Courtesy of Texas Instruments, copyright 1976.)

## EXAMPLE 12-8

Given the number 974 in BCD form, convert it to binary using **74184**. Show the input and output of each chip.

## SOLUTION

The solution is shown in Fig. 12-9. Because the input is a three-digit number, the circuit of Fig. 12-8*b* is used. The BCD equivalent of 974, 100101110100, is written on the top line. The inputs to IC chip 1 are taken directly from the given BCD number, as shown, and its outputs determined from the function table. Now the inputs to chips 2 and 3 are determined from the original number and the output of chip 1. Proceeding down the circuit and writing the inputs and outputs of each IC, we find **1111001110** on the bottom line. This is 974 in binary.

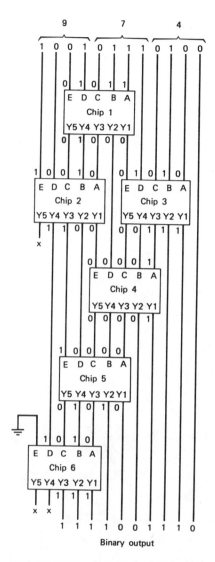

FIGURE 12-9   Conversion of 974 in BCD to binary.
*Notes:* (1) × = irrelevant (unconnected); (2) All chips **74184.**

## 12-5.2   Binary-to-BCD Conversion Using the 74185

The 74185 is a companion IC to the 74184 and is used for binary-to-BCD conversion. It is also a ROM with the same pin layout as the 74184. The function table for the 74185 is given in Fig. 12-10. It performs the code conversion by using a variant of the shift and add algorithm discussed in Sec. 12-4.4. Circuits using the 74185 to convert from binary-to-BCD are

**FUNCTION TABLE**

| BINARY WORDS | INPUTS BINARY SELECT E | D | C | B | A | ENABLE G | OUTPUTS Y8 | Y7 | Y6 | Y5 | Y4 | Y3 | Y2 | Y1 |
|---|---|---|---|---|---|---|---|---|---|---|---|---|---|---|
| 0 · 1 | L | L | L | L | L | L | H | H | L | L | L | L | L | L |
| 2 · 3 | L | L | L | L | H | L | H | H | L | L | L | L | L | H |
| 4 · 5 | L | L | L | H | L | L | H | H | L | L | L | L | H | L |
| 6 · 7 | L | L | L | H | H | L | H | H | L | L | L | L | H | H |
| 8 · 9 | L | L | H | L | L | L | H | H | L | L | L | H | L | L |
| 10 · 11 | L | L | H | L | H | L | H | H | L | L | H | L | L | L |
| 12 · 13 | L | L | H | H | L | L | H | H | L | L | H | L | L | H |
| 14 · 15 | L | L | H | H | H | L | H | H | L | L | H | L | H | L |
| 16 · 17 | L | H | L | L | L | L | H | H | L | L | H | L | H | H |
| 18 · 19 | L | H | L | L | H | L | H | H | L | L | H | H | L | L |
| 20 · 21 | L | H | L | H | L | L | H | H | L | H | L | L | L | L |
| 22 · 23 | L | H | L | H | H | L | H | H | L | H | L | L | L | H |
| 24 · 25 | L | H | H | L | L | L | H | H | L | H | L | L | H | L |
| 26 · 27 | L | H | H | L | H | L | H | H | L | H | L | L | H | H |
| 28 · 29 | L | H | H | H | L | L | H | H | L | H | L | H | L | L |
| 30 · 31 | L | H | H | H | H | L | H | H | L | H | H | L | L | L |
| 32 · 33 | H | L | L | L | L | L | H | H | L | H | H | L | L | H |
| 34 · 35 | H | L | L | L | H | L | H | H | L | H | H | L | H | L |
| 36 · 37 | H | L | L | H | L | L | H | H | L | H | H | L | H | H |
| 38 · 39 | H | L | L | H | H | L | H | H | L | H | H | H | L | L |
| 40 · 41 | H | L | H | L | L | L | H | H | H | L | L | L | L | L |
| 42 · 43 | H | L | H | L | H | L | H | H | H | L | L | L | L | H |
| 44 · 45 | H | L | H | H | L | L | H | H | H | L | L | L | H | L |
| 46 · 47 | H | L | H | H | H | L | H | H | H | L | L | L | H | H |
| 48 · 49 | H | H | L | L | L | L | H | H | H | L | L | H | L | L |
| 50 · 51 | H | H | L | L | H | L | H | H | H | L | H | L | L | L |
| 52 · 53 | H | H | L | H | L | L | H | H | H | L | H | L | L | H |
| 54 · 55 | H | H | L | H | H | L | H | H | H | L | H | L | H | L |
| 56 · 57 | H | H | H | L | L | L | H | H | H | L | H | L | H | H |
| 58 · 59 | H | H | H | L | H | L | H | H | H | L | H | H | L | L |
| 60 · 61 | H | H | H | H | L | L | H | H | H | H | L | L | L | L |
| 62 · 63 | H | H | H | H | H | L | H | H | H | H | L | L | L | H |
| ALL | X | X | X | X | X | H | H | H | H | H | H | H | H | H |

FIGURE 12-10  Function table for the **74185**. (From the *TTL Data Book for Design Engineers,* 2nd ed., Texas Instruments, Inc. Courtesy of Texas Instruments, copyright 1976.)

given in Fig. 12-11. Note that 4, 5, or 6 outputs may be used, depending upon the position of the IC in the circuit, and that all E inputs not shown on the figure are grounded (0).

---

## EXAMPLE 12-9

Convert the binary number 100111011 to BCD using **74185**s.

## SOLUTION

The given binary number contains 9 bits. Since the circuit of Fig. 12-11c is the simplest circuit that accommodates 9 bits, it should be used. This circuit is redrawn in Fig. 12-12. The 5 inputs to chip 1, 10011, are taken from the 5 most-significant-bits (MSBs) of the given binary number. The 6-bit output, 011100 is obtained from

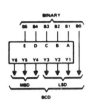

(a) 6-bit binary to BCD converter

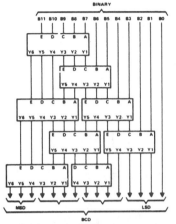

(d) 12-bit binary to BCD converter (see note B)

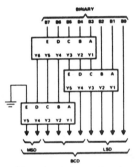

(b) 8-bit binary to BCD converter

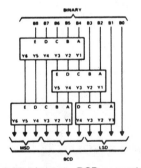

(c) 9-bit binary to BCD converter

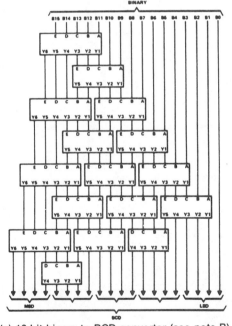

(e) 16-bit binary to BCD converter (see note B)

MSD—Most significant decade
LSD—Least significant decade
*Notes.* A. Each rectangle represents an SN54185A or an SN74185A.
B. All unused E inputs are grounded.

FIGURE 12-11 Circuits for binary-to-BCD conversion using **74185.** (From the *TTL Data Book for Design Engineers,* 2nd ed., Texas Instruments, Inc. Courtesy of Texas Instruments, copyright 1976.)

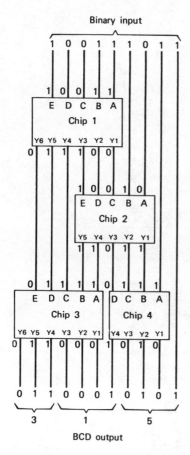

**FIGURE 12-12** Conversion of binary to BCD using **74185.** *Note:* All chips are **74185**s.

the function table (Fig. 12-10) on the line with inputs HLLHH (line labeled 38-39). The three least-significant-bits (LSBs) of chip 1's output and next 2 bits of the given binary number are applied to chip 2, and its output determined from the function table. We proceed down the circuit in this manner until the final output, **01100010101,** is achieved. This is **315** in BCD, and **315** is the value of the given binary number.

## 12-6 INDICATING LIGHTS

Indicating lights are used to display the state of an electronic system. If a single bit, or a group of bits, is to be monitored, as on a computer front panel, a lamp is connected to each bit and its status (ON or OFF) corresponds to the value of the bit (1 or 0). These lights can be read by an engineer or programmer who can determine whether the system is operating properly.

Often these lamps are low current, low voltage incandescent lamps driven by a buffer driver (see Sec. 5-6.2). Recently, however, light-emitting diodes (LEDs) have been gaining popularity.

## 12-6.1 Light-Emitting Diodes (LEDs)

LEDs are small diodes that emit a red glow when current is passed through them. When a LED is ON, it develops a forward voltage of 1.6 V, regardless of its current. Typical LEDs are specified to operate at 20 mA, but they will glow over a wide current range. For use with TTL circuits, a resistor (generally 150 $\Omega$) is placed in series with the diode to limit the current and drop the excess voltage. Although the guaranteed output of a standard TTL inverter is only 16 mA, it is often used to drive a LED.

The circuit of an inverter driving a LED is shown in Fig. 12-13. When the inverter input is HIGH, current is driven through the resistor, the LED, and the lower transistor of the inverter's totem pole output. Typically the resistor absorbs 3 V, the LED 1.6 V, and the inverter output voltage is 0.4 V. Together these add up to the 5 V of the supply. If the LED is ON, it indicates that the input to the inverter is a 1.

When the input to the inverter is a 0, the upper totem pole transistor of the inverter turns ON. Now there is no current path through the LED and it remains dark. In this way the LED monitors the logic level at the inverter input, and the inverter is used as a LED driver.

## 12-6.2 Seven-Segment Displays

A 7-segment display, which is designed to display a single decimal digit, is built of 7 LEDs. An example of converting a BCD decade to its corresponding 7-segment display digit has already been given in Sec. 3-8.3. Fortunately, ICs designed to act as drivers for 7-segment displays are already

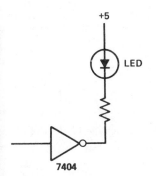

FIGURE 12-13   A TTL inverter driving an LED.

in wide use. These decoder/drivers accept a BCD decade as input and convert it to outputs that illuminate the proper segments of a 7-segment display. Current limiting resistors are still needed with decoder/drivers.

### 12-6.3 The 7447 Decoder/Driver

The most popular and versatile decoder/driver for 7-segment displays is the 7447, whose pin layout, numerical designation, and function table are given in Fig. 12-14. Inputs ABCD form a BCD decade and the seven outputs, a–g, are meant to be connected to a 7-segment display. Current limiting resistors must be connected between each display segment and the **7447** output. Each output is an open-collector transistor. When it is ON a current path exists and illuminates the segment connected to it. For example, if inputs DCBA are 0001, which is the BCD code for 1, only segments b and c glow. They form a single rightmost vertical line and give the appearance of the numeral 1. If a number greater than 9 is applied to the BCD inputs, an unrecognizable symbol appears on the output, as shown in outputs 10 through 15 of the resultant display (Fig. 12-14c).

---

**EXAMPLE 12-10**

A 7447 has inputs of 1001. Show precisely how it drives a 7-segment display, and how the decimal number is displayed.

**SOLUTION**

The result is shown in Fig. 12-15. With a BCD 9 (1001) applied to the inputs, the function table of the **7447**, Fig. 12-14d, shows that all segments except d and e are ON. For each of the remaining 5 segments a current path exists through the segment, the current limiting resistor, and the open-collector transistor within the **7447**. These segments glow as shown shaded on Fig. 12-15 and form the image of a 9.

---

### 12-6.4 Ripple Blanking

Ripple blanking is a feature of decoder/drivers that allows them to blank out leading 0s. If a 6-digit display is used, for example, most people prefer to see 2047, rather than 002047. The number 2047 can be displayed if the first two 7-segment displays are completely OFF. Note that while it is desirable to blank *leading* 0s, *imbedded* 0s (such as the 0 in 2047) cannot be blanked.

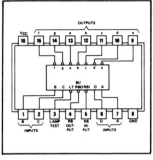

(a) Pin layout

(b) Segment identification

(c) Numerical designations and resultant displays

| DECIMAL OR FUNCTION | INPUTS | | | | | | BI/RBO† | OUTPUTS | | | | | | | NOTE |
|---|---|---|---|---|---|---|---|---|---|---|---|---|---|---|---|
| | LT | RBI | D | C | B | A | | a | b | c | d | e | f | g | |
| 0 | H | H | L | L | L | L | H | ON | ON | ON | ON | ON | ON | OFF | 1 |
| 1 | H | X | L | L | L | H | H | OFF | ON | ON | OFF | OFF | OFF | OFF | 1 |
| 2 | H | X | L | L | H | L | H | ON | ON | OFF | ON | ON | OFF | ON | |
| 3 | H | X | L | L | H | H | H | ON | ON | ON | ON | OFF | OFF | ON | |
| 4 | H | X | L | H | L | L | H | OFF | ON | ON | OFF | OFF | ON | ON | |
| 5 | H | X | L | H | L | H | H | ON | OFF | ON | ON | OFF | ON | ON | |
| 6 | H | X | L | H | H | L | H | OFF | OFF | ON | ON | ON | ON | ON | |
| 7 | H | X | L | H | H | H | H | ON | ON | ON | OFF | OFF | OFF | OFF | |
| 8 | H | X | H | L | L | L | H | ON | ON | ON | ON | ON | ON | ON | |
| 9 | H | X | H | L | L | H | H | ON | ON | ON | OFF | OFF | ON | ON | |
| 10 | H | X | H | L | H | L | H | OFF | OFF | OFF | ON | ON | OFF | ON | |
| 11 | H | X | H | L | H | H | H | OFF | OFF | ON | ON | OFF | OFF | ON | |
| 12 | H | X | H | H | L | L | H | OFF | ON | OFF | OFF | OFF | ON | ON | |
| 13 | H | X | H | H | L | H | H | ON | OFF | OFF | ON | OFF | ON | ON | |
| 14 | H | X | H | H | H | L | H | OFF | OFF | OFF | ON | ON | ON | ON | |
| 15 | H | X | H | H | H | H | H | OFF | OFF | OFF | OFF | OFF | OFF | OFF | |
| BI | X | X | X | X | X | X | L | OFF | OFF | OFF | OFF | OFF | OFF | OFF | 2 |
| RBI | H | L | L | L | L | L | L | OFF | OFF | OFF | OFF | OFF | OFF | OFF | 3 |
| LT | L | X | X | X | X | X | H | ON | ON | ON | ON | ON | ON | ON | 4 |

(d) '46A, '47A, 'L46, 'L47 Function table

FIGURE 12-14 The **7447** display driver. (From the *TTL Data Book for Design Engineers*, 2nd ed., Texas Instruments, Inc. Courtesy of Texas Instruments, copyright 1976.)

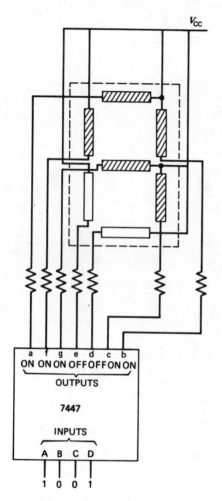

FIGURE 12-15  A **7447** driving a 7-segment display.

The **7447** includes provision for *ripple blanking*. It has a RIPPLE BLANK-ING INPUT (RBI, pin 5) and a RIPPLE BLANKING OUTPUT (RBO, pin 4). If RBI and the BCD inputs ABCD are *all* LOW (see the RBI line in the function table), the RBO output is LOW and the display is blanked (all segments are OFF). If the RBO output of one stage is connected to the RBI input of the next stage, ripple blanking is accomplished. The next display is also blanked if its BCD input is 0. The display can also be blanked by placing a 0 on the BI/RBO output. This is the direct blanking input (BI) feature of the **7447**. Whenever the BCD input to a **7447** is not 0, RBO is HIGH, regardless of RBI and blanking terminates.

## EXAMPLE 12-11

Show how the number 2047 can be displayed in a 6-digit display by making use of ripple blanking.

## SOLUTION

The solution is shown in Fig. 12-16. The RBI of the first chip is grounded. For succeeding chips the RBO is connected to the RBI of the next IC. Consequently, the first two ICs, which have an input of 0 and an RBI of 0, blank their displays by turning all the segments OFF. Since the third IC has an input of 2, it produces 2 on the display and causes RBO to be HIGH. The remaining ICs all receive a high RBI and display the digits on their ABCD inputs including the 0 in 2047.

### 12-6.5 Common Cathode 7-segment Displays and the 7448

It is now common practice to use LEDs with the cathode grounded. A single LED can be driven reasonably well from a TTL gate as shown in Fig. 12-17. When the output of the TTL gate is HIGH, current through the upper transistor of the totem pole lights the LED. A LOW output shorts the LED and keeps it off. The resistor in the upper part of the totem pole limits the current so no pull-up is needed.

Seven-segment common cathode displays, where each segment has its cathode tied to a common ground, are also becoming more popular. They can be driven by common cathode display drivers such as the **7448**, which is similar to the **7447** except that its outputs are reversed, so that it can drive common cathode instead of common anode displays.

A **7448** is shown driving a 7-segment display in Fig. 12-18. If a 0 is to be displayed, for example, the **7448** outputs for segments a through f are HIGH. This forces the current through the resistors to flow through the

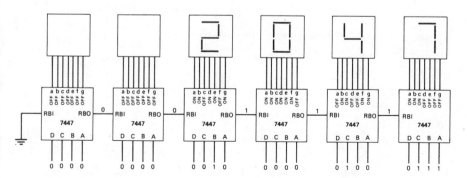

FIGURE 12-16 Display of the number 2047 using ripple blanking.

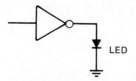

FIGURE 12-17 A TTL gate driving an LED.

LEDs. The output of segment g is LOW, however, so current through the g resistor flows into the 7448 instead of the LED and segment g remains dark.

Figure 12-18 also shows that the 7448 can only absorb a small current ($R_x \geq 650\ \Omega$), so it is limited to driving 7-segment displays that require only a small current. For common LEDs that require 20 mA for reasonable brightness, an IC with a higher $I_{OL}$, such as the **DM8857** made by National Semiconductor, should be used.

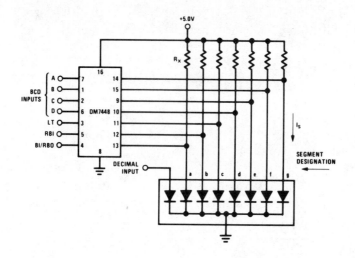

$R_x$ MAY BE CALCULATED USING THE FOLLOWING EQUATION

$$R_x = \frac{5.0 - V_{LED}}{I_s - 1.6}\ k\Omega = \frac{3.3}{I_s - 1.6}\ k\Omega \quad \left[\begin{array}{l} V_{LED} = 1.7V\ @\ 5.0\ mA \\ R_x \geq 650\Omega \end{array}\right]$$

WHERE:

$R_x$ = PULL-UP RESISTOR VALUE

$I_s$ = CURRENT PER SEGMENT IN mA

EXAMPLE:

$I_s$ = 5.0 mA

$R_x$ = 970Ω

FIGURE 12-18 Nonmultiplex application of the DM7448. (From National Semiconductor Tech Note AN-99. Copyright 1974, National Semiconductor Corporation.)

## 12-7  MULTIPLEXED  DISPLAYS

It has recently become common to multiplex 7-segment displays so that many displays are driven by the same driver. In multiplexed displays each display is only driven for a portion of the time, but because of the persistence of the human eye, it appears to be on constantly.

A typical multiplexing scheme is shown in Fig. 12-19. The digit drivers turn on each driver in turn. While this is happening the multiplexing circuitry presents the proper input to each of the segment drivers.

---

**EXAMPLE  12-12**

Inputs from four registers are to be shown on four common cathode 7-segment displays using a **7448.** Design the circuit.

**SOLUTION**

The circuit is shown in Fig. 12-20 and operates as follows.

1.  The four input digits are multiplexed by the **74153**s and passed on to the **7448.**
2.  The counter (the two **74107**s) periodically changes the SELECT inputs to the multiplexer and simultaneously enables each digit driver in turn so the correct digit is shown on the corresponding display. For clarity, only the driving circuit for segment *e* of digit 1 is shown.

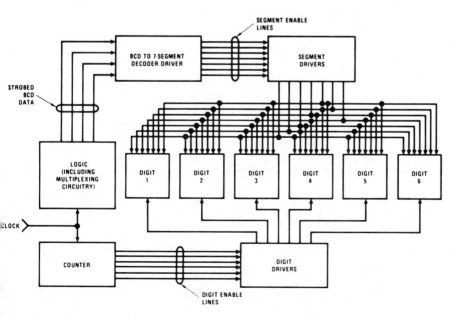

FIGURE  12-19   A typical multiplexing scheme. (From National Semiconductor Tech Note AN-99. Copyright 1974, National Semiconductor Corporation.)

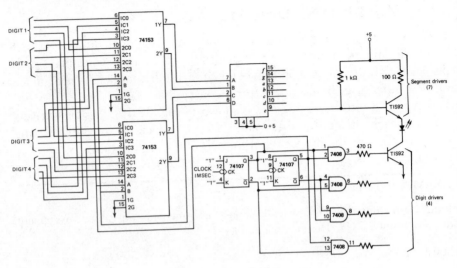

FIGURE 12-20 A circuit for multiplexing displays.

3. Because of the current limitations of the **7448**, an emitter follower is used to provide the current necessary to drive the LED segment.

4. A transistor is also required in the digit driver circuit. The collector of that transistor is connected to the common cathode of the LED. Thus, if all segments are ON and drawing 20 mA, the current through the lower transistor can be 140 mA (160 mA if a decimal point is also used), which is far greater than a TTL output can sink.

5. As shown, segment $e$ of display 1 lights only if the $e$ output of the **7448** is HIGH, turning on the emitter follower, and pin 3 of the **7408** is HIGH, providing base current to turn on the first digit driver.

6. This circuit requires 11 transistors, 7 individual segment drivers, and 4 digit drivers. A higher power decoder could eliminate the segment drivers. Current only flows through each LED one-quarter of the time, but we have found experimentally that the current need only be slightly in excess of 20 mA, and the resistance values shown in Fig. 12-20 give a bright and pleasing display.

---

Microprocessor kits, often used in schools, also use multiplexed displays. The display is driven basically as shown in Ex. 12-12, but there are slight differences in the implementation. They are discussed in Sec. 19-11.

## 12-8 LIQUID CRYSTAL DISPLAYS

Liquid crystal displays (LCDs) operate by polarizing light so that it does or does not reflect ambient light. Like LEDs, they have been organized as 7-segment displays for numerical readout and can be multiplexed.

LCDs have one outstanding advantage over LEDs: they essentially act as a capacitor and consume almost no power. For this reason they are now being used in watches and as readouts for hand-held computers and electronic games. This is a significant advantage because in many circuits LEDs consume more power than all the ICs in the rest of the circuit. Unfortunately, LCDs reflect, rather than generate, light and must be in a well-lit environment to be seen clearly. They must also be driven by an a.c. voltage, but this can be generated by EXCLUSIVE-OR circuits (see Sec. 13-8).

## SUMMARY

In this chapter two algorithms and a hardware method for the conversion of binary-to-BCD and BCD-to-binary are presented. The use of hardware is the fastest way of converting, but many computer manufacturers prefer to use one of the algorithms and utilize the computational capabilities of their computer. In some devices these capabilities may not be available and the use of hardware is necessary.

Commonly used indicators (LEDs and 7-segment displays) and the methods of driving these indicators are also discussed. The advantage of remote blanking is explained and an example using remote blanking is presented. Methods for driving multiplexed displays, where each display is driven by the same decoder and is only on for a part of the time, were also presented.

## GLOSSARY

**BCD.** Binary coded decimal.

**Decade.** A group of 4 bits that represent a single digit in BCD format.

**Algorithm.** A method or procedure for performing a mathematical operation.

**ROM.** Read only memory.

**Current limiting resistor.** A resistor used to limit the current flow in a circuit.

**LED.** Light-emitting diode.

**RBI.** Ripple blanking input to a 7-segment display decoder/driver.

**RBO.** Ripple blanking output.

**BI.** Direct blanking input to a 7-segment display decoder/driver.

## REFERENCES

George K. Kostopoulos, *Digital Engineering*, Wiley, New York, 1975.

Morris and Miller, *Designing With TTL Integrated Circuits*, McGraw-Hill, New York,

Dan I. Porat and Arpad Barna, *Introduction to Digital Techniques*, Wiley, New York, 1979.

*The TTL Data Book for Design Engineers*, Texas Instruments, Inc., 1976.

## PROBLEMS

12-1.    Express the following numbers in BCD:
   (a) 37
   (b) 509
   (c) 360,095
12-2.    Find the decimal equivalent of the following BCD numbers:
   (a) 10001001
   (b) 0010100100000101
   (c) 011101101001100100000001
   (d) 1100011001
12-3.    Are the following numbers in BCD? If so, what numbers are they? If not in BCD, why not?
   (a) 011110011001
   (b) 101000100001
   (c) 010011000010
12-4.    Convert the following decimal numbers to BCD and then to binary, using the three methods, respectively, given below:
   (a) 79
   (b) 442
   (c) 2707

1.   Use the addition algorithm.
2.   Use the shift and subtract algorithm.
3.   Use a **74184** circuit.

12-5.    Convert the last four digits of your social security number to binary by using the shift and subtract algorithm.
12-6.    Convert the following binary numbers to BCD and find their decimal equivalent:
   (a) 1101111
   (b) 101110011

1.   Use the subtraction algorithm.
2.   Use the shift and add algorithm.
3.   Use **74185s**.

12-7.    The input and output of each IC in Fig. P12-7 is given. Two of the ICs are faulty.
   (a) Find the defective ICs, correct them, and demonstrate that the circuit is working properly.
   (b) One of the defective ICs may be working correctly and the problem could be caused by a defective wire. Locate the defective wire and explain.
12-8.    If the inputs to a 7-segment display are 0110, show how it displays a number by drawing a figure similar to Fig. 12-15.
12-9.    Given the binary number 110110001011, which is available as a computer output, design a circuit to display its decimal equivalent. Include remote blanking.

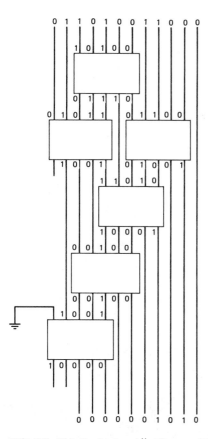

FIGURE P12-7  *Note:* All ICs are **74184**s.

12-10.   Repeat Problem 10-9 for the binary number 000010110001.

12-11.   A 6-digit display uses remote blanking. In the event that all 6 digits are 0, only a single 0 the least significant digit should be displayed. Design the circuit.

12-12.   A circuit must display any number from 0 to 108,000. The input number is available in 6 BCD decades, but only five 7-segment displays and 7447s are available. Design a circuit to display each number unambiguously (i.e., the reader must be able to distinguish between 50 and 100,050).

After attempting the above problems, the student should return to Sec. 12-2. If any of the questions still cause difficulty, he or she should review the appropriate sections of the chapter to find the answers.

# 13

# EXCLUSIVE OR CIRCUITS

## 13-1 INSTRUCTIONAL OBJECTIVES

This chapter demonstrates the use of the EXCLUSIVE OR (XOR) gate, and ICs based on XOR gates, in building comparators, parity checkers, and code converters. After reading it, the student should be able to:

1. Build comparison circuits using XOR gates or 7485s.
2. Build parity checkers and generators using XOR gates.
3. Build parity checkers and generators using 74180s.
4. Parity check long words by cascading 74180s.
5. Design and build a Gray code-to-binary converter.
6. Design an LRC generator and checker.

## 13-2 SELF-EVALUATION QUESTIONS

Watch for the answers to the following questions as you read the chapter. They should help you to understand the material presented. When you have finished the chapter, return to this section and be sure you can answer all of the questions.

1. Explain the differences between XOR and equality gates. What is their relationship?
2. What is the advantage of a wire-AND output for a comparison circuit?
3. Why is parity useful?

4.  Explain the difference between parity generation and parity checking.
5.  Why does parity generation always add one bit to the length of a word?
6.  How can a single **74180** generate a tenth odd parity bit if a 9-bit word is supplied?
7.  Give one advantage and one disadvantage of the Gray code compared to the binary code.
8.  How are XOR gates used to generate a.c. for LCDs?

## 13-3  COMPARISON CIRCUITS

The EXCLUSIVE OR (XOR) gate was introduced in Sec. 5-10. Some of the circuits using XOR gates are described in this chapter. This section considers the use of XOR gates and other circuits to compare two binary numbers.

### 13-3.1  Comparison Circuits Using XOR Gates

The use of discrete XOR gates to build a 4-bit comparator was discussed in Sec. 5-10 (see Ex. 5-9). To briefly review, the single XOR gate can only have two inputs. Its output is HIGH if the two inputs are unequal and LOW if the two inputs are equal. (Both are 0 or both are 1.)

The **7486** quad XOR is currently the most popular XOR IC. The comparator of Fig. 5-24 was designed by making a bit-by-bit comparison of the registers, using **7486** gates. A LOW output indicated that both inputs to the gate were equal. *If the outputs of all the 7486 gates were LOW, it indicated that each pair of bits in the registers were equal and, therefore, the numbers in both registers were equal.*

### 13-3.2  The EXCLUSIVE NOR Gate

The EXCLUSIVE NOR or EQUALITY gate has two inputs and produces a HIGH output whenever the two inputs are equal. The symbol $\odot$ is used to indicate equality. The basic equation is:

$$Y = A \odot B = AB + \overline{A}\overline{B} \tag{13-1}$$

which indicates $Y$ is 1 if the two inputs, $A$ and $B$, are equal (both 0 or both 1).

The **74LS266** is a QUAD X-NOR gate with open collector outputs so that many bits can be wire-ANDed together for a multiple bit comparison. It is a low power Schottky device and is not as commonly used or as readily available as the more popular **7486** XOR.

## EXAMPLE 13-1

Show that an equality gate can be constructed by inverting the output of a 7486, as in Fig. 13-1.

## SOLUTION

### Proof 1. DeMorgan's Theorem

The inverter complements the function by DeMorgan's theorem:

$$
\begin{aligned}
Y &= \overline{(A \oplus B)} \\
&= \overline{(A\bar{B} + B\bar{A})} \\
&= (\bar{A} + B)(\bar{B} + A) \\
&= \bar{A}\bar{B} + AB \text{ or } A \odot B
\end{aligned}
$$

### Proof 2. Pure Reasoning

With two inputs only two possibilities exist: either the inputs are equal or they are not. Since the XOR produces a HIGH output when the two inputs are unequal, its complement must produce a HIGH output when the inputs *are* equal. Note that if the inverter had been placed on one of the inputs to the 7486, instead of the output, the result would have been the same (see Problem 5-15).

## EXAMPLE 13-2

Given two *n*-bit words, A and B, design a comparison circuit using 74LS266 EQUALITY gates to produce a HIGH output when the two words are equal.

## SOLUTION

The solution is shown in Fig. 13-2. Note the symbol for the EQUALITY gate. Each pair of bits is compared by a **74LS266** gate and the outputs are wire-ANDed together. If the inputs to any gate are unequal, the output of that gate is 0, and it causes the wire-ANDed output to go LOW. When each pair of inputs is equal, all the **74LS266** gates produce a HIGH output, allowing the final output to remain HIGH. The wire-ANDing feature of the **74LS266** simplifies the comparison circuit (see Problem 13-2).

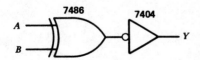

FIGURE 13-1  An equality gate made by inverting the output of an XOR gate. Y is HIGH only if A = B.

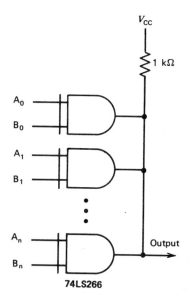

FIGURE 13-2 An $n$-bit comparator using **74LS266** equality (EXCLUSIVE-NOR) gates.

### 13-3.3 The 7485 4-Bit Comparator

The 7485 is an MSI IC specifically designed to compare two binary numbers. The pin layout and function table are shown in Fig. 13-3. The inputs are:

1. Four A inputs, $A_3A_2A_1A_0$.
2. Four B inputs, $B_3B_2B_1B_0$.
3. Three cascading inputs: $A > B$, $A < B$, and $A = B$.

The function of the IC is to compare two 4-bit numbers, A and B. Three outputs are provided, $A > B$, $A < B$, and $A = B$, to give the result of the comparison.

The function table shows that comparison starts with the most significant bits, $A_3$ and $B_3$. If they are unequal, the comparison is already determined (i.e., if $A_3 = 1$ and $B_3 = 0$, then A is larger than B *regardless* of the values of the less significant bits). If $A_3 = B_3$ ($A_3$ and $B_3$ are either both 0 or both 1), the results of the comparison depends on $A_2$ and $B_2$. If these are also equal, the comparison depends on $A_1$ and $B_1$ and finally on $A_0$ and $B_0$. If all A and B bits, respectively, are equal, the two numbers are equal and the output $A = B$ should be HIGH. This occurs if the cascading inputs, $A > B$ and $A < B$ are grounded and the $A = B$ input is tied HIGH, as shown by the eleventh line of the function table.

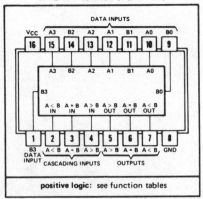

'85, 'S85
J OR N DUAL-IN-LINE OR
W FLAT PACKAGE (TOP VIEW)

(a) Pin Layout

| COMPARING INPUTS | | | | CASCADING INPUTS | | | OUTPUTS | | |
|---|---|---|---|---|---|---|---|---|---|
| A3, B3 | A2, B2 | A1, B1 | A0, B0 | A > B | A < B | A = B | A > B | A < B | A = B |
| A3 > B3 | X | X | X | X | X | X | H | L | L |
| A3 < B3 | X | X | X | X | X | X | L | H | L |
| A3 = B3 | A2 > B2 | X | X | X | X | X | H | L | L |
| A3 = B3 | A2 < B2 | X | X | X | X | X | L | H | L |
| A3 = B2 | A2 = B2 | A1 > B1 | X | X | X | X | H | L | L |
| A3 = B3 | A2 = B2 | A1 < B1 | X | X | X | X | L | H | L |
| A3 = B3 | A2 = B2 | A1 = B1 | A0 > B0 | X | X | X | H | L | L |
| A3 = B3 | A2 = B2 | A1 = B1 | A0 < B0 | X | X | X | L | H | L |
| A3 = B3 | A2 = B2 | A1 = B1 | A0 = B0 | H | L | L | H | L | L |
| A3 = B3 | A2 = B2 | A1 = B1 | A0 = B0 | L | H | L | L | H | L |
| A3 = B3 | A2 = B2 | A1 = B1 | A0 = B0 | L | L | H | L | L | H |
| A3 = B3 | A2 = B2 | A1 = B1 | A0 = B0 | X | X | H | L | L | H |
| A3 = B3 | A2 = B2 | A1 = B1 | A0 = B0 | H | H | L | L | L | L |
| A3 = B3 | A2 = B2 | A1 = B1 | A0 = B0 | L | L | L | H | H | L |

(b) Function table

FIGURE 13-3 The **7485** 4-bit comparator. (From the *TTL Data Book for Design Engineers,* 2nd ed., Texas Instruments, Inc. Courtesy of Texas Instruments, copyright 1976.)

## EXAMPLE 13-3

Design a 4-bit comparator using a single 7485. Show exactly how it works if $A = 0100$ and $B = 0110$.

## SOLUTION

The solution is shown in Fig. 13-4. Note that the cascading inputs are connected to produce a HIGH on the $A = B$ output if the two inputs are equal.

For the numbers given, $A_3 = B_3$, $A_2 = B_2$, and $A_1 < B_1$. Therefore, the sixth

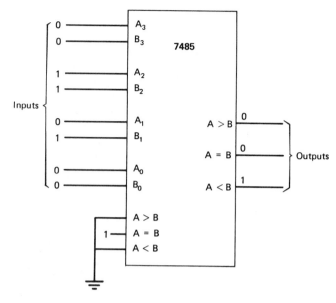

FIGURE 13-4   A **7485** comparing 4 to 6 as specified in Ex. 13-3.

line of the function table is used. $A_0$, $B_0$, and the cascading inputs do not matter. The A < B output goes HIGH, which is correct since A = 4 and B = 6.

## 13-3.4   Cascading 7485s

7485s can be cascaded to compare numbers of any length. Figure 13-5 shows how two 7485s can be used to compare two 8-bit numbers ($X_7$ − $X_0$ to $Y_7$ − $Y_0$). Longer numbers can be compared by connecting the A > B, A = B, and A < B outputs of one **7485** to the corresponding inputs of the *next more significant* set of four bits. The output must be taken from the **7485** whose inputs are the *most significant bits* of the comparison.

### EXAMPLE 13-4

Design a circuit to compare an 8-bit number to $(195)_{10}$.

### SOLUTION

$(195)_{10}$ = 11000011. This is an 8-bit number so the circuit of Fig. 13-5 can be used. If the Y inputs ($Y_7$ − $Y_0$) are connected to 11000011, respectively, and the unknown number is brought into the X inputs, the **7485**s will compare the numbers and produce the proper output.

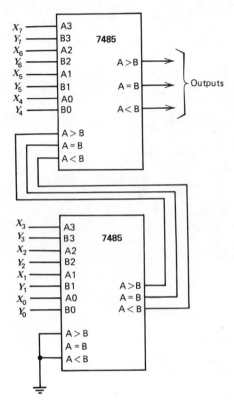

FIGURE 13-5   Cascading two **7485**s to compare 8-bit numbers.

A clever circuit for comparing two 24-bit words is shown in Fig. 13-6. Although this circuit does not reduce the IC count, it is faster. When comparing two 24-bit words, an extension of the circuit of Fig. 13-5 would require six **7485**s and six **7485** delays; whereas the circuit of Fig. 13-6 would also require six **7485**s, but only take two gate delays to make the comparison.

## 13-4   PARITY CHECKING AND GENERATION

*Parity is the addition of a bit to a binary word, to help insure the integrity of the data.* The concept is best made clear by an example, so consider the writing and reading of a magnetic tape. Typically data are written on the tape by a set of *write heads*, which magnetize the tape in accordance with the bits being written. Data are read by the *read heads* that detect the magnetization of the tape and convert it to 1s and 0s. Data are written on tape in character format. If there are eight bits to a character, eight write heads are required and all eight bits are written simultaneously across the

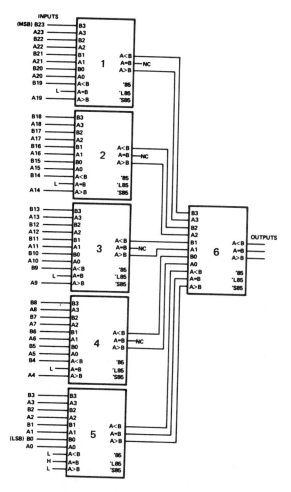

**COMPARISON OF TWO 24-BIT WORDS**

FIGURE 13-6 Cascading **7485**s to produce a 24-bit comparator. (From the *TTL Data Book for Design Engineers,* 2nd ed., Texas Instruments, Inc. Courtesy of Texas Instruments, copyright 1976.)

tape. The tape continues on and the next character is written. Character format is illustrated in Fig. 13-7.

Often data are written on a tape and then the tape is stored until needed, perhaps as long as several years. Parity is a means of checking the data so that we may be confident *the data read are the same as the data originally written* to the tape.

To insure the integrity of the data, a *parity bit* is added to each character and is written on the tape along with the bits that determine the character.

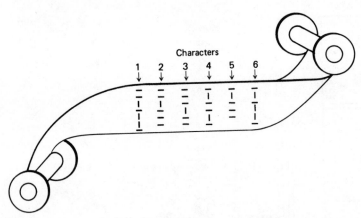

FIGURE 13-7 Characters written on a reel of magentic tape.

There are two types of parity: odd and even. *To generate odd parity, a bit is added to each character so that the number of 1s in the character is odd.* If, for example, a character, as supplied from a computer or other device, consists of 7 bits, an eighth bit is added to make the number of 1s in the 8-bit character odd. This 8-bit, odd parity character is then written to the tape.

Even parity is the complement of odd parity. If even parity is used, the number of 1s in each character is even. This process of adding a bit to make the parity either even or odd is called *parity generation.*

---

**EXAMPLE 13-5**

Characters A, B, C, and D are received from a computer. Generate an eighth bit to maintain:

1. Odd parity.
2. Even parity.

|   |         |
|---|---------|
| A | 0000000 |
| B | 0101011 |
| C | 1100010 |
| D | 1111111 |

**SOLUTION**

1. Characters A and B, as received, contain an even number of 1s. Character A contains no 1s and character B contains four 1s. Characters C and D contain an odd number of 1s (3 and 7, respectively). To generate odd parity, 1s are added to

characters A and B and 0s are added to characters C and D. The 8-bit words written on the tape look like this:

| | |
|---|---|
| 0000000 | 1 |
| 0101011 | 1 |
| 1100010 | 0 |
| 1111111 | 0 |

Original characters          Parity bit

Note that the number of 1s in each 8-bit character are 1, 5, 3, and 7, respecitvely. Thus, each character contains an odd number of 1s.
2.   For even parity, a bit is added to make the number of 1s in each character even. The word written to tape looks like this:

| | |
|---|---|
| 0000000 | 0 |
| 0101011 | 0 |
| 1100010 | 1 |
| 1111111 | 1 |

Original characters          Parity bit

Note that the number of 1s in each character are now 0, 4, 4, and 8, respectively. Each character contains an even number of 1s.

## 13-4.1   The Parity Karnaugh Map

Consider the function $f(A,B,C,D) = \Sigma(1,2,4,7,8,11,13,14)$. Its Karnaugh map is shown in Fig. 13-8 and resembles a checkerboard. No 1s can be combined into subcubes. The function is:

$$f(A,B,C,D) = \bar{A}\bar{B}\bar{C}D + \bar{A}\bar{B}C\bar{D} + \bar{A}B\bar{C}\bar{D} + \bar{A}BCD + A\bar{B}\bar{C}\bar{D}$$
$$+ A\bar{B}CD + AB\bar{C}D + ABC\bar{D} \qquad (13\text{-}2)$$

It requires eight 4-input NAND gates and an 8-input NAND gate if NAND implementation is used.

FIGURE 13-8   Karnaugh map for $f(A,B,C,D) = \Sigma(1,2,4,7,8,11,13,14)$.

Fortunately, there is a better way to implement Eq. (13-2). Functions that produce *checkerboard* Karnaugh maps are easily implemented using XOR gates. Equation (13-2) reduces to:

$$f(A,B,C,D) = A \oplus B \oplus C \oplus D$$

as shown in Appendix C, and the circuit to implement it is shown in Fig. 13-9.

Considering $f(A,B,C,D)$ further, we discover that the number of 1s in each term is odd. This leads to the conclusion that there is a direct relationship between parity circuits and XOR gates. It can be shown (Appendix D) that if *all* the outputs of a register are XORed together, the output of the XOR circuit is HIGH *only* if the number of 1s in the input word is *odd*.

## 13-4.2 Parity Checking Using XOR Gates

*Parity checking* is the process of examining all *n* bits of a word to determine if the number of 1s in the *n* bits is odd or even, and reporting an error if the parity is wrong.

---

### EXAMPLE 13-6

Design an odd parity checker for 8-bit words.

### SOLUTION

An 8-bit parity checker can be built by XORing the 8 inputs together. Two circuits to accomplish this are shown in Fig. 13-10. In each case the output is HIGH *only* if there is an *odd* number of 1s among the 8 inputs. Therefore, any LOW output indicates an error because we are checking for odd parity.

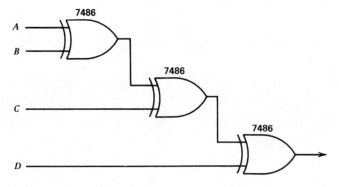

FIGURE 13-9  Implementation of Eq. (13-2) using XOR GATES.

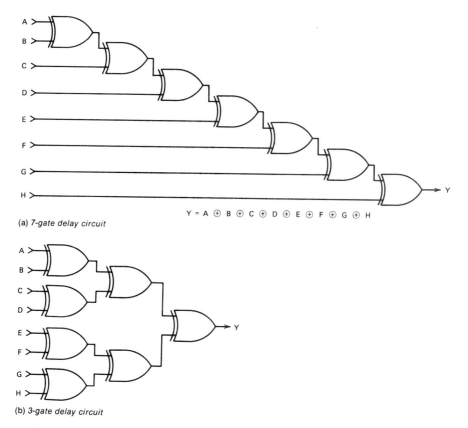

(a) 7-gate delay circuit

$$Y = A \oplus B \oplus C \oplus D \oplus E \oplus F \oplus G \oplus H$$

(b) 3-gate delay circuit

FIGURE 13-10  Two ways of using XOR gates as parity checker/generators. (From Morris and Miller, *Designing with TTL Circuits*, copyright 1971. Courtesy of Texas Instruments, Inc.)

Although *both* circuits of Fig. 13-10 contain the *same* number of gates, Fig. 13-10b is more commonly used. It is faster because it contains only 3-gate delays, whereas Fig. 13-10a has 7-gate delays.

## 13-4.3 Parity Generation Using XOR Gates

Parity generation involves adding an extra bit to an $n$-bit word in order to produce the proper parity in the $(n + 1)$-bit word. The concept was explained in Sec. 13-4 and Ex. 13-5. A circuit to generate the proper parity bit can be built using XOR gates. If, for example, *odd* parity is *required*, the $n$-bit word is checked for odd parity. If the parity of the $n$ bits is odd, a 0 is written in the $n + 1$ bit (the parity bit) and the *odd* number of 1s in the word is still preserved. If the parity of the $n$ bits is even, a 1 is written as the parity bit so that the parity of the $(n + 1)$-bit word is odd.

## EXAMPLE 13-7

(a) Design a parity generating circuit to add an eighth bit to a 7-bit word so that the word has odd parity.

(b) Show how it operates if the 7-bit input word is 1101111.

## SOLUTION

(a) The original 7 bits are XORed together to form a 7-bit odd parity checker. If the parity checker output is HIGH, the eighth bit should be a 0 because there are an odd number of 1s in the original 7-bit word, but if the parity checker output is LOW, indicating even parity, the eighth bit must be a 1 to cause odd parity. Therefore, the eighth bit is generated by *inverting* the output of the 7-bit parity checker.

(b) The parity generating circuit is similar to the parity checker of Fig. 13-10, and is shown in Fig. 13-11. The 1s and 0s specified in the example are also written on the figure. Note that the 8-bit output consists of the 7 input bits plus bit H, which causes the parity of the 8-bit output to be odd.

### 13-4.4 Parity Checking in Microprocessors

Some microprocessors (the **8080, 8085,** and **Z80**) check the parity of an 8-bit word after each arithmetic operation. They set a *parity bit* in their

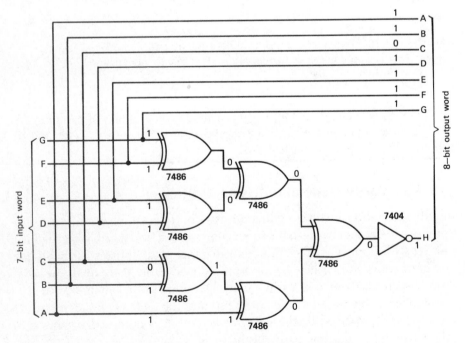

FIGURE 13-11 An 8-bit parity generator. All XORs are **7486**s.

*condition code register* to indicate whether the parity of the word is even or odd. The parity bit, or flag, can then be used to control conditional jump instructions so that the proper action is taken depending upon the parity of the word.

## 13-5 PARITY CHECKING AND GENERATION USING THE 74180

The 74180 is a 9-bit parity generator/checker, which can check a 9-bit input for parity, and generate a tenth parity bit when a 9-bit word is supplied. The layout of the 74180 and its function table are shown in Fig. 13-12. There are 8 inputs for the bits whose parity is to be checked (inputs A through H), plus an EVEN input and an ODD input. The EVEN and ODD inputs can be used either to accommodate a ninth bit or to cascade 74180s for long words. Note that the EVEN and ODD inputs must be *complements* of each other in order for the IC to check parity.

The 74180 has two outputs, an ΣODD output and an ΣEVEN output, which indicate the parity of the inputs. The outputs are always complementary. Once a circuit has been designed, it is easiest to assume all inputs are 0, to determine what parity the output levels indicate.

---

**EXAMPLE 13-8**

   (a) Design a 7-bit parity checker using a single 74180.
   (b) How could this circuit be used to generate an eighth bit for odd parity?

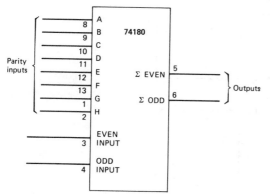

| INPUTS | | | OUTPUTS | |
|---|---|---|---|---|
| Σ OF H's AT A THRU H | EVEN | ODD | Σ EVEN | Σ ODD |
| EVEN | H | L | H | L |
| ODD | H | L | L | H |
| EVEN | L | H | L | H |
| ODD | L | H | H | L |
| X | H | H | L | L |
| X | L | L | H | H |

H = high level, L = low level, X = irrelevant

(a) Pin layout                       (b) Function table

FIGURE 13-12 The **74180.** (From the *TTL Data Book for Design Engineers,* 2nd. ed., Texas Instruments, Inc. Courtesy of Texas Instruments, copyright 1976.)

## SOLUTION

(a) The solution is shown in Fig. 13-13. The seven inputs are applied to the A through G inputs of the **74180**. Since input H is not required, it is grounded so it does not affect the parity. The EVEN input is tied HIGH and the ODD input is grounded to make them complementary. If all the A through G inputs are 0, line 1 of the function table applies. Therefore, if the parity is even, the ΣEVEN output is HIGH and the ΣODD output is LOW. For odd parity, the outputs reverse and the ΣODD output is HIGH.

(b) If all 7 input bits are 0, the parity of the incoming word is even and the ΣEVEN output is HIGH. Therefore, if the 7 input bits and the ΣEVEN output are combined, they form an 8-bit, odd parity word that contains seven 0s and one 1 (the ΣEVEN output). Consequently, the 7-bit parity checker also acts as a parity generator. The 8-bit output word is shown in dashed lines on Fig. 13-13.

## EXAMPLE 13-9

(a) Design a 9-bit parity checker using a single **74180**.

(b) Show how it works if the inputs are 101101100.

## SOLUTION

(a) The first 8-bits are applied to the A through H inputs. The 9th bit is applied to the EVEN input and its complement is applied to the ODD input through an inverter, as shown in Fig. 13-14. The input 1s and 0s are also shown on Fig. 13-

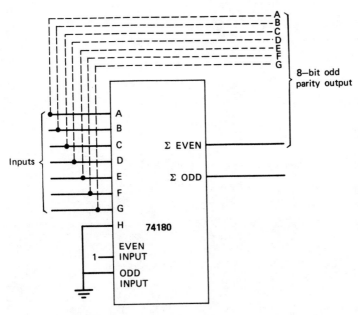

FIGURE 13-13   A 7-bit parity checker.

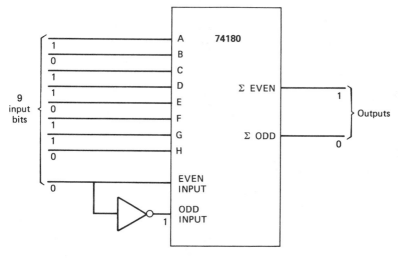

FIGURE 13-14 A 9-bit parity checker.

14. The ninth bit, which is a 0, causes the EVEN input to be 0 and the ODD input to be 1.

(b) Since the A through H inputs contain odd parity (five 1s), line 4 of the function table applies and the ΣEVEN output is HIGH. Therefore, this circuit has its ΣEVEN output HIGH whenever the parity of the input word is odd.

## 13-5.1 Cascading 74180s

To check the parity of words longer than 9 bits, **74180s** can be cascaded by connecting the ΣEVEN output of the first **74180** to the EVEN input of the second **74180** and the ΣODD output of the first to the ODD input of the second. The first **74180** can accept 9 inputs and the second **74180** can accept 8 additional input bits. (Its EVEN and ODD inputs are already used.) Therefore, two **74180s** can check a 17-bit word. Each additional **74180** added in this manner increases the checking capability by 8 bits.

### EXAMPLE 13-10

(a) Design a 14-bit odd parity checker using two **74180s**.
(b) Show how your parity checker works if the input bits are 01101110110101.

### SOLUTION

(a) Perhaps the most straightforward way to build this parity checker is shown in Fig. 13-15. The first 8 inputs are applied to the first **74180**, which has its EVEN

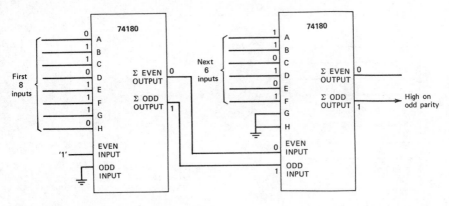

FIGURE 13-15 A 14-bit parity checker.

input tied HIGH and its ODD input LOW. The outputs are tied to the corresponding inputs of the second **74180**, which also receives the remaining 6 bits. The two unused inputs are grounded.

To determine the state of the outputs, assume all 14 input bits are 0. From the **74180** function table, Fig. 13-12*b*, the ΣEVEN output of the first chip is HIGH, and the ΣEVEN output of the second chip is also HIGH. This, of course, is even parity (zero 1s). For odd parity, the ΣODD output will be HIGH.

(b) The 1s and 0s for the given inputs are also shown in Fig. 13-15. For the first **74180**, the input parity is odd (five 1s) and the second line of the function table applies, making the ΣEVEN output LOW. For the second **74180** the input parity is even (four 1s), and the third line of the function table applies because the EVEN input is LOW and the odd input is HIGH. Consequently, the final ΣODD output is HIGH, indicating odd parity (there are nine 1s in the 14-bit word).

## 13-6 MORE SOPHISTICATED ERROR-CORRECTING ROUTINES

Odd or even parity checking schemes are highly effective, but they have two drawbacks: they will not detect two errors in the same word or character, and they provide no capability for correcting an error if one is found. When high reliability is required, additional checks such as *longitudinal redundancy checking* (LRC) or *cyclical redundancy checking* (CRC) are used.

Often a group or block of words is being read (the words stored on one sector of a disk is an example) and the user is concerned that every one of the words is correct. It is now common practice to *add* an additional character or two to insure the integrity of the data. These are the longitudinal or cyclical redundancy characters. They contain no new information and cause more bits to be transmitted for the same message, but their value in checking the data compensates for the additional block length in most systems.

## 13-6.1  Longitudinal Redundancy Checking

Longitudinal redundancy checking is accomplished by adding an LRC character to each block of data. Bit 1 of the LRC character is a 1 if all the bit 1s of the transmitted block taken together contain an *odd* number of 1s. The other bits are similarly determined.

The situation is shown in Fig. 13-16 where, for simplicity, a 6-word-by-4-bit data block is assumed. Vertical or odd parity is added to each of the 6 words, and the 5-bit LRC character is calculated by looking horizontally across the 6 bits and determining if the number of 1s is odd or even.

As with vertical parity, *LRCs are generated when writing* and appended as the last character of a block. The LRC character must be *checked* whenever the block is read.

---

**EXAMPLE 13-11**

Design a circuit to generate an LRC character.

**SOLUTION**

A circuit using J-K FFs is shown in Fig. 13-17. For an N-bit word it requires N FFs. At the start of transmission the CLEAR line is pulsed LOW, clearing all the FFs. The data bit to be written is also connected to the J and K inputs of the corresponding FF. The word clock occurs as each word is transmitted. If the data bit is a 1, the FF toggles; otherwise it does not change. When the entire block is transmitted, the LRC character is on the outputs of the FFs. This character must then be transmitted.

When reading the data, the LRC can be checked by a circuit similar to that of Fig. 13-17. The clock that clocks in the received data bits is also used to toggle the

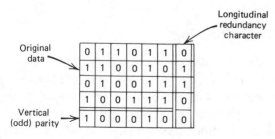

FIGURE 13-16  Addition of vertical parity and an LRC to a 6-word-by-4-bit data block.

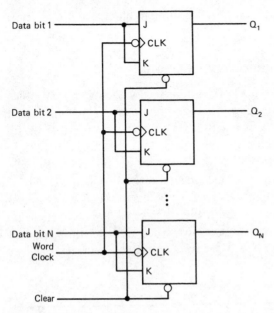

FIGURE 13-17  A circuit for generating an LRC.

N FFs if the received data bit is a 1. When the LRC is received, it can be compared to the FFs and a match indicates that a valid data block has been received. It is simpler, however, to allow the LRC character to be clocked in and toggle the FFs also. If the block has been received correctly, the FF outputs after the LRC has been received will all be 0s.

## 13-6.2  Error Correction

If a block is received with only a single bit in error, the LRC scheme of Fig. 13-16 allows us to find and correct the error. If only one bit is wrong, it will cause one vertical parity error and one bit in the LRC word to be wrong. The bit in error is at the intersection of the vertical and horizontal parity errors and should be complemented (see Problem 13-18).

## 13-6.3  Cyclical Redundancy Checking

This checking is achieved by appending a cyclical redundancy check character (CRC) to each data block. The CRC is calculated by taking the data and dividing it by a given polynomial. This division can be accomplished in hardware by a circuit using a shift register and XOR gates for feedback. CRC circuits are also available as ICs from several manufacturers. Unfortunately, the theory of CRC is too long and complex to be presented here. The reader should consult other books for more information.

| Decimal Number | Binary Equivalent | | | | Gray Code Equivalent | | | |
|:---:|:---:|:---:|:---:|:---:|:---:|:---:|:---:|:---:|
| | $B_3$ | $B_2$ | $B_1$ | $B_0$ | $G_3$ | $G_2$ | $G_1$ | $G_0$ |
| 0 | 0 | 0 | 0 | 0 | 0 | 0 | 0 | 0 |
| 1 | 0 | 0 | 0 | 1 | 0 | 0 | 0 | 1 |
| 2 | 0 | 0 | 1 | 0 | 0 | 0 | 1 | 1 |
| 3 | 0 | 0 | 1 | 1 | 0 | 0 | 1 | 0 |
| 4 | 0 | 1 | 0 | 0 | 0 | 1 | 1 | 0 |
| 5 | 0 | 1 | 0 | 1 | 0 | 1 | 1 | 1 |
| 6 | 0 | 1 | 1 | 0 | 0 | 1 | 0 | 1 |
| 7 | 0 | 1 | 1 | 1 | 0 | 1 | 0 | 0 |
| 8 | 1 | 0 | 0 | 0 | 1 | 1 | 0 | 0 |
| 9 | 1 | 0 | 0 | 1 | 1 | 1 | 0 | 1 |
| 10 | 1 | 0 | 1 | 0 | 1 | 1 | 1 | 1 |
| 11 | 1 | 0 | 1 | 1 | 1 | 1 | 1 | 0 |
| 12 | 1 | 1 | 0 | 0 | 1 | 0 | 1 | 0 |
| 13 | 1 | 1 | 0 | 1 | 1 | 0 | 1 | 1 |
| 14 | 1 | 1 | 1 | 0 | 1 | 0 | 0 | 1 |
| 15 | 1 | 1 | 1 | 1 | 1 | 0 | 0 | 0 |

FIGURE 13-18   Gray code and binary equivalents of decimal numbers.

## 13-7   THE GRAY CODE

The Gray code is a code for representing numbers that is often used when analog-to-digital conversions are required. The Gray code has the advantage that *only one bit in the numerical representation changes between successive numbers.* Unfortunately, it is difficult to perform arithmetic operations on numbers in Gray code, so Gray-to-binary and binary-to-Gray conversions are often performed.

### 13-7.1   Construction of the Gray Code

Figure 13-18 shows the Gray code equivalent of the first 15 numbers. The Gray code table was obtained by using the following algorithm:

1.   For $G_0$ write one 0 followed by two 1s, two 0s, two 1s, and so on.
2.   For $G_1$ write two 0s followed by alternating groups of four 1s and four 0s.
3.   In general, for the $G_n$ column, start by writing $2^n$ 0s. Then write alternating groups of $2^{n+1}$ 1s and 0s.

Note that as the count progresses only a single bit in the Gray code representation changes. When a count goes from 7 to 8, for example, all 4 bits in the binary representation change, but only $G_3$ in the Gray code representation changes.

### 13-7.2 Uses of the Gray Code

One of the common uses of the Gray code is to determine the position of a rotating shaft. Typically, this is done by attaching a code wheel to the shaft as shown in Fig. 13-19. The code wheel contains transparent and opaque segments. Light transmitted through the code wheel is detected by phototransistors or other photosensitive sensors. The position of the shaft is determined by the 1s and 0s (transparent and opaque segments) detected by the photoelectric sensors. This scheme is often used on rotating drum line printers, where the code wheel output determines when the hammers strike the drum and print the character.

A 4-bit binary code wheel is shown in Fig. 13-20a and a 4-bit Gray code wheel is shown in Fig. 13-20b. Each code wheel produces 16 codes, so the shaft position resolution is 360/16 = 22.5 degrees. For greater resolution, additional concentric segments must be added.

If a radius is drawn straight up on the binary code wheel of Fig. 13-20a, the segments to the left of the radius are shaded and all the segments to the right are clear. *When the detector or sensor is placed in exactly this middle position, it could produce either a 0 or a 1 output for each sector.* Consequently, *any number between 0 and 15 might appear on the output.* This ambiguity is avoided by the Gray code wheel, where *only one* segment changes at a time.

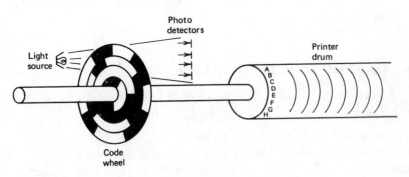

FIGURE 13-19 A code wheel used on a drum printer.

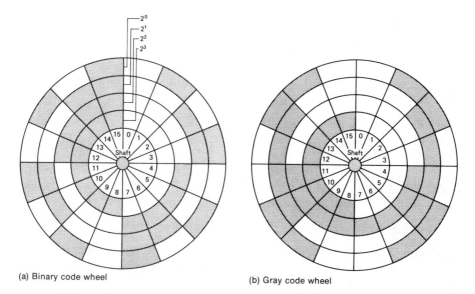

(a) Binary code wheel        (b) Gray code wheel

FIGURE 13-20   Binary and Gray code wheels. (From Barna and Porta, *Integrated Circuits in Digital Electronics*. Copyright John Wiley & Sons, Inc. 1973. Reprinted by permission of John Wiley & Sons, Inc.)

---

**EXAMPLE 13-12**

(a) Assume a sensor is right on the division between 11 and 12 and produces a random output for each sector that changes. What numbers might it select if the code wheel is binary?

(b) Repeat for a Gray code disk.

**SOLUTION**

(a) On the binary code wheel the three least significant (outermost) sectors change between 11 and 12. Only the innermost sector (the MSB) is a 1 on both sides of the line. Consequently, any number between 8 and 15 may appear as an output.

(b) On the Gray code disk only the second sector changes between 11 and 12. Consequently, the output is either 1110 (the Gray equivalent of 11) or 1010 (the Gray equivalent of 12). This improves the shaft resolution considerably.

---

## 13-7.3   Gray-to-Binary Conversion

In determining shaft positions, it is advantageous to pick the position off a disk in Gray code and convert the Gray code output to binary for numerical processing within a computer. Fortunately, Gray-to-binary converters are very simple to design and build and consist entirely of XOR gates.

## EXAMPLE 13-13

Design a converter to change a 4-bit Gray code to its equivalent 4-bit binary representation.

## SOLUTION

Code conversion problems were introduced in Sec. 3-8.4, and the methods developed there are applicable to this problem. The inputs consist of the 4 Gray code bits $(G_3G_2G_1G_0)$ and there are 4 separate binary outputs $(B_3B_2B_1B_0)$. The solution is shown in Fig. 13-21. The decimal numbers are listed in column 1 of the truth table (Fig. 13-21a) and the corresponding Gray code entries are listed in column 2, using the procedure developed in Sec. 13-7.1. The Gray code entries, which are the inputs, do not progress in the normal manner of truth table entries; they do not increment. Therefore, column 3 was written. *The numbers in column 3 are the decimal equivalent of the Gray code entries of column 2, read as binary numbers.* This is done to allow us to develop the equations and Karnaugh maps. The corresponding binary outputs are listed in column 4.

In Fig. 13-21b, the equations for the binary outputs are developed. *These equations are obtained for each binary bit by listing the numbers in column 3 that correspond to the 1s in each binary output column.*

Once the equations are written, Karnaugh maps can be drawn for each bit, as shown in Fig. 13-21c, and the equations relating the binary bits to the Gray code bits developed. The map for $B_0$ is a checkerboard. Fortunately, checkerboard Karnaugh maps were encountered in Sec. 13-4.1 and since the maps are the same, the solution is the same:

$$B_0 = G_3 \oplus G_2 \oplus G_1 \oplus G_0$$

For $B_1$ we have, by combining subcubes,

$$B_1 = \bar{G}_3\bar{G}_2G_1 + \bar{G}_3G_2\bar{G}_1 + G_3G_2G_1 + G_3\bar{G}_2\bar{G}_1$$

which reduces (with a little inspiration) to

$$\begin{aligned} B_1 &= \bar{G}_3(\bar{G}_2G_1 + G_2\bar{G}_1) + G_3(G_2G_1 + \bar{G}_2\bar{G}_1) \\ &= \bar{G}_3(G_2 \oplus G_1) + G_3(\overline{G_2 \oplus G_1}) \\ &= G_3 \oplus G_2 \oplus G_1 \end{aligned}$$

The maps for $B_2$ and $B_3$ are even simpler.

The final step, the design of the converter, becomes very simple, as shown in Fig. 13-21d. The 4 Gray code input bits are converted to the corresponding binary bits simply by using 3 XOR gates.

| Column | 1 | 2 | | 3 | 4 | |
|---|---|---|---|---|---|---|
| | Decimal Number | Gray Code | | | Binary Output | |
| | | $G_3$ $G_2$ $G_1$ $G_0$ | | | $B_3$ $B_2$ $B_1$ $B_0$ | |
| | 0 | 0 0 0 0 | 0 | | 0 0 0 0 | $B_3 = \Sigma\ (12,\ 13,\ 15,\ 14,\ 10,\ 11,\ 9,\ 8)$ |
| | 1 | 0 0 0 1 | 1 | | 0 0 0 1 | |
| | 2 | 0 0 1 1 | 3 | | 0 0 1 0 | $B_2 = \Sigma\ (6,\ 7,\ 5,\ 4,\ 10,\ 11,\ 9,\ 8)$ |
| | 3 | 0 0 1 0 | 2 | | 0 0 1 1 | |
| | 4 | 0 1 1 0 | 6 | | 0 1 0 0 | $B_1 = \Sigma\ (3,\ 2,\ 5,\ 4,\ 15,\ 14,\ 9,\ 8)$ |
| | 5 | 0 1 1 1 | 7 | | 0 1 0 1 | |
| | 6 | 0 1 0 1 | 5 | | 0 1 1 0 | $B_0 = \Sigma\ (1,\ 2,\ 7,\ 4,\ 13,\ 14,\ 11,\ 8)$ |
| | 7 | 0 1 0 0 | 4 | | 0 1 1 1 | (b) Equations |
| | 8 | 1 1 0 0 | 12 | | 1 0 0 0 | |
| | 9 | 1 1 0 1 | 13 | | 1 0 0 1 | |
| | 10 | 1 1 1 1 | 15 | | 1 0 1 0 | |
| | 11 | 1 1 1 0 | 14 | | 1 0 1 1 | |
| | 12 | 1 0 1 0 | 10 | | 1 1 0 0 | |
| | 13 | 1 0 1 1 | 11 | | 1 1 0 1 | |
| | 14 | 1 0 0 1 | 9 | | 1 1 1 0 | |
| | 15 | 1 0 0 0 | 8 | | 1 1 1 1 | |

(a) Truth table

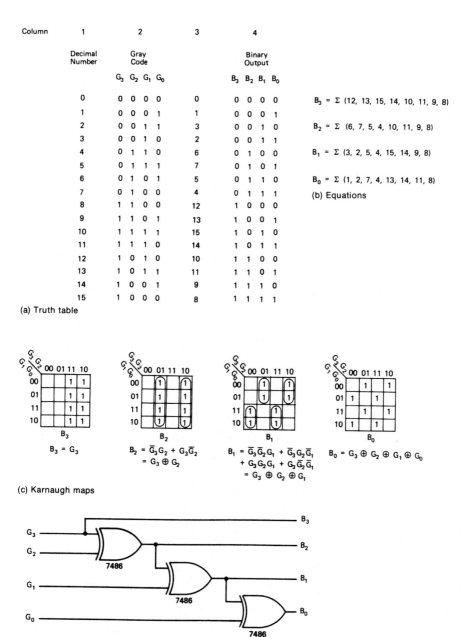

(c) Karnaugh maps

$B_3 = G_3$

$B_2 = \overline{G_3}G_2 + G_3\overline{G_2}$
$\phantom{B_2} = G_3 \oplus G_2$

$B_1 = \overline{G_3}\,\overline{G_2}G_1 + \overline{G_3}G_2\overline{G_1}$
$\phantom{B_1} + G_3G_2G_1 + G_3\overline{G_2}\,\overline{G_1}$
$\phantom{B_1} = G_3 \oplus G_2 \oplus G_1$

$B_0 = G_3 \oplus G_2 \oplus G_1 \oplus G_0$

(d) Circuit

FIGURE 13-21  A Gray code to binary convertor.

## 13-8 LIQUID CRYSTAL DISPLAYS

Liquid Crystal Displays (LCDs) were introduced in Sec. 12-8. They require an a.c. drive voltage because a d.c. component of more than 50 mV will tend to shorten their lives. Fortunately, the a.c. voltage is between the backplane and the LCD segment.

Figure 13-22 shows a 7-segment LCD being driven by a **4511** latch and driver. The a.c. signal required for LCD can be generated using **4070B** CMOS XOR gates. TTL XOR gates should not be used in this circuit because they may produce a d.c. voltage greater than 50 mV.

The backplane drive signal is connected to a 50 percent duty cycle clock (40 Hz is a suggested frequency), as shown in Fig. 13-23. Assume the segment a output of the **4511** is a 1. Then, when the backplane voltage is HIGH, the voltage to segment a is LOW, and vice versa. Thus an a.c. voltage of 3 to 5 V is created between segment a and the backplane of the LCD, turning the segment ON. This is shown in Fig. 13-23*b*. If the segment a output of the **4511** is LOW, the voltages applied to segment a and the backplane will be in phase as shown in Fig. 13-23*a*. Consequently, there will be no effective voltage between them and the segment will remain OFF. Seven-segment LCD displays may also be multiplexed. The reader should consult more advanced literature (see Sec. 13-11) for details.

### SUMMARY

In this chapter, the uses of XOR gates and ICs that depend on them were introduced. The **7485** was presented as a TTL comparator and the **74180** was presented as the

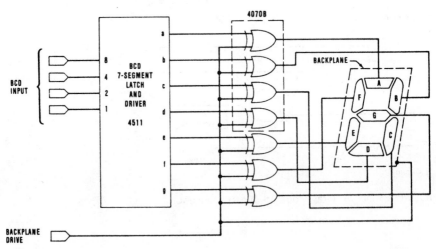

**FIGURE 13-22** Driving a 7-segment LCD. (Courtesy of Beckman Instruments, Inc.)

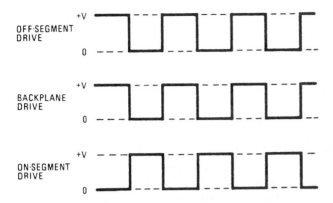

FIGURE 13-23   Phase shifting an LCD. (Courtesy of Beckman Instruments, Inc.)

TTL parity checker—generator. Circuits illustrating the use of these ICs were also presented.

The Gray code was also introduced and its use in shaft position sensing explained. A binary-to-Gray code converter was designed.

## GLOSSARY

**EQUALITY gate.** A 2-input gate that produces a high output when the two inputs are equal.

**EXCLUSIVE NOR.** Synonym for EQUALITY gate.

**Comparator.** A circuit that compares two numbers and produces an output indicating whether they are equal. It may also indicate which number is greater if they are unequal.

**Odd parity.** A word has odd parity if the number of 1s in the word is odd.

**Even parity.** A word has even parity if the number of 1s in the word is even.

**Parity checker.** A circuit that checks the number of 1s in a word to determine if the parity is proper.

**Parity generator.** A circuit that causes the number of 1s in a word to be even or odd (depending on which parity is chosen) by adding a bit to a word.

**Gray code.** A code where only one bit changes between successive numbers.

**Code wheel.** A wheel consisting of transparent and opaque segments, used to give an encoded representation of a shaft position.

## REFERENCES

**Beckman Instruments, Inc.**, Scottsdale, Arizona, *Interfacing Liquid Crystal Displays in Digital Systems*, July 1980.

Steven A. Ciarcia, *BYTE Magazine*, "Make Liquid-Crystal Displays Work for You," Oct. 1980.

Morris and Miller, *Designing With TTL Integrated Circuits*, McGraw-Hill, New York, 1971.

*The TTL Data Book for Design Engineers*, Texas Instruments, Inc., 1976.

## PROBLEMS

13-1.   Given two $n$-bit registers A and B, composed of FFs, show that for the $i$th stage, $Q_A \oplus \overline{Q}_B$ produces a high output when the $i$th bits are equal. Use this principle to design an $n$-bit comparator without using EQUALITY gates.

13-2.   Design a 5-bit comparator circuit using hypothetical EQUALITY gates having totem-pole outputs.

13-3.   Design a 5-bit comparator using only a single 7485 and some SSI gates.

13-4.   Suppose someone, using the circuit of Ex. 13-4, erroneously connected the outputs of the more significant 7485 to the inputs of the less significant 7485, and took the outputs from the less significant 7485. Find a number that would give an incorrect result when compared to $(195)_{10}$.

13-5.   Design a circuit to determine if a 10-bit binary number is greater than, equal to, or less than 400.

13-6.   Two 24-bit numbers differ only in that $A_6 = 1$ and $B_6 = 0$. Show how the comparator of Fig. 13-6 would compare them.

13-7.   Repeat Problem 13-6 if $A_9 = 0$ and $B_9 = 1$.

13-8.   Show how the parity checkers of Fig. 13-10 operate, by showing all the 1s and 0s, if the input word is:

   (a) 11011001
   (b) 11001111

13-9.   Show how the parity generator of Fig. 13-11 operates if the input word is:

   (a) 0001000
   (b) 1101100
   (c) 1111111

13-10.   Show how the 7-bit parity checker and 8-bit parity generator of Fig. 13-13 work for the inputs of Problem 13-9.

13-11.   How could the circuit of Fig. 13-14 be modified to make the ΣODD output high on odd parity. Use no additional gates.

13-12.   How can the circuit of Fig. 13-14 be used to generate a tenth bit for odd parity?

13-13.   Design a 10-bit odd parity checker, using:

   (a) Two 74180s.
   (b) One 74180 and additional gates (no more than 4).

For each case, show how your checker works if the inputs are:

1001100110

PROBLEMS **401**

| | | | | | | |LRC|
|---|---|---|---|---|---|---|---|
|0|0|1|0|0|1|0|0|
|1|0|1|0|0|0|1|0|
|0|0|1|1|1|1|1|1|
|1|0|0|1|1|0|0|1|
|1|1|1|1|1|1|1|1|

Vertical parity

FIGURE P13-18

13-14. Given 15 input bits, design a circuit to produce a 16-bit, odd parity word.

13-15. Design a parity checker to check a 15-bit word and produce a high output on odd parity. Show how your circuit works if the 15 input bits are:

| Bit | 1 | 2 | 3 | 4 | 5 | 6 | 7 | 8 | 9 | 10 | 11 | 12 | 13 | 14 | 15 |
|---|---|---|---|---|---|---|---|---|---|---|---|---|---|---|---|
| | 0 | 1 | 1 | 0 | 1 | 1 | 1 | 0 | 0 | 1 | 1 | 0 | 1 | 0 | 1 |

13-16. You are given 8-input bits and a clock. When the clock is HIGH, the data are valid and must be checked for odd parity. A LOW output should indicate an error (even parity). When the clock is LOW, the data are changing and must not be checked. Consequently, the error output must always be HIGH. Design the circuit. (*Hint:* It can be designed using only a single **74180** and no external gates.)

13-17. Design a 17-bit parity checker using only two **74180**s.

13-18. The data of Figure P13-18 were received and consist of a 7-word by 4-bit data block with odd parity and an LRC appended. One of the received bits is wrong. Find the vertical parity error, the LRC error, and the erroenous bit.

13-19. A disk is being read. The data block consists of 128 words of data (each word consists of 7 bits plus an eighth for odd parity) and an LRC word. The disk presents a START pulse and 129 data clocks. Data are valid during the positive portion of the data clock. Design a circuit to check the vertical and horizontal parity of the received data, to set a FF after the entire block has been received, and to set a second FF if there is an error in the block.

13-20. Construct a table similar to Fig. 13-18 for 5-bit numbers.

13-21. Design a binary-to-Gray converter for 4-bit numbers.

When you have finished the problems, return to Sec. 13-2 and reread the self-evaluation questions. If you have difficulty answering them, return to the appropriate sections of the chapter to find the answers.

CHAPTER

# 14

# ARITHMETIC CIRCUITS

## 14-1 INSTRUCTIONAL OBJECTIVES

This chapter discusses the most common arithmetic circuits using digital ICs. Circuits to add, subtract, and multiply are presented. In addition, the arithmetic-logic unit, a single IC that can perform many arithmetic and logic operations, is introduced. After reading the chapter, the student should be able to:

1. Design binary adders and subtracters.
2. Subtract by complementing the subtrahend and adding.
3. Convert negative binary numbers to their 2's complement or hexadecimal form.
4. Add and subtract numbers in 2's complement form.
5. Build an adder/subtracter using 7483s.
6. Design overflow and underflow detectors for 2's complement adder/ subtracters.
7. Design BCD adders and subtracters.
8. Utilize arithmetic-logic units such as the 74181.
9. Utilize look-ahead carry.
10. Build a multiplier circuit.

# 14-2  SELF-EVALUATION QUESTIONS

Look for the answers to the following questions as you read the chapter. They should help you understand the material presented.

1.  If the inputs to an adder are $n$-bit numbers, how many bits must the *sum register* contain?
2.  What is the difference between a *half-adder* and a *full-adder?*
3.  How does a subtracter circuit react if the difference is negative (the subtrahend is greater than the minuend)?
4.  In 9's complement subtraction, how does an end-around-carry indicate whether the result is positive or negative? What action must be taken in each case to obtain the correct answer?
5.  How is the sign of a 2's complement number determined from inspection?
6.  What is the major advantage of 2's complement notation?
7.  How can the positive equivalent of 2's complement negative numbers or hex numbers be found?
8.  In a binary adder/subtracter, how is the operand 2's complemented for subtraction and presented unchanged for addition?
9.  In 2's complement arithmetic, what is the criteria for overflow or underflow?
10.  What is the significance of the carry in a BCD adder?
11.  How does an arithmetic-logic unit select the function to be performed?
12.  What is the advantage of look-ahead carry? When should it be used?
13.  If an $m$-bit number is multiplied by an $n$-bit number, how long is the product register?

# 14-3  THE BASIC ADDER

The simplest arithmetic circuit is the *basic adder*. For binary numbers it accepts two $n$-bit binary numbers as inputs, and produces an $(n + 1)$-bit binary number as the sum. Essentially it consists of a half-adder, $n - 1$ full-adder circuits, and the $n + 1$ stage (the MSB) that only receives the carry from the $n$th stage.

## 14-3.1  The Half-Adder

The basic adder stage accepts inputs from the *addend*, *augend*, and a *carry-in* from the less significant stage. It generates a *sum* output *and* a *carry* output, which it sends to the next more significant (succeeding) stage. A *half-adder* is simpler. It only accepts addend and augend inputs. Because

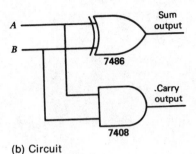

| Inputs | | Outputs | |
|:---:|:---:|:---:|:---:|
| A | B | Sum | Carry |
| 0 | 0 | 0 | 0 |
| 0 | 1 | 1 | 0 |
| 1 | 0 | 1 | 0 |
| 1 | 1 | 0 | 1 |

**(a) Truth table**      **(b) Circuit**

FIGURE 14-1   The half-adder.

it need *not* accept a carry-in, it can be used for the least significant stage of an adder, where there is never a carry input.

The truth table and circuit for the basic half-adder are shown in Fig. 14-1. The sum is 1 if *either* the addend or augend (A or B) are 1. If both A and B are 1, it is adding 1 plus 1 and the result is 10(2). The 0 is the LSB of the sum output and the 1 is the carry output. From the truth table it is apparent that the sum is merely an XOR gate while the carry output is generated by an AND gate.

## 14-3.2   The Full-Adder

The *full-adder* accepts the *i*th bit of the augend and addend ($A_i$ and $B_i$) and produces the *i*th bit of the sum. It must also accept a carry-in from the $i - 1$ stage, and generate a carry-out to the $i + 1$ stage.

The truth table, Karnaugh maps, and circuit of the full adder are shown in Fig. 14-2. The truth table (Fig. 14-2*a*) is constructed by considering the addition of three binary bits, $A_i$, $B_i$, and the carry-in.[1] The sum is 1 if only one or all three of the inputs are 1. This is the same as a 3-bit odd parity checker (Sec. 13-4.1). The carry output must be 1 if two or three of the inputs are 1.

The Karnaugh maps for the sum and carry are shown in Fig. 14-2*b*. The map for the sum is a checkerboard (as it would be for an odd-parity checker) and

$$S = A \oplus B \oplus C_{IN}$$

[1]The reader may wish to refer to the binary addition example presented in Sec. 1-7.1.

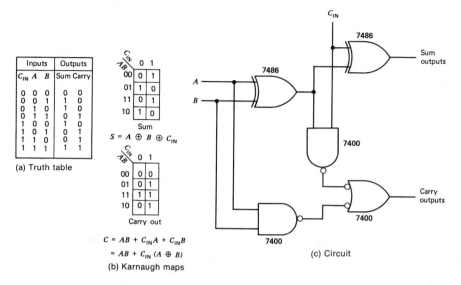

| Inputs | | | Outputs | |
|---|---|---|---|---|
| $C_{IN}$ | A | B | Sum | Carry |
| 0 | 0 | 0 | 0 | 0 |
| 0 | 0 | 1 | 1 | 0 |
| 0 | 1 | 0 | 1 | 0 |
| 0 | 1 | 1 | 0 | 1 |
| 1 | 0 | 0 | 1 | 0 |
| 1 | 0 | 1 | 0 | 1 |
| 1 | 1 | 0 | 0 | 1 |
| 1 | 1 | 1 | 1 | 1 |

(a) Truth table

| $C_{IN}$ / AB | 0 | 1 |
|---|---|---|
| 00 | 0 | 1 |
| 01 | 1 | 0 |
| 11 | 0 | 1 |
| 10 | 1 | 0 |

Sum

$S = A \oplus B \oplus C_{IN}$

| $C_{IN}$ / AB | 0 | 1 |
|---|---|---|
| 00 | 0 | 0 |
| 01 | 0 | 1 |
| 11 | 1 | 1 |
| 10 | 0 | 1 |

Carry out

$C = AB + C_{IN}A + C_{IN}B$

$= AB + C_{IN} (A \oplus B)$

(b) Karnaugh maps

(c) Circuit

FIGURE 14-2   The full-adder.

The map for the carry-out yields:

$$
\begin{aligned}
C_{OUT} &= AB + C_{IN}A + C_{IN}B \\
&= AB + C_{IN}A(B + \bar{B}) + C_{IN}B(A + \bar{A}) \\
&= AB + C_{IN}AB + C_{IN}A\bar{B} + C_{IN}AB + C_{IN}B\bar{A} \\
&= AB(1 + C_{IN} + C_{IN}) + C_{IN}(A\bar{B} + B\bar{A}) \\
&= AB + C_{IN}(A \oplus B)
\end{aligned}
$$

The latter form is generally preferred to the form

$$
C_{OUT} = AB + C_{IN}(A + B)
$$

because the term $A \oplus B$ can be used in *both* the sum and carry outputs. A full-adder circuit designed on the basis of these equations is shown in Fig. 14.2c.

---

**EXAMPLE 14-1**

Design an adder for 3-bit numbers. Show how your adder works if $A = 7$ and $B = 6$.

**SOLUTION**

The adder circuit is shown in Fig. 14-3. It can be understood best by referring to the block diagram, Fig. 14-3a, which shows that it is composed of a half-adder

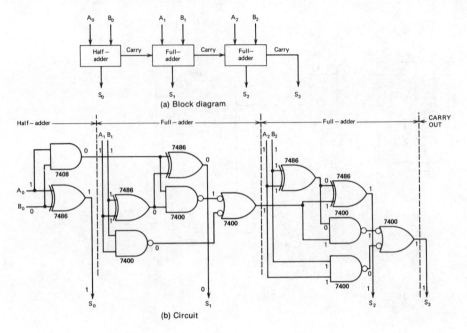

FIGURE 14-3 A 3-bit adder.

and two full-adders. *The MSB of the output is simply the carry of the second full-adder.*

The circuit is shown in Fig. 14-3*b*, where the half- and full-adders are taken from Figs. 14-1 and 14-2. The 1s and 0s for inputs of 7 and 6, respectively (A = 111, B = 110) are also shown on the figure. By continuing to follow the logic, we see that the output is **1101** or **13.**

Adders for larger numbers can be built in this manner simply by adding a a full-adder stage for each additional bit.

## 14-4 SUBTRACTION

*Binary subtraction* is the process of finding the *difference* (A minus B) between two binary numbers, A and B. It can be performed in the same manner as binary addition, by using a *half-subtracter* for the least-significant stage, followed by a series of full-subtracters. Each subtracter stage generates a *borrow-out* if it must borrow from a more-significant stage. Each stage except the least significant stage (the half-subtracter) may receive a borrow-in signal from the preceding stage.

The truth table for a full-subtracter can be constructed by considering the rules for binary subtraction (Sec. 1-7.2). The inputs to the full-subtracter are A, B, and borrow-in, and the full-subtracter generates a difference output and a borrow-out. The truth table is shown in Fig. 14-4. The entries are obtained by considering the conditions on each line of the truth table.

The subtraction circuit for $A - B$ can be designed on the basis of the truth table in a manner similar to the design of the adder circuit.

## 14-4.1 Subtraction by Complementation

*Subtraction by complementation* is a method of performing subtraction by adding, and works well for decimal numbers. In subtraction by complementation, the subtrahend must be replaced by its 9's complement, which is obtained by taking each decimal digit and replacing it by the difference between itself and 9. The 9's complement of 2, for example, is 7, and the 9's complement of 0 is 9.

---

**EXAMPLE 14-2**

Find the 9's complement of the decimal number 399704.

**SOLUTION**

The 9's complement is obtained by replacing each digit with its 9's complement as shown:

$$399704 \qquad \text{(original number)}$$
$$600295 \qquad \text{(9's complement)}$$

Note that each digit plus its 9's complement adds to 9.

---

| Line number | Inputs | | | Outputs | |
|:-----------:|:-----:|:--:|:--:|:--:|:---:|
| | $B_{IN}$ | $A$ | $B$ | $D$ | $B_{OUT}$ |
| 0 | 0 | 0 | 0 | 0 | 0 |
| 1 | 0 | 0 | 1 | 1 | 1 |
| 2 | 0 | 1 | 0 | 1 | 0 |
| 3 | 0 | 1 | 1 | 0 | 0 |
| 4 | 1 | 0 | 0 | 1 | 1 |
| 5 | 1 | 0 | 1 | 0 | 1 |
| 6 | 1 | 1 | 0 | 0 | 0 |
| 7 | 1 | 1 | 1 | 1 | 1 |

FIGURE 14-4  Truth table for a full-subtracter.

Decimal subtraction can be performed by using the following procedure:

1. Take the 9's complement of the subtrahend and add it to the minuend.
2. Remove the most significant 1, and add it to the least-significant digit. This is known as an *"end-around-carry."*

---

**EXAMPLE 14-3**

Subtract 19,307 from 28,652.

**SOLUTION**

The 9's complement of 19,307 is 80,692. Adding this to the minuend, we obtain

$$
\begin{array}{ll}
\text{(Minuend)} \quad 28{,}652 & \text{(Original number)} \\
\text{(Subtrahend)} \quad \underline{80{,}692} & \text{(9's complement)} \\
\phantom{\text{(Subtrahend)} \quad} 109{,}344 &
\end{array}
$$

Removing the most significant 1 and adding it to the least significant digit yields

$$
\begin{array}{l}
①09{,}344 \\
\underline{\phantom{09{,}3}+ 1} \quad \text{(end-around-carry)} \\
\phantom{0}9{,}345
\end{array}
$$

This is the correct result.

---

## 14-4.2  Negative Results

If the subtrahend is greater than the minuend, there is no end-around-carry. The **absence** *of an end-around-carry indicates a* **negative** *result.* In this case, the 9's complement of the sum is taken to get the correct magnitude of the answer.

---

**EXAMPLE 14-4**

Subtract 9934 from 807.

**SOLUTION**

| | |
|---|---:|
| Minuend | 807 |
| 9's complement of subtrahend | + 0065 |
| Result (note no end-around-carry) | 0872 |
| Correct answer (9's complement of the result) | − 9127 |

## 14-5   2's COMPLEMENT ARITHMETIC

When building hardware to accommodate binary numbers, two problems arise:

1.  The number of bits in a hardware register is finite.
2.  Negative integers must also be represented.

These problems do not arise in conventional pencil-and-paper arithmetic. If additional bits are needed, the number can always be extended to the left and negative numbers can always be represented by a minus sign.

Since a hardware register consists of a finite number of bits, the range of numbers that can be represented is finite. An $n$-bit register can contain one of $2^n$ numbers. If positive binary numbers are used, the $2^n$ numbers that can be represented are 0 through $2^n - 1$ (a string of $n$ 1s represents the number $2^n - 1$).

The range of numbers that can be represented by a single computer word is also restricted by the *word length* of the computer word. This is one reason why *larger* computers have *longer* word lengths. The IBM 370 uses 32-bit words while mini-computers, such as the PDP-11 or the Nova, use 16-bit words. A single PDP-11 word can represent one of $2^{16}$ or 65,536 numbers.

The simplest approach to the problem of designating negative integers is to *use the MSB to denote the sign of the number*. Normally an MSB of 0 indicates a positive number, and an MSB of 1 indicates a negative number. The remaining bits denote the magnitude of the number. This is called *sign-magnitude* representation.

---

**EXAMPLE 14-5**

What range of positive and negative numbers can be represented in sign-magnitude notation by a 16-bit computer?

**SOLUTION**

Since the 16th bit is reserved for the sign, the largest number that can be represented is a string of fifteen 1s, which is $2^{15} - 1$, or 32,767.

Therefore, 32,767 positive and 32,767 negative numbers can be represented. Zero, of course, can also be represented for a total range of 65,535 *different* numbers. There are two representations for 0; a positive 0 (all 0s) and a negative 0 (an MSB or sign bit of 1 followed by all 0s).

---

Some computers use sign-magnitude representation. The vast majority, however, use the 2's *complement* method of representing numbers. It does not have the double representation of 0, and has other advantages that will soon become clear. For the remainder of this section, 2's complement representation of binary numbers is discussed.

## 14-5.1 2's Complementing Numbers

As in sign-magnitude representation, the *MSB of a 2's complement number denotes the sign* (0 means the number is positive; 1 means the number is negative), but the *MSB is also part of the number. In 2's complement notation, positive numbers are represented as simple binary numbers with the restriction that the MSB is 0.* Negative numbers are somewhat different. To obtain the representation of a negative number, use the following algorithm:

1. Represent the number as positive binary number.
2. Complement it. (Write 0s where there are 1s and 1s where there are 0s in the positive number.)
3. Add 1.
4. Ignore any carries out of the MSB.

---

### EXAMPLE 14-6

Given 8-bit words, find the 2's complement representation of:
  (a) 25
  (b) −25
  (c) −1

### SOLUTION

(a) The number +25 can be written at 11001. Since 8 bits are available, there is room for three leading 0s, making the MSB 0.

$$+25 = 00011001$$

(b) To find −25, complement +25 and add 1:

$$
\begin{array}{r}
+25 = 00011001 \\
\overline{(+25)} = 11100110 \\
+1 \\
\hline
-25 = 11100111
\end{array}
$$

Note that the MSB is 1.

(c) To write −1, take the 2's complement of +1.

$$
\begin{array}{r}
+1 = 00000001 \\
\overline{(+1)} = 11111110 \\
+1 \\
\hline
-1 = 11111111
\end{array}
$$

From this example, we see that a solid string of 1s represents the number $-1$ in 2's complement form.

To determine the magnitude of any *unknown negative number*, simply take its 2's complement as described above. The result is a *positive number whose magnitude equals that of the original number.*

---

**EXAMPLE 14-7**

What decimal number does 11110100 represent?

**SOLUTION**

Complementing the given number, we obtain

$$
\begin{array}{r}
00001011 \\
+1 \\
\hline
00001100
\end{array}
$$

Adding 1

This is the equivalent of $+12$. Therefore, $11110100 = -12$.

---

## 14-5.2   The Range of 2's Complement Numbers

The maximum positive number that can be represented in 2's complement form is a single 0 followed by all 1s, or $2^{n-1} - 1$ for an $n$-bit number. The most negative number that can be represented has an MSB of 1 followed by all 0s, which equals $-2^{n-1}$. Therefore, an $n$-bit number can represent any one of $2^{n-1} - 1$ positive numbers, plus $2^{n-1}$ negative numbers, plus 0, which is $2^n$ total numbers. Every number has a unique representation.

Other features of 2's complement arithmetic are:

1.   Even numbers (positive or negative) have an LSB of 0.
2.   Numbers divisible by 4 have the 2 LSBs equal to 0 (see Ex. 14-8).
3.   In general, numbers divisible by $2^n$ have $n$ LSBs of 0.

---

**EXAMPLE 14-8**

What range of numbers can be represented by an 8-bit word using 2's complement representation?

**SOLUTION**

The most positive number is $01111111 = 127$.
The most negative number in 8 bits is $10000000 = -128$.

Therefore, any number between $+127$ and $-128$ can be represented by an 8-bit number in 2's complement form. There are 256 numbers in this range, as expected, since $2^8 = 256$. Note also that the 7 LSBs of $-128$ are 0, as required, since $-128$ is divisible by $2^7$.

## 14-5.3 Adding 2's Complement Numbers

Consider the simple equation $C = A + B$. Although it seems clear enough, we cannot immediately determine whether an addition or subtraction operation is required. If A and B are both positive, addition is required. But if one of the operands is negative and the other is positive, a subtraction operation must be performed.

The major advantage of 2's complement arithmetic is:

*If an addition operation is to be performed, the numbers are added regardless of their signs. The answer is in 2's complement form with the correct sign.* Any carries out of the MSB are meaningless and should be ignored.

### EXAMPLE 14-9

Express the numbers 19 and $-11$ as 8-bit, 2's complement numbers, and add them.

### SOLUTION

The number $+19$ is simply 00010011. To find $-11$, take the 2's complement of 11.

$$11 = 00001011$$
$$(\overline{11}) = 11110100$$
$$-11 = 11110101$$

Now $+19 + (-11)$ equals:

$$\begin{array}{r} 00010011 \\ +\,11110101 \\ \hline 00001000 \end{array}$$

Note that there is a carry out of the MSB, which is ignored. The 8-bit answer is simply the number $+8$.

### EXAMPLE 14-10

Add $-11$ and $-19$.

## SOLUTION

First $-19$ must be expressed as a 2's complement number:

$$19 = 00010011$$
$$(\overline{19}) = 11101100$$
$$-19 = 11101101$$

Now the numbers can be added:

$$
\begin{array}{rr}
(-19) & 11101101 \\
+ (-11) & 11110101 \\
\hline
\text{Answer} \quad (-30) & 11100010
\end{array}
$$

Again, a carry out of the MSB has been ignored.

## 14-5.4 Subtraction of Binary Numbers

*Subtraction* of binary numbers in 2's complement form is also very simple and straightforward. *The 2's complement of the subtrahend is taken and added to the minuend.* This is essentially subtraction by changing the sign and adding. As in addition, the signs of the operands and carries out of the MSB are ignored.

## EXAMPLE 14-11

Subtract 30 from 53. Use 8-bit numbers.

## SOLUTION

Note 30 is the subtrahend and 53 the minuend.

$$53 = 00110101 \quad \text{(Minuend)}$$
$$30 = 00011110 \quad \text{(Subtrahend)}$$

Taking the 2's complement of 30 and adding, we obtain

$$
\begin{array}{rl}
(\overline{30}) = & 11100001 \\
-30 = & 11100010 \\
+53 = & 00110101 \\
\hline
& 00010111 = 23
\end{array}
$$

## EXAMPLE 14-12

Subtract $-30$ from $-19$.

## SOLUTION

Here $-19 = 11101101$ (see Ex. 14-10).

$\quad -30 = 11100010$ (Subtrahend)

*Note:* $-30$ is the subtrahend. 2's complementing $-30$ gives $+30$ or $00011110$.

$$
\begin{array}{rl}
-19 & 11101101 \\
+30 & \underline{00011110} \\
& \overline{00001011} = +11
\end{array}
$$

The carry out of the MSB is ignored and the answer, $+11$, is correct.

---

## 14-6  HEXADECIMAL ARITHMETIC

A problem associated with binary arithmetic should now be apparent: there are too many 1s and 0s, and a shorthand notation for expressing them is needed. In most literature and documentation today the convention of using *hexadecimal notation* has been adopted. The manufacturers of all the modern microprocessors use hexadecimal arithmetic in their literature to condense the 1s and 0s required to express addresses or data.

The hexadecimal system is a *base 16* arithmetic system. Since such a system requires 16 different digits, the letters A through F are added to the 10 decimal digits (0–9). The advantage of having 16 hexadecimal digits is that each digit can represent a unique combination of 4 bits, and that any combination of 4 bits can be represented by a single hex[2] digit. Table 14-1 gives both the decimal and binary value associated with each hexadecimal digit.

In hex representation of numbers the terms *nibble* and *byte* are often used. A nibble is 4 bits of a single hex digit. A byte is two nibbles or 8 bits. A byte is the word size for many microprocessors, and memories (see Chapter 15) are often described by the number of bytes they contain.

### 14-6.1  Conversions between Hexadecimal and Binary Numbers

To convert a binary number to hexadecimal, start at the least significant bit (LSB) and divide the binary number into groups of four bits each. Then replace each 4-bit group with its equivalent hex digit obtained from Table 14-1.

---

[2]The word hex is often used as an abbreviation for hexadecimal.

TABLE 14-1 TABLE OF HEXADECIMAL DIGITS

| Hexadecimal Digit | Decimal Value | Binary Value |
|---|---|---|
| 0 | 0 | 0000 |
| 1 | 1 | 0001 |
| 2 | 2 | 0010 |
| 3 | 3 | 0011 |
| 4 | 4 | 0100 |
| 5 | 5 | 0101 |
| 6 | 6 | 0110 |
| 7 | 7 | 0111 |
| 8 | 8 | 1000 |
| 9 | 9 | 1001 |
| A | 10 | 1010 |
| B | 11 | 1011 |
| C | 12 | 1100 |
| D | 13 | 1101 |
| E | 14 | 1110 |
| F | 15 | 1111 |

## EXAMPLE 14-13

Convert the binary number 11000010111111101 to hex.

## SOLUTION

We start with the LSB and divide the number into 4-bit nibbles. Each nibble is then replaced with its corresponding hex digit as shown:

$$0011 \quad 0000 \quad 0101 \quad 1111 \quad 1101$$
$$3 \quad\quad 0 \quad\quad 5 \quad\quad F \quad\quad D$$

When the most significant group has less than 4 bits, as in this example, leading 0s are added to complete the 4-bit nibble.

To convert a hex number to binary, simply replace each hex digit by its 4-bit binary equivalent.

## EXAMPLE 14-14

Convert the hex number 1CB09 to binary.

## SOLUTION

We simply expand the hex number:

$$\begin{array}{ccccc} 1 & C & B & 0 & 9 \\ \underline{0001} & \underline{1100} & \underline{1011} & \underline{0000} & \underline{1001} \end{array}$$

Thus the equivalent binary number is:

$$11100101100001001$$

It is not necessary to write the leading 0s.

## 14-6.2 Conversion of Hex Numbers to Decimal Numbers

The hex system is a base 16 system; therefore, any hex number can be expressed as:

$$H_0 \times 1 + H_1 \times 16 \times H_2 \times 16^2 + H_3 \times 16^3 \cdots$$

where $H_0$ is the least significant hex digit, $H_1$ the next, and so on. This is similar to the binary system of numbers discussed in Sec. 1-5.

### EXAMPLE 14-15

Convert 2FC to decimal.

### SOLUTION

The least significant hex digit, $H_0$, is C or 12. The next digit ($H_1$) is F or 15. This must be multiplied by 16 giving 240. The next digit, $H_2$, is 2, which must be multiplied by $16^2$, or 256. Hence, 2FC = 512 + 240 + 12 = 764.

An alternate solution is to convert 2FC to the binary number 1011111100 and then perform a binary to decimal conversion.

Decimal numbers can be converted to hex by repeatedly dividing them by 16. After each division, the remainder becomes one of the hex digits in the final answer.

### EXAMPLE 14-16

Convert 9999 to hex.

## SOLUTION

Start by dividing by 16 as shown in the table below. After each division, the quotient becomes the number starting the next line and the remainder is the hex digit with the least significant digit on the top line.

| Number | Quotient | Remainder | Hex Digit |
|--------|----------|-----------|-----------|
| 9999 | 624 | 15 | F |
| 624 | 39 | 0 | 0 |
| 39 | 2 | 7 | 7 |
| 2 | 0 | 2 | 2 |

This example shows that $(9999)_{10} = (270F)_{16}$. The result can be checked by converting 270F to decimal, as shown in Ex. 14-15. By doing so we obtain:

$$(2 \times 4096) + (7 \times 256) + 0 + 15 = 9999$$
$$8192 \quad + \quad 1792 \quad + 0 + 15 = 9999$$

## 14-6.3  Hexadecimal Addition

When working with $\mu$Ps, it is often necessary to add or subtract hex numbers. They can be added by referring to hexadecimal addition tables, but we suggest the following procedure.

1. Add the two hex digits (mentally substituting their decimal equivalent).
2. If the sum is 15 or less, it can be directly expressed in hex.
3. If the sum is greater than or equal to 16, subtract 16 and carry 1 to the next position.

The following examples should make this procedure clear.

---

**EXAMPLE 14-17**

Add D + E.

## SOLUTION

D is the equivalent of decimal 13 and E is the equivalent of decimal 14. Together they sum to 27 = 16 + 11. The 11 is represented by B and there is a carry. Therefore, D + E = **1B**.

**EXAMPLE 14-18**

Add B2E6 and F77.

## SOLUTION

The solution is shown below.

| Column | 4 3 2 1 |
|---|---|
| Augend | B 2 E 6 |
| Addend | F 7 7 |
| Sum | C 2 5 D |

- ☐ **Column 1** $6 + 7 = 13 = D$. The result is less than 16 so there is no carry.
- ☐ **Column 2** $E + 7 = 14 + 7 = 21 = 5 +$ a carry because the result is greater than 16.
- ☐ **Column 3** $F + 2 + 1$ (the carry from Column 2) $= 15 + 2 + 1 = 18 = 2 +$ a carry.
- ☐ **Column 4** $B + 1$ (the carry from Column 3) $= C$.

Like addition, *hex subtraction* is analogous to decimal subtraction. If the subtrahend digit is larger than the minuend digit, one is borrowed from the next most significant digit. If the next most significant digit is 0, a 1 is borrowed from the next digit and the intermediate digit is changed to an F.

## EXAMPLE 14-19

Subtract 32F from C02.

## SOLUTION

The subtraction proceeds as follows:

| Column | 3 2 1 |
|---|---|
| Minuend | C 0 2 |
| Subtrahend | 3 2 F |
| Difference | 8 D 3 |

- ☐ **Column 1** Subtracting F from 2 requires a borrow. Because a borrow is worth 16, it raises the minuend to 18. Column 1 is therefore $18 - F = 18 - 15 = 3$.
- ☐ **Column 2** Because Column 2 contains a 0, it cannot provide the borrow out for Column 1. Consequently, the borrow out must come from Column 3, while the minuend of Column 2 is changed to an F. Column 2 is therefore $F - 2 = 15 - 2 = 13 = D$.
- ☐ **Column 3** Column 3 can provide the borrow out needed for Column 1. This reduces the C to a B and $B - 3 = 8$.

As in decimal addition, the results can be checked by adding the subtrahend and difference to get the minuend.

## 14-6.4 Negating Hex Numbers

The negative equivalent of a positive hex number can always be found by converting the hex number to binary and taking the 2s complement of the result (Sec. 14-5). A shorter method exists, however.

1. Add to the least significant hex digit the hex digit that makes it sum to 16.
2. Add to all other digits the digits that make it sum to 15.
3. If the least significant digit is 0, write 0 as the least significant digit of the answer and start at the next digit.
4. The number written is the negative equivalent of the given hex number.

This procedure works because the sum of the original number and the new number is always 0.

---

**EXAMPLE 14-20**

Find the negative equivalent of the hex number 20C3.

**SOLUTION**

The least significant digit is 3. To make 16, D must be added to 3. The other digits are 2, 0, and C. To make 15 in each case, we add D, F, and 3, respectively. The negative equivalent of 20C3 is therefore **DF3D**. This example can be checked by adding the negative equivalent to the positive number. Since X plus $-X$ always equals 0, the result should be 0.

$$
\begin{array}{r}
2\,0\,C\,3 \\
+\ \ D E\,3D \\
\hline
0\,0\,0\,0
\end{array}
$$

The carry out of the most significant digit is ignored.

---

## 14-6.5 Octal Notation

Octal notation is an alternate form of concise notation used by many mini-computer manufacturers. It divides binary numbers into 3-bit groups and each group represents the particular digit (0 through 7) that corresponds to the value of the 3-bit group.

Octal notation is a base 8 system. The rules for conversions and addition and subtraction in octal are analogous to those for hexadecimal.

Octal is not as convenient as hex when expressing two bytes as a bit word. For example, compare these three ways to describe the same two bytes of information:

| Octal | 2 | 6 | 5 | 3 | 0 | 6 |
|---|---|---|---|---|---|---|
| Binary | 1 0 1 | 1 0 1 0 1 | | 1 1 0 0 0 1 1 0 | | |
| Hexadecimal | B | 5 | | C | 6 | |

When expressed as a 16-bit address, these same bits would be:

$$132706 \quad \text{in octal}$$
$$\text{or} \quad B5C6 \quad \text{in hex}$$

It is obvious that the hex version is much easier to use when combining groups of 8-bit words because the hex notation is simply the linking of the two 8-bit notations, while octal requires conversion to different numbers to express the 16-bit value. Because hex is far more popular among $\mu$P users, it is used exclusively throughout this book.

## 14-7 THE 7483 4-BIT ADDER

The 7483[3] is essentially a hexadecimal adder. It accepts two 4-bit numbers (A and B) and a carry-in ($C_0$) as inputs. The 4 bits of the A input are connected to the A1, A2, A3, and A4 inputs on the 7483. The 4 bits of the B input are similarly connected. The 7483 produces a 4-bit sum output ($\Sigma$) and a carry output ($C_4$). A functional drawing of the 7483 is shown in Fig. 14-5.

The IC operates as follows:

1. If the sum of the two inputs plus the carry-in is between 0 and 15, the sum appears in the $\Sigma$ outputs and the carry-out (labeled $C_4$) is 0.
2. If the sum is between 16 and 31, carry out $C_4$ is 1 and the $\Sigma$ outputs are 16 less than the sum. Note that in a 4-bit adder, the carry-out has a value or weight of 16.

[3]For new designs, the 74283, if available, is preferable to the 7483. The 74283 is identical to the 7483, except that power and ground are on the pins 16 and 8, respectively.

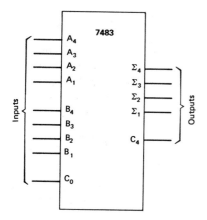

FIGURE 14-5 The **7483** 4-bit adder. Note that $V_{CC}$ is on pin 5 and GND is on pin 12.

**EXAMPLE 14-21**

A **7483** has the following inputs:

$$A_4A_3A_2A_1 = 0111 \quad (7)$$
$$B_4B_3B_2B_1 = 1010 \quad (A)$$
$$C_0 = 1$$

Find the outputs.

**SOLUTION**

The inputs are seen to be 7 and 10, with a carry input. The sum of the inputs is therefore 18. According to the above rules, a carry-out is produced and the $\Sigma$ outputs are 2 (18 minus 16). Therefore, $C_4 = 1$ and $\Sigma_4\Sigma_3\Sigma_2\Sigma_1 = 0010$ (2).

## 14-7.1 Cascading 7483s

To build an adder for words longer than 4 bits, 7483s can be cascaded simply by tying $C_4$ of one **7483** to the $C_0$ input of the next most significant **7483**, as shown in Fig. 14-6. Note that the least-significant stage is placed on the right and the addition proceeds from *right* to *left*. Although this is opposite to the usual directions of data flow, it allows us to read the input numbers and the output in the normal manner, from *left* to *right*.

**EXAMPLE 14-22**

Add 202 and 231 using two 7483s.

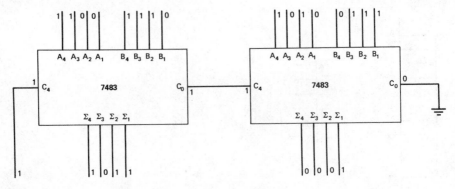

FIGURE 14-6 An 8-bit adder using two **7483**s.

**SOLUTION**

The 1s and 0s for this example are also shown in Fig. 14-6. First 202 and 231 are converted to binary or hex:

$$202 = 11001010 = CA$$
$$231 = 11100111 = E7$$

These become the A and B inputs to the **7483**s. The four LSBs of each number are connected to the least significant **7483**, which sees inputs of A and 7 and no carry. Its sum output is 1 and it produces a carry-out. The more significant **7483** sees inputs of C and E and a carry input. Therefore, it produces an output of B, or 1011 plus a carry output. If the carry output plus the 8-bit output of the two **7483**s are read as a single binary number, we have

$$\textbf{Sum} = \textbf{1B1} = \textbf{110110001} = \textbf{(433)}_{10}$$

## 14-7.2   The 2's Complement Adder/Subtracter

An *adder/subtracter* for 2's complement numbers can be built from **7483**s, as shown in Fig. 14-7. The mode of the adder/subtracter is controlled by a toggle switch or an add-subtract line. When the line is LOW, the circuit is an adder. The carry-in is 0 and the XORs act as straight-through gates. The circuit simply sums the A and B inputs.[4]

In subtract mode, $C_0 = 1$ and the output of the XORs is the inversion of the B inputs. Hence, B is complemented and 1 is added to it, effectively 2's complementing the operand. Consequently, if B is the subtrahend, subtraction is accomplished by 2's complementing and adding B to A.

---

[4]The reader should not confuse the inputs to the **7483** that are designated A and B with the hex numbers A and B.

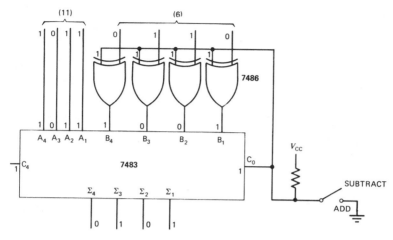

FIGURE 14-7 A 4-bit ADDER/SUBTRACTER.

## EXAMPLE 14-23

Subtract 6 from 11 (B) using the circuit of Fig. 14-7.

### SOLUTION

The 1s and 0s for this circuit are shown on Fig. 14-7. The switch is shown in the subtract position and the A and B inputs are 11 and 6. Because of the inversion caused by the XORs, the **7483** receives inputs of 11, 9, and a carry. Since these sum to 21, the **7483** produces a carry-out and a sum output of 5. The result of **5** is correct and the carry-out is ignored.

In this and many other subtractor circuits, a carry-in or carry-out of 1 indicates *no borrow*. A carry-in or carry-out of 0 indicates that the subtrahend is greater than the minuend, giving a negative result and requiring a borrow.

## EXAMPLE 14-24

Add and subtract 89 from −35 using a 2's complement adder/subtracter.

### SOLUTION

Since the operands take more than 4 bits, a 2-stage adder/subtracter is required. The solution is shown in Fig. 14-8, where the A input is −35 (11011101) and the B input is +89 (01011001). The numbers in boldtype indicate the addition operation. Here the least significant IC, which is on the right, sees inputs of 13 and 9. They sum to 22, or 6 plus a carry-out. The more significant chip sees inputs of

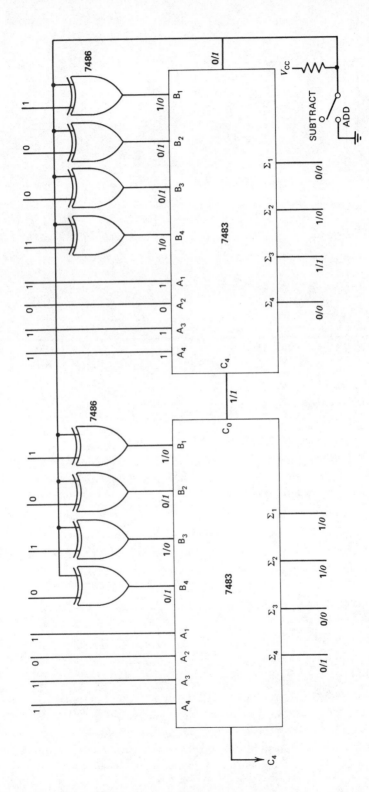

FIGURE 14-8  An 8-bit ADDER/SUBTRACTER.

13 and 5, plus a carry-in, which sums to 19, or 3 plus a carry-out. The result of the addition is **00110110** or $(54)_{10}$, which is correct.

The numbers in italics apply to the subtraction of 89 from $-35$. The least significant chip sees 13 plus 6 plus a carry-in and produces a sum of 4 and a carry-out. The more significant chip sees inputs of 13, 10, and a carry-in. It, therefore, produces an output of 8 (1000) and a carry-out. The results of the subtraction operation are **10000100**, or $-124$, which is correct.

This problem can be shortened considerably by using hex notation. Inspecting the binary numbers above, we see that $(-35)_{10} = DD$ and $(+89)_{10} = 59$. Their sum is $DD + 59 = (36)_{16}$ or $(54)_{10}$ and their difference is $DD - 59 = (84)_{16}$ or $(-124)_{10}$. The hex results, 36 and 84, are the outputs of the two 7483s in addition and subtraction.

Clearly this circuit can be extended to handle longer words. It is ideal for use in computers where the numbers are stored in memory in 2's complement form and only need to be brought out and passed through the adder/subtracter to achieve the desired results.

## 14-8 OVERFLOW AND UNDERFLOW IN 2s COMPLEMENT ARITHMETIC

*Overflow* and *underflow* are two problems that have been ignored up to now. *Overflow occurs when the result of an arithmetic operation produces a number larger than the register can accommodate. Underflow occurs when the result produces a number smaller than the register can accommodate.*

The limitations on the numbers that can be handled by an $n$-bit register are, as shown in Sec. 14-5, $2^{n-1} - 1$ positive numbers, and $2^{n-1}$ negative numbers. An 8-bit register is restricted to numbers between $+127$ and $-128$.

To illustrate overflow, consider the number 100 expressed as an 8-bit number, 01100100. If an attempt is made to add 100 plus 100, the result is 200 (11001000). Unfortunately, considered as a 2's complement number, it equals $-56$. This ridiculous result occurred because the answer, $+200$, was *beyond the range* of numbers that could be handled by an 8-bit register.

### 14-8.1 Overflow and Underflow in Addition and Subtraction

If overflow or underflow can be a problem, circuits must be built to detect this condition. Microprocessors detect overflow or underflow by setting a FF called the V *flag*. The V flag or bit is called the overflow bit but it actually detects both overflow and underflow. It is SET when any out-of-range result occurs in an addition or subtraction operation.

There are two criteria for overflow and underflow in microprocessors:

1. For *addition* instructions the basic Boolean equation for overflow is

$$V = \overline{A_7}\,\overline{B_7}R_7 + A_7 B_7 \overline{R_7} \tag{14-1}$$

where it is assumed that the operation is A plus $B \rightarrow R$ and $A_7$ is the MSB of A (the augend), $B_7$ is the MSB of B (the addend), and $R_7$ is the MSB of the result. The plus sign in the equation indicates the logical OR.

If the first term of the equation is 1, it indicates that two positive numbers have been added (because $A_7$ and $B_7$ are both 0) and the result is negative (because $R_7 = 1$). This possibility has been illustrated in the preceding paragraph.

The second term indicates that two negative numbers have been added and have produced a positive result.

---

### EXAMPLE 14-25

Show how the hex numbers 80 and C0 are added.

### SOLUTION

80 + C0 = 40 plus a carry (see Sec. 4-3.3). Note that 80 and C0 are both negative numbers, but their sum (as contained in a single byte) is positive. This corresponds to the second term of equation 1. Fortunately, this addition sets the V bit to warn the user that overflow (in this case underflow) has occurred.

---

2. For *subtraction* operations, the Boolean equation is

$$V = A_7 \overline{B_7}\,\overline{R_7} + \overline{A_7} B_7 R_7 \tag{14-2}$$

The assumption here is that $A - B \rightarrow R$. The first term indicates that a positive number has been subtracted from a negative number and produced a positive result. The second term indicates that a negative number has been subtracted from a positive number and produced a negative result. In either case, the overflow bit is set to warn the user.

Inspection of Eqs. (14-1) and (14-2) reveals that if two numbers of *unlike sign* are added, there can *never* be overflow in a 2's complement system. Similarly, there can never be overflow if two numbers of like sign are subtracted.

## 14-9 BCD ARITHMETIC

For certain applications, such as calculators, it is advantageous to perform all arithmetic operations in BCD format. This eliminates the need for conversions. Unfortunately, BCD arithmetic circuits are more complex and costly than binary circuits. This fact should be carefully considered before using BCD circuits.

### 14-9.1 The BCD 4-Bit Adder

A BCD 4-bit adder accepts a single BCD digit from the A operand, a single BCD digit from the B operand and a carry-in. *It produces a single BCD digit as the sum output. It must also produce a carry-out if the sum is greater than 9.*

Perhaps the simplest way to design a BCD adder is to base its operation on the 7483. If the sum of the three inputs (the A digit, the B digit, and the carry-in) is 9 or less, the 7483 produces the correct output. If the sum is greater than 9, a carry-out must be generated and the sum corrected.

A single-decade BCD adder is shown in Fig. 14-9. It consists of three parts:

1. The basic adder.
2. The carry detection circuit.
3. The correction circuit.

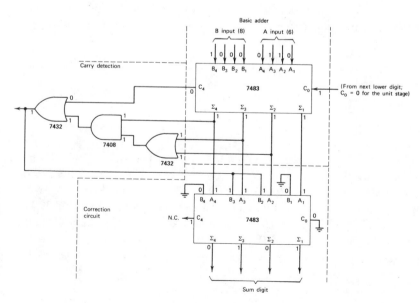

FIGURE 14-9 The BCD adder.

The basic adder is simply a 4-bit *binary* adder. Note, however, that the A and B inputs can never be greater than 9 because they are in BCD form, and the output of the 7483 can never be greater than 19.

The carry detection circuit produces a carry-out if the sum is greater than 9. Sums between 10 and 15 are detected by the logic gates and sums between 16 and 19 produce a carry-out of the 7483. The equation is

$$\text{Carry-out} = C_4 + \Sigma_4(\Sigma_3 + \Sigma_2)$$

which is implemented by the logic gates.

A second 7483 is used in the correction circuit, and the final sum is the output of this IC. If the sum of the inputs is 9 or less, there is no carry-out and the lower 7483 adds 0 to the output of the upper 7483, giving the correct sum. If the sum is greater than 9, the carry-out has a weight of 10, and 10 must be subtracted from the sum. This is equivalent to adding the 2's complement of 10 (i.e., 6) to the output of the first adder. Therefore, whenever a carry-out is produced, 6 is added to the first sum to correct and give the proper sum output in BCD form.

Most microprocessors are essentially hex computers, but they can do BCD addition by using the DAA (Decimal Adjust Accumulator) instruction. This instruction functions much like the circuit of Fig. 14-9 and adds 6 whenever an adjustment must be made to convert a hex sum into a BCD sum.

BCD adders can be cascaded to add numbers several digits long simply by tying the carry-out of a stage to the carry-in of the next stage. Of course, the least-significant stage never has a carry-in, so its carry input is grounded.

---

### EXAMPLE 14-26

The inputs to a BCD adder are 6, 8, and a carry-in. Show how it performs the addition.

### SOLUTION

The 1s and 0s for this example are shown on Fig. 14-9. The upper 7483 accepts the inputs and produces an output of 15. The carry circuit produces a carry-out and causes the lower 7483 to add 6 to the 15 it receives from the upper 7483. The $\Sigma$ output of the lower 7483 is 5 (0101). Thus the circuit has successfully added the inputs by producing an output of 15 (a sum output of 5, plus a carry output).

## 14-9.2　The BCD Subtracter

BCD subtraction is more difficult (and painful) than BCD addition, especially if the possibility of a negative result is allowed. Most BCD subtracters work on the complement-and-add principles developed in Sec. 14-4.1. The simplest BCD subtracter that assumes a 0 or positive result is shown in Fig. 14-10. Figure 14-10*a* shows a single stage of the subtracter and Fig. 14-10*b* shows how the stages can be cascaded for numbers of several digits.

The circuit functions by adding the 15's complement of the *B* input (produced by the 4 inverters) to the *A* input. A correction of 10, when necessary, is made by the lower **7483**. Note that a carry of 1 into a stage indicates that the previous stage is *not* attempting to borrow, and *a stage generates a carry-out of 0 when it must borrow from the next stage.* Therefore, the least significant stage has its carry-in tied to 1, which indicates that *there is no borrow into it.* It also implicitly assumes that the end-around carry is 1, giving a positive result.

---

### EXAMPLE 14-27

Show how the circuit of Fig. 14-10*b* operates when subtracting 3627 from 8353.

### SOLUTION

Since the given numbers contain 4 digits, 4 subtracter stages are required. It requires a great deal of work to draw the 4 stages individually and indicate the 1s and 0s on each gate. The work can be reduced using a table, shown in Fig. 14-11, where $A$ and $B$ are the minuend and subtrahend, and $\bar{B}$ is the 15's complement of $B$.

The table of Fig. 14-11 is developed as follows:

1. Start at the least-significant digit (stage 1), where the carry-in is known to be 1.
2. The inputs to the upper adder are 3, 8, and 1 (the carry-in). It produces a sum of 12 and no carry-out.
3. The inputs to the lower adder are 12 and 10, because the sum of the upper adder is connected to the A inputs of the lower adder, and the B inputs to the lower adder are either 10 (when there is no carry) or 0 (when carry-out = 1).
4. The lower adder produces a result of 12 + 10 = 22 = 6 + carry-out. The difference digit is 6.
5. The inputs to the second stage are now seen to be 5, 13, and 0 (no carry-in from the first stage). It produces a sum of 2 and a carry-out.
6. The rest of the table is constructed by continuing with the procedure developed here.

*Note:* The most significant stage must have a carry-out of 1 for a positive result.

The table gives the numbers at all the *intermediate* stages of the subtracter and simplifies debugging.

---

It is possible to build BCD adder/subtractors by modifying and expanding upon the designs presented here. These circuits become complex, however,

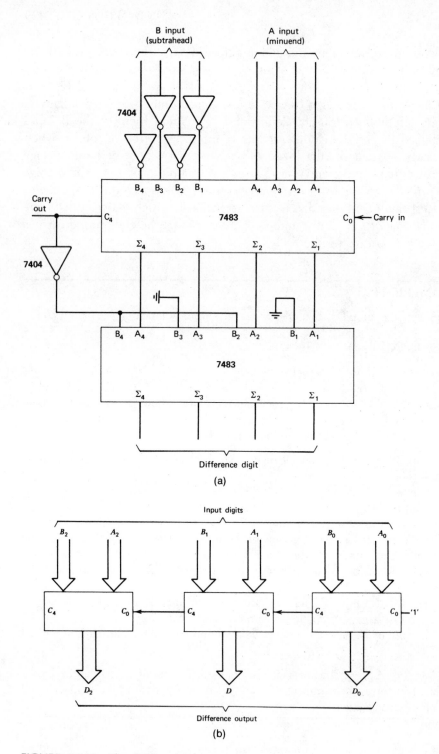

FIGURE 14-10 The BCD subtracter.

| | Stage | | | |
|---|---|---|---|---|
| | 4 | 3 | 2 | 1 |
| A | 8 | 3 | 5 | 3 |
| B | 3 | 6 | 2 | 7 |
| $\overline{B}$ | 12 | 9 | 13 | 8 |
| $C_{IN}$ | 0 | 1 | 0 | 1 |
| $\Sigma$ | 4 | 13 | 2 | 12 |
| $C_{OUT}$ | 1 | 0 | 1 | 0 |
| $B''$ | 0 | 10 | 0 | 10 |
| Result | 4 | 7 | 2 | 6 |

FIGURE 14-11   Subtracter table for Ex. 14-27.

and the reader is advised to investigate commercially available BCD arithmetic units in IC form before attempting to build these circuits using SSI and MSI chips.

## 14-10   ARITHMETIC/LOGIC UNITS

The discussions of the previous sections have shown a need for larger scale circuits that perform complex arithmetic operations. Such units are called *arithmetic/logic* units.

### 14-10.1   The 74181

The 74181 is the basic arithmetic/logic unit (ALU) in the 7400 TTL series. A simplified functional layout of the 74181 is shown in Fig. 14-12a and the function table is given in Fig. 14-12b.

The 74181 accepts two 4-bit words, A and B, as *data inputs* and a CARRY-IN, which acts as an *inverted carry* during addition operations because it is LOW when a carry-in occurs. There are also *five control inputs* that determine the *operation* performed on the inputs. The MODE input determines whether the output is a logical or arithmetic function of the inputs. The carry-in does not affect the logic functions. The four SELECT lines select any 1 of 16 possible logic operations or arithmetic operations.

Data
inputs

Outputs

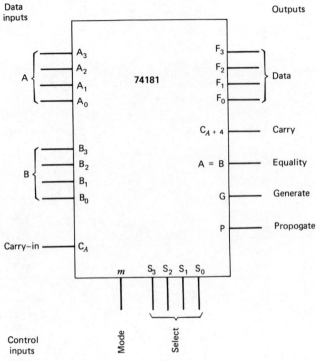

(a) Functional layout

**TABLE 1**

| SELECTION | M = H | M = L; ARITHMETIC OPERATIONS | |
|---|---|---|---|
| | LOGIC | $C_n$ = H | $C_n$ = L |
| S3 S2 S1 S0 | FUNCTIONS | (no carry) | (with carry) |
| L L L L | F = $\bar{A}$ | F = A | F = A PLUS 1 |
| L L L H | F = $\overline{A+B}$ | F = A + B | F = (A + B) PLUS 1 |
| L L H L | F = $\bar{A}B$ | F = A + $\bar{B}$ | F = (A + $\bar{B}$) PLUS 1 |
| L L H H | F = 0 | F = MINUS 1 (2's COMPL) | F = ZERO |
| L H L L | F = $\overline{AB}$ | F = A PLUS A$\bar{B}$ | F = A PLUS A$\bar{B}$ PLUS 1 |
| L H L H | F = $\bar{B}$ | F = (A + B) PLUS A$\bar{B}$ | F = (A + B) PLUS A$\bar{B}$ PLUS 1 |
| L H H L | F = A $\oplus$ B | F = A MINUS B MINUS 1 | F = A MINUS B |
| L H H H | F = A$\bar{B}$ | F = A$\bar{B}$ MINUS 1 | F = A$\bar{B}$ |
| H L L L | F = $\bar{A}$ + B | F = A PLUS AB | F = A PLUS AB PLUS 1 |
| H L L H | F = $\overline{A \oplus B}$ | F = A PLUS B | F = A PLUS B PLUS 1 |
| H L H L | F = B | F = (A + $\bar{B}$) PLUS AB | F = (A + $\bar{B}$) PLUS AB PLUS 1 |
| H L H H | F = AB | F = AB MINUS 1 | F = AB |
| H H L L | F = 1 | F = A PLUS A* | F = A PLUS A PLUS 1 |
| H H L H | F = A + $\bar{B}$ | F = (A + B) PLUS A | F = (A + B) PLUS A PLUS 1 |
| H H H L | F = A + B | F = (A + $\bar{B}$) PLUS A | F = (A + $\bar{B}$) PLUS A PLUS 1 |
| H H H H | F = A | F = A MINUS 1 | F = A |

*Each bit is shifted to the next more significant position.

(b) Function table

FIGURE 14-12  The **74181** arithmetic/logic unit. (From The *TTL Data Book for Design Engineers*, 2nd ed., Texas Instruments, Inc. Courtesy of Texas Instruments, copyright 1976.)

The outputs of the **74181** include the 4-bit result (the F outputs), a CARRY-OUT labeled $C_{n+4}$, an $A = B$ output, and the GENERATE and PROPAGATE outputs. The outputs are determined by carefully following the function table. Note that the + sign means logical OR and the word *plus* means the sum of the inputs.

The advantage of the **74181** is that it can perform the operations of addition, subtraction, shifting (one place), AND, OR, XOR, and many others on the input variables simply by changing the SELECT and MODE inputs.

For addition operations, a 0 on the carry line indicates a carry. Any plus operation performed by the **74181** produces a LOW carry-out if the sum is greater than 15.

In subtraction, a 0 carry-out indicates a positive or 0 result, and a HIGH on the carry line indicates a negative result or a borrow. *If the result of a minus operation is negative, it is presented as a 4-bit, 2's complement number.* For example, if the result is $-4$, the F outputs read 1100 and the carry-out is HIGH.

A table of the outputs of a **74181**, when $A = 9$ (1001) and $B = 10$ (1010) is shown in Fig. 14-13 to help explain its operation. Using the given

| Line | Selection $S_3\ S_2\ S_1\ S_0$ | M = 1 Logic Functions $F_3\ F_2\ F_1\ F_0$ | M = 0 Arithmetic Operations $C_{IN} = 1$ (no carry) $F_3\ F_2\ F_1\ F_0$ | $C_{n+4}$ | $C_{IN} = 0$ (with carry) $F_3\ F_2\ F_1\ F_0$ | $C_{n+4}$ |
|---|---|---|---|---|---|---|
| 0 | 0 0 0 0 | 0 1 1 0 | 1 0 0 1 | 1 | 1 0 1 0 | 1 |
| 1 | 0 0 0 1 | 0 1 0 0 | 1 0 1 1 | 1 | 1 1 0 0 | 1 |
| 2 | 0 0 1 0 | 0 0 1 0 | 1 1 0 1 | 1 | 1 1 1 0 | 1 |
| 3 | 0 0 1 1 | 0 0 0 0 | 1 1 1 1 | 1 | 0 0 0 0 | 0 |
| 4 | 0 1 0 0 | 0 1 1 1 | 1 0 1 0 | 1 | 1 0 1 1 | 1 |
| 5 | 0 1 0 1 | 0 1 0 1 | 1 1 0 0 | 1 | 1 1 0 1 | 1 |
| 6 | 0 1 1 0 | 0 0 1 1 | 1 1 1 0 | 1 | 1 1 1 1 | 1 |
| 7 | 0 1 1 1 | 0 0 0 1 | 0 0 0 0 | 0 | 0 0 0 1 | 0 |
| 8 | 1 0 0 0 | 1 1 1 0 | 0 0 0 1 | 0 | 0 0 1 0 | 0 |
| 9 | 1 0 0 1 | 1 1 0 0 | 0 0 1 1 | 0 | 0 1 0 0 | 0 |
| 10 | 1 0 1 0 | 1 0 1 0 | 0 1 0 1 | 0 | 0 1 1 0 | 0 |
| 11 | 1 0 1 1 | 1 0 0 0 | 0 1 1 1 | 0 | 1 0 0 0 | 0 |
| 12 | 1 1 0 0 | 1 1 1 1 | 0 0 1 0 | 0 | 0 0 1 1 | 0 |
| 13 | 1 1 0 1 | 1 1 0 1 | 0 1 0 0 | 0 | 0 1 0 1 | 0 |
| 14 | 1 1 1 0 | 1 0 1 1 | 0 1 1 0 | 0 | 0 1 1 1 | 0 |
| 15 | 1 1 1 1 | 1 0 0 1 | 1 0 0 0 | 0 | 1 0 0 1 | 0 |

FIGURE 14-13 Outputs at a **74181** if $A = 9$ and $B = 10$.

inputs and the function table (Fig. 14-12$b$), the outputs presented in the table can be calculated.

---

### EXAMPLE 14-28

Explain how the outputs of line 14, Fig. 14-13, were determined.

### SOLUTION

(a) The logic output is given in Fig. 14-12$b$ as $F = A + B(A$ or $B)$ for line 14. Since $A = 1001$ (9) and $B = 1010$ (10), $A + B = 1011$ as shown.

(b) With $M = 0$ and no carry, $(C_{IN} = 1)$, $F = (A + \bar{B})$ plus A. Here $A + \bar{B}$ is A, 1001, ORed with $\bar{B}$, 0101, which equals 1101, or numerical 13. This is added to the number A, as the *plus* operation specfies, and the result is 22, or 6 and a carry-out. Because the sum is greater than 15, the carry-out is LOW and is shown as a 0 on the $C_{n+4}$ line.

(c) With a LOW carry-in $(C_n = 0)$, $F = (A + \bar{B})$ plus A plus $1 = 23$. The results as shown in Fig. 14-13 are 7 and a LOW carryout.

---

### EXAMPLE 14-29

Design an adder/subtracter using cascaded **74181**s. Show how it works if $A = 57$ and $B = 28$ under:

(a) Addition.
(b) Subtraction.
(c) Repeat using the hex equivalent of the given numbers.

### SOLUTION

The **74181** can be cascaded in the normal manner, by tying the carry-out of a stage to the carry-in of the succeeding stage.

(a) Addition can be performed by setting S to 9, M to 0, and the carry-in of the least significant stage to 1.

$$\text{If } A = 57 = 00111001$$
$$\text{and } B = 28 = 00011100$$

The least significant **74181** sees inputs of 1001, 1100, and a HIGH carry-in. Its output, as specified by the function table, is 9 plus $12 = 5$. Since the sum is greater than 15, the carry-out is LOW.

The second stage sees inputs of $A = 3$, $B = 1$ and a LOW carry-in. Its output is A plus B plus $1 = 5$. The final output is, therefore, $(01010101)_2 = 85$, which is the correct answer.

(b) Subtraction can be performed by setting S to 6, M to 0, and the carry-in of the least significant stage to 0. Then the least significant **74181** sees inputs of 9,

12, and a LOW carry-in. Its output is $-3$ or 1101 (the 4-bit 2's complement of 3). It also produces a HIGH carry-out because the result is negative.

The next stage sees inputs of 3, 1, and a HIGH carry-in. The results are A minus B minus $1 = 3 - 1 - 1 = 1$. The final answer is, therefore, 00011101, or 29, which is also correct.

(c) Using the hex equivalent $A = 39$ and $B = 1C$. Their sum is $(55)_{16}$ or $(85)_{10}$. For subtractions we first have $9 - C = -3 = D$ (the 4-bit 2's complement of D). This also produces a borrow out so the most significant nibble becomes $3 - 1 - 1 = 1$, and the final result is $(1D)_{16} = (29)_{10}$. It is wise to use hex arithmetic because the 74181s essentially operate in hex mode and it greatly simplifies debugging should a problem arise.

## 14-11 LOOK-AHEAD CARRY[5]

In the adders discussed thus far, the output of each stage depended upon its inputs and the carry-in from the previous stage. These adders are relatively slow because the carry must ripple through all the stages before the result is correct. *Look-ahead carry* is a method of speeding up the carry generation on an adder. It requires additional hardware and should only be used when the *speed* of the adder circuit is extremely critical.

The equation for the carry output of an adder stage was developed in Sec. 14-3.2

$$C_{OUT} = AB + C_{IN}(A \oplus B)$$

The $i$th stage always produces a carry-out if both $A_i$ and $B_i$ are 1. This term is defined as $G_i$ (i.e., $G_i = A_iB_i$), the GENERATE term for the $i$th stage. The $i$th stage also produces a carry output if $A \oplus B = 1$ and there is a carry-in from the $i - 1$ stage. The term $(A_i \oplus B_i)$ is called the carry PROPAGATE term.[6]

Consider an adder starting with the least significant stage. Equations for the carry-out of each stage can be written as follows:

$C_0 = G_0$     $C_{IN} = 0$ to this stage. Therefore, the PROPAGATE term is irrelevant.

$C_1 = G_1 + G_0P_1$

$C_2 = G_2 + G_1P_2 + G_0P_2P_1$

$C_3 = G_3 + G_2P_3 + G_1P_3P_2 + G_0P_3P_2P_1$

---

[5]This section contains advanced material and may be omitted on first reading.
[6]Some authors define $P_i = A_i + B_i$. For look-ahead carry circuits, the OR and XOR are equivalent.

In general,

$$C_n = G_n + G_{n-1}P_n + G_{n-2}P_nP_{n-1} \cdots \qquad \text{14-3}$$

$$\text{or} \quad C_n = G_n + P_nC_{n-1}$$

---

**EXAMPLE 14-30**

Given the binary numbers A = 110110001, and B = 11011011, identify by inspection which stages produce carry outputs and which stages produce GENERATE and PROPAGATE outputs.

**SOLUTION**

Arrange the numbers as shown below:

| Stage | 8 | 7 | 6 | 5 | 4 | 3 | 2 | 1 | 0 |
|-------|---|---|---|---|---|---|---|---|---|
| A | 1 | 1 | 0 | 1 | 1 | 0 | 0 | 0 | 1 |
| B |   | 1 | 1 | 0 | 1 | 1 | 0 | 1 | 1 |

Carry GENERATION occurs when both A and B are 1, or on stages 0, 4, and 7. PROPAGATION occurs when $A \oplus B$ is 1 or on stages 1, 3, 5, 6, and 8. By applying Eq. 14-3, we find carries out of all stages except stages 2 and 3. Consequently, there are no carries into stages 0, 3, and 4.

---

The advantage of look-ahead carry is speed. It only takes three gate delays to generate the carry input (one gate to produce the GENERATE or PROPAGATE output, and two gates delays for the NAND realization of Eq. (14-3). A 4-bit adder circuit using look-ahead carry is shown in Fig. 14-14.

### 14-11.1 Multibit Adders

Look-ahead carry can also be used to speed up multibit adders. The **7483** 4-bit adder for example, already incorporates look-ahead carry in its internal circuitry, but adders using several 7483s (16- or 32-bit adders are examples) are faster if look-ahead carry is used. The equations for look-ahead carry still apply except that for a 4-bit adder, GENERATE is 1 if the sum of the inputs is 16 or more, and PROPAGATE is 1 if the sum of the inputs is 15.

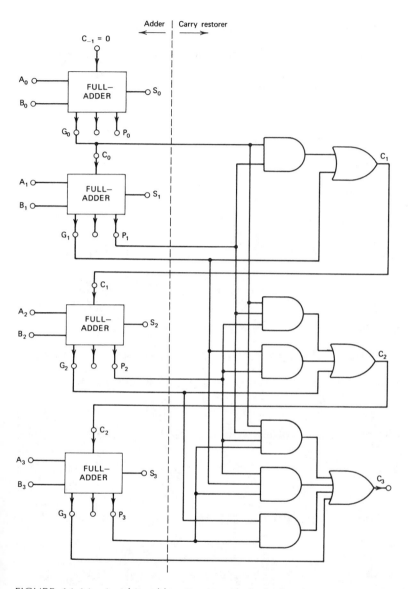

FIGURE 14-14  A 4-bit adder circuit with look-ahead carry. (Barna and Porat, *Integrated Circuits in Digital Electronics*. Copyright John Wiley & Sons, Inc. 1973. Reprinted by permission of John Wiley & Sons, Inc.)

### 14-11.2 The 74182 Look-Ahead Carry Generator

The 74182 is an IC that contains all the logic required for look-ahead carry, and should be used whenever look-ahead carry is necessary. The schematic of the 74182 is shown in Fig. 14-15$a$. The equations for the 74182 are

$$C_{n+x} = \bar{G}_0 + \bar{P}_0 C_n$$
$$C_{n+y} = \bar{G}_1 + \bar{P}_1 G_0 + \bar{P}_1 \bar{P}_0 C_2$$
$$C_{n+z} = \bar{G}_2 + \bar{P}_2 \bar{G}_1 + \bar{P}_2 \bar{P}_1 \bar{P}_0 C_n$$
$$G = \bar{G}_3 (\bar{P}_3 + \bar{G}_2)(\bar{P}_3 + \bar{P}_2 + \bar{G}_0)(\bar{P}_3 + \bar{P}_2 + \bar{P}_1 + \bar{C}_0)$$
$$\bar{P} = \bar{P}_3 \bar{P}_2 \bar{P}_1 \bar{P}_0$$

Essentially, the GENERATE and PROPAGATE terms must be inverted before being applied to the 74182. The carry restorer portion of Fig. 14-14 could be replaced by a 74182 if the GENERATE and PROPAGATE terms are inverted.

The 74182 is designed to work directly with the 74181 and increases the speed of the adder if the word length is greater than 8 bits. The connections between the 74181 and the 74182 are shown in Fig. 14-15$b$.

### 14-12 BINARY MULTIPLICATION

The ability to perform binary multiplication and division is not as important as the ability to add, subtract, and logically manipulate data. Many mini-computers do not multiply and divide (these operations are performed by a program, or "in software") or else hardware MULTIPLY/DIVIDE (which is faster than software) is supplied as an option, at additional cost. In this section, a multiplier for 7-bit words is designed using the shift-and-add algorithm developed below.

### 14-12.1 The Multiplication Algorithm

Multiplication starts with a multiplier, multiplicand, and PRODUCT REGISTER. The PRODUCT REGISTER, which is initially 0 and eventually contains the product, must be as long as the number of bits in the multiplier and multiplicand added together. In computers, the multiplier and multiplicand are typically one $n$-bit word each, and two $n$-bit words must be reserved for the product.

Multiplication can be performed in accordance with the following algorithm:

1. Examine the least significant bit of the multiplier. If it is 0, shift the PRODUCT REGISTER one bit to the right. If it is 1, add the multiplicand to the MSB of the PRODUCT REGISTER and then shift.
2. Repeat step 1 for each bit of the multiplier.

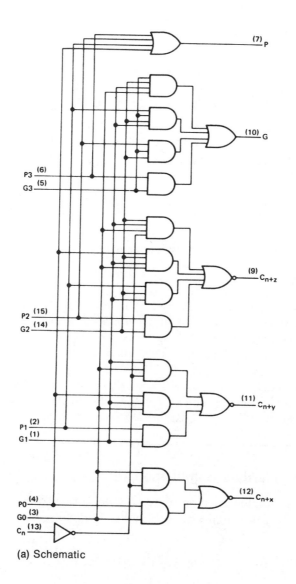

(a) Schematic

## TYPICAL APPLICATION DATA

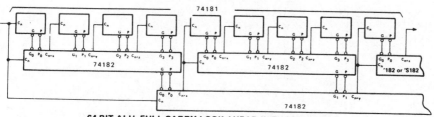

**64-BIT ALU, FULL-CARRY LOOK-AHEAD IN THREE LEVELS**
A and B inputs and F outputs of 74181 are not shown.

(b) Typical application

**FIGURE 14-15** The **74182** look-ahead carry generator. (From the *TTL Data Book for Design Engineers*, 2nd ed., Texas Instruments, Inc. Courtesy of Texas Instruments, copyright 1976.)

At the conclusion the product should be in the PRODUCT REGISTER.

It is common to shift the multiplicand one bit *left* for each multiplier bit and then add it to the PRODUCT REGISTER. This is analogous to multiplication as taught in grade school. The algorithm presented here, however, holds the *position* of the multiplicand *constant* but shifts the PRODUCT REGISTER *right*. This algorithm is more easily implemented in hardware (see Sec. 14-12.2).

---

**EXAMPLE 14-31**

Multiply 22 × 26 using the above algorithm.

**SOLUTION**

The solution is shown in Fig. 14-16. The multiplier bits are listed in a column with the LSB on top. The multiplication proceeds in accordance with these bits as the leftmost column shows.

Consider line 5 as an example. The multiplier bit is 1, so the 5-bit multiplicand (26) is added to the 5 MSBs of the PRODUCT REGISTER that appear on line 4 as 13. The result, 39, appears on line 5. The product moves steadily to the right. The final product appears on line 9 and is a 10-bit number in this case. Note that the MSB of the PRODUCT REGISTER is reserved for carries that may result from the additions.

---

## 14-12.2 Hardware Implementation

A hardware implementation of a multiplier for two 7-bit words is shown in Fig. 14-17. It uses many circuits that were described in earlier sections of this book.

Multiplier = 22 = 1 0 1 1 0

Multiplicand = 26 = 1 1 0 1 0

| Line Number | | | | Multiplier Bit | |
|---|---|---|---|---|---|
| 1 | | 0 0 0 0 0 | | | Initial product |
| 2 | | 0 0 0 0 0 | 0 | 0 | Shift product register |
| 3 | | 1 1 0 1 0 | 0 | 1 | Add multiplicand |
| 4 | | 0 1 1 0 1 | 0 0 | | Shift product register |
| 5 | 1 | 0 0 1 1 1 | 0 0 | 1 | Add multiplicand |
| 6 | | 1 0 0 1 1 | 1 0 0 | | Shift product register |
| 7 | | 0 1 0 0 1 | 1 1 0 0 | 0 | Shift product register |
| 8 | 1 | 0 0 0 1 1 | 1 1 0 0 | 1 | Add multiplicand |
| 9 | | 1 0 0 0 1 | 1 1 1 0 0 | | Shift – final product |

FIGURE 14-16 Multiplying 22 × 26.

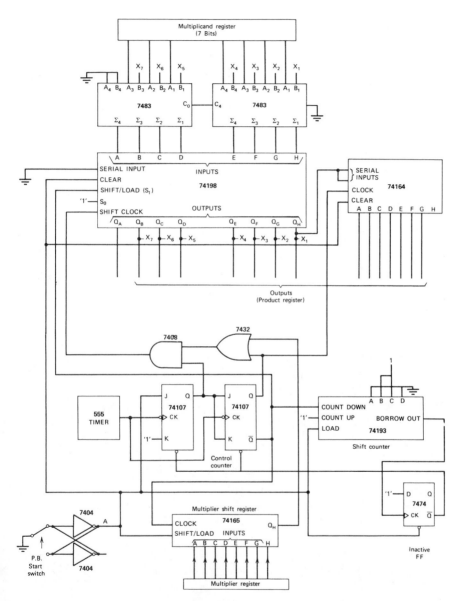

FIGURE 14-17 A 7-bit multiplier.

The basic parts of the circuit are:

1. **The MULTIPLICAND REGISTER.** This contains the multiplicand.
2. **The adder.** It consists of **7483**s and adds the multiplicand register and the product register.
3. **The PRODUCT REGISTER.** It is built of a **74198** and a **74164**. The more significant stages of the register must have both parallel load and parallel output

capability; hence a **74198** is chosen. Two 4-bit shift registers such as **74195**s could also be used. The most significant bit of the PRODUCT REGISTER must be available to hold any carry-out of the adder (see lines 5 and 8 of Fig. 14-16). Since the less significant stages of the PRODUCT REGISTER only need serial inputs, a **74164** parallel-output shift register is selected.

4.   **The 555 timer.** This produces a continuous clock. Other oscillators discussed in Chapter 7 would serve as well.

5.   **The CONTROL COUNTER.** The more significant stage of this counter controls the SHIFT/LOAD input to the **74198**, and changes the mode on alternate pulses. The clock pulses to the **74198** are supplied by the least significant stage.

6.   **The gating of the 7432 and 7408.** This gating supplies a clock to the **74198** on each pulse of the CONTROL COUNTER *if it is in shift mode or $Q_H$ of the* MULTIPLIER SHIFT REGISTER is 1. It eliminates LOAD clocks when $Q_H$ of the MULTIPLIER REGISTER is 0.

7.   **The SHIFT COUNTER.** This is a **74193** that is loaded with the number of shifts required. For a 7-bit multiplier, the **74193** is loaded with the number 6, as shown.

8.   **The MULTIPLIER SHIFT REGISTER.** This register is loaded with the multiplier ($Q_H$ is the LSB) that is shifted out during its operation.

9.   The INACTIVE FF. After the proper number of shifts the BORROW-OUT of the **74193** sets the INACTIVE FF, which then holds the CONTROL COUNTER clear and stops the circuit from changing further. This retains the answer in the PRODUCT REGISTER until the next start pulse occurs.

10.   **The START switch.** Depressing the debounced START switch initializes the circuit. When it is released, the multiplication begins.

The operation of the circuit is briefly described:

1.   Depressing the START switch places a 0 at point A. This clears the PRODUCT REGISTER and the INACTIVE FF, loads the MULTIPLIER SHIFT REGISTER with the multiplier, and the SHIFT COUNTER with the count. It prevents the CONTROL COUNTER from counting by holding J LOW.

2.   When the START switch is released, the CONTROL COUNTER starts. It clocks the MULTIPLIER SHIFT REGISTER and SHIFT COUNTER and provides the proper SHIFT and LOAD pulses to the PRODUCT REGISTER. The LOAD pulses load the sum of the multiplicand and the PRODUCT REGISTER into the PRODUCT REGISTER, effectively implementing the add-and-shift algorithm.

3.   When the SHIFT COUNTER counts down the proper number of pulses, it produces a BORROW OUT. This sets the INACTIVE FF and stops the circuit from pulsing further.

---

**EXAMPLE 14-32**

Show how the circuit of Fig. 14-17 multiplies 75 by 127.

## SOLUTION

The solution is shown by the timing chart and data table of Fig. 14-18. The shift clocks occur at $t = 3, 7, 11$, and so on. Load clocks occur at $t = 1, 5,\ldots$ There are no load clocks at $t = 9, 17$, and 21 because the MULTIPLIER SHIFT REGISTER output is 0.

The data table shows the PRODUCT REGISTER contents as the multiplication progresses. At $t = 1$, for example, the PRODUCT REGISTER is loaded with the multiplicand, and is shifted at $t = 3$. At $t = 5$, the PRODUCT REGISTER is loaded with the sum of its own contents plus the multiplicand, or $63 + 127 = 190$. The shifts and loads continue until the final product occurs at $t = 27$.

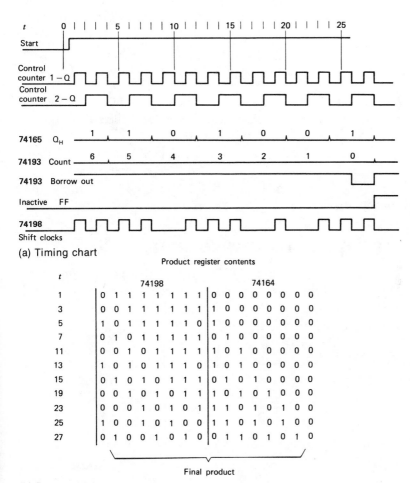

(a) Timing chart

Product register contents

| $t$ | 74198 | | | | | | | | 74164 | | | | | | | |
|---|---|---|---|---|---|---|---|---|---|---|---|---|---|---|---|---|
| 1 | 0 | 1 | 1 | 1 | 1 | 1 | 1 | 1 | 0 | 0 | 0 | 0 | 0 | 0 | 0 | 0 |
| 3 | 0 | 0 | 1 | 1 | 1 | 1 | 1 | 1 | 1 | 0 | 0 | 0 | 0 | 0 | 0 | 0 |
| 5 | 1 | 0 | 1 | 1 | 1 | 1 | 1 | 0 | 1 | 0 | 0 | 0 | 0 | 0 | 0 | 0 |
| 7 | 0 | 1 | 0 | 1 | 1 | 1 | 1 | 1 | 0 | 1 | 0 | 0 | 0 | 0 | 0 | 0 |
| 11 | 0 | 0 | 1 | 0 | 1 | 1 | 1 | 1 | 1 | 0 | 1 | 0 | 0 | 0 | 0 | 0 |
| 13 | 1 | 0 | 1 | 0 | 1 | 1 | 1 | 0 | 1 | 0 | 1 | 0 | 0 | 0 | 0 | 0 |
| 15 | 0 | 1 | 0 | 1 | 0 | 1 | 1 | 1 | 0 | 1 | 0 | 1 | 0 | 0 | 0 | 0 |
| 19 | 0 | 0 | 1 | 0 | 1 | 0 | 1 | 1 | 1 | 0 | 1 | 0 | 1 | 0 | 0 | 0 |
| 23 | 0 | 0 | 0 | 1 | 0 | 1 | 0 | 1 | 1 | 1 | 0 | 1 | 0 | 1 | 0 | 0 |
| 25 | 1 | 0 | 0 | 1 | 0 | 1 | 0 | 0 | 1 | 1 | 0 | 1 | 0 | 1 | 0 | 0 |
| 27 | 0 | 1 | 0 | 0 | 1 | 0 | 1 | 0 | 0 | 1 | 1 | 0 | 1 | 0 | 1 | 0 |

Final product

(b) Data table

FIGURE 14-18   Multiplying 75 × 127 using the circuit of Fig. 14-17.

### 14-12.3   Practical Considerations

The multiplier of Fig. 14-17 was designed so that the two halves of the PRODUCT REGISTER do *not* shift at the same time. The **74164** shifts at $t = 2, 6, 10, \ldots$ while the **74198** shifts at $t = 3, 7, 11, \ldots$. This prevents race and sneak problems that may arise if the same clock drives both shift registers and a clock delay occurs somewhere in the circuit. It also simplifies the circuitry.

The circuit of Fig. 14-17 is very easy to test in the laboratory. It is best tested by replacing the start switch with a PULSE GENERATOR; this provides repetitive short negative pulses. Multiplication occurs during the long positive portion of the wave. It would also be wise to synchronize the oscillator with the PULSE GENERATOR (possibly using the reset on the **555**). Operation of the multiplier can be observed on a CRO and any faults corrected.

### 14-12.4   Signed Multiplication

The *multiplication* of *signed* numbers is accomplished by converting both operands to positive numbers, multiplying them and then converting, if necessary, so the product has the proper sign.

If the numbers are in 2's complement form and a 2's complement answer is required, the *Booth algorithm*[7] can be used to produce the result directly. The circuit of Fig. 14-17 can be converted to a *Booth algorithm multiplier* with minor modifications.

### 14-13   ARITHMETIC PROCESSING UNITS

The Arithmetic Processing Unit (APU) is an IC that can perform many complex arithmetic functions. One such processing unit is the **Am9511** produced by Advanced Micro Devices, Sunnyvale, California. This 24-pin IC can perform many operations, such as:

ADD
SUBTRACT
MULTIPLY
DIVIDE
SQUARE ROOT
SINE, COSINE, TANGENT
INVERSE SINE, COSINE, TANGENT
LOGARITHMS

[7]For a discussion of the Booth algorithm, see Bartee (References).

The APU can perform these operations on fixed or floating point data and on 16- or 32-bit words. The APU is driven by a clock (just as the multiplier circuit of Fig. 14-17). Each operation takes a number of clock cycles and more complex operations take more clock cycles.

The APU is designed to get data and commands from a microprocessor. When the APU receives a command, it becomes busy for the number of clock cycles required to complete the arithmetic operation. It then outputs an $\overline{END}$ signal to signify the completion of the operation and its ability to accept another command.

APUs are too advanced and complex to be discussed fully in this book. The reader who needs further information should consult the manufacturer's literature. (See References.)

## SUMMARY

This chapter discussed various circuits for performing binary arithmetic. It began with half-adders, full-adders, and subtracter circuits built from SSI gates. More sophisticated methods for performing arithmetic operations, particularly subtraction by complementation and 2's complement notation were covered. 2's complement notation was emphasized because of its wide usage.

Larger scale integrated (LSI) circuits for performing binary arithmetic, specifically the 7483, 74181, and 74182, were introduced. Use of the 7483 in 2's complement and BCD ADDER/SUBTRACTERS was considered. Use of the 74181 ALU to perform arithmetic and logic functions was described in some detail. Finally, look-ahead carry, with and without the 74182, and a binary multiplication circuit were presented.

At this point, the reader should be able to design circuits to perform any simple arithmetic or logical operations he or she requires. More sophisticated arithmetic operations are generally done by computers, using programmed algorithms, or by arithmetic processing units when high speed is required.

## GLOSSARY

**Half-adder.** A circuit that accepts 2 binary bits as inputs and produces their sum and carry as outputs.

**Full-adder.** A circuit that accepts 2 binary bits and a carry-in as inputs and produces their sum and carry as outputs.

**Half-subtracter.** A circuit that accepts 2 inputs, computes their difference, and determines whether there is a borrow out.

**Full-subtracter.** A circuit that accepts 2 inputs and a borrow-in and produces their difference and a borrow as outputs.

**Complementation.** The process of taking the difference between a given number and an input number.

**Sign-magnitude notation.** A representation of numbers where the MSB represents the sign of the number.

**2's complement.** A representation of numbers where negative numbers are obtained by complementing their positive equivalent and adding 1.

**ADDER/SUBTRACTER.** A circuit that can either add or subtract 2 operands, depending on the mode.

**Overflow.** Overflow occurs when the result of an arithmetic operation is a larger number than the output register can accommodate.

**Underflow.** Underflow occurs when the result of an arithmetic operation is a more negative number than the output register can accommodate.

**BCD arithmetic.** Arithmetic involving decimal digits, expressed in BCD form.

**Arithmetic/logic unit (ALU).** A circuit capable of performing a variety of arithmetic or logical functions.

**Look-ahead-carry.** Using special circuitry to rapidly generate the carry-in of an adder circuit.

**End-around-carry.** Removing an MSB of 1 and adding it to the LSB to complete a subtraction by the complement-and-add method.

**PRODUCT REGISTER.** The register used to hold the product in a multiplier.

## REFERENCES

**Arpad Barna** and **Dan I. Porat,** *Integrated Circuits in Digital Electronics,* Wiley, New York, 1973.

**Thomas C. Bartee,** *Digital Computer Fundamentals,* 5th ed., McGraw-Hill, New York, 1980.

**E.B. Croson, F.H. Carlin,** and **J.A. Howard,** "Integrated Arithmetic Processing Unit Enhances Processor Execution Times," *Computer Design,* April 1981.

**George K. Kostopoulos,** *Digital Engineering,* Wiley, New York, 1975.

**Morris** and **Miller,** *Designing With TTL Integrated Circuits,* McGraw-Hill, New York, 1971

**Hermann Schmid,** *Decimal Computation,* Wiley, New York, 1974.

**Ronald J. Tocci,** *Digital Systems,* Prentice-Hall, Englewood Cliffs, N.J., 1980.

*The Am9511 Arithmetic Processing Unit,* Advanced Micro Devices, Inc., Sunnyvale, California, 1978.

## PROBLEMS

14-1.    Add 5 and 3 using the circuit of Fig. 14-3. Show all your 1s and 0s.

14-2.    Design a subtractor circuit to subtract 3-bit binary numbers. Show how it works if the numbers are 6 − 5.

14-3.    Perform the following subtractions using complement-and-add techniques:

(a)    73
        $-28$

(d)    843
        $-966$

(b)    8193
        $-6095$

(e)    512
        $-16096$

(c)    14312
        $-857$

14-4    Find the 8-bit 2's complement of the following numbers:
(a) 99
(b) $-7$
(c) $-102$

14-5.    Determine by inspection which of the following 2's complement numbers are divisible by 4:
(a) 11011010
(b) 10011100
(c) 01111000
(d) 00001010
(e) 01000001
(f) 01010100

14-6.    A PDP-8 is a computer with a 12-bit word length that uses 2's complement arithmetic. What range of numbers can be expressed by a single PDP-8 word?

14-7.    Express each of the following numbers in 9-bit, 2's complement form, and add them:

(a)    85
        $+37$

(c)    $-85$
        $+37$

(b)    85
        $+(-37)$

(d)    $-85$
        $+(-37)$

14-8.    Do the following subtractions after expressing the operands in 11-bit, 2's complement notation:

(a)    36
        $-(23)$

(c)    $-450$
        $-(-460)$

(b)    835
        $-(214)$

(d)    310
        $-(-579)$

14-9.    A number in 2's complement form can be inverted by subtracting it from $-1$.
(a) Invert 25 using this procedure and 8-bit numbers.
(b) Show that there can never be overflow or underflow using this procedure.

14-10.  Convert the following binary numbers to hexadecimal.
   (a)  11111011
   (b)  1011001
   (c)  10000011111100
   (d)  10010101100011101

14-11.  Convert the following hex numbers to binary:
   (a)  129
   (b)  84C5
   (c)  5CF035
   (d)  ABCDE2F

14-12.  Convert the numbers in Problem 14-11 to decimal numbers:

14-13.  Convert the following numbers to hex:
   (a)  139
   (b)  517
   (c)  2,000
   (d)  105,684

14-14.  Perform the following hex additions:

(a)     99            (b)     CB
     + 89                    DD

(c)  15F02            (d)  2CFB4D
     3C3E                  5DC98B

14-15.  Perform the following hex subtractions:

(a)  59               (b)  1CC
      F                    DE

(c)  1002             (d)  5F306
     5F8                   135CF

14-16.  Find the negative equivalent of the following hex numbers:
   (a)  23
   (b)  CB
   (c)  500
   (d)  1F302
   (e)  F5630

14-17.  Use the circuit of Fig. 14-8 to perform the following operations:
   (a)  65 + 57
   (b)  65 − 57
   (c)  − 35 + 42
   (d)  −35 − 42
   (e)  −35 − (−42)

14-18.  For the circuit of Figure P14-18:
   (a) What are we trying to do? (Consider your inputs as 2's complement numbers).
   (b) Find the two defective circuits.
   (c) Re-draw the circuit to show it working without defective gates.

14-19.  Design a circuit to detect overflow in addition (see Sec. 14-8).

14-20.  Add 32 and 69 using a BCD adder. Show all the 1s and 0s.

14-21.  Subtract 676 from 2432 by using the BCD subtracter of Fig. 14-10.

14-22.  Explain line 5 of Fig. 14-13.

14-23.  Perform the following operations using 74181s (all given numbers are decimal numbers):

(a)     $-87$      (b)     $-87$      (c)     $-43$      (d)     $-43$
     $+(+43)$        $-(+43)$        $+(-17)$        $-(-17$

(e)     $+65$      (f)     $+65$      (g)     $+65$      (h)     $+65$
     $+(-37)$        $-(-37)$        $+(-100)$        $-(-100)$

14-24.  Which results in Problem 14-23 produce overflow?

14-25.  How can 74181s be used to subtract 1 from a number? Show the 1s and 0s if the number is 80.

14-26.  Add 1 to 271 by using 74181s. Show the 1s and 0s.

14-27.  Design a single circuit to perform the following functions on two 8-bit operands using two 74181s.
   (a) Addition.
   (b) Subtraction.
   (c) AND
   (d) OR
   (e) NAND
   (f) XOR
Use a rotary switch to select the function.

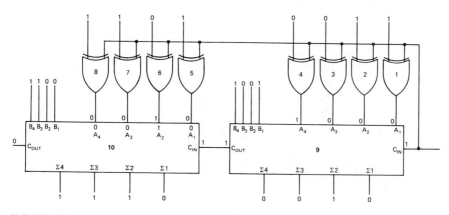

FIGURE P14-18

14-28. Two **74181**s are connected together. Finish the following table: For any irrelevant answers, use X.

| | $A_0$ | $B_0$ | $F_0$ | $C_0$ | $A_1$ | $B_1$ | $F_1$ | $C_1$ |
|---|---|---|---|---|---|---|---|---|
| S = 12<br>M = 0<br>$C_N$ = 0 | B | C | | | 5 | 3 | | |
| S = 6<br>M = 0<br>$C_N$ = 0 | A | D | | | 6 | A | | |
| S = 9<br>M = 1<br>$C_N$ = 1 | A | D | | | 3 | 9 | | |
| S = 15<br>M = 0<br>$C_N$ = 1 | 0 | 7 | | | 5 | 7 | | |

14-29. The circuit of Figure P14-29 can be used to determine if all 8 bits of A are 0 (if A = 0). Explain how it works. Where is the output taken from?

14-30. Design a 12-bit adder using **7483**s and look-ahead carry. For the look-ahead carry circuit use:
   (a) SSI gates.
   (b) A **74182**.

14-31. Multiply $37 \times 45$ using the algorithm of Sec. 14-12.1.

14-32. Multiply $63 \times 107$ using the circuit of Fig. 14-17. Produce a timing chart and data table similar to that of Fig. 14-18.

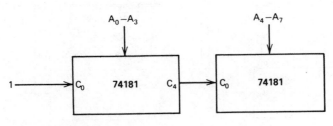

For the **74181**s: S = 15 (F)
              M = 0

FIGURE P14-29

14-33.    Design a multiplier for 8-bit words. (*Hint:* Modify Fig. 14-7 by adding a FF and some SSI gates.)

After working the problems, return to the self-evaluation questions of Sec. 14-2. If you cannot answer some of them, review the pertinent sections of the chapter to find the answers.

CHAPTER

# 15

# MEMORIES

## 15-1 INSTRUCTIONAL OBJECTIVES

Electronic systems are often required to retain or remember many bits of information. When the amount of information exceeds several bits, this information is stored in a device called a *memory* rather than in individual FFs or a register. This chapter introduces the student to core and semiconductor memories, which are used in today's computers. After reading it he or she should be able to:

1.   Explain the steps involved in reading or writing a memory word.
2.   Explain the function of each wire that goes through a core.
3.   Use a semiconductor RAM to read or write information at a specific location.
4.   Assemble several smaller RAMs together to make a larger RAM.
5.   Utilize both MOS and TTL RAMs.
6.   Utilize dynamic RAMs and be able to build refresh circuits for them.
7.   Design and use diode and NAND gate ROMs for small memories.
8.   Understand and use ROMs and PROMs.

## 15-2 SELF-EVALUATION QUESTIONS

Watch for the answers to the following questions as you read the chapter. They should help you understand the material presented.

1. In memory terminology, the symbol K does not mean 1000. What number does it represent and why?

2. What are MAR and MDR? What factor determines their size?

3. What is the difference between a ROM and a RAM? What input and output signals are connected to each?

4. Why are memory request signals unnecessary in a static RAM?

5. Define access time and cycle time. Why are they important?

6. What are the advantages and disadvantages of a bipolar static RAM compared to an MOS static RAM? For what applications should each be used?

7. In a dynamic RAM, why are refresh cycles necessary? How often must they be performed?

8. What is the difference between a PROM and a ROM?

9. Why is an erasable PROM the best choice when developing a microprocessor program?

## 15-3 MEMORY CONCEPTS

Many sophisticated electronic circuits, up to and including computers, require the ability to retain large amounts of information. This information is stored in the device's *memory* in the form of binary bits, and is used, as needed, to control the operation of the device.

Flip-flops, introduced in Chapter 6, function as *1-bit memories*; they remember whether they were CLEARED or SET last. When FFs are grouped into registers their capability increases; $n$ FFs can remember $2^n$ bits.

Even the smallest computers, however, require several thousand bits of memory. It is impractical to build memories this large from discrete FFs. The basic element of large memories are either large-scale, integrated circuits (LSI), ICs (chips), or magnetic cores.

Whether they are built from cores or chips, memories are subdivided into groups of bits called *words*. A *word consists of the bits involved in each data transfer*. A *memory* is composed of *many* words. Each word is stored at a *selected address*. Data are written into the memory one word at a time and preserved for future use. It is read at a later time when the information is needed. During the interval between writing and reading, other words may be stored or read at other locations. Typical *word sizes* for minicomputers are 12 and 16 bits, and most registers in a computer are the *same* size as the memory word.

Computers required many words of memory. A typical memory size is 4K words. Unlike standard engineering terminology, where K is an abbreviation for kilo, or 1000, $K = 2^{10}$, or 1024, when applied to memories.

This value of K is used because normal memory design leads to sizes that are even powers of 2.

Each word in memory is given an *address* to locate it. Thus each memory word has *two parameters*; its *address*, which locates it within the memory, and the *data*, which are retained within the word.

Memories also require *two registers*, one associated with *address* and one with *data*. The *memory address register* (MAR) holds the address of the word currently being accessed, and the *memory data register* (MDR) holds the data being written into or read out of the addressed memory location. The words stored within the memory are *not* immediately accessible to the system, but the MAR and MDR are available. In some memories the MAR and MDR are *not* part of the internal memory. In this case, they must be part of the user's circuitry. The MDR serves as a temporary storage area (buffer) for transferring words between the memory and the external system.

A block diagram of a basic 4K-word-by-16-bit core memory is shown in Fig. 15-1. The 12 bits in the MAR select one of the 4096 words in memory. The data transfer is between the selected word and the MDR.

---

**EXAMPLE 15-1**

In the 4K-by-16-bit memory:
   (a) How many bits are required in the MAR?
   (b) How many bits are required in the MDR?
   (c) How many data bits are contained within the entire memory?

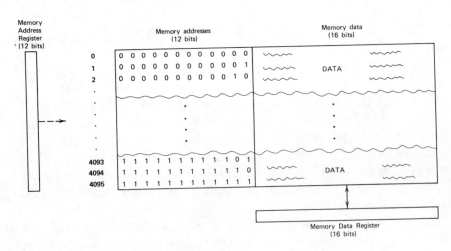

FIGURE 15-1   Block diagram of 4K × 16 bit memory.

## SOLUTION

(a) The specification 4K-by-16 means 4K words of 16 bits each. Therefore, the memory contains $4K = 4096 = 2^{12}$ words. The MAR must hold one of $2^{12}$ possible locations which requires **12 bits** (see Theorem 1, Sec. 1-4). Note that the address of each memory location corresponds to the 12 bits in the MAR. Therefore, memory address numbering starts at 0, and the 4096 addresses range from 0 to 4095.

(b) Each memory location contains 16 bits. Consequently, the MDR must be **16 bits** long to accommodate an entire data word.

(c) A 4K-by-16-bit memory contains $2^{12}$ words $\times$ $2^4$ bits/word $= 2^{16}$ or **65,536 bits.** For a computer, this is a relatively *small* memory.

---

## 15-3.1 Reading Memory

Memories operate in two basic modes: READ or WRITE. A *memory is read when the information at a particular address is required by the system.* To read a memory:

1. The location to be read is loaded into the MAR.
2. A READ command is given.
3. The data are transferred from the addressed word in memory to the MDR, where it is accessible to the system.

Normally the word being read must *not* be altered by the READ operation, so that the word in a particular location can be read many times. The process of reading out of a location without changing it is called *nondestructive readout.*

## 15-3.2 Writing Memory

A *memory location is written when the data must be preserved for future use.* In the process of writing, the previous information in the specified location is destroyed (overwritten). To write a memory:

1. The address is loaded into the MAR.
2. The data to be written are loaded into the MDR.
3. The WRITE command is then given, which transfers the data from the MDR to the selected memory location.

## 15-4 CORE MEMORIES

A core memory uses small magnetic cores as the basic memory element. For many years, core was the dominant memory technology. While IC memories are replacing core in most applications, there are still enough core memories used to warrant an introdction to them. At present, many large, main-frame computers use core memory; mini-computers give their users a choice of core or semiconductor memory; and microprocessors use semiconductor memory almost exclusively.

## 15-4.1 Basic Magnetic Core Operation

A single core is a very small toroid of magnetic material. Typical modern cores have outer diameters of 0.030 or 0.022 in. (0.762 or 0.559 mm), which makes them hard to see and handle. Very thin wires (AWG #36 or 38 is often used) are strung through the center of a core and are very brittle. Stringing cores is a meticulous and difficult job, requiring great patience and skill.

The core is activated by a wire through its center, as shown in Fig. 15-2. If a current pulse is applied to the wire, the resultant magnetic field around the wire magnetizes the high-retentivity core. When the current is removed, the magnetic flux remains within the magnetized core, as shown in Fig. 15-2a. If the current is reversed, as shown in Fig. 15-2b, the flux direction within the core reverses and remains reversed until another pulse of forward current is applied.

The information stored in a magnetic core (as 1 or 0) is determined by the direction of flux within the core. *The memory content (magnetism) is not destroyed even if power to the system is removed.* The ability to retain

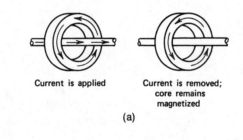

Current is applied     Current is removed;
core remains
magnetized

(a)

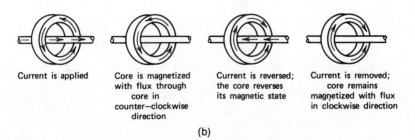

Current is applied    Core is magnetized    Current is reversed;    Current is removed;
with flux through    the core reverses    core remains
core in    its magnetic state    magnetized with flux
counter–clockwise               in clockwise direction
direction

(b)

FIGURE 15-2 The magnetic core. (From T. Bartee, *Digital Computer Fundamentals,* copyright 1981, 5th ed. Used with permission of McGraw-Hill Book Company.)

information without requiring power is called *nonvolatility*. This is a major advantage of core memories. *Semiconductor memories are volatile* and must be rewritten whenever power is turned on. Microprocessor programs, for example, must be preserved in Read Only Memories (See Sec. 15-9) disks, or cassettes so they can be rewritten rapidly after power has been turned off.

Core operation can best be understood by referring to the hysteresis loop of Fig. 15-3. The *hysteresis loop* is a plot of B, the *magnetic flux density within the core*, versus H, the *magnetic field intensity*, which is proportional to the *current* in the wire. *The variation of flux density in a core resulting from a current pulse depends on the magnitude of the current pulse and the previous state of the core.* For optimum use in memories, the hysteresis loop of the core should be as square (or rectangular) as possible, and considerable manufacturing research and effort has resulted in production of "square-loop" cores.

## 15-4.2 Writing and Reading a Core

For the core whose hysteresis loop is shown in Fig. 15-3, a *positive* current pulse *writes* the core; a *negative* current pulse *reads* the core. If a WRITE

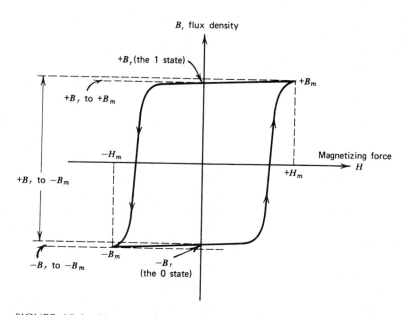

FIGURE 15-3 Hysteresis loop for a magnetic material. Note: The magnetic flux H is caused by and is proportional to the winding current I. (From T. Bartee, *Digital Computer Fundamentals*, copyright 1981, 5th ed. Used with permission of McGraw-Hill.)

current pulse, which produces a magnetic field equal to $H_m$, is applied to a core, its flux density goes to $+ (B_{max})$. When the pulse is removed, $H$ becomes 0, but the flux density follows the hysteresis loop to $+ (B_{residual})$. This indicates that the core is storing a 1.

To read the core, a negative current pulse is applied and the flux density goes to $-B_m$. After the read pulse ends, the core flux follows the hysteresis loop to $-B_r$, which indicates it is in the 0 state. *After any READ operation, the core is always in the 0 state. Core reads are* always *destructive*; the data are destroyed in the process of reading it.

## 15-4.3 Sensing the Core

If a READ pulse occurs when the core is in the 1 state, the flux density changes from $+B_r$ to $-B_m$. This is a relatively *large* change of flux. If a READ pulse occurs when the core is in the 0 state, however, the flux change is only from $-B_r$ to $-B_m$, a small change. During the READ pulse a *large flux change* indicates the core was in the *1 state* and a *small flux change* indicates the core was in the 0 *state*.

To sense the change of flux and determine the state of the core, another wire called the *sense winding* is strung through the center of the core. A flux change induces a voltage in the sense winding $[e = N(d\phi/dt)]$. *The sense winding is connected to a SENSE AMPLIFIER*, as shown in Fig. 15-4, which differentiates between a large and small flux change and determines whether the output is a 0 or a 1. SENSE AMPLIFIERS also have a STROBE input that disables them, except during the portion of the READ pulse when a flux change must be detected. This prevents noise from creating a spurious SENSE AMPLIFIER output.

---

**EXAMPLE 15-2**

The pulses on a core line are WRWRRRWRWRR. (W = WRITE, R = READ). What data are read on each READ pulse?

**SOLUTION**

The WRITE current pulses write a 1, but each READ current pulse returns the core to the 0 state $(-B_r)$. A 1 is read if a WRITE pulse precedes a READ pulse, but *not on the second consecutive READ pulse*. The data read in this example are, therefore, **1100110**.

---

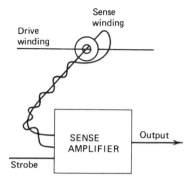

FIGURE 15-4 The sense winding and SENSE AMPLIFIER.

### 15-4.4 Core Memory Organization

An $n$-bit-by-$m$-word core memory is organized into $n$ core planes. Each core plane contains $m$ bits, which are one bit of each of the $m$ words in the memory. Core plane 0, for example, contains bit 0 of each word of the memory.

In practice, each core has four wires strung through it; an X and Y drive wire, a SENSE wire and an INHIBIT wire. Details of core memory organization and operation are presented in other books (see References).

## 15.5 INTRODUCTION TO SEMICONDUCTOR MEMORIES

Semiconductor memories are now the dominant memory technology. Like ICs, they come in DIP (dual in line) packages. They are small, easy to install, inexpensive, and consume little power. They have the disadvantage of losing information when power is turned off, but manufacturers have designed circuits using batteries to back up the power source in case of power failure to preserve the information in the memory. Nonvolatile semiconductor memories are available at additional cost.

### 15-5.1 Bipolar and MOS Memories

Semiconductor memories are built primarily of bipolar or MOS gates. Both types of memories store their information in FFs or on capacitors within the IC. The internal structure of the FFs and gates comprising a memory are beyond the scope of this text. Instead, we will concentrate on the input-output (I-O) characteristics, which must be understood if one is to use a memory.

## 15-5.2 RAMs and ROMs

Semiconductor memories are available as either RAMs (random access memories) or ROMs (read only memories). ROMs are discussed in Sec. 15-9. The term RAM means that each word in memory is *equally* accessible. It also means, implicitly, that the memory is capable of being written into or read from. A RAM implies a *read-write* memory.

## 15-5.3 Static and Dynamic RAMs

RAMs are divided into two catagories, *static* and *dynamic*. *Static* RAMs store their information in FFs and are very simple to use. *Dynamic* RAMs store their information on capacitors that lose their charge ultimately and must be refreshed periodically. The remainder of this section considers static RAMs. (Dynamic RAMs are discussed in Sec. 15-8.)

## 15-5.4 Interfacing with a RAM Memory

Like core memories, semiconductor memories are arranged in matrices of $n$ words by $m$ bits. A semiconductor RAM normally has pins for the following inputs and outputs:

1. $m$ output bits.
2. $m$ input bits.
3. $n$ address bits (for $2^n$ words).
4. A READ/WRITE input.
5. A CHIP SELECT or ENABLE input.

Because semiconductor ICs are often paralleled to build larger memories, most outputs are either open collector or 3-state. When the IC is *not* selected, as determined by the CHIP SELECT signal, it cannot be written into, and all its outputs are turned OFF, or in the high impedance state.

There are no memory request signals in a static semiconductor memory. To read a location, the READ/WRITE line is simply set to the READ level. The memory then presents the data at the addressed word and continues to present this data until the address, READ/WRITE command or CHIP SELECT signals change. *Reading is nondestructive;* the state of the internal FFs at the selected address is simply brought to the outputs.

A memory is written into by setting the READ/WRITE line to WRITE. Whatever data are on the input lines are entered into the memory *at the addressed location.* One must be careful about changing the input data while in WRITE mode. Any change of data is gated into the memory and *overwrites* the previous contents of the memory, which are lost.

## 15-5.5   Memory Timing

For high speed systems, *memory timing* is very important. A READ cycle is limited by the access time of a memory and a WRITE cycle is limited by the CYCLE TIME.[1]

*Access time* is the time required for the memory to present valid data *after* the address and select signals are firm. *Cycle time* or WRITE *time* is the length of time the address and data must be held constant in order to write to the memory.

---

**EXAMPLE 15-3**

A 256-bit memory is organized into 64 words of 4 bits each. What input and output lines are required?

**SOLUTION**

This memory requires:

1.  Six address bits to select 1 of 64 words.
2.  Four data input bits, 1 for each word.
3.  Four data output bits, 1 for each word.
4.  One READ/WRITE command.
5.  One CHIP SELECT or ENABLE.

These signals and the power inputs are required to operate the memory.

---

## 15-6   BIPOLAR RAMs

Most memories are built of bipolar (TTL) or MOS FFs. Bipolar memories are all static and are discussed in this section. (MOS static RAMs are discussed in Sec. 15-7.)

The basic memory element within bipolar RAMs is the TTL type FF. Consequently, these memories are very fast and are used in high-speed systems. They also have the advantage of requiring only one standard + 5-V power supply. Many MOS memories require two or more different voltages.

---

[1] A more complete discussion of memory timing is given in Sec. 9-3 Luecke, Mize, and Carr, of *Semiconductor Memory Design and Application*. New York, McGraw-Hill, 1973. (See References.)

## 15-6.1   Scratch-Pad Memories

Small memories, with capacities of less than 256 bits, called *scratch-pad* memories, are used to store small amounts of information that must be changed frequently. The general-purpose registers or accumulators of a computer provide one use for scratch-pad memories.

The smallest TTL scratch-pad memories have 16 bits, arranged as a 16-word-by-1-bit memory (the 7481, for example) or as a 4-word-by-4-bit memory like the 74170.

The 74170 is a typical scratch-pad memory. Its pin layout, function table and functional block diagram are shown in Fig. 15-5. From the pin layout (Fig. 15-5a) we note that there are:

1.   Four data inputs (D1–D4), one for each bit in a word.
2.   Four outputs (Q1–Q4), one for each bit in a word. The outputs are open collector to allow for parallel operation.
3.   Two READ SELECT inputs ($R_A$ and $R_B$) to select 1 of the 4 words to be read.
4.   Two WRITE SELECT inputs ($W_A$ and $W_B$) to select 1 of the 4 words to be written to.
5.   One READ ENABLE line ($G_R$).
6.   One WRITE ENABLE line ($G_W$).

The functional block diagram (Fig. 15-5c) shows that the 74170 consists of 16 latch-type FFs and associated gates for encoding and decoding the input and output signals. The 16 FFs form 4 words horizontally, and each word consists of a vertical column of 4 bits.

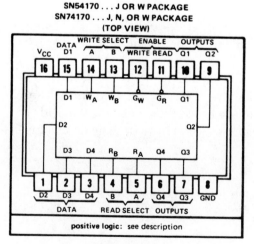

(a) Pin layout

| WRITE FUNCTION TABLE (SEE NOTES A, B, AND C) | | | | | | |
|---|---|---|---|---|---|---|
| WRITE INPUTS | | | WORD | | | |
| $W_B$ | $W_A$ | $G_W$ | 0 | 1 | 2 | 3 |
| L | L | L | Q = D | $Q_0$ | $Q_0$ | $Q_0$ |
| L | H | L | $Q_0$ | Q = D | $Q_0$ | $Q_0$ |
| H | L | L | $Q_0$ | $Q_0$ | Q = D | $Q_0$ |
| H | H | L | $Q_0$ | $Q_0$ | $Q_0$ | Q = D |
| X | X | H | $Q_0$ | $Q_0$ | $Q_0$ | $Q_0$ |

| READ FUNCTION TABLE (SEE NOTES A AND D) | | | | | | |
|---|---|---|---|---|---|---|
| READ INPUTS | | | OUTPUTS | | | |
| $R_B$ | $R_A$ | $G_R$ | Q1 | Q2 | Q3 | Q4 |
| L | L | L | W0B1 | W0B2 | W0B3 | W0B4 |
| L | H | L | W1B1 | W1B2 | W1B3 | W1B4 |
| H | L | L | W2B1 | W2B2 | W2B3 | W2B4 |
| H | H | L | W3B1 | W3B2 | W3B3 | W3B4 |
| X | X | H | H | H | H | H |

NOTES:  A. H = high level, L = low level, X = irrelevant.
B. (Q = D) = The four selected internal flip-flop outputs will assume the states applied to the four external data inputs.
C. $Q_0$ = the level of Q before the indicated input conditions were established.
D. W0B1 = The first bit of word 0, etc.

(b) Function table

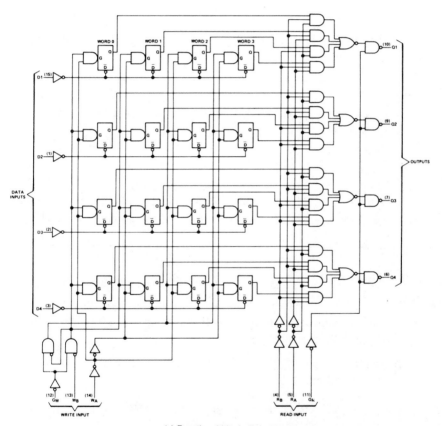

(c) Functional block diagram

FIGURE 15-5  The **74170** 16-bit memory. (From the *TTL Data Book for Design Engineers,* 2nd ed., Texas Instruments, Inc. Courtesy of Texas Instruments, copyright 1976.)

The operation of the **74170** can be determined from the function table (Fig. 15-5*b*) or the functional block diagram. To write, the $G_W$ input must be LOW. If $G_W$ is HIGH, the G (ENABLE) inputs to all FFs are LOW and the data is locked within them. With $G_W$ LOW, the $W_A$ and $W_B$ inputs select 1 word (column) and the levels on the D inputs are written into the 4 FFs that comprise the selected word.

The output of a word is read when READ ENABLE, $G_R$, is LOW. This enables the output gates shown on the functional block diagram. The $R_A$ and $R_B$ inputs select 1 of the 4 inputs to each of the 4 internal AOI gates. The data at the selected word appear at the output. If $G_R$ is HIGH, the output open collector NAND gates cut off and the outputs are all HIGH.

---

**EXAMPLE 15-4**

If the inputs to a 74170 are $D_1 = D_2 = D_3 = D_4 = G_W = W_B = W_A = G_R = 0$ and $R_B = R_A = 1$, what is happening to the memory?

**SOLUTION**

Since $G_R = G_W = 0$, both READ and WRITE are enabled. With $W_A$ and $W_B$ both 0, word 0 is being written with the data on the D lines (all 0s in this example). With $R_A$ and $R_B$ both 1, word 3 is being read and the contents of word 3 are available at the outputs.

---

## 15-6.2   The 7489

The **7489** is a 64-bit RAM whose pin layout, function table and functional block diagram are shown in Fig. 15-6. It is arranged as 16 words of 4 bits each. The inputs to the **7489** consist of:

1. Four data input bits.
2. Four data output bits.
3. Four select bits (to select 1 of the 16 words).
4. Two control lines, MEMORY ENABLE (ME) and WRITE ENABLE (WE).

As the function table shows, this allows the designer four options:

1. READ
2. WRITE
3. Invert data
4. Do nothing

---

FIGURE 15-6 (*opposite*)   The **7489** 64-bit memory. (From the *TTL Data Book for Design Engineers*, 2nd ed., Texas Instruments, Inc. Courtesy of Texas Instruments, copyright 1976.)

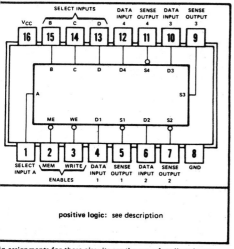

positive logic: see description

in assignments for these circuits are the same for all packages.

(a) J or N dual-in-line or W flat package (top view)†

| ME | WE | OPERATION | CONDITION OF OUTPUTS |
|----|----|-----------|----------------------|
| L | L | Write | Complement of Data Inputs |
| L | H | Read | Complement of Selected Word |
| H | L | Inhibit Storage | Complement of Data Inputs |
| H | H | Do Nothing | High |

(b) Function table

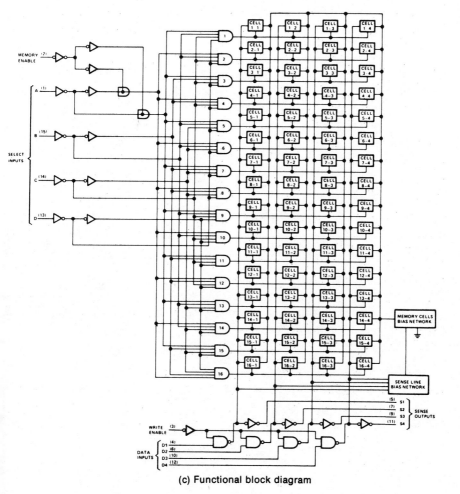

(c) Functional block diagram

The *do nothing* option is important because it *deselects* the IC. Since the 7489 has open collector outputs, another 7489 can be paralleled with it and selected. This increases the size of the memory by 16 words for each additional IC.

---

**EXAMPLE 15-5**

Construct a 64-word-by-4-bit memory using 7489s. Show how to address, READ, and WRITE the memory.

**SOLUTION**

A memory of this size requires four 7489s. The signals necessary to operate it are:
    (a) **Six SELECT lines** (to select 1 of the 64 words in memory).
    (b) **Four DATA INPUT lines,** which hold the 4 bits to be written to the selected word.
    (c) **Four DATA OUTPUT lines,** to read the 4 bits of the selected word.
    (d) **A READ/WRITE line,** which determines whether the READ or the WRITE operation is being performed.
    One way to construct the memory is shown in Fig. 15-7. The 4 DATA INPUT lines and 4 of the 6 SELECT lines drive all 7489s in parallel. The 4 output lines are paralleled by tying each output bit to a common pull-up resistor, which is needed because of the open collector output.
    Since the four 7489s are essentially in parallel, three of them must be *deselected*, and the remaining 7489 must receive the proper READ or WRITE command. This is accomplished by using a 74155 demultiplexer functioning as a dual 2-line-to-4-line decoder, which is controlled by the two MSBs of the select (address) inputs and the READ/WRITE line. The section 1 outputs of the 74155 are tied to the ME inputs, causing the ME input of the selected 7489 to be LOW and all other ME inputs to be HIGH. The section 2 outputs are tied to the WE lines and the READ/WRITE line controls the data input of section 2. If the READ/WRITE line is in WRITE mode (LOW), the selected 7489 receives LOW inputs on both the ME and WE lines, and writes the data into the memory in accordance with the function table. If the READ/WRITE input is HIGH, WE stays HIGH, and the selected IC is read. Note that all unselected 7489s have both ME and WE HIGH. They are in the *do-nothing* state. All their open collector output transistors are OFF, and they cannot be written into.

---

**EXAMPLE 15-6**

What inputs are needed to write word 29 of the 64-word memory of Ex. 15-5?

**SOLUTION**

Word 29 is word 13 of the 7489-1. (7489-0 holds words 0–15, 7489-1 holds words 16–31, etc.) The 6 SELECT lines must contain a binary 29, 011101. The 4 LSBs

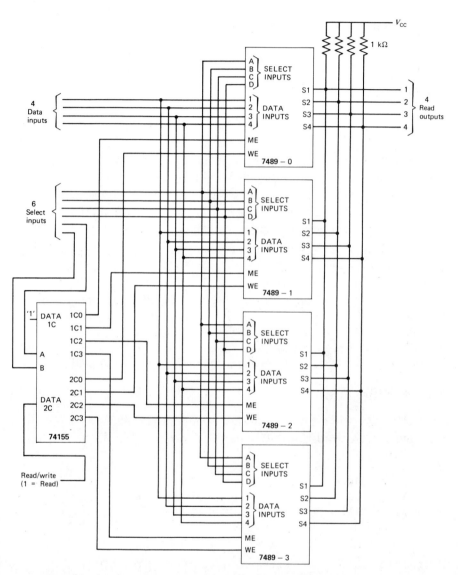

FIGURE 15-7 A 64 word by 4-bit memory using **7489**s.

(1101) are connected to all the **7489** SELECT inputs and select word 13 on each IC. The two MSBs (01) select outputs 1C1 and 2C1 of the **74155**. If the READ/WRITE line is LOW, outputs 1C1 and 2C1 are *both* LOW. They are tied to WE and ME of **7489**-1. This is the *only* IC receiving the WRITE command. It writes the data on the 4 input lines into its 13th word, which is the 29th word in the memory.

### 15-6.3   Larger Bipolar Memories

Larger bipolar memories such as the **74LS215**, which is a 1K-word-by-1-bit memory, are being manufactured. These are generally only used where the high speed of TTL logic is required. Most large memories currently use MOS technology.

## 15-7   MOS MEMORIES

Large static semiconductor memories are built of thousands of MOS (metal-oxide-semiconductor) FFs. Although MOS memories are considerably slower than bipolar (typical access time for a bipolar memory is 40 ns, while MOS memories have access times of about 300 ns), MOS memories have the following important advantages:

1.   MOS memories (especially CMOS) consume less power. A bipolar memory requires about three times as much power per bit as an MOS memory. CMOS memories require even less power.
2.   A higher packaging density can be achieved using MOS circuitry. The smaller area required inside the chip, for an MOS FF, results in more efficient use of "silicon real estate" and produces ICs containing more bits.
3.   MOS memories are generally less expensive. They are presently available at between 0.1 to 0.15 cs per bit. Bipolar memories cost about 2 cs per bit.

### 15-7.1   The 2102

The **2102** is a widely used and very popular 1024-bit static RAM. It is arranged as a 1K-word-by-1-bit memory, as shown in Fig. 15-8. The inputs to the **2102** are:

1.   10 address lines.
2.   DATA IN
3.   READ/WRITE
4.   DATA OUT
5.   CHIP ENABLE

The **2102** has 1 output, DATA OUT, which is 3-stated when the chip is not enabled ($\overline{CE} = 1$). Larger memories can be built using several **2102**s, as Ex. 15-7 shows.

---

**EXAMPLE 15-7**

Design a 4K-byte (4096-words-by-8-bit) memory using **2102**s.

**SOLUTION**

The solution proceeds as follows:

1.   The number of **2102**s required must be determined. Since the total memory

## PIN CONFIGURATION

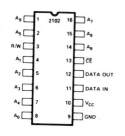

## LOGIC SYMBOL

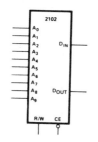

## PIN NAMES

| $D_{IN}$ | DATA INPUT | $\overline{CE}$ | CHIP ENABLE |
|---|---|---|---|
| $A_0 - A_9$ | ADDRESS INPUTS | $D_{OUT}$ | DATA OUTPUT |
| R/W | READ/WRITE INPUT | $V_{CC}$ | POWER (+5V) |

FIGURE 15-8 The **2102** 1024-bit static RAM memory. (From the *Intel Semiconductor Memory Book*. Copyright John Wiley & Sons, Inc. 1978. Reprinted by permission of John Wiley & Sons, Inc.)

requires 4K words by 8 bits/word = 32K bits and each **2102** has 1K bit, thirty-two **2102**s are needed.

2. For a memory of this size, the computer or system driving the memory must provide 12 address bits, 8 data input bits, and a read/write line. It must be capable of accepting 8 data outputs from the **2102**s.

3. The thirty-two **2102**s are arranged in an 8-by-4 matrix, as shown in Fig. 15-9, to provide 4K words of 8 bits.

4. The 10 LSBs of the address are connected to the 10 address inputs of each **2102**. The two remaining address bits become the inputs to a decoder (**74155** in Fig. 15-9) that selects which row of **2102**s will be enabled. Note that the decoder provides a single LOW output and $\overline{CE}$ must be LOW for the IC to be enabled, so the decoders can be directly connected to the $\overline{CE}$ inputs.

5. The 8 data inputs are each connected to one column of **2102**s.

6. The 4 data outputs in each column are all connected together. Pull up resistors are not needed since only one of the four **2102**s is enabled at any time and the rest are disabled, presenting a high impedance to the bus.

7. The read/write line is connected to all **2102**s.

8. To read the memory the two high-order address bits select a row of **2102**s and the 10 low-order address bits select one of the 1024 words within the **2102**s. After the addresses are firm for the specified access time ($\approx$300ns), the 8 data output bits appear on the output lines.

9. Writing is similar. When a WRITE pulse is applied to the R/W line the 8 data inputs are written into the **2102**s. Note that *writing destroys the previous contents*

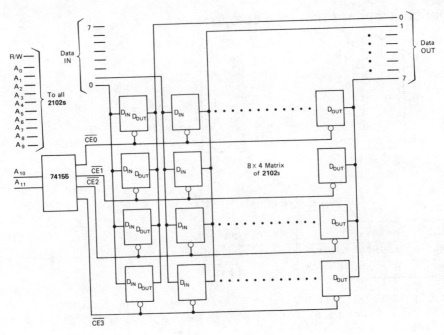

FIGURE 15-9 A 4K-by-8-bit memory using **2102**s.

*of the memory* at the selected address, so the R/W line should always be kept in READ mode, except when actually trying to write.

10. The 10 address lines and the R/W line drive all 32 ICs in Fig. 15-9. Fortunately, these are MOS ICs with high input impedances, so there is no loading or fanout problem.

## 15-7.2 The 2114

The **2114** is a 1K-by-4-bit static RAM that is being used increasingly in place of the **2102** because it contains four times as many bits. The **2114** is shown in Fig. 15-10. It comes in an 18-pin package and requires:

- □ 10 address inputs
- □ 4 data lines
- □ 1 chip select ($\overline{CS}$)
- □ 1 read/write input ($\overline{WE}$)

The four data lines are *bidirectional*; the data go from or go to the **2114** depending on whether the memory is being read ($\overline{WE} = 1$) or written ($\overline{WE} = 0$).

In a typical application the **2114** is connected to a microprocessor that controls the addresses and read/write line, as shown in Fig. 15-11. If $\overline{CS} = 1$, the **2114** is deselected; it presents a high impedance to the bus and

## BLOCK DIAGRAM

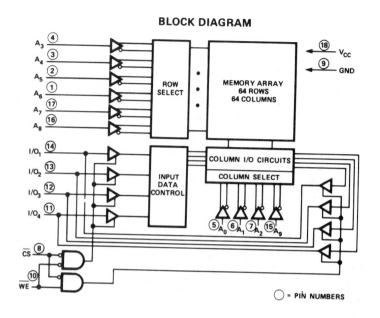

## PIN CONFIGURATION

## LOGIC SYMBOL

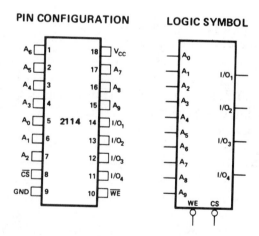

## PIN NAMES

| | | | |
|---|---|---|---|
| $A_0 - A_9$ | ADDRESS INPUTS | $V_{CC}$ POWER (+5V) | |
| $\overline{WE}$ | WRITE ENABLE | GND GROUND | |
| $\overline{CS}$ | CHIP SELECT | | |
| $I/O_1 - I/O_4$ | DATA INPUT/OUTPUT | | |

FIGURE 15-10   The **2114** static RAM. (From the *Intel Semiconductor Memory Book.* Copyright John Wiley & Sons, Inc. 1978. Reprinted by permission of John Wiley & Sons, Inc.)

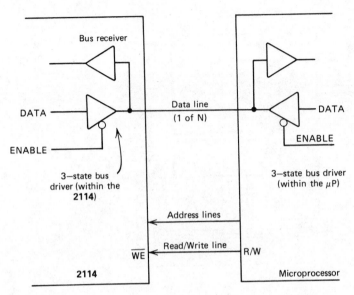

FIGURE 15-11 A microprocessor connected to a **2114**.

cannot be written into. In a read operation ($\overline{\text{CS}} = 0$, $\overline{\text{WE}} = 1$), the 3-state bus drivers within the **2114** turn on and the data at the selected address are sent from the **2114** to the $\mu$P. During a write operation the $\mu$P first selects the **2114** (causing $\overline{\text{CS}}$ to be LOW). It then sends a WRITE pulse, which causes $\overline{\text{WE}}$ on the **2114** to go LOW. This disables the **2114** bus drivers. At this time the bus drivers within the $\mu$P are enabled, and the data to be written flow from the $\mu$P to the **2114**..

---

**EXAMPLE** 15-8

Describe how a 4K-by-8-bit memory can be made of **2114**s.

**SOLUTION**

The solution is similar to the **2102** memory shown in Fig. 15-9, except that:

1. Only eight **2114**s arranged in a 4-row-by-2-column matrix are required.
2. The DATA-IN and DATA-OUT lines are merged to become a single bi-directional bus.

## 15.8 DYNAMIC RAMs

Dynamic RAMs also use MOS gates, but they store their information as charges on internal capacitors, rather than in FFs. This results in a greater saving of silicon real estate and larger ICs (more bits per IC) can be built. Consequently, dynamic RAMs are inexpensive. At present, they can be obtained for as low as .025 cents per bit. Unfortunately, data stored in the form of a charge on a capacitor leaks off as time progresses. Consequently, dynamic RAMs must be *refreshed* periodically to restore the information. This requirement complicates the external circuitry of a dynamic RAM (Sec. 15-8.3).

The **4116** is a popular and typical dynamic RAM. It is a 16K-by-1-bit memory so it requires 14 address lines, DATA-IN, DATA-OUT, and a read/write ($\overline{\text{WRITE}}$) input.

A simplified drawing of the operation of the **4116** is shown in Fig. 15-12. The 16K-memory bits are stored on a square matrix of capacitors (128 × 128). The 7 low-order address bits select an entire row. The contents of this row are driven into the 128 sense/refresh amplifiers. The 128 high-order address bits select one of the sense amplifiers and place that bit on the DATA-OUT line. At the same time the information in all 128 bits is refreshed (the capacitors are recharged) and rewritten to the proper row in memory. *Thus every read or write cycle refreshes all 128 bits in the selected row.*

## 15-8.1 $\overline{\text{RAS}}$ and $\overline{\text{CAS}}$

The pin configuration for the **4116** is shown in Fig. 15-13. Notice that there are only 7 address pins on the chip. The 14 system addresses are *multiplexed* onto the chip by using the $\overline{\text{RAS}}$ (Row Address Select) and $\overline{\text{CAS}}$ (Column Address Select) signals. This saves pins and is the way most modern dynamic RAMs operate.

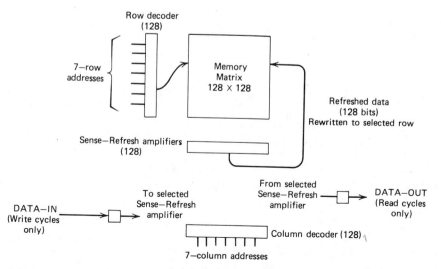

FIGURE 15-12  A simplified diagram of the **4116**.

PIN ASSIGNMENT

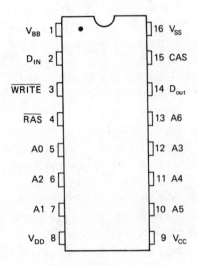

PIN NAMES

A0–A6  Address Inputs
$\overline{\text{CAS}}$  Column Address Strobe
$D_{in}$  Data in
$D_{out}$  Data out
$\overline{\text{RAS}}$  Row Address Strobe
WRITE  Read/Write Input
$V_{SS}$  Power (−5 V)
$V_{CC}$  Power (+5 V)
$V_{DO}$  Power (+ 12 V)
$V_{SS}$  Ground

FIGURE 15-13  Pin assignments for the **4116**. (Courtesy of Motorola Integrated Circuits Division.)

The timing for Read, Write, and Refresh cycles is shown in Fig. 15-14. For Read or Write cycles the **4116** output is 3-stated (high impedance) when $\overline{\text{RAS}}$ and $\overline{\text{CAS}}$ are both HIGH. To start a cycle, the row addresses (typically the 7 LSBs of the 14 address bits) are put on the **4116**'s address inputs and $\overline{\text{RAS}}$ goes LOW. The **4116** then latches in the row address. After about 60 $\mu$s (see the manufacturer's specifications for precise times), the row addresses

FIGURE 15-14 (*opposite*)  Read, write, and refresh cycle timing for the **4116**. (From the *Intel Semiconductor Memory Book*. Copyright John Wiley & Sons, Inc. 1978. Reprinted by permission of John Wiley & Sons, Inc.)

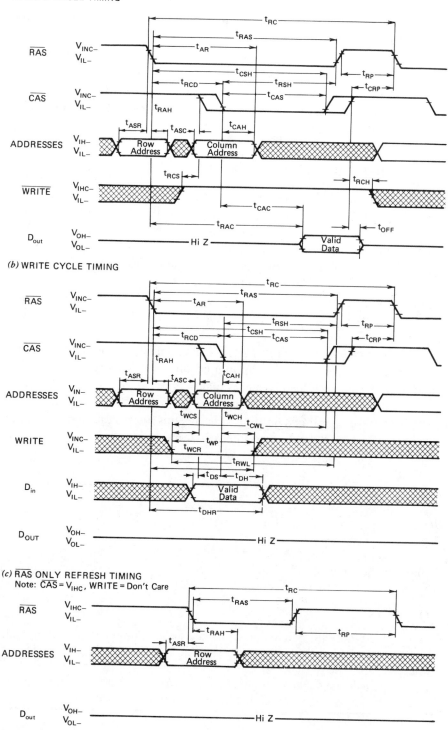

(a) READ CYCLE TIMING

(b) WRITE CYCLE TIMING

(c) $\overline{RAS}$ ONLY REFRESH TIMING
Note: $\overline{CAS}$ = $V_{IHC}$, WRITE = Don't Care

are replaced by the 7 MSBs of the address inputs and $\overline{CAS}$ goes LOW. Now the **4116** has the 14 address bits it requires.

For a Read cycle, valid data appear at DATA-OUT after the access time, measured from the start of $\overline{CAS}$ ($t_{CAC}$ on Fig. 15-14) and remains valid until $\overline{RAS}$ and $\overline{CAS}$ go HIGH. For a Write cycle, DATA-IN must be valid when $\overline{CAS}$ goes LOW and $\overline{WRITE}$ should go LOW slightly before $\overline{CAS}$ goes LOW.

## 15-8.2 Refresh Cycles

Because dynamic RAMs store their data on capacitors, the charge on these capacitors must be refreshed periodically. For the **4116** and most other dynamic RAMs, each location must be refreshed once every 2 ms. Fortunately, refreshing can be accomplished by reading a row of data into the Sense-Refresh Amplifiers and then rewriting them to memory. Thus 128 locations are refreshed during each refresh cycle.

The timing for a refresh cycle is shown in Fig. 15-14c. Note that only $\overline{RAS}$ is required for a refresh cycle. It causes the data to be transferred to and from the Sense-Refresh Amplifiers.

To refresh all 128 rows during each 2-ms interval, a 7-bit *refresh counter* is required. The outputs of this counter become the memory address during each refresh cycle and the counter is incremented at the end of each cycle.

There are two modes of refresh: *burst* and *periodic*. In burst mode all 128 rows are refreshed at the beginning of each 2-ms interval; in periodic mode refresh cycles occur at approximately equal intervals.

---

**EXAMPLE 15-9**

If each refresh cycle takes 500 ns, how is the time divided between refresh and Read/Write memory cycles in:
  (a) burst mode
  (b) periodic mode

**SOLUTION**

  (a) In burst mode the 128 refresh cycles take 64 $\mu$s. Thus in each 2-ms interval, the first 64 $\mu$s are used for refresh, leaving the remaining 1936 $\mu$s available for Read/Write cycles.
  (b) In periodic mode there is one refresh cycle every 15.6 $\mu$s. Thus the memory periodically takes .5 $\mu$s out of each 15.6-$\mu$s interval for a refresh cycle.

---

## 15-8.3 A Refresh Controller

A Refresh Controller for a RAM like the **4116** is shown in Fig. 15-15. This controller multiplexes $\overline{RAS}$, $\overline{CAS}$, and the address counter. It also assumes that Refresh requests come periodically, and that Data requests occur as required by the computer system controlling the memory. Cycles in progress cannot be interrupted so Data requests must wait if they occur when a Refresh cycle is in progress and vice versa.

The circuit of Fig. 15-15 operates as follows.

1. Both the Data cycle and Refresh cycle one-shots are set for the cycle time of the memory. When they are up, $\overline{RAS}$ is set LOW. While either one of them is up, the other cannot be SET.

2. When either of them SETS, it triggers the $\overline{CAS}$ delay one-shot and removes the CLEAR from the $\overline{CAS}$ FF.

3. When the $\overline{CAS}$ DELAY one-shot expires, it SETS the $\overline{CAS}$ FF. This sends $\overline{CAS}$ to the memory and also changes the SELECT input to the RAS/CAS multiplexer so that the column addresses are sent to the memory.

4. When the Data or Refresh cycle one-shot expires, it causes $\overline{RAS}$ to go HIGH. It also resets the $\overline{CAS}$ FF, and $\overline{CAS}$ goes HIGH. The memory is now waiting for the next cycle.

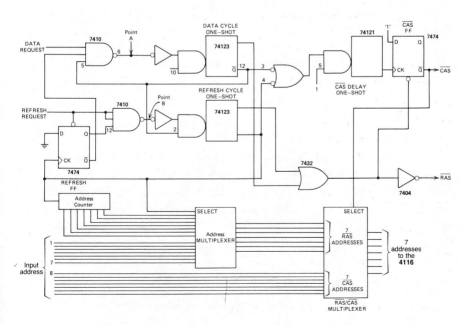

FIGURE 15-15   A refresh circuit for a dynamic RAM.

5.   A refresh oscillator (not shown) provides REFRESH REQUEST, a 300-ns negative pulse that occurs at the refresh rate required by the memory (once every 15.6 $\mu$s for a **4116**). The oscillator can be built from a single **74123** (see Sec. 7-6.2).

6.   When $\overline{\text{REFRESH REQUEST}}$ goes LOW, it sets the REFRESH FF, which prevents any further DATA REQUESTS from triggering the DATA CYCLE one-shot.

7.   After 300 ns, $\overline{\text{REFRESH REQUEST}}$ returns HIGH and triggers the RE-FRESH CYCLE one-shot, provided the DATA CYCLE one-shot is not set.

8.   During refresh cycles, the REFRESH CYCLE one-shot, which controls the ADDRESS MULTIPLEXER, places the contents of the ADDRESS COUNTER on the LSBs of the memory address lines. The other addresses, bits $A_8$–$A_{14}$, are irrelevant during a refresh cycle. During normal memory cycles, the REFRESH CYCLE one-shot is CLEAR and the entire input address is sent to the **4116**s via the $\overline{\text{RAS}}$/$\overline{\text{CAS}}$ multiplexer.

9.   Data requests cannot trigger the DATA CYCLE one-shot from the start of REFRESH REQUEST until the end of the REFRESH CYCLE because the REFRESH FF is set and cuts off the **7410**.

10.   The REFRESH CYCLE ends when the REFRESH CYCLE one-shot expires. This clears the REFRESH FF and increments the address counter, so the next REFRESH CYCLE occurs at the next row.

11.   A system memory request is initiated when DATA REQUEST goes HIGH. While the DATA CYCLE one-shot is HIGH, it prevents the REFRESH CYCLE one-shot from being triggered, and also cuts off the input gate to itself. (This assures us that $\overline{\text{RAS}}$ is LOW for the required time.)

## 15-8.4   IC Dynamic RAM Controllers

Manufacturers now provide dynamic RAM controllers in a DIP package. One such controller is the **i3242**, manufactured by Intel, Inc., and shown in Fig. 15-16.

The **3242** has 14 address inputs and 7 address outputs. Figure 15-16 shows that its outputs are a series of AND-OR-INVERT (AOI) gates. ROW ENABLE determines whether the low-order or high-order 7 bits are fed through to the output, thus implementing $\overline{\text{RAS}}$ and $\overline{\text{CAS}}$. REFRESH EN-ABLE disconnects the 14 system addresses and causes the internal counter to be placed on the outputs for a Refresh cycle. $\overline{\text{COUNT}}$ increments the counter.

FIGURE 15-16 (*opposite*)   The Intel® **3242**, a Schottky bipolar address multiplexer and refresh counter to complement the **2104**A and **2116** dynamic RAMs. (From the *Intel Semiconductor Memory Book*. Copyright John Wiley & Sons, Inc. 1978. Reprinted by permission of John Wiley & Sons, Inc.)

# LOGIC DIAGRAM

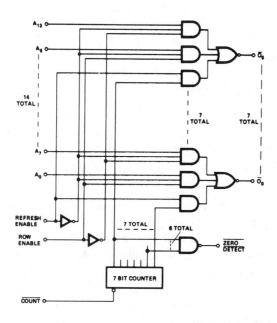

## PIN CONFIGURATION

| | | | | |
|---|---|---|---|---|
| COUNT | 1 | | 28 | V_CC |
| REFRESH ENABLE | 2 | | 27 | A_6 |
| ROW ENABLE | 3 | | 26 | A_13 |
| N.C. | 4 | | 25 | A_5 |
| A_1 | 5 | | 24 | A_12 |
| A_5 | 6 | | 23 | A_4 |
| A_2 | 7 | 3242 | 22 | A_11 |
| A_3 | 8 | | 21 | A_3 |
| A_0 | 9 | | 20 | A_10 |
| A_7 | 10 | | 19 | $\overline{O}_6$ |
| $\overline{O}_0$ | 11 | | 18 | $\overline{O}_3$ |
| $\overline{O}_2$ | 12 | | 17 | $\overline{O}_4$ |
| $\overline{O}_1$ | 13 | | 16 | $\overline{O}_5$ |
| GND | 14 | | 15 | ZERO DETECT |

NOTE: $A_0$ THROUGH $A_6$ ARE ROW ADDRESSES.
$A_7$ THROUGH $A_{13}$ ARE COLUMN ADDRESSES.

## TRUTH TABLE AND DEFINITIONS

| REFRESH ENABLE | ROW ENABLE | OUTPUT |
|---|---|---|
| H | X | REFRESH ADDRESS (FROM INTERNAL COUNTER) |
| L | H | ROW ADDRESS ($A_0$ THROUGH $A_6$) |
| L | L | COLUMN ADDRESS ($A_7$ THROUGH $A_{13}$) |

COUNT – ADVANCES INTERNAL REFRESH COUNTER.
ZERO DETECT – INDICATES ZERO IN THE FIRST 6
SIGNIFICANT REFRESH COUNTER
BITS (USED IN BURST REFRESH MODE)

The **3242** contains the equivalent of the $\overline{RAS}/\overline{CAS}$ multiplexer, the address multiplexer, and address counter of Fig. 15-16. The problems of keeping Data and Refresh requests separate and proper timing must still be handled by external circuits. (See Problem 15-12.)

## 15-9 READ ONLY MEMORIES

*Read only memories* (ROMs) are *prewritten* in some *permanent* or *semipermanent form. They are not written during the course of normal device operation.* ROMs are used to perform code conversions, table look-up, and to control special purpose programs for computers. They usually run the programs in microprocessors.

The concept of an ROM is extremely simple; the user supplies an address and the ROM provides the data output of the word prewritten at that address. As with RAMs, ROMs are organized on an $n$-word-by-$m$-bit basis. Supplying the proper address results in an $m$-bit output. Access time for a ROM is the time between the address input and the appearance of the resulting data word.

### 15-9.1 Diode Matrices

A small, simple ROM can be made from a multiplexer and a diode matrix. The placement of the diodes determines the output of the ROM, which can easily be changed by moving a diode.

The first step in building any ROM is to construct a truth table relating the inputs (addresses) to the required outputs. For example, consider the table of Fig. 15-17, which has a 4-bit input, a 4-bit output and can be used to convert an angle between 0 and 90 degrees to its cosine. The angle is represented by a 4-bit binary input. The output represents the cosine of the angle. With 4 bits, 16 outputs are allowed. An output of 0000 means the cosine of the angle is between 0.0000 and 0.0625; an output of 0001 is for cosines between 0.0625 and 0.1250, and so on.

The truth table of Fig. 15-17 can be implemented as an ROM by using a **74154** decoder and a diode matrix, as shown in Fig. 15-18. The inputs are applied to the SELECT lines of the **74154**. The corresponding line goes LOW and causes outputs that are connected to the selected line by diodes to go LOW. The diodes used in this matrix should have a small forward voltage drop so the sum of the diode drop and $V_{OL}$ of the decoder does not exceed 0.8 V.

---

**EXAMPLE 15-10**

What output does the ROM of Fig. 15-18 produce if the input is 48 degrees?

| Angle in Degrees | Input in Binaty | | | | Output (cos $\theta$) | | | | |
|---|---|---|---|---|---|---|---|---|
| 0 | 0 | 0 | 0 | 0 | 1 | 1 | 1 | 1 | 1 |
| 6 | 0 | 0 | 0 | 1 | 1 | 1 | 1 | 1 | 1 |
| 12 | 0 | 0 | 1 | 0 | 1 | 1 | 1 | 1 | 1 |
| 18 | 0 | 0 | 1 | 1 | 1 | 1 | 1 | 1 | 1 |
| 24 | 0 | 1 | 0 | 0 | 1 | 1 | 1 | 0 | |
| 30 | 0 | 1 | 0 | 1 | 1 | 1 | 0 | 1 | |
| 36 | 0 | 1 | 1 | 0 | 1 | 1 | 0 | 0 | |
| 42 | 0 | 1 | 1 | 1 | 1 | 0 | 1 | 1 | |
| 48 | 1 | 0 | 0 | 0 | 1 | 0 | 1 | 0 | |
| 54 | 1 | 0 | 0 | 1 | 1 | 0 | 0 | 1 | |
| 60 | 1 | 0 | 1 | 0 | 1 | 0 | 0 | 0 | |
| 66 | 1 | 0 | 1 | 1 | 0 | 1 | 1 | 0 | |
| 72 | 1 | 1 | 0 | 0 | 0 | 1 | 0 | 0 | |
| 78 | 1 | 1 | 0 | 1 | 0 | 0 | 1 | 1 | |
| 84 | 1 | 1 | 1 | 0 | 0 | 0 | 0 | 1 | |
| 90 | 1 | 1 | 1 | 1 | 0 | 0 | 0 | 0 | |

FIGURE 15-17   A cosine conversion table.

## SOLUTION

An input of 48 degrees results in a 1000 input to the SELECT lines of the **74154** and line 8 of the decoder goes LOW. Since output lines 0 and 2 are connected to line 8 via diodes, the output is 1010. This corresponds to the function table of Fig. 15-17 for the ROM, and indicates that the cosine of 48 degrees is between 0.625 and 0.6875.

## 15-9.2   NAND Gate ROMs

Small ROMs can be built using a decoder and a NAND gate for each output bit. It may be more convenient to mount NAND gates in DIP packages, than to mount diodes, especially if IC panels are used. A NAND gate ROM is built by creating a LOW input to each output NAND gate whenever a 1 is required on the output. NAND gates with as many inputs as possible should be used.

## EXAMPLE 15-11

(a) Design the cosine generator ROM using NAND gates.
(b) If the angle is 30 degrees, how are the outputs produced?

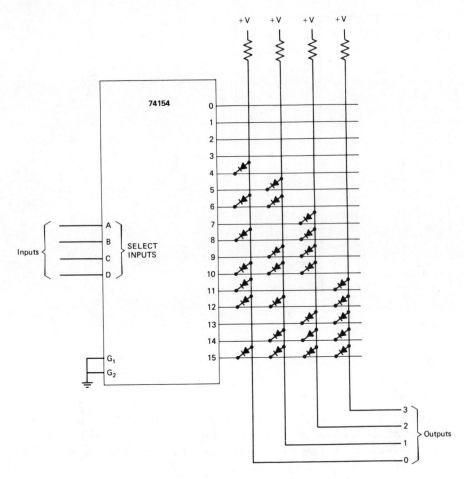

FIGURE 15-18 Diode matrix implementation of a cosine ROM.

## SOLUTION

(a) The solution is shown in Fig. 15-19. Again a **74154** 4-line-to-16-line decoder is used. The **74133** NAND gate was chosen because it has 13 inputs. When the decoder input is applied, the corresponding line goes LOW and causes the NAND gate outputs connected to it to go HIGH.

(b) A 30-degree angle produces an input of 0101 (see Fig. 15-17). This causes line 5, which is connected to gates 3, 2, and 0 to go LOW. The outputs of gates 3, 2, and 0 are therefore HIGH and the result is 1101, which indicates the cosine of 30 degrees is between 0.8125 and 0.8750.

Note that the outputs of this ROM can be changed simply by rerouting the wires between the decoder and the NAND gates.

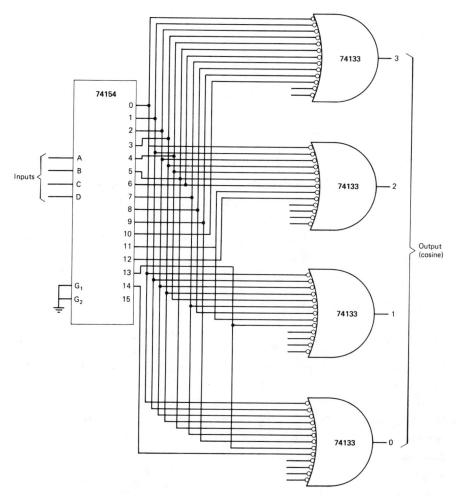

FIGURE 15-19 A NAND gate ROM. *Note:* All unused **74133** inputs connected to 1.

## 15-10 IC ROMs

The diode matrix and NAND gate ROMs discussed in Sec. 15-9 are easily changed, but they are only used for very small memories. Most ROMs used today are DIP ICs that are readily available from many manufacturers.

Figure 15-20 shows the various types of ROMs available. They divide into MOS and bipolar and subdivide into Mask ROMs and various types of PROMs. Bipolar ROMs are faster, but MOS ROMs contain more bits per IC and are less expensive. Consequently, most applications use MOS ROMs exclusively.

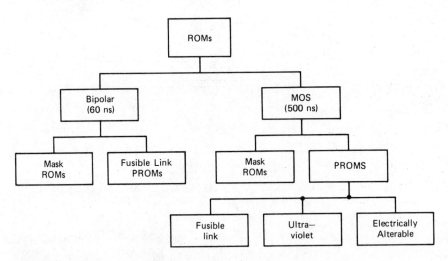

FIGURE 15-20 Types of ROMs.

### 15-10.1 Mask ROMs

Mask programmed ROMs are built by the manufacturer. The ROM is made by including or excluding a small semiconductor jumper to produce a 1 or 0 for each bit. The user must supply the manufacturer of the ROM with the code he or she wishes each word to contain. Because of the custom programming involved, a mask is made that costs about $1000. The user should be sure of the inputs before he or she incurs the masking charge because the ROM is usually useless if a single bit is wrong. Once the mask is made, identical ROMs can be produced inexpensively.

Some ROM patterns are very commonly used and are available at low cost. The Basic interpreters, which control microprocessor systems such as the Apple or TRS-80 computers use mask programmed ROMs. Character generators for CRT displays, such as the **2513**, are another example of commonly used and readily available mask programmed ROMs (see Sec. 20-4).

### 15-10.2 Fusible Link PROMs

A PROM (Programmable Read Only Memory) is a ROM whose program is written by the user. This saves the user the mask cost and the turn-around time—the time between sending a mask program to the manufacturer and the delivery of the ROM. The user must enter the program into the PROM at the facility before it can be used.

The fusible link PROM comes with a small nichrome or silicon fuse in each bit position. All outputs are initially 0. The user can then drive an excessive current through the fuses that blows the fuses open and changes the output to a 1. Specifications on the programming pulse (the required voltages, currents, and pulse durations) are supplied by the manufacturer. Many users will purchase a device called a PROM *programmer*, or PROM blaster. This facilitates blowing the fuses, and means the user does not have to design the necessary circuitry for blowing the fuses.

## 15-10.3   EPROMs

One problem with both mask and fusible line ROMs is that a single bad bit can make the memory useless. EPROMs (Erasable Programmable Read Only Memories) that can be both programmed and erased at the user's facility are a solution to this problem.

EPROMs contain a floating gate, electrically insulated from the rest of the circuit, as part of each bit. The presence or absence of charge on the gate determines whether the bit is a 1 or 0. The charge is deposited during the programming cycle.

EPROMs can be erased by exposing them to ultraviolet light. The ICs are equipped with a transparent lid to allow the ultraviolet light to enter. In normal operation this lid is often closed off by a piece of tape to prevent ambient ultraviolet light from entering and causing erasures.

One commonly used EPROM is the **2708**, whose pin configuration is shown in Fig. 15-21. The memory is 1K words by 8 bits. The **2708** contains:

- □   10 address pins
- □   8 data pins
- □   4 voltage pins ($+12$, $+5$, $-5$, and GND)
- □   1 $\overline{CS}$/WE pin
- □   1 program pin

The **2708** operates in three modes: Deselected, Read, or Program. If $\overline{CS}$/WE is a logic 1, the **2708** is deselected, and all its outputs are high impedance. If $\overline{CS}$/WE is a logic 0, the IC is in normal read mode, and the contents of the address appear on the data outputs.

To program the **2708**, the data for a location must be placed on the data (O) lines and $\overline{CS}$/WE raised to $+12$ V. Then a 26-V pulse must be applied to the PROGRAM input. Actually, all inputs must be programmed sequentially. We recommend the reader consult the manufacturers literature for details on programming.

Programmers for the **2708** and similar EPROMs are available. They also contain a source of ultraviolet light for erasure.

## 15-10.4   EAROMs

EAROMs (Electrically Alterable Read Only Memories) are the latest advance in ROMs. They can be erased and rewritten while still in the circuit.

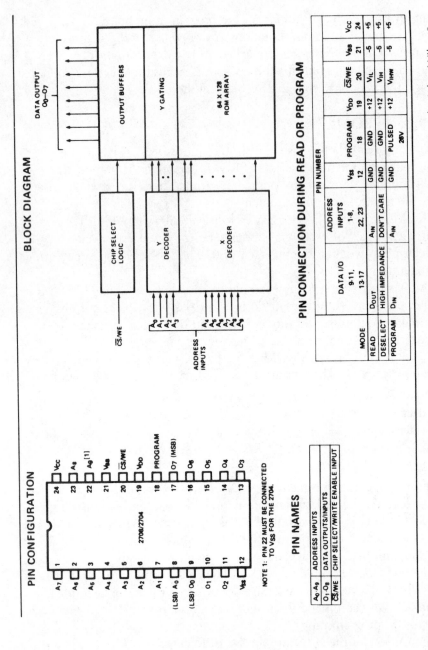

FIGURE 15-21 The **2708** EPROM. (From the *Intel Semiconductor Memory Book*. Copyright John Wiley & Sons, Inc. 1978. Reprinted by permission of John Wiley & Sons, Inc.)

It takes about 10 ms to erase data and it can be rewritten at a rate of 1 ms per byte. This is more convenient and faster than any other type of EPROM.

One such EAROM is the **ER3400,** manufactured by General Instruments. It is controlled by two mode input lines that allow four modes of operation: Block erase, Word erase, Read, and Write.

This EAROM has one drawback: it can only be read $2 \times 10^{11}$ times. This is not sufficient for holding a computer program, but it is useful for storing tables that are rarely changed and not read continuously. The reader is advised to consult the manufacturer's specifications for a detailed description of any EAROM.

## SUMMARY

This chapter started with an introduction to memory concepts, such as reading, writing, the MAR, and the MDR. Core memories, still used in many main-frame computers, were then introduced.

The rest of the chapter concentrated on semiconductor memories. Static RAMs, both bipolar and MOS, were introduced first. This was followed by a section on dynamic RAMs, including the use of the dynamic RAM controller. Finally, the various types of ROMs and PROMs that are currently used were presented.

## GLOSSARY

**Semiconductor memory.** A memory whose basic element is a semiconductor FF or gate.

**RAM.** Random access memory.

**ROM.** Read only memory.

**Scratch-pad memory.** A small memory generally less than 256 bits.

**Bipolar.** Circuits that are essentially TTL.

**Static memory.** A memory where the data are stored in FFs.

**Dynamic memory.** A memory where the data are stored on capacitors, and must be refreshed periodically.

**Access time.** The time required to read valid data after the addresses and chip enable are firm.

**Cycle time.** The time required to write data into a static RAM (or to read and rewrite a dynamic RAM).

**Refresh.** The act of restoring the data on the capacitors within the memory.

**Masking.** The process of building a custom ROM.

**PROM.** Programmable read only memory, a ROM that can be programmed by the user.

**EPROM.** Erasable Programmable ROM.

**EAROM.** Electrically Alterable ROM.

## REFERENCES

Steve Ciarcia, "Add Nonvolatile Memory to Your Computer," *Byte Magazine*, Dec. 1979.

Robert Glaser, "Program Those 2708s," *Byte Magazine*, April 1980.

Joseph D. Greenfield, *Practical Digital Design Using ICs*, Wiley, New York, 1977.

Gerald Luecke, Jack P. Mize, and William N. Carr, *Semiconductor Memory Design and Application*, McGraw-Hill, New York, 1973.

*Intel—The Semiconductor Memory Book*, Wiley Interscience, New York, 1978.

*The Semiconductor Memory Data Book*, Texas Instruments, 1975.

## PROBLEMS

15-1.   A memory module is 16K by 32 bits. Determine:
(a) How many bits are in the MAR?
(b) How many bits are in the MDR?
(c) How many planes are in the memory?
(d) How many cores are in the memory?
(e) How many SENSE AMPLIFIERS and SENSE WINDINGS are required?
(f) How many cores are threaded by each SENSE WINDING?
(g) How many INHIBIT DRIVERS are required?
(h) How many cores are threaded by each INHIBIT winding?
(i) How many X and Y drive lines are there?

15-2.   Construct the equivalent of a 7489 using 74170s.

15-3.   Design a 32-word-by-8-bit memory using 7489s.

15-4.   In Fig. 15-7 what is happening if the SELECT lines are 101011 and the READ/WRITE line is LOW?

15-5.   Place 1s and 0s on Fig. 15-7 if the SELECT lines are 110101, the READ/WRITE line is HIGH, the data input is 1000, and the data being read are 1011.

15-6.   An 8K-by-12-bit memory is made up of 2102s.
(a) How many data inputs are required?
(b) How many data outputs are required?
(c) How many address bits are required?
(d) How many read/write lines are required?
(e) How many different chip select lines are required?
(f) Draw that portion of the circuit that selects particular chip select lines.

15-7.   Repeat Problem 15-6 using 2114s.

15-8.   A 6800 $\mu$P requires 2K of bytes of memory.
(a) What input and output lines are required to the memory?
(b) Design it using 2102 RAMs, which are 1K by 1. The 2102 has one $\overline{CS}$ line and is selected when $\overline{CS}$ is low.
(c) Design it using 6810 RAMs, which are 128 words by 8 bits. Assume the 6810 also has one $\overline{CS}$ line.

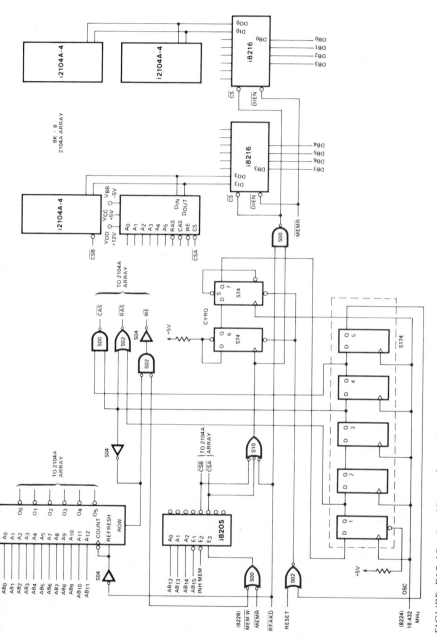

FIGURE P15-12  An 8K-word × 8-bit synchrononous refresh memory system using the Intel® 2104A dynamic RAM. (From the *Intel Semiconductor Memory Book*. Copyright John Wiley & Sons, Inc. 1978. Reprinted by permission of John Wiley & Sons, Inc.)

15-9. Assume a dynamic RAM controller exists for a 4K dynamic RAM with multiplexed addresses.

(a) How many input addresses are required? Where do they come from?

(b) How many output addresses are on the chip? Where do they go?

(c) What are the address outputs during $\overline{RAS}$?

(d) What are the address outputs during $\overline{CAS}$?

(e) What are the address outputs during a refresh cycle?

15-10. Design the REFRESH REQUEST oscillator for the circuit of Fig. 15-15.

15-11. Using the refresh controller circuit of Fig. 15-15, draw a timing chart to show what happens if a REFRESH REQUEST occurs at $t = 0$ and a DATA REQUEST occurs at $t = 0.1$ $\mu$s. Assume the DATA and REFRESH CYCLE one-shots are set for 400 ns, and the CAS delay one-shot is set for 200 ns.

15-12. Figure P15-12 shows an array of **2104** memories. Each **2104** is a 4K-by-1-bit dynamic RAM.

(a) How many **2104**s are in the 8K-by-8-bit array?

(b) The **8205** is a decoder. What memory addresses cause it to select $\overline{CSA}$ and $\overline{CSB}$?

(c) What is the function of the **8216**s?

(d) What is the function of FFs 6 and 7? (*Hint:* See Sec. 6-16.)

(e) The **3242** has no connections to inputs A6 or A13. Why?

(f) Draw a timing chart showing the response of the circuit to a $\overline{MEM\ W}$ request. Show $\overline{RAS}$, $\overline{CAS}$, $\overline{WE}$, and the outputs of each of the 7 FFs.

15-13. Why are diodes necessary in Fig. 15-18? (Why can't the connections be made by wire?)

15-14. The 4-bit binary and Gray-code equivalents are given in Fig. 13-18. Design a binary-to-Gray-code converter using an ROM.

(a) Use a diode matrix.

(b) Use NAND gates.

15-15. Explain how you would design a diode matrix ROM to drive a common cathode 7-segment display for a HEX output. Specifically state what hardware you would use. Draw the diode configuration for the line where the inputs are 1 1 0 0.

After attempting the above problems, return to Sec. 15-2 and review the self-evaluation questions. If any of them are still unclear, reread the appropriate sections of the text to find the answers.

# 16

# THE BASIC COMPUTER

## 16-1  INSTRUCTIONAL OBJECTIVES

The object of this chapter is to show the reader how to build a rudimentary minicomputer using ICs. Two computer designs are presented: a 16-bit computer to best develop the ideas and concepts behind the design of a computer, and a very inexpensive 8-bit computer. The 8-bit design was built by students at RIT and is used to illustrate the practical considerations involved in designing and building computers.

After reading this chapter the student should be able to:

1. List each basic part of the computer and describe its function.
2. Write a simple computer program.
3. Draw a flowchart.
4. Explain how an arithmetic or logic instruction is executed.
5. Explain how a BRANCH or JUMP instruction is executed.
6. Design a rudimentary computer.
7. Modify the computers presented here to add or delete specific instructions.

## 16-2  SELF-EVALUATION QUESTIONS

Watch for the answers to the following questions as you read the chapter. They should help you understand the material presented.

1. What is an Op code?
2. Is there any distinction between data and instructions when they are in memory? How does the computer tell the difference?
3. What is a decision box? How many entries and exits does it have? What instructions does it correspond to?
4. What is a self-modifying program? What are its drawbacks?
5. What is the function of the FETCH/EXECUTE FF?
6. Why is the MAR loaded during both FETCH and EXECUTE cycles?
7. Why does the small computer of Sec. 16-9 execute conditional skips rather than conditional jumps?

## 16-3 INTRODUCTION TO THE COMPUTER

A computer consists of four basic parts, as shown in Fig. 16-1.

1. The memory
2. The arithmetic-logic unit (ALU)
3. The input-output system
4. The control unit

Fortunately, the ALU and memory have been covered in Chapters 14 and 15, respectively, and the input-output system is only required so the computer can communicate with external devices such as TTYs, disks, and printers. These will be discussed in Chapters 17 and 18. This chapter concentrates on the Control Unit that regulates the operation of the computer.

Unfortunately, the programs that run the computer (the software) are not shown in the diagram of Fig. 16-1. It is necessary, therefore, to present an introduction to programming before we can consider the Control Unit further.

### 16-3.1 Introduction to Programming

Computers are controlled by the *programs* written for them. This makes computers very versatile because *their operation can be changed simply by changing the program.*

A program is a *group of instructions* that cause the computer to perform a given task. Each instruction causes the computer to do something specific. Most instructions affect the *memory* and the *accumulator*, a register within the control unit that contains the basic operand used in each instruction. When an arithmetic or logical instruction that requires two operands is

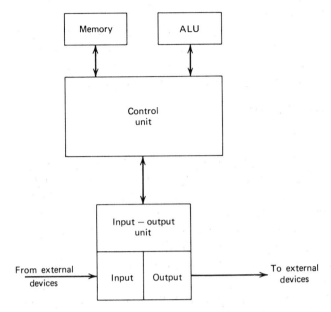

FIGURE 16-1  Block diagram of a bsic computer. (From Greenfield and Wray. *Using Microprocessors and Microcomputers: The* **6800** *Family.* Copyright John Wiley & Sons, Inc. 1981. Reprinted by permission of John Wiley & Sons, Inc.)

executed, one of the operands is in the accumulator and the other operand is in memory. The results are generally placed in the accumulator.

As an introduction, consider that each instruction consists of two parts: an Op code and an address.[1] The *Op code* tells the computer *what to do* when it encounters that instruction. The *address* is the *memory location involved.*

The most common instructions are listed in Table 16-1. The symbol (M) means the *contents of memory* at the address contained in the instruction. If location 500 contains the number 25, for example, the instruction LOAD 500 causes the memory to be read at address 500 and puts the 25 it finds there in the accumulator.

---

**EXAMPLE 16-1**

If the accumulator contains 35 and 500 contains 25, what will the accumulator contain after the instruction ADD 500 is executed?

---

[1]In any commercial computer, each instruction has some additional sections.

TABLE 16-1   SOME COMMON INSTRUCTIONS

| Instruction | Symbol | Explanation |
|---|---|---|
| LOAD | $(M) \rightarrow (A)$ | The contents of memory (at the address specified in the instruction) are written to the accumulator. |
| STORE | $(A) \rightarrow (M)$ | The contents of the accumulator are written to memory. |
| ADD | $(A) + (M) \rightarrow (A)$ | The contents of memory are added to the accumulator. The results go to the accumulator. |
| SUB | $(A) - (M) \rightarrow (A)$ | The contents of memory are subtracted from the accumulator. The results go to the accumulator. |

### SOLUTION

An ADD instruction adds two numbers. One is the number already in the accumulator (35) and the other is the number in memory at location 500 (25). After the instruction the accumulator contains 60.

## 16-3.2   A Simple Introductory Program

To make a computer do something useful, a program must first be written and entered into memory. The group of instructions that comprise the program *reside in memory*. Thus a computer's memory is divided into two areas: a *program area* that contains the instructions, and a *data area* that contains the data used by the program.[2] To run the program, the instructions are generally executed sequentially. Figure 16-2 is a sample program to add the numbers 5, 6, and 7, and store the result in location 23. The program proceeds as follows.

1. First the program area had to be selected. For this simple program it was arbitrarily chosen as locations 4 to 7.
2. Next the data area was chosen at memory locations 20 to 23. Note that program and data occupy different areas of memory. Figure 16-3 shows the contents of memory for this problem.
3. The data (the numbers 5, 6, and 7) are written into locations 20, 21, and 22, respectively. Address 23 is reserved for the result.

[2]Most modern computers also contain a stack area, but that is beyond the scope of this book.

| | Instruction | |
|---|---|---|
| Location | Op code | Address |
| 4 | LOAD | 20 |
| 5 | ADD | 21 |
| 6 | ADD | 22 |
| 7 | STORE | 23 |

FIGURE 16-2  A program to add the contents of locations 20, 21 and 22.

4.   The program is placed in locations 4 through 7 of memory, and execution is started at location 4. .

5.   The first instruction LOAD 20 causes the contents of location 20 (5) to be placed in the accumulator.

6.   The next instruction adds the contents of 21 (6) to the accumulator. It now contains 11.

7.   Next the contents of location 22 are added to the accumulator.

8.   The last instruction stores the contents of the accumulator into location 23. At the end of the program it contains 18, the sum of the numbers in 20, 21, and 22.

**Memory**

| Address | Data | | |
|---|---|---|---|
| • | | | |
| • | | | |
| • | | | |
| 4 | LOAD | 20 | |
| 5 | ADD | 21 | Program |
| 6 | ADD | 22 | area |
| 7 | STORE | 23 | |
| • | | • | |
| • | | • | |
| • | | • | |
| 20 | | 5 | |
| 21 | | 6 | Data |
| 22 | | 7 | area |
| 23 | | * | |
| • | | | |
| • | | | |
| • | | | |

*Reserved for result.

FIGURE  16-3  Memory contents for the program of Fig. 16-2.

## 16-4 FLOWCHARTS

Flowcharts are used by programmers to show the progress of their programs graphically. They are a clear and concise method of presenting the programmer's approach to a problem. They are often used as a part of programming documentation, where the program must be explained to those unfamiliar with it. Since good documentation is essential for proper use of any computer, the rudiments of flowcharts are presented in this section.

### 16-4.1 Flowchart Symbols

The flowchart symbols used in this book are shown in Fig. 16-4.

1. The *oval* symbol is either a *beginning* or *termination* box. It is used simply to denote the start or end of a program.
2. The *rectangular block* is the *processing or command block*. It states what must be done at that point in the program.
3. The *parallelogram* is an *input-output block*. Such commands as READ or WRITE, especially from an external device such as disk or card reader, are flowcharted using these boxes.

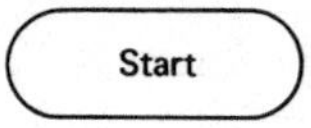

Start

**(a) Beginning or termination block**

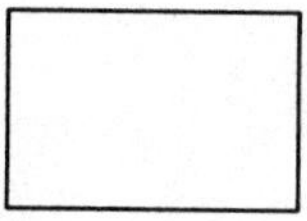

**(b) Processing or command block**

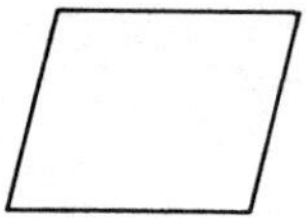

**(c) Input/Output block.**

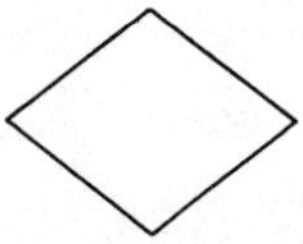

**(d) Decision block**

FIGURE 16-4 The most common standard flowchart symbols. (From Greenfield and Wray. *Using Microprocessors and Microcomputers: The* **6800** *Family.* Copyright John Wiley & Sons, Inc. 1981. Reprinted by permission of John Wiley & Sons, Inc.)

4.  The *diamond box* is a *decision box*. It usually contains a question within it. There are typically two output paths; one if the answer to the question is yes, and the other if the answer is no. Sometimes when a comparison between two numbers is made, there might be three exit paths corresponding to the greater than, less than, and equal, possibilities.

---

**EXAMPLE  16-2**

---

Draw a flowchart to add the numbers 1, 4, 7, and 10 together.

**SOLUTION**

The solution is shown in Fig. 16-5. It consists simply of a start box, four command boxes, and a stop box. This is an example of straight-line programming since no decisions were made. It was also assumed that the numbers 1, 4, 7, and 10 were available in the computer's memory and were not read from an external device.

---

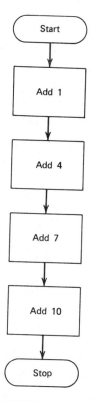

FIGURE 16-5  Flowchart for Example 16-2. (From Greenfield and Wray. *Using Microprocessors and Microcomputers: The* **6800** *Family.* Copyright John Wiley & Sons, Inc. 1981. Reprinted by permission of John Wiley & Sons, Inc.)

## 16-5 BRANCH INSTRUCTIONS AND LOOPS

The program for the flowchart of Ex. 16-2 can be written very simply as:

```
LOAD  20
ADD   21
ADD   22
ADD   23
HALT
```

where locations 20, 21, 22, and 23 contain the numbers 1, 4, 7, and 10, respectively. This program is extremely simple; indeed, the user can compute the answer more quickly than he or she can write the program. Suppose, however, the program is expanded so that we are required to add the numbers 1, 4, 7, 10,. . ., 10,000.[3] Conceptually, this could be done by expanding the program and flowchart of Sec. 16-4. However, the program and data areas would then require 3333 locations each and just writing them would become very tedious. Obviously, something else must be done.

### 16-5.1 Branch Instructions

Branch instructions provide the solution to the above problem. Normally instructions are executed one after the other as the program proceeds through the list of instructions. A *branch or jump instruction alters the normal sequence of program execution* and is used to create *loops* that allow the same sequence of instructions to be executed many times.

An example of a branch instruction is:

BRANCH 500

where BRANCH is the Op code and 500 is the branch address. It causes the branch address to be written into the PC (Program Counter. See Sec. 16-6.2). Thus the location of the next instruction to be executed is the branch address (500 in this case), rather than the sequential address.

There are two types of branch instructions—unconditional and conditional.

*The unconditional branch always causes the program to jump to the branch address.* It is written as BRA, which stands for BRANCH ALWAYS.

*The conditional branch causes the program to branch only if a specified*

---

[3]Since the object of this section is to teach programming and not mathematics, ignore the fact that this is an arithmetic progression whose sum is given by a simple formula.

condition is met. For this introductory chapter two conditional branch instructions are used:

BPL—branch on positive accumulator (0 is considered as a positive number).
BMI—branch on negative accumulator.

Therefore, should the computer encounter one of these instructions, it can simply test the MSB of the accumulator. This determines whether the number in the accumulator is positive (since we are using a 2s complement $\mu$P) and whether the branch should be taken.

---

**EXAMPLE 16-3**

A computer is to add the numbers 1, 4, 7, 10,. . ., 10,000.[+] Draw a flowchart for the program.

**SOLUTION**

The flowchart is shown in Fig. 16-6. We recognize that we must keep track of two quantities. One is the number to be added. This has been labeled N in the flowchart and progresses 1, 4, 7, 10,. . . . The second quantity is the sum S, which progresses 1, 1 + 4, 1 + 4 + 7,. . . or 1, 5, 12,. . . . The first box in the flowchart is an initialization box. It sets N to 1 and S to 0 at the beginning of the program. The next box ($S = N + S$) sets the new sum equal to the old sum plus the number to be added. The number to be added is then increased by 3. At this point the flowchart loops around to repeat the sequence. This is accomplished by placing an unconditional branch instruction in the program. The quantities S and N will progress as specified.

---

## 16-5.2 Decision Boxes and Conditional Branches

The reader has probably already realized that there is a serious problem with Example 16-3; the loop never terminates. Actually, putting a program into an endless loop is one of the most common programming mistakes.

There are two common ways to determine when to end a loop: *loop counting* and *event detection*. Either method requires the use of decision boxes in the flowchart and corresponding conditional branch instructions in the program. Loop counting is considered first.

Loop counting is done by determining the number of times the loop should be traversed, counting the actual number of times through and comparing the two.

---

[+]For the introductory problems of this section we have ignored the fact that decimal 10,000 cannot be contained within a single byte.

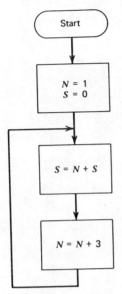

FIGURE 16-6 Flowchart for Example 16-3. (From Greenfield and Wray. *Using Microprocessors and Microcomputers: The* **6800** *Family.* Copyright John Wiley & Sons, Inc. 1981. Reprinted by permission of John Wiley & Sons, Inc.)

## EXAMPLE 16-4

Improve the flowchart of Fig. 16-6 so that it terminates properly.

## SOLUTION

The program should terminate not when N = 10,000 but when 10,000 is added to the sum. For the flowchart of Fig. 16-6, N is increased to 10,003 immediately after this occurs. At the end of the first loop N = 4, the second loop N = 7, and so on. It can be seen here that N = 3L + 1, where L is the number of times through the loop. If N is set to 10,003, L = 3334. The loop must be traversed 3334 times.

The correct flowchart is shown in Fig. 16-7. The loop counter L has been added and set initially to − 3334. It is incremented each time through the loop and tested to see if it is positive. After 3334 loops, it becomes 0. Then the YES path from the decision box is taken and the program halts.

## EXAMPLE 16-5

Write the code for the flowchart of Fig. 16-7.

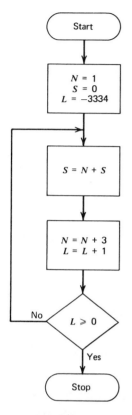

FIGURE 16-7 Flowchart for Example 16-4. (From Greenfield and Wray. *Using Microprocessors and Microcomputers: The* **6800** *Family*. Copyright John Wiley & Sons, Inc. 1981. Reprinted by permission of John Wiley & Sons, Inc.)

## SOLUTION

The program is arbitrarily started at location 10 and the data area is set at location 80. In the data area three variables N, S, and L are needed. These should be initialized to 1, 0, and $-3334$, respectively, before the program starts. In addition, two constants, 1 and 3, are needed during the execution of the program. Before the program starts, the data should look like this:

| Location | Term | Initial value |
|----------|----------|---------------|
| 80 | N | 1 |
| 81 | S | 0 |
| 82 | L | $-3334$ |
| 83 | Constant | 3 |
| 84 | Constant | 1 |

The program can now be written directly from the flowchart.

| Locations | Instruction | Comments |
|---|---|---|
| 10 | LOAD 81 | |
| 11 | ADD 80 | $S = N + S$ |
| 12 | STORE 81 | |
| 13 | LOAD 80 | |
| 14 | ADD 83 | $N = N + 3$ |
| 15 | STORE 80 | |
| 15 | LOAD 82 | |
| 16 | ADD 84 | $L = L + 1$ |
| 17 | STORE 82 | |
| 18 | BMI 10 | |
| 19 | HALT | |

Note that the instructions follow the flowchart. The decision box has been implemented by the BMI instruction. The program loops as long as $L$ remains negative.

*Check:* As a check on the program, we can write the contents of $N$, $S$, and $L$ at the end of each loop in the following table.

| Times through the Loop | $N$ | $S$ | $L$ |
|---|---|---|---|
| 1 | 4 | 1 | $-3333$ |
| 2 | 7 | 5 | $-3332$ |
| 3 | 10 | 12 | $-3331$ |

The chart shows that each time around the loop:

$$\frac{N - 1}{3} + |L| = 3334$$

Therefore, when $N = 10{,}003$, $L$ indeed equals 0 and the loop terminates.

*Event detection terminates a loop when an event occurs that should make the loop terminate.* In Example 16-5, that event could be the fact that $N$ is greater than 10,000. Using event detection, locations 15 through 18 could be replaced by:

```
15   LOAD N
16   SUBTRACT 105
17   Not needed (NOP)
18   BMI 10
```

where location 105 contains 10,001. This program branches back until

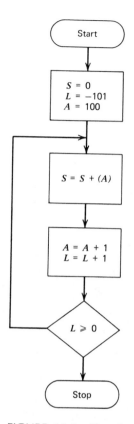

FIGURE 16-8 Flowchart for adding together the contents of locations 100 to 200. (From Greenfield and Wray. *Using Microprocessors and Microcomputers: The* **6800** *Family.* Copyright John Wiley & Sons, Inc. 1981. Reprinted by permission of John Wiley & Sons, Inc.)

$N = 10{,}003$. In this problem, the use of event detection is conceptually simpler and saves one instruction.

As a final introductory programming example, consider the problem of adding together all the numbers in memory between locations 100 and 200. This problem has practical applications. If, for example, a firm has 101 debtors, it can store the amount owed by each debtor in one of the locations. The sum of these debts is the firm's accounts receivable.

The flowchart for this example is shown in Fig. 16-8. At the start the sum is set to 0 and the loop counter to 101 because there are 101 addresses between 100 and 200.[5] The loop counter and the addresses are then incremented. Note the box $S = S + (A)$. The parenthesis around $A$ means that the contents of location $A$ are being added to the sum.

[5]In this introductory problem, decimal addresses are assumed for simplicity.

The difficult part of this problem is to find a way to increment the addresses. One way to do this is to increment the ADD instruction. This is done in Ex. 16-6.

___

**EXAMPLE 16-6**

Write the code for the flowchart of Fig. 16-8.

**SOLUTION**

Again, the program is started at location 10. Because locations 100 to 200 contain data, the constant area is set up at 300. It looks like this:

| Location | Term | Initial Value |
|----------|------|---------------|
| 300 | $S$ | 0 |
| 301 | $L$ | $-101$ |
| 302 | Constant | 1 |

The coding can then proceed as follows.

| Location | Instruction | | Comments |
|----------|-------------|------|----------|
| 10 | LOAD | 300 | |
| 11 | ADD | 100 | $S = S + (A)$ |
| 12 | STORE | 300 | |
| 13 | LOAD | 11 | |
| 14 | ADD | 302 | Increment address portion of instruction |
| 15 | STORE | 11 | 11 and resume |
| 16 | LOAD | 301 | |
| 17 | ADD | 302 | $L = L + 1$ |
| 18 | STORE | 301 | |
| 19 | BMI | 10 | If minus, branch back to 10 |
| 20 | Halt | | |

Note particularly instructions 13, 14, and 15. These instructions treat location 11, which contains an instruction, as data. The assumptions are that the instructions consist of an Op code and address, and the address occupies the LSBs of the instruction, and that adding 1 to the instruction increments the address without affecting the Op code or other parts of the instruction. These assumptions are valid for most computers.

___

This method of solving this problem is not generally used, and *we do not recommend it*, because more powerful methods, such as indexing, are preferable. It was presented, however, to emphasize that *instructions occupy memory space and can be treated as data.* In this example, the instruction in location 11 is treated as data when the instructions in locations 13, 14, and 15 are executed. This is a *self-modifying program*, which means it *changes its own contents as it executes.* Most programmers *avoid* self-modifying programs. They are complex and difficult to document or understand, and the program cannot be run successively without reentering the initial values between each run. Also, they cannot be used in ROM.

## 16-6 THE CONTROL UNIT

The control unit consists of the timing circuits and registers necessary to control the operation of the computer.

### 16-6.1 FETCH, EXECUTE, and Timing

The execution of a computer instruction generally requires two cycles: a FETCH cycle and an EXECUTE cycle. During the FETCH cycle the computer reads memory to get the instruction; during the EXECUTE cycle it actually executes the instruction. The computer then starts the next instruction by performing another FETCH cycle. Thus a computer operates basically by alternately doing FETCH, EXECUTE, FETCH, EXE-CUTE, . . . until it steps through the entire program and encounters a HALT instruction.[6]

The FETCH and EXECUTE cycles are subdivided into time slots. For the first computer to be described we will assume four time slots or phases, labeled $T_0$, $T_1$, $T_2$, and $T_3$, occur during each cycle.

### 16-6.2 A Rudimentary Computer

The diagram of a rudimentary computer is shown in Fig. 16-9. The dimensions of the memory are assumed to be 4K words by 16 bits. An instruction word is assumed to consist of a 4-bit Op code and a 12-bit address as shown in Fig. 16-10. The memory in Fig. 16-9 is shown containing a simple program to add the numbers in location 100 and 101 and store the results in 101. It also shows the data are at 100 and 101.

Besides the memory, the computer consists of the following parts.

1. **The FETCH/EXECUTE FF.** This FF determines whether the computer is in FETCH or EXECUTE mode.

---

[6]The front panel of many computers such as the PDP-8 or HP have FETCH and EXECUTE lights on them. When operated in single cycle these lights indicate which cycle will be performed next.

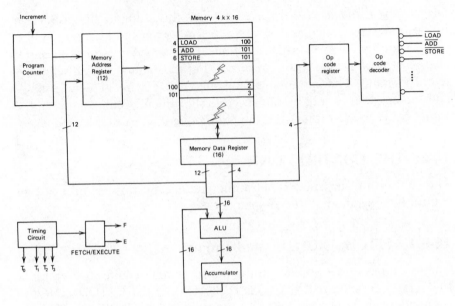

FIGURE 16-9   A rudimentary computer.

2.   **The timing generator.** This develops the four-phase clock used to synchronize the operations during the FETCH and EXECUTE cycles.

3.   **The MAR (Memory Address Register).** This contains the memory address (see Sec. 15-3).

4.   **The MDR (Memory Data Register).** This contains the data going to or from the memory (see Sec. 15-3).

5.   **The Program Counter (PC).** This register is as long as the MAR. It contains the *address* of the next instruction to be executed. Before each FETCH cycle begins the contents of the PC are transferred to the MAR so that the next instruction can be fetched from memory. The PC is incremented during each instruction so that the next instruction will be fetched from the next sequential location.

6.   **The Op code register.** This register retains the Op code portion of the instruction word.

7.   **The Op code decoder.** This IC decodes the information in the Op code register and presents a single active line (LOAD, ADD, STORE, etc.) corresponding to the particular instruction being executed.

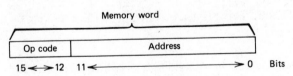

FIGURE 16-10   Division of an instruction word into OP code and address portions.

8. **The accumulator.** This register holds data, one of the operands used by the instruction. Since it contains data, it must be as long as the MDR.

9. **The Arithmetic-Logic Unit (ALU).** The ALU generally gets an operand from memory, performs an operation (ADD, SUBTRACT, AND, OR, etc.) on the memory data, and returns the results to the accumulator.

The interconnecting wires or busses between the various components are also shown in Fig. 16-9.

---

**EXAMPLE 16-7**

For the above computer how many bits are in:
  (a) The MAR
  (b) The MDR
  (c) The PC
  (d) The Op code register
  (e) The accumulator
  (f) The ALU

**SOLUTION**

1. Because the memory is 4K by 16, the MAR must contain 12 bits ($2^{12} = 4K$) and the MDR must contain 16 bits.

2. The PC essentially contains an address (the address of the instruction to be executed). It contains as many bits as the MAR (12).

3. Because the Op code portion of the instruction word is 4 bits, the Op code register only needs 4 bits.

4. Both the accumulator and the ALU contain data. Because a data word is as long as a memory word, the accumulator and ALU both contain 16 bits.

---

## 16-6.3 The FETCH Cycle

Because the instructions that comprise a program are retained in memory, but must be executed in the control unit, instruction execution starts with a FETCH cycle that reads (or FETCHes) the instruction from memory to the control unit. In many computers, including our sample computer, all FETCH cycles are identical. A FETCH cycle occurs while the FETCH FF is HIGH. Remember there are four time slots or phases during each cycle. A FETCH cycle proceeds as follows:

1. The MAR is loaded with the contents of the PC. This happens before or just at the beginning of the FETCH cycle.

2. During the $F \cdot T_0$ (F means FETCH mode or the FETCH FF is HIGH and $T_0$ means the first time slot in the cycle) the memory is read. This brings the instruction into the MDR.

3. The Op code portion of the instruction (bits 12–15 in our computer) is loaded into the Op code register at $F \cdot T_1$. The output of the Op code register goes to the Op code decoder, which activates one line that corresponds to the instruction being executed.

4. At $F \cdot T_2$ the Program Counter is incremented.

5. At $F \cdot T_3$ the address portion of the instruction (bits 0–11) is placed in the MAR. Table 16-2 shows the events that occur during the FETCH.

At the end of the FETCH cycle the proper output of the Op code decoder is activated and the operand address is in the MAR. If the instruction was LOAD 500, for example, the LOAD line out of the decoder would be active and the address (500) would be in the MAR.

## 16-6.4   The EXECUTE Cycle

When the FETCH cycle ends, the FETCH/EXECUTE FF toggles and the EXECUTE cycle begins. What occurs during the EXECUTE cycle depends primarily on the instruction decoded by the Op code decoder. The events that occur during the LOAD, ADD, and STORE instructions are shown in Table 16-3. The LOAD is actually accomplished by a CLEAR and ADD procedure; the accumulator is first *cleared*, then it is *added* to the contents of memory. The results are the memory contents and they are placed in the accumulator. Note that the ADD instruction is identical to the LOAD instruction, except that the accumulator is not cleared at $E \cdot T_0$.

## 16-6.5   Execution of a Simple Program

In this section we will consider how our rudimentary computer can perform the simple program shown in Fig. 16-9.

$$
\begin{array}{lll}
4 & \text{LOAD} & 100 \\
5 & \text{ADD} & 101 \\
6 & \text{STO} & 101 \\
\end{array}
$$

TABLE 16-2   THE FETCH PORTION OF AN INSTRUCTION

| | |
|---|---|
| $F \cdot T_0$ | Memory is read to FETCH the instruction. |
| $F \cdot T_1$ | Op code portion of the instruction goes to the Op code register. |
| $F \cdot T_2$ | Program Counter is incremented to point to the address of the next instruction. |
| $F \cdot T_3$ | The address portion of the instruction word is sent to the MAR. |

TABLE 16-3   EXECUTION OF THE LOAD, ADD, AND STORE INSTRUC-
TIONS

| | |
|---|---|
| **LOAD** | |
| $E \cdot T_0$ | Accumulator is cleared. ALU commanded to ADD. |
| $E \cdot T_1$ | ALU adds memory contents to accumulator contents (0 because accumulator is CLEAR). |
| $E \cdot T_2$ | Sum goes into the accumulator. |
| $E \cdot T_3$ | (PC) goes into MAR. |
| | |
| **ADD** | |
| $E \cdot T_0$ | ALU commanded to ADD. |
| $E \cdot T_1$ | ALU adds memory contents to accumulator contents. |
| $E \cdot T_2$ | Sum goes into the accumulator. |
| $E \cdot T_3$ | (PC) goes into the MAR. |
| | |
| **STORE** | |
| $E \cdot T_0$ | Accumulator output sent to MDR. |
| $E \cdot T_2$ | Memory placed in WRITE mode. |
| $E \cdot T_3$ | (PC) goes into the MAR. |

Table 16-4 has been prepared to show the contents of the registers in the control unit during each step in the program. An X in a register indicates the data are irrelevant or unknown at that time.

The program starts with the PC and the MAR locating the first instruction. At $F \cdot T_1$ the 4 bits indicating LOAD are transferred to the Op code register. The PC is incremented to 5 at $F \cdot T_2$ and the instruction address, 100, is transferred to the MAR at $F \cdot T_3$.

The accumulator is cleared at $E \cdot T_0$ of the LOAD instruction. The ALU then adds the 2 (read from location 100) to the 0 in the accumulator and puts the sum in the accumulator at $E \cdot T_2$. At $E \cdot T_3$ the PC contents, now 5, are transferred to the MAR, so the computer is ready to fetch the next instruction.

The execution of the ADD and STORE instruction is similar. The FETCH cycles are identical. A WRITE pulse, not shown on the table, occurs at $E \cdot T_2$ of the STORE instruction.

## 16-7   THE HARDWARE DESIGN OF A COMPUTER

Section 16-6 described the operation of a computer *conceptually*. In this section we will go through the design of a computer in greater detail, we will identify the ICs and other components that comprise each part of the computer, and finish with a rudimentary but viable computer.

TABLE 16-4   STEP-BY-STEP EXECUTION OF THE PROGRAM OF SEC-
TION 16-6.5

| Time | PC | MAR | MDR | Op Code | Accum. | ALU |
|------|----|-----|-----|---------|--------|-----|
| **LOAD 100** | | | | | | |
| $F \cdot T_0$ | 4 | 4 | LOAD 100 | X | X | X |
| $F \cdot T_1$ | 4 | 4 | LOAD 100 | LOAD | X | X |
| $F \cdot T_2$ | 5 | 4 | LOAD 100 | LOAD | X | X |
| $F \cdot T_3$ | 5 | 100 | LOAD 100 | LOAD | X | X |
| $E \cdot T_0$ | 5 | 100 | 2 | LOAD | 0 | X |
| $E \cdot T_1$ | 5 | 100 | 2 | LOAD | 0 | 2 |
| $E \cdot T_2$ | 5 | 100 | 2 | LOAD | 2 | 2 |
| $E \cdot T_3$ | 5 | 5 | 2 | LOAD | 2 | X |
| **ADD 101** | | | | | | |
| $F \cdot T_0$ | 5 | 5 | ADD 101 | LOAD | 2 | X |
| $F \cdot T_1$ | 5 | 5 | ADD 101 | ADD | 2 | X |
| $F \cdot T_2$ | 6 | 5 | ADD 101 | ADD | 2 | X |
| $F \cdot T_3$ | 6 | 101 | ADD 101 | ADD | 2 | X |
| $E \cdot T_0$ | 6 | 101 | 3 | ADD | 2 | 5 |
| $E \cdot T_1$ | 6 | 101 | 3 | ADD | 2 | 5 |
| $E \cdot T_2$ | 6 | 101 | 3 | ADD | 5 | 5 |
| $E \cdot T_3$ | 6 | 6 | 3 | ADD | 5 | X |
| **STORE 101** | | | | | | |
| $F \cdot T_0$ | 6 | 6 | STO 101 | ADD | 5 | X |
| $F \cdot T_1$ | 6 | 6 | STO 101 | STO | 5 | X |
| $F \cdot T_2$ | 7 | 6 | STO 101 | STO | 5 | X |
| $F \cdot T_3$ | 7 | 101 | X | STO | 5 | X |
| $E \cdot T_0$ | 7 | 101 | X | STO | 5 | X |
| $E \cdot T_1$ | 7 | 101 | 5 | STO | 5 | X |
| $E \cdot T_2$ | 7 | 101 | 5 | STO | 5 | X |
| $E \cdot T_3$ | 7 | 7 | 5 | STO | 5 | X |

## 16-7.1   The Timing Circuit

The basic timing of most computers and microprocessors is derived from a high-frequency clock. Often a crystal controlled clock is used to keep the timing precise. The timing for our computer is shown in Fig. 16-11. The clock feeds a 2-bit counter, which goes to a **74155** decoder to produce the required 4-phase clock. The leading edge of $\overline{T}_0$ also toggles the FETCH-EXECUTE FF so the computer alternates between FETCH AND EXECUTE cycles. The times are expressed as $\overline{T}_0$, $\overline{T}_1$, $\overline{T}_1$, and $\overline{T}_3$ because they are LOW active. In some figures we occasionally use uncomplemented times such as $T_1$. $T_1$ is simply the inverse of $\overline{T}_1$.

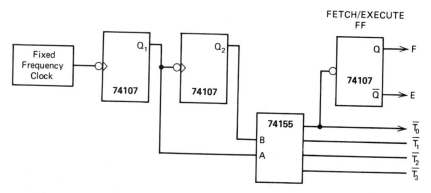

FIGURE 16-11   The basic timing for the computer.

---

**EXAMPLE 16-8**

Draw a timing chart for the basic computer timing.

**SOLUTION**

The solution is shown in Fig. 16-12. Note that each time is LOW for one quarter of each cycle because the active output of a decoder is LOW.

---

## 16-7.2   The Memory, MAR, and MDR

The memory is shown in Fig. 16-13. For simplicity, 2102s (see Sec. 15-7.1) were chosen as the basic memory IC. A 4K-word-by-16-bit memory requires 64 of them in a $4 \times 16$ matrix. A 12-bit MAR is also required. The MAR is made up of two **74174**s. Ten of its outputs go directly to the

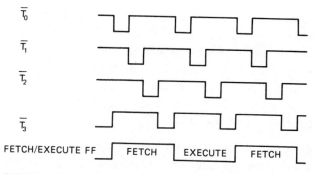

FIGURE 16-12   Timing waveforms for the computer.

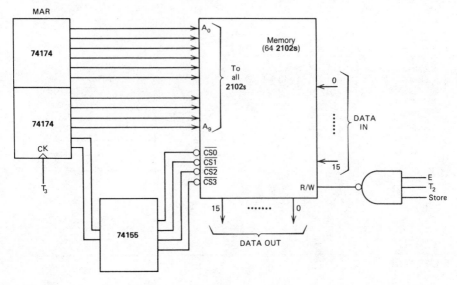

FIGURE 16-13   The memory and MAR.

2102s. The other two addresses go into a decoder, which selects one of the 4 rows of memory by driving the proper CS line LOW.

Because 2102s are used, the MDR really consists of the DATA OUT and DATA IN lines. The memory is almost always in READ mode, so that its contents are not changing. The DATA OUT lines contain the memory contents of its selected address.

The only time the memory is written is at $E \cdot T_2$ of a STORE instruction. The DATA IN lines are driven by the accumulator to provide the data to be stored in memory.

### 16-7.3   The Op Code Register and Decoder

The Op code register, shown in Fig. 16-14, only needs 4 bits and is a single 74174 or 74175. Its inputs are the Op code, the 4 MSBs of the MDR, and are clocked in at $F \cdot T_1$. The four outputs go to the SELECT lines on a 74154, which functions as the Op code decoder. The single LOW output from the 74154 is the instruction being executed.

### 16-7.4   The Accumulator and the ALU

The accumulator, the ALU, and its interconnections are shown in Fig. 16-15. The accumulator is made of three 74174s (see Sec. 6-7). The instruction inputs ($\overline{LOAD}$, $\overline{ADD}$, etc.) come directly from the Op code decoder. The accumulator is cleared at $E \cdot T_0$ by CLEAR and LOAD in-

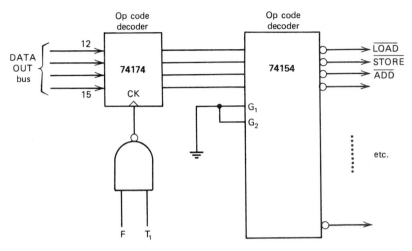

FIGURE 16-14   The Op code register and decoder.

structions. It is loaded at E · T₂ by LOAD instructions and all the arithmetic or logic instructions. Those instructions that do not change the accumulator, such as STORE or BRANCH, are not inputs to the NAND gate controlling the accumulator clock.

The A inputs to the ALU come directly from the accumulator, and the B inputs come from the memory. The ALU outputs go to the D inputs of the accumulator where they are clocked in the E · T₂ of those instructions that affect the accumulator.

The ALU is controlled by the six NAND gates shown on the right side of Fig. 16-15. They function as a NAND gate ROM (see Sec. 15-9.2), and send the proper S, M, and carry inputs to the 74181s.

---

**EXAMPLE 16-9**

Explain how the following instructions control the ALU:
   (a) ADD
   (b) AND

**SOLUTION**

(a) To ADD, the ALU must have a 9 on its S inputs, $C_{in}$ must be HIGH, and M must be LOW. An ADD instruction causes the Op code decoder to set $\overline{ADD}$ LOW, leaving all other outputs HIGH. Because $\overline{ADD}$ is connected to the $S_0$, $S_3$, and $C_{in}$ inputs, these will be HIGH and the other S and M inputs will be LOW. This combination of inputs causes the ALU to add.

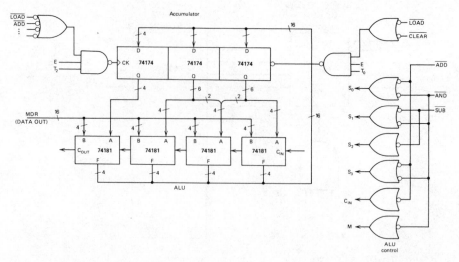

FIGURE 16-15 The accumulater and ALU.

(b) Because AND is a logic instruction, Fig. 14-12 shows that M, $S_0$, S, and $S_3$ must be HIGH. This occurs if $\overline{\text{AND}}$ goes LOW. The CARRY IN is irrelevant for logic instruction and is not connected for simplicity.

In most computers other instructions that affect the accumulator will also be used, and they will provide additional inputs to the NAND gates (see Problem 16-6).

## 16-8 THE COMPLETE COMPUTER

The complete computer is shown in Fig. 16-16. The parts of this computer that were not discussed in Sec. 16-7 are:

1. **The program counter.** The PC is a 12-bit counter. Because it must be capable of being incremented and parallel loaded (see Sec. 16-8.3), **74193** presettable counters were chosen.

2. **The address multiplexer.** Because the MAR must be loaded from both the PC and the data bus, a multiplexer is required. We used the **74157** quad 2 line to 1 line multiplexer. The SELECT input to the **74157** is controlled by the FETCH-EXECUTE FF so that when the computer is in FETCH mode, the DATA OUT inputs from the **2102s** (bits 0–11) are on the multiplexer's output and during EX-ECUTE mode the PC is fed through the multiplexer to the MAR. The reasons for this are explained in Sec. 16-8.1.

3. **The branch logic.** This is discussed in Sec. 16-8.3.

### 16-8.1 The FETCH Cycle

The FETCH cycle for this computer is as previously discussed. The following events occur during the FETCH cycle.

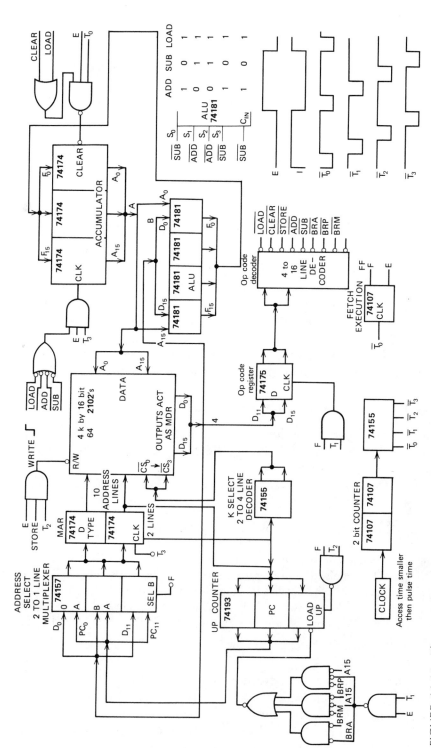

FIGURE 16-16 The complete computer.

1.  The Op code register is loaded at $F \cdot T_1$. It retains the instruction for the rest of the FETCH cycle and throughout the EXECUTE cycle.
2.  The PC is incremented at the trailing edge of $F \cdot T_2$ by the NAND gate connected to the COUNT-UP input of the **74193s**.
3.  The MAR is loaded with the address on the DATA-OUT lines at the beginning of $T_3$. This happens because in FETCH mode the Data bus comes through the Address Multiplexer.

Note that the MAR is loaded at the start of every $T_3$ time slot. During EXECUTE mode it is loaded with the PC contents; this causes the memory to read the next instruction for the following FETCH cycle.

## 16-8.2    Execution of Arithmetic and Logic Instructions

When the Op code decoder detects an arithmetic or logic instruction, it sets the S, M, and $C_{in}$ inputs to the ALU. This occurs at $F \cdot T_1$. At $F \cdot T_3$ the operand address is clocked into the MAR. After one memory access time, the memory operand appears at the B inputs of the ALU. All instructions that cause the accumulator to change are brought to the NAND gate connected to the accumulator CLOCK. Thus at $E \cdot T_2$, after the ALU has had time to perform the proper operation between the memory data and the accumulator, the results are clocked into the accumulator.

## 16-8.3    BRANCH Instructions

The BRANCH instruction logic is shown in the lower left corner of Fig. 16-16. A BRANCH instruction causes the PC to be loaded with the address contained in the instruction.

BRA 500, for example, is an unconditional branch that causes the next instruction to be executed at location 500. It is executed as follows:

1.  At $F \cdot T_1$, $\overline{\text{BRA}}$ is decoded by the Op code decoder.
2.  At $F \cdot T_3$, the MAR is loaded with the address in the instruction (500). The contents of memory at location 500 will be placed on the DATA OUT lines, but nothing will be done with it during this instruction. The MAR outputs, however, are also connected to the parallel load inputs to the **74193s**. At $E \cdot T_1$ a LOAD pulse is generated that loads the MAR contents (500) into the PC. At $E \cdot T_3$ the contents of the PC (now 500) are loaded into the MAR and the next FETCH occurs with the MAR containing 500. Thus the instruction in location 500 is executed next.

Conditional branches are executed similarly. The condition is tested. If it is satisfied, the PC is loaded with the memory address. Otherwise the PC is not loaded and the computer executes the next sequential instruction. The logic for two conditional branches, BRM and $\overline{\text{BRP}}$ is shown in Fig. 16-16. $\overline{\text{BRM}}$ from the Op code decoder is gated with $\overline{\text{A}}_{15}$. If $A_{15}$ is a 1, the number in the accumulator is negative, $\overline{A}_{15} = 0$, and a LOAD pulse is produced. The BRP instruction is similar except that $\overline{BRP}$ is gated with $A_{15}$.

---

### EXAMPLE 16-10

How can a BRZ (branch on zero accumulator) be added to this computer?

### SOLUTION

The simple solution is to use a 16-input AND gate to determine if all bits of the accumulator are 0. However, a more elegant solution that requires fewer gates exists:

1. Set up the ALU so that $S = 15$ (the function is $F = A$ minus 1), the CARRY-IN is 1 and $M = 0$.
2. The CARRY OUT of the ALU will be 1 only if all bits of the accumulator are 0 (see Problem 16-7).
3. The CARRY OUT can be inverted and gated with the $\overline{BRZ}$ output of the OP code decoder to provide a load pulse to the PC. This occurs only if the accumulator contains a 0.

---

Other instructions, in addition to the BRZ, can be added to this computer as required. The length of the Op code register and decoder, however, limit it to 16 different instructions.

## 16-9 BUILDING THE COMPUTER

Using the ideas developed in the previous sections, a computer was designed, built, and tested by the students at RIT.[7] We felt this was the best way to learn how computers actually work, and it also presented the students with nontrivial debugging problems.

Unfortunately, this computer makes demands on two of the students most limited resources: time and money.[8] Both constraints dictate that the computer be designed with as few ICs as possible. The first decision, consequently, was to go to an 8-bit computer with a 3-bit Op code and a 5-bit address. Two **2114**s were used to give a 1K-by-8-bit memory.

### 16-9.1 Reading and Writing Memory

The entire computer is shown in Fig. 16-17 through 16-21. The computers discussed previously had no front panel; no way of reading or writing memory was included for simplicity.

---

[7]Each student or lab group of two students built their own computer.
[8]At RIT the computer is supposed to be built during the students' co-op block and be ready when the course starts.

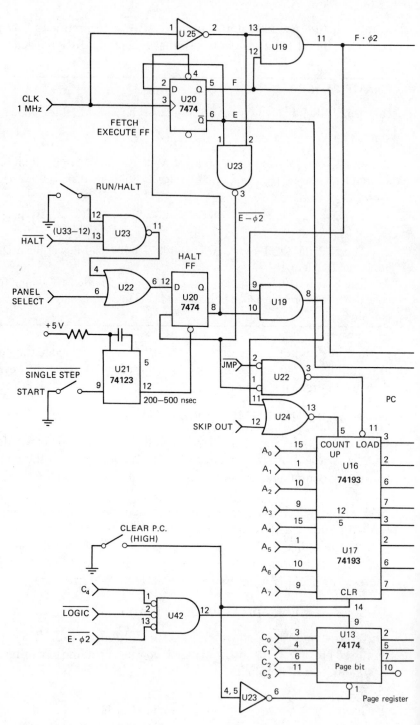

FIGURE 16-17  The control section of the R.I.T. computer.

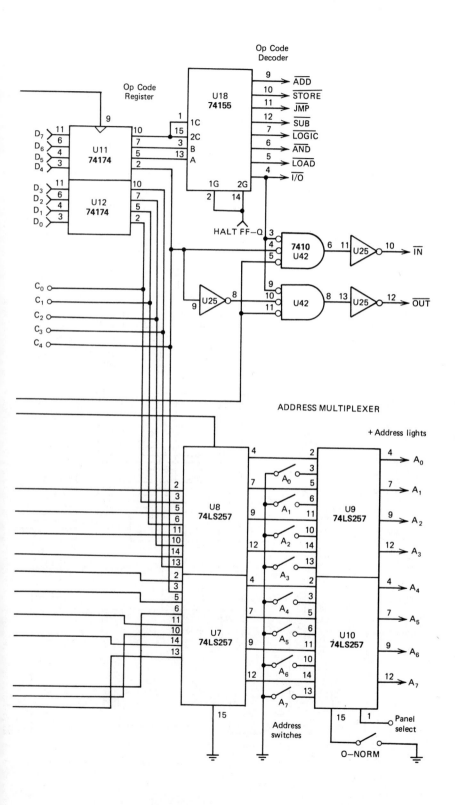

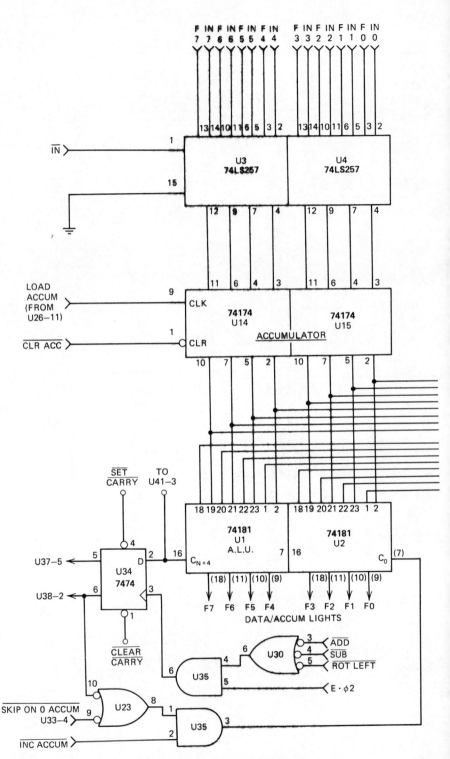

FIGURE 16-18  The accumulator and ALU of the R.I.T. computer.

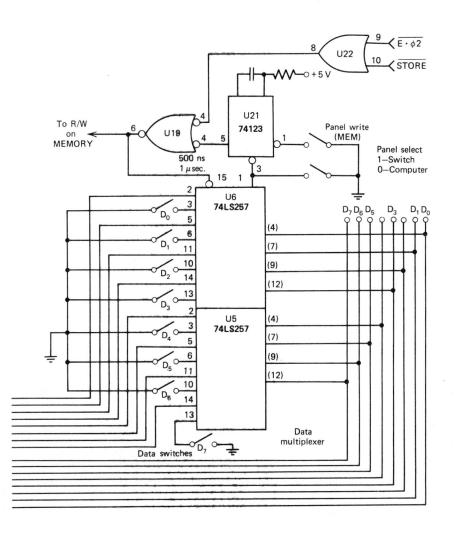

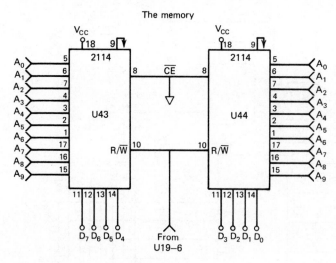

FIGURE 16-19 The memory.

Reading or writing memory requires address switches to set the memory address, data switches, and a way to generate a WRITE pulse. It also requires multiplexers to select the switch data or the computer data and send it to the memory bus.

The **74LS257** was chosen as the multiplexer. It is like a **74157**, except that it has 3-state outputs that are compatible with the bi-directional data bus of the **2114**s. The Data Multiplexer (U5 and U6) multiplexes the accumulator or data switches onto the memory data bus.

There is a two-level address multiplexer (Fig. 16-17). U7 and U8 are the PC/instruction Address Multiplexer (see Sec. 16-8). U9 and U10 then multiplex the computer address or the address switches onto the **2114**'s address inputs.

The multiplexers are controlled by the Panel Select switch (Fig. 16-18). In PANEL SELECT the address and data switches control the memory; otherwise it gets its data and addresses from the computer.

In PANEL SELECT the memory is written by flipping the debounced Panel Write switch, which triggers the one-shot (U21) to provide a WRITE pulse. This writes the switch data into memory at the address on the address switches.

When the computer is running, memory is written only by a STORE instruction. Normally the outputs of U5 and U6 are disabled (high impedance) so the memory outputs control the data bus. Any WRITE pulse, however, comes through U19 pin 6, drives the R/W line on the **2114**'s

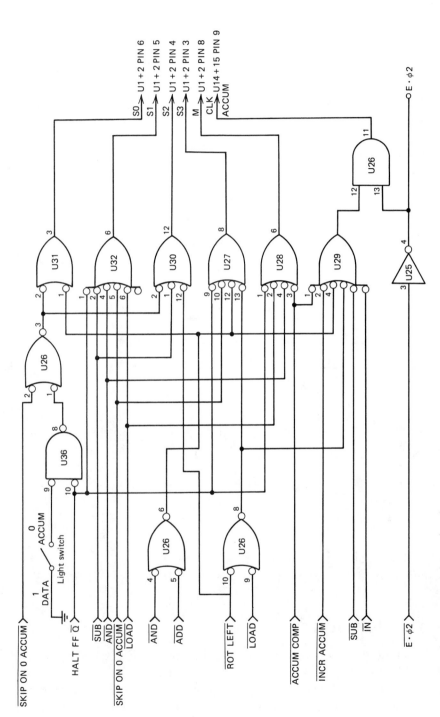

FIGURE 16-20 Control logic for the ALU.

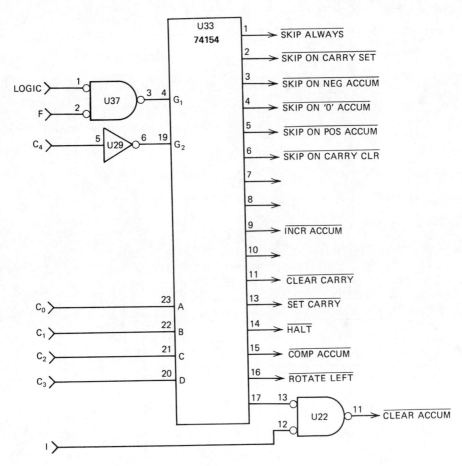

FIGURE 16-21 Logic instruction decoder.

LOW, and enables the outputs of the Data Multiplexer. This allows the memory to be written to, or memory is read via the ALU (see Sec. 16-9.5).

## 16-9.2 The FETCH/EXECUTE FF

The operation of the computer is controlled by the FETCH/EXECUTE FF at U20 (Fig. 16-17). This FF and the clock driving it produce the timing pulses $F \cdot \phi_2$ and $\overline{E} \cdot \phi_2$ that provide the time slots for the computer. At $F \cdot \phi_2$ the PC is incremented and the instruction register (U11 and U12) loaded. At $\overline{E} \cdot \phi_2$ the page register, the accumulator, and the PC are loaded for those instructions that require it.

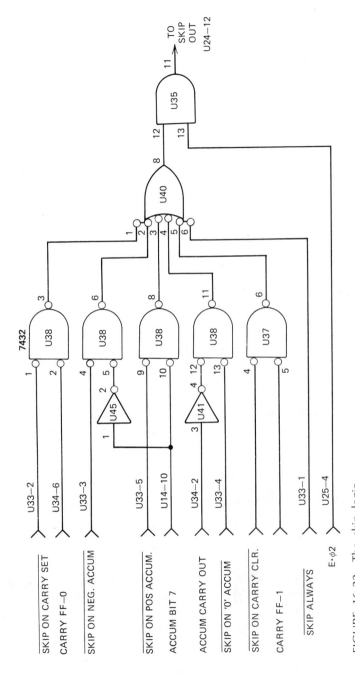

FIGURE 16-22 The skip logic.

The operation of the FETCH/EXECUTE FF is affected by the HALT FF (U20). When it is SET the FETCH/EXECUTE FF is held in the FETCH state so the next instruction to be executed appears on the readouts. The HALT FF is set by:

1. The RUN/HALT switch being placed in HALT mode.
2. The Panel Select switch.
3. A HALT instruction.

It is cleared by depressing the START switch. The computer can be run in single step mode, executing one instruction at a time, by putting the RUN/HALT switch in HALT and depressing the START switch each time a single instruction is to be executed.

### 16-9.3   The Program Counter (PC) and Page Register

The PC uses two 74193s in U16 and U17. This limits the computer to 256 instructions. Each instruction only has 5 bits for address. This effectively subdivides the memory into 32-word *pages*. To allow greater memory usage, the memory address is augmented by the 4-bit Page Register. The Page Register (U13) is set by an instruction whose 4 MSBs are 1000. The 4 LSBs of that instruction are loaded into the Page Register. The Memory address during EXECUTE is determined by the 5 LSBs of the instruction word and the 4 bits of the Page Register. The 5 LSBs of the instruction and the 3 LSBs of the Page Register are inputs to the Address Multiplexer (U7 and U8).

### 16-9.4   The Op Code Decoder

The first-level Op code decoder (U18) decodes the 3 MSBs of the instruction. It provides the instructions shown at the outputs of U18. All of these outputs, except the $\overline{\text{LOGIC}}$ and $\overline{\text{I/O}}$, use memory addresses that are a combination of the 5 LSBs of the instruction word and the Page Register.

Any instruction whose MSBs are 100 is a $\overline{\text{LOGIC}}$ instruction and does not involve memory. Instructions that start 1000 are SET PAGE REGISTER instructions. The Page Register is set to the 4 LSBs of the instruction.

---

**EXAMPLE  16-11**

How can this computer perform a JUMP to location $(56)_{16}$.

**SOLUTION**

$(56)_{16}$ looks like:

$$0 \quad 010 \quad 1 \quad 0110$$

Page          Instruction
Register      bits
bits

To get the proper address the Page Register must first be set to a 2 by a 1000 0010 instruction. The 1000 in the MSB indicates SET PAGE REGISTER and the 0010 sets the 2 into the Page Register. This is followed by a JMP $(16)_{16}$ instruction:

$$010 \quad 1 \quad 0110$$

JUMP          16
Op code       Jump Address

---

If the LOGIC instruction starts 1001, the second level Op code decoder at U33 (Fig. 16-21) provides for 16 additional instructions (not all implemented). None of these instructions may involve a memory operand. As seen at the output of U33, they are the SKIP instructions (see Sec. 16-9.6), SET and CLEAR the CARRY FF, INCREMENT, COMPLEMENT and ROTATE the accumulator, and HALT.

## 16-9.5    The Accumulator and ALU

The accumulator (U14 and U15) and the ALU (U1 and U2) (Fig. 16-18) function as described in Sec. 16-7.4. A set of NAND gates (Fig. 16-20) provides the control logic for the ALU. The $\overline{\text{LOAD}}$ instruction is executed in this computer by setting the ALU to the logic function F = B. As the memory data bus is connected to the B inputs, it becomes the accumulator output and is clocked into memory at $\overline{\text{E}} \cdot \phi_2$.

The output of the ALU is also meant to be connected to a set of LEDs or 7-segment displays, through buffers if necessary. The DATA/ACCUMULATOR SWITCH (Fig. 16-20), connected to U36, allows the lights to indicate either what is in the accumulator or on the data bus when the computer is halted. In ACCUMULATOR position, the switch places the ALU in the F = A mode, and in Data position it places the ALU in the F = B position. Thus the *same set of lights can read either the Data Bus or the Accumulator* depending on the switch setting.

The CARRY FF (U34, Fig. 16-18) is also associated with the ALU. On ADD, SUBTRACT, and ROTATE instructions it retains the final CARRY output. It also forms part of the control for the CARRY-IN to the least significant stage of the accumulator. This makes multiple precision arithmetic possible. Unfortunately, the CARRY FF must be SET before performing a single precision ADD and CLEARED before performing a single precision subtraction. The CARRY FF, however, allows the user to control the input to the LSB on ROTATE instructions.

## 16-9.6   SKIP instructions

A SKIP instruction causes the computer to skip the next instruction. There are no conditional branches in this computer, but there are a number of *conditional skips*. These instructions cause the SKIP OUT signal to be generated at U35 pin 11 (Fig. 16-22). SKIP OUT is applied to the COUNT-UP input of the PC at $\overline{E} \cdot \phi_2$ and increments the PC a second time, causing one instruction to be skipped.

---

**EXAMPLE 16-12**

A program must jump to location $(74)_{16}$ only if the accumulator is 0. How can this be done?

**SOLUTION**

The program should look like this:

$$
\begin{array}{ll}
1000\ 0011 & \text{Set Page Register to 3} \\
1001\ 0011 & \text{Skip on 0 accumulator} \\
1001\ 0000 & \text{Skip always} \\
0101\ 0100 & \text{Jump 14}
\end{array}
$$

If the accumulator is 0, the JUMP instruction will be executed; otherwise the SKIP ALWAYS instruction will be executed and skip over the JUMP instruction.

---

## 16-9.7   IN and OUT Instructions

The I-O instructions are designed to allow the computer to exchange information with external devices. The 4 MSBs of an IN instruction are 1110. This causes an $\overline{\text{IN}}$ pulse at U25 pin 10. The external data must be on the inputs to the IN Multiplexer (U3 and U4). A LOW $\overline{\text{IN}}$ signal gates the external data through the multiplexer and it is clocked into the accumulator at $E \cdot \phi_2$. At all other times $\overline{\text{IN}}$ is HIGH and the ALU outputs come through the multiplexer. The $\overline{\text{IN}}$ pulse can also be used as an *acknowledge* to let the external device know that the computer has accepted its data.

OUT instructions are generally used to transfer data from the computer's memory to an external device. The word to be transferred is loaded into the accumulator and an OUT instruction given. The data bus to the external device must be connected to the accumulator outputs (this is not shown in Fig. 16-17), and the external device clocks in the data when it receives the $\overline{OUT}$ pulse.

The 4 LSBs in an IN or OUT instruction are used as *port* designations. With proper decoding of the LSBs of the address bus, this computer can send or receive data to or from 16 external devices.

## 16-9.8   Retrospective

The ICs in this computer, wire wrap sockets, and a perforated board to hold them were purchased from discount suppliers and sold to the students at $50 per computer. They added some switches, LEDs, or 7-segment displays, and often a chassis. The cost, especially if split between two students, is still reasonable. We feel the depth of understanding of computers gained by this project adequately compensates the student for his or her cost and efforts.

It is unfortunate that this computer has to be programmed using self-modifying programs. A longer word length, index registers, indirect addressing, and subroutine jumps are highly desirable, but increase the cost and size. Of course, the computer as presented here can be augmented to include these features. For a few dollars more, anything is possible.

### SUMMARY

The object of this chapter was to show the reader how to design a rudimentary computer. The reader was first introduced to the basic instructions used in a computer and simple programs built from them. Flowcharts to help document and clarify program flow were discussed.

Then each basic part of the computer was discussed. The operations during the Fetch and Execute cycles were explained. Finally, two examples of complete computers were shown. The RIT computer was presented to demonstrate the practical problems involved in building a computer.

### GLOSSARY

**Branch instruction.** An instruction that alters the normal course of a program by causing it to branch or jump to another instruction.

**Conditional branch.** An instruction that causes the program to branch only if a certain condition is met.

**Event detection.** Using the occurrence of an event to terminate a loop or program.

**Flow chart.** A graphic method used to outline or show the progress of a program.

**Load.** An instruction that causes data to be brought from memory into an accumulator register.

**Loop.** Returning to the start of a sequence of instructions so that the same instructions may be repeated many times.

**Store.** An instruction that causes data in the accumulator to be moved to memory or a peripheral register.

**Accumulator.** A register or registers in a computer that hold the operands used by the instructions.

**Bus.** A group of interrelated wires; generally a bus carries a set of signals from one digital device to another.

**Control unit.** The internal parts of a computer that control and organize its operations.

**Execute.** The portion of an instruction cycle where the instruction is executed.

**Fetch.** The portion of an instruction cycle where the instruction is sent from memory to the instruction register.

**Increment.** To add 1 to a number.

**Input-Output (I-O) system.** The part of a computer that communicates with external devices.

**Instruction.** A command that directs the computer to perform a specific operation.

**Op code.** The portion of an instruction that tells the computer what to do.

**Program.** A group of instructions that control the operation of a computer.

**Program counter (PC).** A register in a computer that contains the address of the next instruction to be executed.

**Rotate.** A circular shift where the bits from one end of the word move to the opposite end of the word.

**Software.** Refers to the programs used in a computer.

### REFERENCES

Thomas C. Bartee, *Digital Computer Fundamentals*, 5th ed., McGraw-Hill, New York, 1981.

Boillot, Gleason, and Horn, *Essential of Flowcharting*, William C. Brown, Dubuque, Iowa, 1975.

Joseph D. Greenfield and William C. Wray, *Using Microprocessors and Microcomputers: The 6800 Family*, Wiley, New York, 1981.

## PROBLEMS

16-1.   Write a program to add the numbers 1, 6, 11, 16, . . . , 20, 001. Use the standard instructions.

16-2.   There is a set of numbers in locations 100 to 300. Some of them are 22. Write a program to count the number of 22s in the area, and store the result in location 99. Use a self-modifying program and the instructions we discussed.

16-3.   The positive integer N is in location 500. Write a program to store $(N + 2)!$ in location 501. Be sure to identify any constants you use.

16-4.   Write a program to transfer the contents of locations 2000 to 2047 to 2410 to 2457.

16-5.   Assume the memory data register and accumulator are made up of 7474s. Describe how the instruction 'OR 500' is fetched and executed and show a hardware implementation. The instruction ORs the contents of location 500 with the accumulator, the results remain in the accumulator.

16-6.   The 16-bit computer discussed in Sec. 16-8 must be able to execute the following instructions:

> Add
> Subtract
> And
> Or
> Shift (1 Bit)

Draw the memory data outputs, the ALU, the accumulator, and the Op code decoder. Show the connections between them to execute these instructions.

16-7.   The ALU consists of four 74181s. What is a CARRY-IN and CARRY-OUT of each 74181 if S = 15, CARRY-IN to the first IC is 1, M = 0, and the 16 A inputs are $(0000)_{16}$. Repeat if the A inputs are $(0B00)_{16}$.

16-8.   In Fig. 16-17 what is the state of CARRY-IN on:
  (a) An INCREMENT ACCUMULATOR instruction?
  (b) A SKIP ON ZERO ACCUMULATOR instruction?
In both cases, explain why your answer is so.

After attempting to solve the problems, try to answer the self-evaluation questions in Sec. 16-2. If any of them still seem unclear, review the appropriate sections of the chapter to find the answers.

# 17

# COMPUTER INTERFACES

## 17-1 INSTRUCTIONAL OBJECTIVES

This chapter considers the most common ways of communicating with, and interfacing to, a digital computer. The design of digital devices that must work and communicate with computers, such as peripheral controllers, is discussed in detail.

After reading the chapter, the student should be able to:

1. List the signal lines that comprise an I-O bus and explain their function.
2. Design controllers for slow speed devices using only I-O bus communications.
3. List the signal lines on a memory bus and explain their function.
4. Design a DMA channel into a controller.
5. Design a memory controller.
6. Design a controller that can produce interrupts and interrupt vectors.
7. List the steps involved in a high speed data transfer.
8. Utilize A/D and D/A converters.

## 17-2 SELF-EVALUATION QUESTIONS

Watch for the answers to the following questions as you read the chapter. They should help you understand the material presented.

1. What is the advantage of inverted data on an I-O or memory bus?

2. What is the function of terminating resistors on a bus line?

3. Explain how DONE, BUSY, DATA IN, and DATA OUT control data transfers between the processor and its peripheral devices.

4. How are device priorities determined on a memory bus? Which devices generally get top priority?

5. Why are interrupts used?

6. How does the processor determine which device on the I-O bus caused an interrupt?

7. How does a vectored interrupt system work? How are priorities determined?

8. What is the function of A/D and D/A converters? Why are they used?

## 17-3 INTRODUCTION TO COMPUTER INTERFACES

In Chapter 16, where the designs of rudimentary minicomputers were presented, communications between the computer and external devices (TTY, disks, tapes, lineprinters, etc.) was not covered thoroughly. This chapter and Chapter 19 concentrate on the topic of computer input-output (I-O). In this chapter minicomputer I-O is emphasized. Chapter 19 is devoted to microprocessor I-O.

Speaking broadly, computers function in one of two ways: they solve problems or they control processes. The problem-solving computer typically receives inputs and commands in the form of punched cards or paper tape and usually produces an output in the form of a line printer printout, which people can understand and use.

Process control computers[1] usually receive inputs from *sensors* that sense the magnitude of specific physical quantities and convert them into voltages. Temperature, humidity, shaft position, and gasoline consumption are some examples of physical quantities that can be sensed. The sensor output voltage is then converted to digital information by using *analog-to-digital* (A/D) converters and become an input to the computer. The outputs of a process control computer are commands to control the operation of the process and printouts to inform human operators of the system status.

Line printers, card punches, sensors, and so on are called *computer peripherals*. A computer must be able to communicate with its peripherals. Because computer manufacturers don't know which peripherals their customers are using and peripheral manufacturers don't know which computers their products will be tied to, this communication is not automatic. It is done via an *interface* or *controller*, which is a device that communicates with the computer and produces the proper commands to the peripheral.

[1]Process control computers are often called *processors*. The terms processor and computer are used synonymously in this book.

It also regulates data flow between the peripheral and the computer. There is usually one controller for each peripheral device in the computer system. It acts as a translator between the language of the peripheral and the language of the computer.

There are three common modes of communication between a computer and a controller; an *input-output (I-O) bus, interrupts,* and *direct memory accesses (DMA).* The I-O bus is normally used with relatively slow-speed devices, such as teletypes and card readers, while DMAs are normally made by high-speed devices, such as disks and drums. Interrupts are used with both types of data transfer. Data transfers between the computer and external devices are controlled by the computer program. The design of typical controllers is considered here and in Chapter 18.

## 17-4   THE I-O BUS

*The I-O bus is a group of wires that runs between the computer and the external devices it uses.* The path of a typical minicomputer I-O bus is shown in Fig. 17-1. Note that the bus talks to the device controllers rather than to the device itself.

Figure 17-1 shows a *daisy-chained* I-O bus. All devices on the bus receive the same signals. This is the most common type of I-O bus and is used on computers such as PDP-8 and the NOVA. Some computers, such as the HP 2100 use a separate bus to each interface. The daisy-chained bus uses open-collector drivers (see Sec. 5-6) and terminating resistors on the far end.

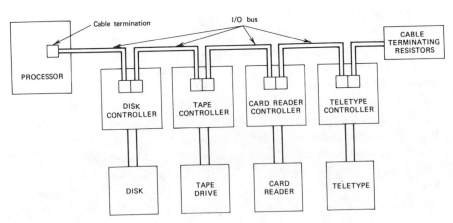

FIGURE 17-1   Routing of the I-O bus between the processor and several peripheral controllers.

The signals on an I-O bus are usually transmitted by *twisted pairs*. A twisted pair is a pair of wires tightly wrapped around each other. One wire carries a signal while the other wire is attached to ground in the devices at both ends of the cable. The advantage of twisted pairs is that the ground lead forms a shield around the signal lead that minimizes noise and *cross talk* (unwanted signal transmission between wires). Twisted pairs wires are available from wire and cable vendors. They can also be made in the laboratory by placing two wires in an electric drill and spinning it. The impedance of a twisted pair line is about 100 to 120 $\Omega$. This impedance should be matched for optimum data transmission.

## 17-4.1 Signals on the I-O Bus

A rudimentary minicomputer I-O bus consists of the following signals.[2]

1.  **Select lines.** These lines determine which of the several devices on an I-O bus is to communicate with the computer. Typical I-O busses have six select lines.
2.  **Data lines.** These lines transfer information between the computer and the external devices. Separate lines may be provided for data to and from the computer, or the data lines may be bi-directional.
3.  **Computer command lines.** These lines from the computer are used to issue *commands* to the controllers. Typical commands are:
    (a) Activate the device (set BUSY).
    (b) Clear the device.
    (c) DATA OUT.
    (d) DATA IN.
4.  **Device sense lines.** These lines transfer *status information* from the device controller to the computer. Typically, these lines contain the status of the BUSY, DONE, and INTERRUPT FFs (Sec. 17-4.6).

Lines on an I-O bus generally carry *inverted data*. A 1 on the I-O bus is represented by a LOW input, usually caused by a saturated transistor. The active level of all command signals should be LOW. This allows a device to be removed without disturbing the rest of the bus. When a device on the bus is removed, or has its power turned off, its output signals provide no path to ground and look like HIGH (inactive) levels.

## 17-4.2 Terminating Resistors

Each device on a daisy-chained bus must have two parallel sets of cable terminations, as shown in Fig. 17-1. One set accommodates the incoming bus; the other is for the cable that is connected to the next device in the chain. The last device on a bus must also have two sets of cable terminations.

[2]Microprocessor busses are considered in Chapter 19.

One set accommodates the cable and the other set is for the terminating resistors.

Terminating resistors are placed at the far end of the I-O bus. A typical set of terminating resistors is shown in Fig. 17-2. All lines are driven by 3-state or open-collector drivers; totem pole outputs are never used in bus lines. The terminating resistors provide TTL level outputs and low impedance. The impedance of the terminating resistors approximately matches that of the line in order to minimize noise and reflections. The best way to mount terminating resistors is on a card that plugs into the last cable connector on the controller at the end of the bus.

---

### EXAMPLE 17-1

There are nine controllers on an I-O bus. How can a tenth controller be added?

### SOLUTION

The terminating resistor card must be unplugged from the cable connector in the ninth controller. A cable is then connected between the ninth and tenth controllers and the terminating resistor card is plugged into the vacant cable connector on the tenth controller.

### EXAMPLE 17-2

For the I-O bus line of Fig. 17-2:
    (a) What is the terminating impedance of the line?
    (b) What voltage is on the line if the processor is sending out a 1?
    (c) What voltage is on the line if the processor is sending out a 0?
    (d) How much current must the line driver absorb?

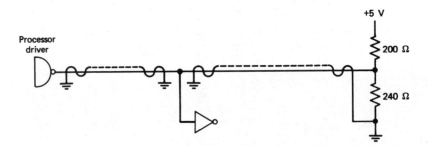

FIGURE 17-2 A single direction, twisted pair, I-O bus line and its terminatory resistor. (Reprinted from *How to Use the Nova Computers* by permission of Data General Corp. © Data General Corp. 1971.)

## SOLUTION

(a) The output of a power supply is usually a large capacitor that makes it effectively ground for AC or noise signals. Therefore, the 240 and 200$\Omega$ resistors are in parallel:

$$R = \frac{200 \times 240}{440} = 109 \; \Omega$$

This is approximately the characteristic impedance of a twisted pair line.

(b) The processor places a 1 on the bus by leaving its open-collector transistor off. Then the resistors function as a voltage divider:

$$V_{line} = \frac{240}{440} \times 5 = 2.7 \; V$$

(c) The processor places a 0 in the line by turning on its open-collector driver. This drives the line to $V_{CE(sat)}$, or about 0.2 V. Note that the logic 1 and 0 levels are TTL compatible.

(d) When the processor turns on it must sink the current through the 200-$\Omega$ resistor.

$$I_C = \frac{V_{CC} - V_{CE(sat)}}{R_A} = \frac{5 - 0.2}{200} = 24 \; mA$$

This is too much current for an ordinary TTL gate and a buffer should be used. ICs used for driving bus lines are 7406s, 7407s, and 7438s (a quad, 2-input, NAND, open-collector buffer/driver). These are all open-collector buffers and are capable of sinking 40 mA.

---

## 17-4.3 Bi-directional Busses

Data lines on an I-O bus are often bi-directional to minimize the number of bus wires. *Bi-directional lines transmit data from a selected controller to the processor or vice versa. The processor controls the direction of the data transfer.*

Each bi-directional line must be terminated in an open-collector (or 3-state) driver and a receiver as shown in Fig. 17-3. These can be TTL gates or special purpose driver/receivers. TTL manufacturers supply ICs specifically designed for line driving and receiving. The drivers have high current capability and the receivers have high impedances to prevent loading down the line. A bi-directional driver is called a *transceiver*. The 75138, shown in Fig. 17-4, is an example of a transceiver.

Bi-directional lines and lines that drive signals from the controllers to the processor are terminated at the end of the bus and within the processor. A bi-directional transmission line for a NOVA computer and its recommended

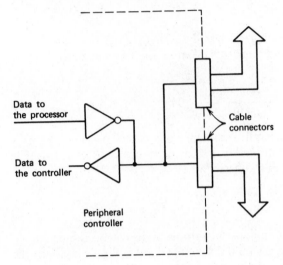

FIGURE 17-3 Termination of a bi-directional I-O bus data line within a controller.

termination is shown in Fig. 17-5. The diodes are to prevent the I-O bus voltages from going negative and also damp out severe "ringing" on the line.

## 17-4.4 DEVICE SELECT Lines

All controllers on the I-O bus receive the same set of signals but *only one must respond.* Each controller is given a *device address* and the I-O bus contains several DEVICE SELECT lines. *The signals on the DEVICE SELECT lines must match the device's address or the controller must not respond to any other signals on the I-O bus.* (Interrupt requests are sometimes an exception). When the signals do match, the controller develops a DEVICE SELECTED signal and responds to the commands on the I-O bus.

---

**EXAMPLE 17-3**

A controller has been assigned an I-O bus address of 30. Design a circuit to generate DEVICE SELECTED.

**SOLUTION**

The I-O bus is assumed to contain 6 device select bits. One solution is shown in Fig. 17-6, where the 6 address lines are simply gated together. The I-O bus is assumed to be carrying inverted data, and 30 is assumed to be binary (011110). DEVICE SELECTED is active when the output of the **7430** is low.

---

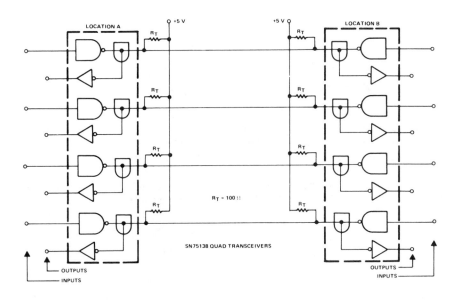

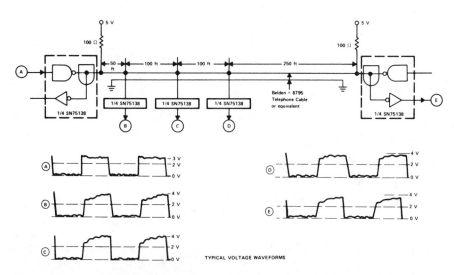

FIGURE 17-4 The **75138** transceiver and its use in a bi-directional bus. (From *Linear and Interface Circuits Applications Manual,* published by Texas Instruments, Inc. Courtesy of Texas Instruments, Inc.)

## 17-4.5 DATA IN and DATA OUT

DATA IN and DATA OUT are signals from the processor that tell the controller in *which direction data is flowing.* The directions of data flow are with respect to the processor. DATA IN means data flow *from the device to the processor,* and DATA OUT means data flow from the processor to

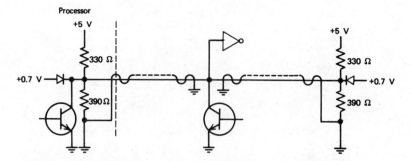

FIGURE 17-5 Typical termination of a bi-directional bus or a bus sending information from a controller to the processor. (Reprinted from *How to Use the Nova Computers* by permission of Data General Corp. © Data General Corp. 1971.)

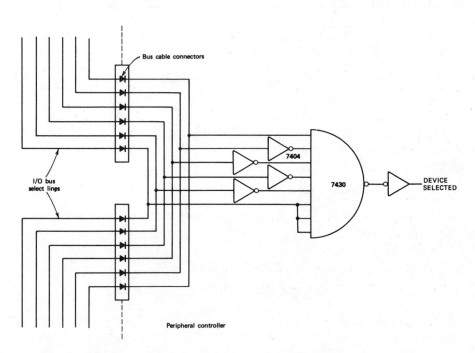

FIGURE 17-6 Decoding a device code from the I-O bus select lines.

the device. The controller accepts data or places them on the bus when it is selected and receives these signals.

---

**EXAMPLE 17-4**

Design the following simple controller:

1. When the processor activates DATA OUT, it places 4 bits on the I-O bus data lines. These 4 bits form a BCD digit which must appear on a 7-segment display.
2. When the processor activates DATA IN, the status of 4 switches within the controller must be sent to the processor via the same 4 data lines.

**SOLUTION**

A simple design is shown in Fig. 17-7. First the DEVICE SELECT circuitry is built as in Ex. 17-3. To receive the BCD digit, DEVICE SELECTED is ANDed with DATA OUT to clock the bus data bits into four 7474 FFs. The outputs of the FFs then drive a 7447 decoder/driver that turns on the 7-segment display.

To transmit information to the processor, the status of the switches is gated with DEVICE SELECT to drive the I-O bus when DATA IN is active. Note that the command and data lines are active (1) when they are LOW.

---

## 17-4.6  I-O Bus Status Lines

The I-O bus is controlled by four status lines:

□ CLEAR
□ START
□ BUSY
□ DONE

Most controllers on the bus are required to have BUSY and DONE FFs. When the processor decides to activate the device it issues a START command to the device controller, which sets the BUSY FF and clears the DONE FF. The controller then executes any commands the processor may issue to the peripheral device. When the execution of the command is finished, the controller clears BUSY and sets DONE.

The controller often effects a data transfer. If a data transfer from the processor is required, the data are strobed into the controller's register by DATA OUT. The controller then transfers the data to the peripheral at a rate determined by the peripheral's ability to accept data. When the transfer is complete, the controller clears BUSY and sets DONE.

If the processor requests a data transfer from the peripheral, the controller assembles a character from the peripheral. When it has the full character, the controller clears BUSY and sets DONE. The processor must recognize

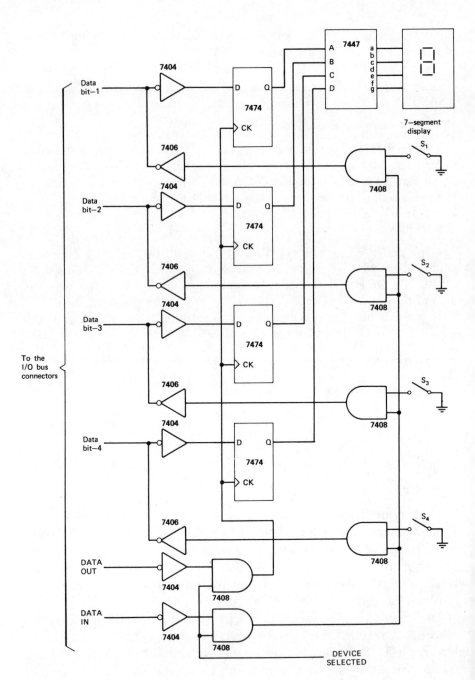

FIGURE 17-7  The controller for Ex. 7-4.

that DONE is SET and return a DATA IN to the controller in order to receive the character.

If interrupts are not used, the controller must periodically monitor the status of the DONE and BUSY FFs. This is accomplished via the DONE and BUSY lines on the I-O bus, which send data from the devices to the processor. Typically, whenever a device is selected, the status of its DONE and BUSY FFs drive the DONE and BUSY lines. To determine the status of a particular controller, the processor need merely send out the controller's address. It then receives that controller's DONE and BUSY status on its I-O lines.

For example, consider the design of a hypothetical controller for a paper tape reader. The paper tape itself has 8 columns of holes. The 8 holes or bits across the tape comprise an 8-bit character, as shown in Fig. 17-8. The 1s and 0s are determined by the presence or absence of a hole. Each character also has a small sprocket hole aligned with it, which is used for synchronization.

Our hypothetical paper tape reader accepts a tape advance signal from the controller, which causes it to move the tape from one sprocket hole to the next, or to advance the tape one character. When the reader detects a sprocket hole, it provides a TAPE CHARACTER READY signal and the 8 data bits corresponding to the holes in the tape for that particular character.

The controller has the following major requirements:

1.  When it receives a START signal from the processor, it must move the tape.
2.  When it receives TAPE CHARACTER READY, it must strobe in the character and inform the computer by setting its DONE FF.
3.  When the controller receives DATA IN on the I-O bus, it must place the character on the I-O bus data lines.

The design of the controller is shown in Fig. 17-9.

1.  The START signal sets the BUSY FF and clears the DONE FF.
2.  Setting the BUSY FF causes the paper tape to advance one character because the BUSY FF generates a TAPE ADVANCE command that is sent from the controller to the tape reader.
3.  After the tape advance, TAPE CHARACTER READY, which indicates that the tape reader has detected a sprocket hole, causes the data to be read into the controller buffer. It also clears BUSY and sets DONE. Clearing BUSY stops the tape advance.
4.  TAPE READER SELECTED goes high whenever the computer places the tape reader address on the I-O bus select lines. Note that it is gated with *all* the input signals. The status of the BUSY and DONE FFs is always placed on the I-O lines whenever TAPE READER SELECTED is HIGH.
5.  When the processor senses that the DONE FF is set, it can issue a DATA IN command. This places the character on the I-O bus.

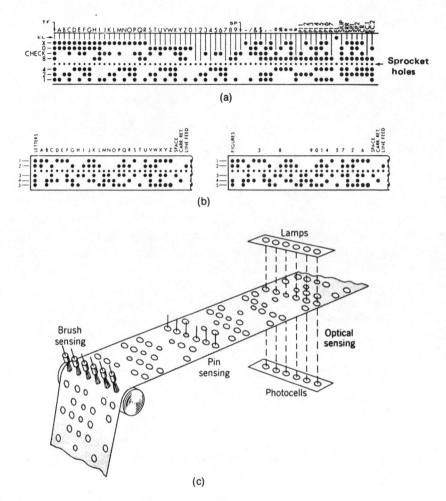

**FIGURE 17-8** Paper table showing (a) eight-channel code, (b) five-channel code, (c) methods of reading punched tape. (From Louis Nashelsky, *Introduction to Digital Computer Technology*, 2nd ed. Copyright 1977 John Wiley & Sons, Inc. Reprinted with permission of John Wiley & Sons, Inc.)

6.    The processor can then ask for another character by issuing another START command, or it can idle the tape reader by issuing a CLEAR command on the I-O bus. This clears the DONE and BUSY FFs and the reader will be idle until the next START command.

We have now designed a complete, albeit simple, controller using only the I-O bus. To design any controller, the engineer should check the signals available on the I-O bus for the particular computer he is using and the signals required by the peripheral (the tape reader in this case). The controller can then be designed by the methods shown in this section.

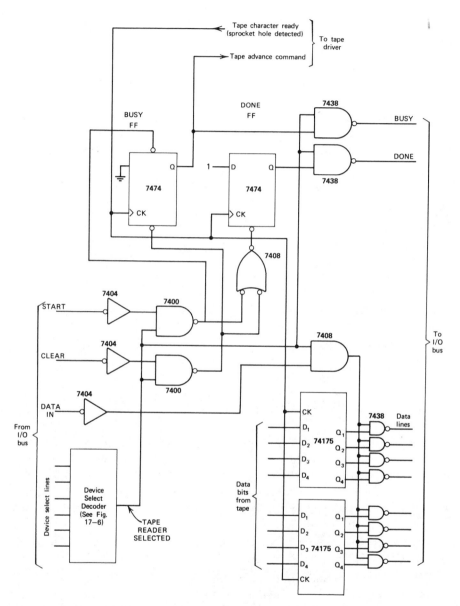

FIGURE 17-9 A tape reader controller.

## 17-5 DIRECT MEMORY ACCESSES

The higher-speed peripherals (e.g., drums and disks) must access a computer's memory frequently. This can be done via the I-O bus, but it is more efficient to grant high-speed peripherals the ability to communicate directly with memory. This is known as *direct memory access* (DMA). The principles of DMA and a rudimentary DMA system are described in this section.

A memory cycle, once started, cannot be interrupted. It must run to completion and send the results to the requesting device. The processor itself is a device that continually requires memory cycles. Under a DMA system, the processor competes with the peripherals for memory cycles. This slows down the processor because it may have to wait several memory cycles before its memory request is granted, if several peripherals are active. Nevertheless, DMA is the fastest way to transfer data because it does not require specific I-O instructions to exchange a word of data between the computer's memory and the peripheral.

Since many devices, including the processor, are competing for memory cycles, a *memory controller* is required. *The function of the memory controller is to regulate traffic and control memory priorities. Memory access from all devices must go through the memory controller.*

### 17-5.1 The Memory Bus

The memory bus runs between the memory, or the memory controller, and the peripherals. One way to design the memory bus is shown in Fig. 17-10.

The memory bus is split into two parts. The common bus runs between the memory and all the peripheral devices. It is "daisy-chained" like an I-O bus and the last device on the line must have terminating resistors. There is also a dedicated bus running between each peripheral and the memory controller. In the simplest case, we assume a static RAM memory and the only direct communication between the memory and the memory controller is the DATA AVAILABLE line.

The peripheral controller making the memory request must supply the memory with the address, the data on a write request, and the direction of data transfer (from memory to controller on a READ cycle, and the reverse on a WRITE cycle).

The common memory bus contains the following signals (which are common to all devices on the memory bus).

1. **Data.** A set of bi-directional lines carrying data between the memory and the controllers. There must be one data line for each bit in a memory word.
2. **Address.** A set of lines carrying the address from the controllers to the memory. There must be one line for each bit in the address.

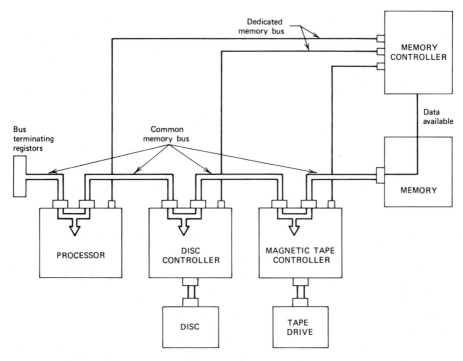

FIGURE 17-10  A DMA memory bus.

3. **Read/write.** A single line that controls the direction of data transfer.
4. **Data available.** This signal is transmitted from the memory to the controllers. Its appearance terminates a memory cycle.

## EXAMPLE 17-5

For a 4K-word-by-16-bit memory, how many lines are required on the memory bus? Assume the lines are twisted pairs.

## SOLUTION

This memory bus requires:

- □  16 bi-directional data lines
- □  12 address inputs (from the device controllers to the memory)
- □  1 READ/WRITE command (from the device controllers to the memory)
- □  1 DATA AVAILABLE (from the memory controller to the device controllers)

This is a total of 30 lines. Since each line is twisted with a companion ground wire to reduce noise and crosstalk, 60 lines are required and the cable connectors at each end of the bus must have at least 60 terminations.

Each dedicated bus, which runs from the memory controller to the device controllers, need have only two lines:

1.   **Memory request *n*,** from device *n* to the memory controller. A LOW level on this line indicates device *n* is making a memory request.
2.   **Memory acknowledge *n*,** from the memory controller to device *n*. This signal indicates device *n* has been granted the memory cycle.

In some cases, it might be advantageous to run a single common memory bus that contains a pair of dedicated lines for each controller on the bus.

## 17-5.2   Memory Bus Protocol

Since all device controllers share the common memory bus, a *bus protocol* must be set up to prevent the devices from interfering with each other.

A simple protocol for the memory bus described here is:

1.   Whenever a device controller needs to communicate with memory, it places its MEMORY REQUEST signal on the dedicated bus.
2.   A device controller may not place any signals on the common bus until it receives MEMORY ACKNOWLEDGE on the dedicated bus.
3.   When a requesting device receives MEMORY ACKNOWLEDGE, it places the memory address, data (if WRITE), and READ/WRITE signals on the line.
4.   When the acknowledged device receives DATA AVAILABLE, it may remove MEMORY REQUEST. It must also use DATA AVAILABLE to strobe in the data.

## 17-5.3   Design of the Device Controller Interface to the Memory Bus

Each device controller must have its own MAR, MDR, and READ/WRITE FF. Through its internal circuitry and its communication with the peripheral device, the controller determines when it requires a memory request. At this time it sets its MEMORY REQUEST FF. The address in the MAR, the data in the MDR, and the READ/WRITE signal must be firm in the device controller's MAR and MDR at this time.

When MEMORY ACKNOWLEDGE is received, the device controller gates its MAR and MDR onto the common bus. When DATA AVAILABLE occurs, it strobes the bus data into the MDR on a READ cycle and also terminates MEMORY REQUEST.

A circuit to accomplish this is shown in Fig. 17-11. For simplicity, only one bit of the MAR and one bit of the MDR are shown. In WRITE mode as determined by the READ/WRITE FF, data from the MDR are placed on the bus when MEMORY ACKNOWLEDGE is received, but in READ

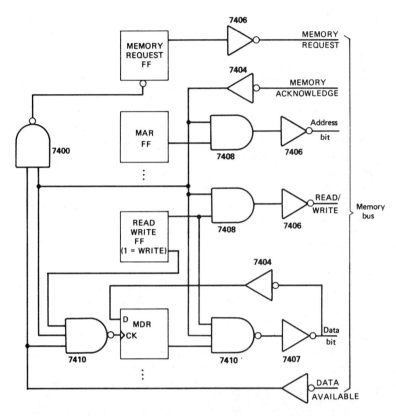

FIGURE 17-11  Device controlled interface. *Note:* Only one bit of the MAR and MDR are shown.

mode, the data from the bus are strobed into the MDR by DATA AVAIL-ABLE. Note that a logic 1 is a LOW level on the bus and the occurrence of MEMORY ACKNOWLEDGE and DATA AVAILABLE reset the MEMORY REQUEST FF.

## 17-5.4  Design of a Memory Controller

A memory controller can be designed to work with the memory bus and controller discussed in Secs. 17-5.2 and 17-5.3. This controller uses the 74148 priority interrupt controller whose pinout and function table are given in Fig. 17-12.

The 74148 can handle memory requests from eight different devices. A device requests memory by driving its MEMORY REQUEST line LOW, as shown in Fig. 17-11. The LOW input to the 74148 causes its GS output to go LOW; it also causes the A outputs to present the number of requesting

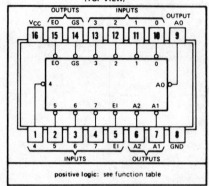

SN54148, SN54LS148 ... J OR W PACKAGE
SN74148, SN74LS148 ... J OR N PACKAGE
(TOP VIEW)

positive logic: see function table

**(a) Pinout**

'148, 'LS148
**FUNCTION TABLE**

| | INPUTS | | | | | | | | OUTPUTS | | | | |
|---|---|---|---|---|---|---|---|---|---|---|---|---|---|
| EI | 0 | 1 | 2 | 3 | 4 | 5 | 6 | 7 | A2 | A1 | A0 | GS | EO |
| H | X | X | X | X | X | X | X | X | H | H | H | H | H |
| L | H | H | H | H | H | H | H | H | H | H | H | H | L |
| L | X | X | X | X | X | X | X | L | L | L | L | L | H |
| L | X | X | X | X | X | X | L | H | L | L | H | L | H |
| L | X | X | X | X | X | L | H | H | L | H | L | L | H |
| L | X | X | X | X | L | H | H | H | L | H | H | L | H |
| L | X | X | X | L | H | H | H | H | H | L | L | L | H |
| L | X | X | L | H | H | H | H | H | H | L | H | L | H |
| L | X | L | H | H | H | H | H | H | H | H | L | L | H |
| L | L | H | H | H | H | H | H | H | H | H | H | L | H |

**(b) Function Table**

FIGURE 17-12 The **74148**. (From the *TTL Data Book for Design Engineers*, 2nd ed., Texas Instruments, Inc. Courtesy of Texas Instruments, copyright 1976.)

inputs. The EI input acts as a strobe; if it is HIGH, the IC is disabled and all outputs are HIGH.

The **74148** *prioritizes* its incoming requests. The function table (Fig. 17-12*b*) shows that input 7 has the highest priority; if it is LOW, all A outputs are LOW regardless of any other input. Input 6 has the next priority, and so on, down to input 0, which will be recognized only if all other inputs are inactive (HIGH).

A memory controller designed using the **74148** is shown in Fig. 17-13. It works as follows.

1. If one or more of the 8 MEMORY REQUEST inputs goes LOW, it drives GS LOW.
2. This sets the MEMORY CYCLE FF, which disables the **74148**, but clocks the A outputs into the ACK Register, which drives the ACK decoder and provides MEMORY ACK to the highest priority requesting peripheral. Note that when the memory cycle FF is set, the decoder is disabled by the HIGH level on its strobe inputs.
3. The MEMORY CYCLE FF also triggers the ACCESS TIME one-shot (O.S.). The time of this one-shot is set equal to the worst case memory access time, plus data transmission times.
4. When the ACCESS TIME O.S. expires, it triggers the DATA AVAILABLE ONE SHOT. This drives the DATA AVAILABLE signal to all devices.
5. When DATA AVAILABLE expires, it clears the MEMORY CYCLE FF. This enables the **74148** to respond to the next memory request.

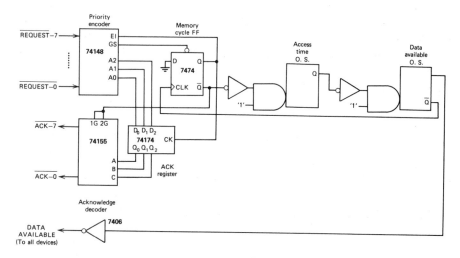

FIGURE 17-13  A memory request controller.

### EXAMPLE 17-6

While processing MEMORY REQUEST-1, MEMORY REQUEST-3 and MEMORY REQUEST-5 arrive. Describe how the memory controller reacts.

### SOLUTION

While processing MEMORY REQUEST-1, the **74148** is disabled by the MEMORY CYCLE FF and the arrival of higher priority requests does not affect the cycle in progress. When the cycle ends the MEMORY CYCLE FF clears. This allows GS to go LOW, setting the MEMORY CYCLE FF again and clocking 010 into the ACK register. This causes MEMORY ACK-5 to go LOW, providing an acknowledgment to device 5. The **74148** gives the highest priority requesting device the acknowledgment. After the request has been processed, MEMORY REQUEST-5 returns HIGH and MEMORY REQUEST-3 is processed, if no higher priority requests were received while MEMORY REQUEST-5 was being processed.

## 17-5.5  Priorities

In Fig. 17-13 the device connected to MEMORY REQUEST-7 gets the highest priority. The system designer must determine which device gets the highest priority. *High speed devices* like a disk *get high priority*. When a disk is being read, words are coming from it every 5 to 10 $\mu$s. The disk controller usually has very little storage capability, and must write the words to memory as quickly as they are received. The devices with the highest word transfer rates are assigned the highest priority in a memory controller.

The processor itself is treated as a device by the memory controller. Generally the processor is assigned low priority because it can afford to wait even though this slows down the progress of the program.

## 17-6 INTERRUPTS

When the processor issues a command to a device controller via the I-O bus, or asks for a series of data transfers via the memory bus, *it must be able to determine when the device has completed the action in order to proceed.* When the device controller has finished its assigned task, the processor may proceed by:

1. Giving it more data if it requires additional data transfers from memory to the device.
2. Acting on the information received or requesting more information if the processor is reading information.
3. Giving the controller another command.

We saw in Sec. 17-4.6 that each device controller has a DONE flag (and perhaps other status bits) that can be tested by the processor to determine if the controller has finished its command. But this requires that the processor periodically test the status bits to see if DONE is set and causes the processor to execute many extra instructions. Device controllers are often required to have *interrupt capability* to alleviate this problem.

*An interrupt is a signal from the device controller to the processor; the signal informs the processor that the controller has completed its command, or has information for it. The processor then takes appropriate action via its program.* A detailed explanation of programming, especially interrupt programming, is beyond the scope of this book and is not absolutely necessary for designing controllers.

### 17-6.1 Interrupts on the I-O Bus

Perhaps the most common way of handling interrupts is to add an INTERRUPT line to the I-O bus. Whenever a device wishes to interrupt, it drives the I-O bus INTERRUPT line LOW. The processor recognizes the interrupt request and tests the status of the device controllers in turn to determine which device caused the interrupt. This allows the processor to establish interrupt priorities. The device that has the highest interrupt priority is tested first. Many device controllers generate an interrupt whenever the DONE FF sets. Other events associated with its peripheral may also cause a controller to generate an interrupt.

In most systems the processor can allow or prevent a controller from interrupting. This is done by having the processor send out a MASK signal and a MASK bit. When the MASK signal is received at the selected controller, it sets an internal INTERRUPT ENABLE FF that depends on the status of the MASK bit.

---

**EXAMPLE 17-7**

A device controller is to interrupt provided:
  (a) Its DONE FF is set.
  (b) The MASK bit was 1 at the leading edge of the last MASK signal.
Design this part of the controller.

**SOLUTION**

The design is shown in Fig. 17-14. An INTERRUPT MASK FF is included within the controller. It sets or clears on the leading edge of the MASK signal depending on the MASK bit. Whenever both the DONE and INTERRUPT MASK FFs are set, the controller drives the interrupt I-O bus line LOW. The processor can terminate the interrupt by clearing the INTERRUPT MASK FF or by servicing the controller and clearing the DONE FF.

---

## 17-6.2 Vectored Interrupts and Interrupt Chaining

The process of querying each device controller to determine which one interrupted also requires significant processor time. The process can be shortened by using *vectored interrupts*.

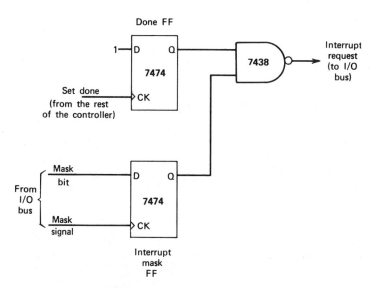

FIGURE 17-14 Generating an interrupt request within a controller.

With vectored interrupts, each controller has its own unique interrupt vector. An interrupt vector is a group of bits that identify the controller to the processor. A device controller places its interrupt vector on the data lines when it is interrupting and it receives an INTERRUPT ACKNOWL- EDGE pulse from the processor.

Since several devices may interrupt simultaneously and the interrupt lines are common, the following protocol is typically set up to determine which device is serviced, and to establish a set of interrupt priorities so that the most critical device controller is serviced first.

1. Each device controller has an INT-ACK-IN (interrupt acknowledge in) and an INT-ACK-OUT (interrupt acknowledge out) line.
2. When the processor receives an INTERRUPT REQUEST, it sends an IN- TERRUPT ACKNOWLEDGE pulse to the first device on the I-O bus. The highest priority device controller is assigned the first position on the I-O bus.
3. The first controller receives the pulse on its INT-ACK-IN line. If it is currently requesting an interrupt, it places its INTERRUPT VECTOR on the data bus, and does not generate INT-ACK-OUT.
4. If the first controller is not requesting an interrupt and it receives INT-ACK- IN, it generates INT-ACK-OUT, which goes to INT-ACK-IN of the second con- troller. In this manner, the INTERRUPT ACKNOWLEDGE pulse is passed from controller to controller, until it is intercepted by the highest priority controller requesting an interrupt.
5. The processor receives the interrupt vector from the highest priority device requesting service and services it. During this process the INTERRUPT REQUEST from the controller should be turned off. Then the next INTERRUPT ACKNOWL- EDGE signal sent from the processor can proceed to another controller.

---

## EXAMPLE 17-8

Design the interrupt system for a device controller. Assume its interrupt vector is 101110.

## SOLUTION

When INT-ACK-IN is received, the controller must put its INTERRUPT VEC- TOR on the data lines if it is interrupting; otherwise, it must generate INT-ACK- OUT. The design is shown in Fig. 17-15. If the DONE and MASK FFs are both set, the controller is requesting an interrupt and point A goes LOW. This cuts off the INT-ACK-OUT gate and enables the INTERRUPT VECTOR gates. The data inputs to the INTERRUPT VECTOR gates are wired to ground or $V_{CC}$ to generate the controller's particular vector. If the controller is not interrupting, the appearance of INT-ACK-IN causes the controller to send INT-ACK-OUT to the next controller.

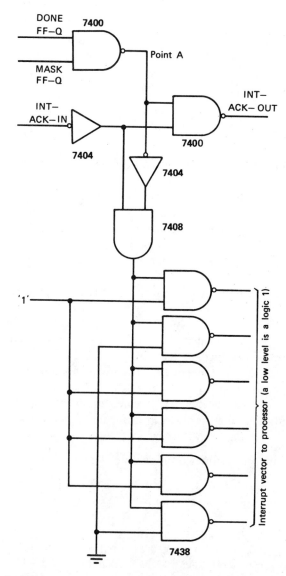

FIGURE 17-15  A circuit for generating an interrupt vector.

## 17-6.3  High Speed Data Transfers

Data transfers between the processor and high speed peripherals such as drums and disks rarely proceed one word at a time. Typically, a single command results in the transfer of a group of words called a *data block*, between the processor and the peripheral. Typical data block sizes are 128 and 256 words. These data transfers, which must proceed via the peripheral controller, often utilize the I-O bus, the memory bus, and interrupts.

For example, assume the processor is to read a block of 256 words from a disk. Before issuing such a command, the processor must set aside a block of 256 words in its memory to receive the data. The data transfer would typically proceed in the following manner.

1.   Using the I-O bus, the processor issues a command to the disk controller. This command must specify:

   (a) The direction of data transfer. (In this case data go from the disk to the processor's memory.)

   (b) The place on the disk (the disk address) where the data to be read exist.

   (c) The address of the first word in memory where the data are to be stored.

2.   The disk controller gives the READ command and the address to the disk.

3.   After the latency time, which is the time required for the disk to rotate so that the specified address is under the READ heads, words start coming off the disk every few microseconds.

4.   When the controller receives the first word, it makes a WRITE request to memory via the DMA port.

5.   When the DMA request is honored, the controller writes the first word into the processor's memory, then it increments its MAR so that the next word from the disk is written to the next sequential memory location.

6.   When the next word is received, it is also written to memory. This continues until all 256 words in the block have been written to memory.

7.   After the controller has completed the transfer, it generates an interrupt to inform the processor that the data are now available in its memory, and that the controller is now idle and may accept another command.

The disk controller generally performs two other functions.

1.   **Reformatting.** The memory words may not contain the same number of bits as a disk word. In this case, the words would have to be *reformatted* to make the two devices compatible.

2.   **Error control.** The disk controller must report to the processor (usually via interrupts) any errors it detects. Two common errors are parity (between the controller and the memory or between the controller and the disk) and DATA LATE. DATA LATE occurs if the disk controller cannot write words to memory as fast as it must read them from the disk (see Problem 17-10). In this case the controller has more information than it can store and must discard some of it. Disk controllers are given high priority on the DMA channel to prevent DATA LATE errors.

---

### EXAMPLE 17-9

Assume the processor's memory contains 16-bit words and each word is transferred to the disk one bit at a time. Assume also that the disk sends a synchronizing signal approximately every 0.5 $\mu$s to request a bit. Describe the function of each part of the disk controller that would be needed for this interface.

### SOLUTION

The necessary parts of the disk controller are:

1. The I-O bus gates.
2. The memory bus gates.
3. The MAR for disk DMA requests. This must be loaded on an I-O command and must be capable of being incremented.
4. The MDR for holding the data being received from or written to the memory.
5. A 16-bit parallel-in, parallel-out, shift register, which shifts the data in from the disk or out from the memory to the disk.
6. The disk address register.
7. Control circuitry for detecting disk address matches and errors and generating interrupts.

### EXAMPLE 17-10

For the controller of Ex. 17-9, describe how the shift register and the MDR work together when the disk is being read.

### SOLUTION

The disk synchronizing circuit produces a pulse every 500 ns, which clocks the data into the shift register. After 16 pulses, the shift register is full and the controller must transfer it to the MDR. The controller now makes a WRITE request via DMA. The processor has 8 $\mu$s in which to write the data into its memory while the next disk word is being shifted in.

---

Although the design of controllers is involved, and requires a detailed knowledge of the I-O signals of the processor and the peripheral, it still comes down to a group of gates and registers and should not be beyond the ability of the reader at this point.

## 17-7 COMMUNICATION BETWEEN THE ANALOG AND DIGITAL WORLD

Computers or special purpose digital controllers are often used to monitor and control processes rather than to solve problems. The inputs and outputs to these computers are generally *analog* quantities. For example, a computer might monitor the ambient temperature in an oven. If it exceeds a certain limit, the computer must reduce the fuel input. Microprocessors monitor many processes. They are being considered for use in automobiles, where they will monitor the engine vacuum, RPM, and so on and deliver (hopefully) the optimum mixture of air and gas to the carburetor.

The input analog quantities, which are used to monitor the process, must be converted to digital information before they can be used by the processor or digital controller. This requires an *analog-to-digital* (A/D) converter. We have already seen one A/D converter in the code wheel (Sec. 13-7.2) that converted a shaft position to digital information.

The digital output quantities must often be converted to analog quantities to control a process. Voltages and currents are typical analog quantities. This requires a *digital-to-analog* (D/A) converter.

### 17-7.1 D/A Converters

A D/A converter accepts a digital input and produces an analog output equal, as nearly as possible, to the value of the digital input. The analog output of a D/A converter is shown in Fig. 17-16.

---

**EXAMPLE 17-11**

A D/A converter has a maximum output of 10 V and accepts 6 binary bits as inputs. How many volts does each analog step represent?

**SOLUTION**

With a 6-bit input, the 10 V can be divided into 64 levels. Therefore, each analog step represents $\frac{10}{64}$ = 0.15625 V. This is the *resolution* of the output. Greater resolution can be obtained by using more binary bits.

---

D/A converters are usually built using weighted resistor networks. Two examples are shown in Fig. 17-17. For the summing network (Fig. 17-17a), the analog output must be at ground potential. For the ladder network, the analog output must have an impedance of 2R to ground.[3] In these circuits, "ground" is usually the input of an operational amplifier.

---

**EXAMPLE 17-12**

Assume the digital input to the summing network of Fig. 17-17a is 100010, where 1 = + 5 V and 0 = ground. How much current flows into the analog output if it is at ground potential?

---

[3]A thorough discussion of the ladder network is given in Barna and Porat, Chapter 10. (See Sec. 17-10.)

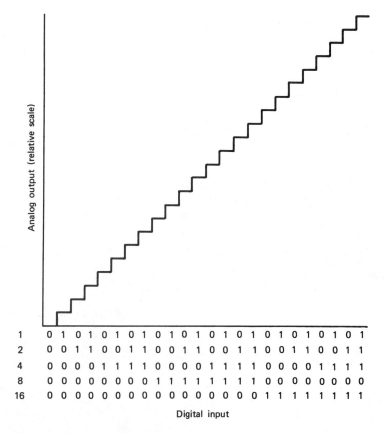

FIGURE 17-16   Analog output versus digital input in a D/A converter. (From George K. Kostopoulos. *Digital Engineering.* Copyright 1975, John Wiley & Sons, Inc. Reprinted by permission of John Wiley & Sons, Inc.)

## SOLUTION

With the given digital input, current flows in the $2^5$ and $2^1$ branches and their sum flows into the analog output.

$$I = \frac{5}{R/32} + \frac{5}{R/2} = \frac{170}{R}$$

The output current is proportional to the binary input number. If $R$ is chosen as 5000 $\Omega$, $I = 34$ mA. For $R = 5000$ $\Omega$, the output current in mA equals the binary value of the input bits.

## 17-7.2   Analog-to-Digital Converters

There are two popular methods of building an A/D converter to convert an analog quantity (typically a voltage) to digital. Both utilize comparators. A

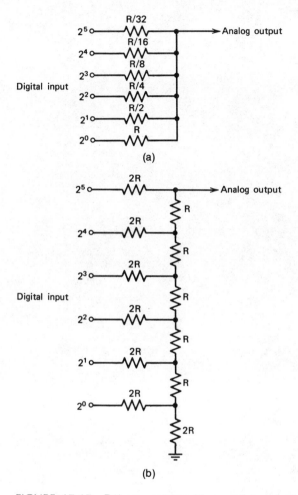

(a)

(b)

FIGURE 17-17   D/A converters: (a) summing network; (b) ladder network.
George K. Kostopoulos, *Digital Engineering.* Copyright 1975, John Wiley &
Sons, Inc. Reprinted by permission of John Wiley & Sons, Inc.)

*comparator* (shown in Fig. 17-18a) *detects a small difference in the input
voltage and produces a digital output.* Figure 17-18b shows that if the voltage
difference between the inputs is $-2$ mV or more, the output is $+3$ V, and
if the input voltage difference is $+2$ mV or more, the output is $-0.5$ V.

One method of A/D conversion is *successive approximations*. To perform
the conversion, a D/A converter and a comparator are used. The inputs to
the comparator are the unknown voltage to be converted and the output of
the D/A converter. Digital inputs are applied to the D/A converter until its
output most closely approximates the analog voltage.

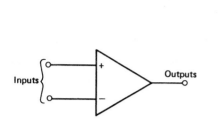

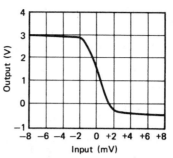

(a) Symbol

(b) Output voltage vs. input voltage

FIGURE 17-18 An analog voltage comparator. (From Barna and Porat, *Integrated Circuits in Digital Electronics.* Copyright 1973, John Wiley & Sons, Inc. Reprinted by permission of John Wiley & Sons, Inc.)

The simplest way to build this converter is to use a counter to increment the digital inputs. The counter stops when the comparator changes states.

---

### EXAMPLE 17-13

(a) Design an A/D converter. Assume the input voltage and the input of the D/A converter can vary between 0 and $+10$ V. Use the comparator of Fig. 17-18a and a pushbutton to clear the counter.

(b) How long does the conversion take if the analog input voltage is 6 V and the clock frequency is 1 MHz?

### SOLUTION

(a) The design is shown in Fig. 17-19. Depressing the pushbutton clears the counter causing the output of the comparator to be HIGH. When the pushbutton is released, the counter increments. The output of the A/D converter increases until it is greater than the analog input. At this time, the output of the comparator goes LOW and stops the clocks to the counter by cutting off the AND gate. The number locked into the counter is the digital equivalent of the analog voltage.

(b) The resolution of this counter is 0.15625 V. If the analog input is 6 V, the counter stops when the A/D converter output becomes greater than 6 V, or after 39 counts ($39 \times 0.15625 = 6.09375$). Therefore, it requires 39 $\mu$s to make the conversion.

---

Successive approximations can be made more quickly by first applying the MSB to the D/A converter, determining if the resultant output is too large, and removing the MSB if it is. Then the next MSB is applied and

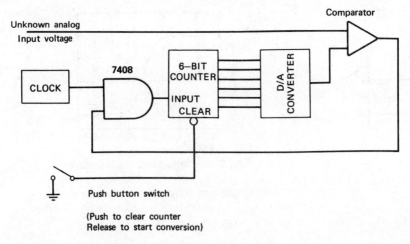

FIGURE 17-19  Design of an A/D converter.

the procedure repeated. This method requires more sophisticated circuitry, but results in a faster conversion.

Another method of A/D conversion, sometimes used in digital voltmeters, is to apply the unknown voltage to one input of a comparator and a RAMP voltage, which increases linearly with time, to the other input. The higher the unknown voltage, the longer the RAMP voltage takes to reach it and trigger the comparator. While the RAMP voltage is rising, a counter is being incremented by a fixed frequency clock. Higher voltages allow more clock pulses before the comparator fires and blocks them. Thus the final count is proportional to the unknown analog voltage.

## 17-7.3  Commercial A/D and D/A Converters

Today, most engineers buy A/D and D/A converters, rather than building them. The important parameters of an A/D converter are:

1.  **Resolution.** The number of binary bits of output.
2.  **Maximum conversion time.** The worst case time of a conversion.
3.  **Output codes.** A/D converters are available with either binary or BCD outputs.
4.  **Analog input range.** The range of analog input voltages that are converted.
5.  **Input impedance.**
6.  **Power dissipation.**
7.  **Size.**
8.  **Price.**

The parameters that apply to a D/A converter are:

1.  **Resolution.** The number of binary input bits.
2.  **Error.** The worst case error output as a percent of full scale.

3.  **Output mode.** Current or voltage outputs are available in D/A converters.
4.  **Settling time.** The time required for the output analog voltage to become firm.
5.  **Input coding.** D/A converters may accept binary or BCD inputs.
6.  **Analog output range.** The range of the analog output voltage or current.
7.  **Power.**
8.  **Size.**
9.  **Price.**

The engineer should consult the manufacturer's catalogs to determine the best D/A or A/D converter for his particular application.

## 17-7.4   The DAC 0800 D/A Converter

The **DAC 0800** is a readily available D/A converter manufactured by the National Semiconductor Corporation (see References). It is shown in Fig. 17-20 as set up in the laboratory to provide approximately a $\pm 5$ V output.

The **DAC 0800** provides complementary current outputs on pins 2 and 4, but the output on pin 2 was connected to ground, and not used. The power supply voltages were $\pm 10$ V. The eight TTL digital inputs were applied to pins 5 to 12. With pin 4 connected to $+5$ V through a 10-K$\Omega$ resistor, the output was found to vary between $\pm 4.9$ V. This limit can be adjusted by changing the voltage on the 10-K$\Omega$ resistor.

The positive reference voltage was applied to pin 14, and the negative reference was applied to pin 15 via the 5K resistors, as the manufacturer

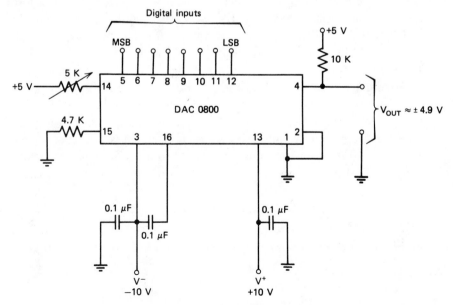

FIGURE 17-20   The **DAC 0800** as set up in the laboratory.

recommends. Pin 15 was grounded and the potentiometer connected to pin 14 was adjusted so that the mid-scale voltage (with the digital inputs at 10000000 or 01111111) was as close to zero as possible. The analog output at pin 4 then precisely followed the digital input.

---

**EXAMPLE 17-14**

Assume the **DAC 0800** is adjusted for a $\pm 5$-V output. What output voltage would we expect if the digital inputs are 10010110 or $(96)_{16}$?

**SOLUTION**

The 10-V analog range is divided into 256 intervals by the 8 digital inputs, so each digital increment corresponds to 10/256 or 0.0390625 V. The digital input corresponds to the decimal number 150 so the voltage output is 150 times 0.0390625 or 5.86 V. This is 5.86 V above the $-5$ V that occurs when the digital inputs are all 0, and the output voltage should be 0.86 V.

---

## 17-7.5 The ADC 0800 A/D Converter

One example of a commercially available A/D converter is the **ADC 0800** (**MM4357B/MM5357B**) 8-bit converter also manufactured by National Semiconductor Corporation. Its block diagram is shown in Fig. 17-21. For TTL compatible outputs $V_{SS} = 5$ V and $V_{GG} = 12$ V. The range of analog voltages to be converted is determined by the voltages applied to the top and bottom of the resistor network (R-NETWORK TOP, pin 15, and R-NETWORK BOTTOM, pin 5). These voltages must be between $V_{SS}$ and $V_{GG}$. Generally they are $+5$ and ground or $+5$ and $-5$ V.

The timing for the **ADC 0800** is shown in Fig. 17-22. The analog voltage to be converted is applied to $V_{IN}$ and a start of conversion (SOC) pulse applied. This causes end-of-conversion (EOC) to go LOW for 40 clock periods while the selection and control logic matches the analog input to the output of the resistor network. When the conversion is complete, EOC goes HIGH and the digital output may be read. A more thorough explanation of the **ADC 0800** is given in the manufacturers notes (see References).

---

**EXAMPLE 17-15**

Explain how continuous conversions can be obtained using the **ADC 0800**.

**SOLUTION**

The EOC output of the **ADC 0800** can be connected to the SOC input. Whenever a conversion is complete EOC and SOC go HIGH. The HIGH on SOC causes

## Block Diagram

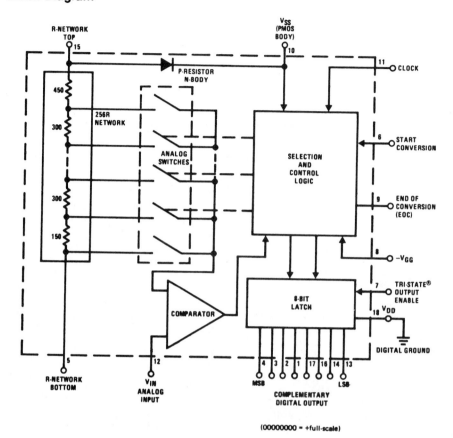

FIGURE 17-21   The **ADC 0800** A/D converter. (Copyright 1980, National Semiconductor.)

the **ADC 0800** to initiate another conversion, which causes EOC and SOC to go LOW. When the conversion is complete the process repeats.

One simple way of testing an A/D converter is to apply a known input voltage to $V_{IN}$ and connect the digital outputs to LEDs. As the input voltage varies, the LEDs should indicate a binary number corresponding to the input.

### SUMMARY

This chapter discussed the design of controllers that handle the communication between a minicomputer and its peripheral devices. It emphasized the communication between the minicomputer and the external world. The I-O bus, memory

## Timing Diagram

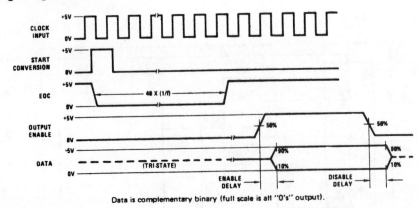

Data is complementary binary (full scale is all "0's" output).

FIGURE 17-22 Timing for the **ADC 0800**. (Copyright 1980, National Semiconductor.)

bus, and interrupt systems were discussed. Examples of designs involving each mode of communication were presented.

In Sec. 17-7, the basic principles of A/D and D/A conversion were explained. The parameters of commercial A/D and D/A converters were also discussed.

### GLOSSARY

**Digital subsystem.** A group of ICs used to perform a specific task.

**Processor.** A term often used to mean the controlling computer.

**Computer peripheral.** A device that receives commands from, or exchanges data with, a computer.

**Controller.** A digital subsystem that handles the transfer of commands and data between a processor and a peripheral device.

**Interface.** Sometimes used as a synonym for controller. It also sometimes refers to connections between devices.

**Bus.** A group of wires or cables carrying a related set of signals.

**I-O bus.** Input-output bus, used for programmed data transfers between a computer and its peripherals.

**Daisy-chain cable.** A cable that goes from one device to another, to another, and so on.

**Twisted pair.** A pair of wires wrapped around each other. One wire carries the signal while the other wire is connected to ground to minimize noise and crosstalk.

**Inverted data.** Data on an I-O bus where a 1 is represented by a LOW voltage.

**Terminating resistors.** Resistors placed at the end of a bus to terminate it in its characteristic impedance.

**Bi-directional line.** A bus line that carries signals in either direction.

**Transceiver.** A circuit capable of both transmitting and receiving signals on a bus.

**Device select lines.** I-O bus lines that specify which device or controller is being addressed.

**BUSY.** A FF that indicates a controller is in the process of executing a command.

**DONE.** A FF that indicates a controller has finished executing a command.

**Mask.** A signal command sent from the processor that allows or prohibits a controller from interrupting.

**Memory controller.** A device that regulates the accesses to memory and determines their priority.

**Memory bus.** A bus that communicates between the controllers and the main memory.

**DMA.** Direct memory access, the mode of operation that allows a peripheral to access the processor's memory without going through the processor itself.

**DATA AVAILABLE.** A signal on the memory bus that tells the controller that the data is available and can be strobed in. DATA AVAILABLE is also used to terminate a memory cycle.

**Interrupt.** A signal that informs the computer that a peripheral device is requesting service. It generally causes the processor to branch to a service routine for the requesting device.

**Interrupt vector.** A group of bits that tells the processor which device is causing an interrupt.

**Sensor.** A device that senses an analog quantity, and typically converts it into a voltage that is proportional to the magnitude of the quantity.

**A/D.** Analog-to-digital converter.

**D/A.** Digital-to-analog converter.

**Comparator.** A circuit that compares two input voltages and produces an output depending on which input is larger.

**RAMP voltage.** A voltage that increases linearly with time.

**SOC.** Start-of-Conversion.

**EOC.** End-of-Conversion.

## REFERENCES

**Arpad Barna** and **Dan I. Porat**, *Integrated Circuits in Digital Electronics*, Wiley, New York, 1973.

George K. Kostopoulos, *Digital Engineering*, Wiley, New York, 1975.
*How To Use The Nova Computers*, Data General Corporation, Southboro, Mass., 1971.

## PROBLEMS

17-1.    In Fig. 17-5:

(a) What is the impedance at each end of the line?

(b) What are the voltage levels on the line?

(c) How much current must a driver be capable of absorbing?

(d) Can this line be driven with a 7406 or 7407?

17-2.    A bus line is to be terminated in 150 $\Omega$ and produces 3 V when not driven to ground. If the source voltage is 5 V, what resistors should be used?

17-3.    In Fig. 17-7, use the clear line on the I-O bus to allow the processor to blank the display. Design the necessary circuitry within the controller.

17-4.    In order to write to a peripheral, a processor must send out a 16-bit word and DATA OUT. The 8 MSBs are the device's address and the 8 LSBs are the character to be written. When the controller receives the DATA OUT, it sends the address to the peripheral and awaits an ADDRESS MATCH signal. When ADDRESS MATCH is received, the controller sends the data to the peripheral. Design the controller.

17-5.    Repeat Problem 17-4. When ADDRESS MATCH is received, the controller must present the character to the peripheral in two 4-bit bursts on the same 4 lines between the controller and the peripheral. The data bursts must be 8 $\mu$s apart.

17-6.    A controller is receiving serial inputs and a 100-KHz clock from its peripheral. Each bit is valid on the negative transition of the clock. It is to group the bits into 8-bit characters. When the eighth bit appears, it must:

(a) Check for odd parity.

(b) Set DONE.

A DATA IN signal from the processor resets DONE. The controller must also set an error FF if:

(a) Even parity is detected when the character is received.

(b) The computer is late. This error occurs when DATA IN occurs after the first shift for the next word.

Design the controller.

17-7.    For a 16K-word-by-32-bit memory, how many lines must be on the common memory bus?

17-8.    A system uses an 8K-by-12-bit memory. How many FFs are required in each controller's DMA port? What is their function?

17-9.    Design a refresh request circuit for a dynamic RAM to use an input port on the memory controller of Fig. 17-13.

17-10.    Assume a disk controller receives a serial input and generates a pulse whenever it has assembled a complete word to be written to memory. Design a

DATA LATE FF that sets only if the previous memory request has not been honored when the next complete word pulse occurs.

17-11. Explain how the MDR and shift register in the disk controller of Ex. 17-10 work when data are being written to the disk.

17-12. A D/A converter has 8 input bits. If its output ranges from 0 to +5 V, how much voltage does each digital step require?

17-13. Repeat Problem 17-12 if the output range is from −7 to +7 V.

17-14. A D/A converter is required to have an output range of 0 to +20 V. It must have an input resolution of 0.01 V or better. How many bits are required on the digital input?

17-15. For the summing network of Fig. 17-17a, what is the current output if R = 10K and the binary input is:
  (a) 37
  (b) 55

17-16. For the ladder network of Fig. 17-17b, if the digital input is 0 or +5 V, what value of R is needed to make the output current in mA equal to the binary value of the input. Remember the output impedance is 2R Ω to ground.

17-17. What is the output current for the network of Fig. 17-17b if the resistances are 120 Ω and the digital input is:
  (a) 20
  (b) 30
  (c) 50

17-18. If the A/D converter of Fig. 17-19 is required to have a worst case settling time of 50 μs, how fast must the clock run?

17-19. Design an A/D converter that first compares the MSB to the input voltage, then the next most significant MSB, and so on. How many steps are involved for a worst case comparison?

17-20. If a DAC 0800 is set up to provide an analog output of ±10 V and its input is 00110011, what is the analog output voltage?

17-21. If an ADC 0800 is set up to convert analog voltages between 0 and +10 V, what is its digital output if the input is 3.5 V?

After attempting the problems, return to Sec. 17-2 and reread the self-evaluation questions. If any of them still seem difficult, review the appropriate sections of the chapter to find the answers.

# 18

# MODEMS AND TELETYPES

## 18-1 INSTRUCTIONAL OBJECTIVES

This chapter considers the uses of synchronous and asynchronous MO-DEMS, and the design of controllers to interface with asynchronous I-O devices. Emphasis is on the most common asynchronous device, the *tele-type*.

After reading the chapter, the student should be able to:

1.  Describe the functions of a MODEM.
2.  Explain how an FSK transmission system works.
3.  Construct an asynchronous transmission pattern for a given group of characters.
4.  List the signals required by a synchronous MODEM and explain the function of each signal.
5.  Design a TTY driver and receiver.
6.  Connect a computer or microprocessor to a serial TTY or printer using a UART.

## 18-2 SELF-EVALUATION QUESTIONS

Watch for the answers to the following questions as you read the chapter. They should help you to understand the material presented.

1. What is meant by the speed of a MODEM? Why is it important?
2. Explain the difference between full duplex, half duplex, and simplex transmission.
3. Why are START and STOP bits necessary for asynchronous data transmission?
4. What synchronizes the transmit and received data in a synchronous MODEM?
5. Explain what time division multiplexing is. Why is it used?
6. What is double-buffering?
7. What are FRAMING and OVERRUN errors?

## 18-3 INTRODUCTION TO MODEMS

The source of data for a computer may be a long distance from the actual computer. An example of this is a time sharing computer system, where the input data originates in each user's office, which is equipped with a teletype machine or other I-O device. The user's input must then be transmitted to the computer, which can be many miles away, and the computer's reply must be transmitted back to the user.

The most common way to transmit digital data is via the already existing telephone network. This requires that the *digital data be transformed into an audio tone that the telephone lines can transmit, and be demodulated back to the original bit stream at the receiving end.* This is accomplished by a MODEM. The term MODEM is an acronym for MODULATOR/DEMODULATOR.

A MODEM can range from a single PC card to a highly sophisticated prepackaged device. The function of a MODEM is to transform a digital bit stream into audio signals compatible with transmission lines on the transmitting end, and reconstruct the original digital bit stream at the receiving end. A typical MODEM transmission system is shown in Fig. 18-1.

### 18-3.1 Modes of Data Transmission

Figure 18-1 shows the most common mode of data transmission, *full duplex*. In full duplex operation, there is a signal line in each direction so that *simultaneous data transmission* can take place. Other modes of operation are *half duplex*, where one side sends and one side receives at any one time (the sides take turns sending and receiving), and *simplex*, where one side is always sending and the other side is always receiving. Simplex is one-way data transmission.

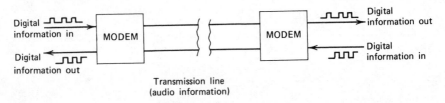

FIGURE 18-1   A full-duplex data transmission system.

## 18-3.2   Speed of MODEMS

MODEMS are generally classified by the number of *bits per second* (BPS) they can transmit. High speed MODEMS transmit between 2000 and 9600 BPS. Low speed MODEMS generally handle data rates between 75 and 1800 BPS.

As the speed of a MODEM increases, its error rate also increases. This adversely affects the actual data rate because errors can cause the retransmission of large quantities of data.

## 18-4   Low Speed MODEMS

Low speed MODEMS with speeds up to 300 BPS are attractive for many applications because they can be used on directly dialed telephone lines. The 103A MODEM, manufactured by the Bell System, is a popular low speed MODEM. These MODEMS often are equipped with cradles for a telephone, so the user can dial the proper phone number, place the handset in the receiver, and start transmitting digital data. A *voice coupler*, often used with teletypes, is a form of low speed MODEM.

The operation of a telephone transmission system is shown in Fig. 18-2. The method of modulation is known as *frequency shift keying* (FSK),

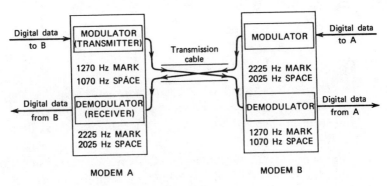

FIGURE 18-2   Data transmission using FSK modems.

where the 1s and the 0s are sent out at different frequencies. All frequencies must be below the 3000 cycle limit on the bandwidth of a telephone line.

As shown in Fig. 18-2, these MODEMS utilize four different frequencies:

1.  The frequency at which they transmit a 1.
2.  The frequency at which they transmit a 0.
3.  The frequency at which they receive a 1.
4.  The frequency at which they receive a 0.

Because the MODEMS operate in pairs, MODEM A must transmit at the frequency MODEM B receives. The receiver for MODEM A and the transmitter for MODEM B work on the alternate pair of frequencies. The standard frequencies are shown in Fig. 18-2. Under this scheme, data can be transmitted over the telephones in both directions (full duplex) without interference, because the data from A to B use different frequencies than the data from B to A.

The terms *mark* and *space* are used in conjunction with data transmission and MODEMS. A mark corresponds to a logic 1, and a space corresponds to a logic 0.

---

**EXAMPLE 18-1**

What occurs when a 1 is being transmitted from MODEM A?

**SOLUTION**

The logic 1 is applied to the digital input of MODEM A, which modulates it and sends out a 1270-Hz signal on the telephone line. The receiver of MODEM B detects the 1270-Hz note and demodulates it, producing a logic 1 on its digital output.

---

## 18-4.1   Asynchronous Transmission

Low speed MODEMS generally utilize *asynchronous* (unclocked) *data transmission*. The most common code for asynchronous data transmission is the 11-bit START-STOP code that is described here. Some systems and devices use other codes that are variants of the 11-bit START-STOP code.

The basic pattern for asynchronous data transmission is shown in Fig. 18-3. When the line is quiescent (transmitting no data), it is constantly in the mark, or 1, state. The start of a character is signaled by the START bit, which drives the line to the 0, or space state, for one bit time.

The 8 bits immediately following the START bit are the data bits of the character. In most systems, the ASCII (American Standard Code for Information Interchange) code is used. This code consists of 7 data bits and

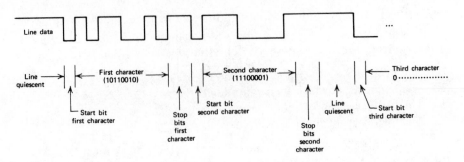

FIGURE 18-3  Asynchronous transmission.

a parity bit. Normally even parity is used because most teletypes generate even parity characters. The ASCII code is given in Appendix E. The LSB of the character is transmitted first, right after the START bit.

After the last datum bit, the transmission line must go HIGH for at least two bit times. These are called the STOP bits. If no further data are to be transmitted, the line simply stays HIGH (marking) until the next START bit occurs.

This data pattern requires 11 bits:

1. One START bit (always a space or 0).
2. Eight data bits.
3. Two STOP bits (always a mark or 1).

---

### EXAMPLE 18-2

How many characters can a 300 BPS line transmit if it is running at its maximum rate?

### SOLUTION

Because each character requires 11 bits, the maximum data transmission rate is

$$\frac{300 \text{ bits/second}}{11 \text{ bits/character}} = 27.27 \text{ characters/second}$$

### EXAMPLE 18-3

(a) What is the bit pattern of the character shown in Fig. 18-4?
(b) Is the parity odd or even?
(c) What ASCII character is it?

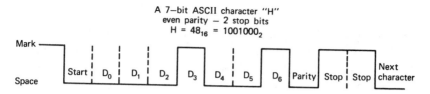

A 7–bit ASCII character "H"
even parity — 2 stop bits
H = $48_{16}$ = $1001000_2$

FIGURE 18-4 Character pattern for Example 18-3. (From Greenfield and Wray. *Using Microprocessors and Microcomputers: The* **6800** *Family.* Copyright John Wiley & Sons, Inc. 1981. Reprinted by permission of John Wiley & Sons, Inc.)

## SOLUTION

(a) After the initial quiescent period, the waveform bits are 00001001011. The first 0 is the START bit and the next 8 are the character. Therefore, the character is 00010010. The two 1s following the last datum bit are the STOP bits.

(b) Since the characters contain an even number of 1s, this is an *even* parity character.

(c) The character bits are 00010010. Remembering that the LSB is transmitted first, this is a $(48)_{16}$. Appendix E shows that 48 is the ASCII code for the letter "H."

## 18-5 HIGH-SPEED MODEMS

MODEMS that operate at speeds between 2000 BPS to 9600 BPS are considered to be high speed MODEMS. They generally operate in accordance with EIA (Electronic Industries Association) standard specification no. RS-232-C.

The major requirements for a high speed MODEM as specified by RS-232-C are:

1. The digital signals that communicate with the MODEM must be between −3 V and −25 V for a 1 and +3 and +25 V for a 0. ICs are available that transform TTL levels to MODEM input levels and vice versa.

2. The MODEMS provide a TRANSMIT CLOCK and a RECEIVE CLOCK. These clocks are used to synchronize the data. *The transition from 0 to 1 is the center of the datum bit.* When transmitting a stream of bits, data should change when the TRANSMIT CLOCK makes a 1 to 0 transition. On the receive side, data should be sampled when the RECEIVE CLOCK makes a 0 to 1 transition because this is the center of the datum bit.

Data transmission on high speed MODEMS is synchronized and controlled by the RECEIVE and TRANSMIT CLOCKS. This is called *synchronous data transmission.* START and STOP bits (which transmit no data and therefore add to the transmission overhead) are not used. Once data transmission is started, the data flow continuously with no apparent demarcation between characters. Consequently, the sending and receiving MODEMS must be carefully *synchronized and remain in step throughout the data transmission.* When no data are available for transmission, the transmitter should send continuous 1s, as in asynchronous transmission.

Dial up telephone lines are incapable of handling high speed MODEM data transfers. Generally special lines are leased from the telephone company and these lines must be *conditioned*, which means that special circuitry is added to monitor and improve the characteristics of the leased lines, so that they can operate at the required speeds. Privately leased lines are an additional expense that must be recovered by the increased rate of data transmission.

## 18-5.1 Connecting to a MODEM

The important signals between a MODEM and a digital device are shown in Fig. 18-5. They are:

1. **TRANSMITTED DATA**—from the digital device to the MODEM. These are the data to be transmitted.
2. **RECEIVED DATA**—from the MODEM to the digital device. These are the data received from the remote MODEM.
3. **REQUEST TO SEND**—from the digital device to the MODEM. This signal should be a 0 whenever data are to be transmitted. It is important in half duplex transmission where it is a 1 when data are being received. In full duplex operation, it generally remains in the 0 state.
4. **CLEAR TO SEND**—from the MODEM to the digital device. This signal is a response to REQUEST TO SEND and indicates that the MODEM can accept data for transmission. In full duplex operation, it is normally always active and presents a 0 level to the digital device.

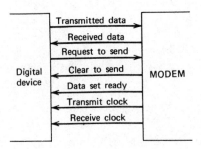

FIGURE 18-5 Interconnections between a MODEM and a digital device.

5.  **DATA SET READY**—from the MODEM to the digital device. A 1 on this line indicates that the data set is not ready, usually because of an abnormal condition. This line must be in the 0 state for transmission or reception.

6.  **TRANSMIT CLOCK**—from the MODEM to the digital device. This clock is provided to synchronize the data to be transmitted.

7.  **RECEIVE CLOCK**—from the MODEM to the digital device. This clock is provided to synchronize the data being received.

Additional signals such as ring indicators are available for use in special situations. A complete listing and specifications of all the available signals are given in specification RS-232-C. Users can provide their own clock to control the bit rate of the MODEM as an option. Most users, however, prefer to synchronize their data with the clocks provided by the MODEM.

Level translators are required between the TTL signals of the digital equipment and the RS-232-C levels required by the MODEM. Two popular level translators are the **MC 1488** and the **MC 1489,** both manufactured by Motorola, Inc. The **MC 1488** is a quad TTL-to-RS-232-C level translator for data going to the MODEM and the **MC 1489** is a quad RS-232-C-to-TTL level translator for data coming from the MODEM to the digital ICs. Unfortunately, level translators require additional power supplies ($+12$ V and $-12$ V are typical) to drive the MODEMS.

---

### EXAMPLE 18-4

Design a circuit to receive data from a MODEM and group it into 8-bit characters.

### SOLUTION

Data are received in bit serial form from a MODEM. A serial-to-parallel shift register is required. A circuit to receive data is shown in Fig. 18-6. The received data from the MODEM become, via the **MC1489** level translator, the serial input to the **74164** SHIFT REGISTER. The RECEIVE CLOCK clocks the data into the SHIFT REGISTER on the 0 to 1 transition, when the received data are not changing. The **74164** groups the bits into 8-bit characters for use by the digital device.

### EXAMPLE 18-5

A computer presents 8 bits to a controller along with DATA OUT. When the controller receives DATA OUT, it is to set BUSY and transmit the data on a high speed MODEM. When no data character is to be transmitted, the MODEM is to transmit 1s. Design this portion of the controller and the CONTROLLER-MODEM interface.

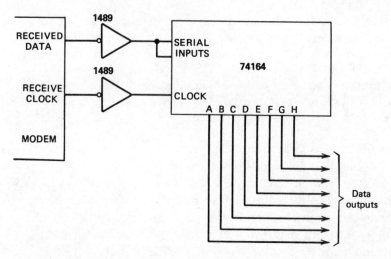

FIGURE 18-6 Receiving data from a MODEM.

**SOLUTION**

After considering the problem, we realize the following are needed.

1.   A SHIFT REGISTER because the data are being presented in parallel, but must be shifted out serially.
2.   Level translators to go from TTL to RS-232-C logic levels.
3.   A counter to determine when the last bit has been shifted out and set DONE.

A circuit that satisfies these requirements is shown in Fig. 18-7. The concurrence of DATA OUT and DEVICE SELECTED causes 8 bits on the I-O bus to be loaded into the **74165** SHIFT REGISTER. It also sets the BUSY FF.

If CLEAR TO SEND and DATA SET READY are both 0s (HIGH levels within the MODEM), the data are clocked through the SHIFT REGISTER by the TRANSMIT CLOCK. Note that the positive edge of the shift clock occurs when the TRANSMIT CLOCK is making a 1 to 0 (LOW to HIGH) transition. Note also that all MODEM signals go to RS-232-C-to-TTL level translators (**1488**s or **1489**s). The driving circuit for REQUEST TO SEND is not shown; for full duplex transmission it could be continuously grounded.

The clock signals also increment the **7493** counter. After 8 counts, the **7493** clears BUSY and sets DONE. The serial input to the **74165** is tied to 1 so that any data transmitted after the character is shifted out are 1s or marks.

## 18-5.2   Time Division Multiplexing

*Time division multiplexing* (TDM) *is the process of giving each of several input lines a time slot on a transmission line or bus for its data.* In this way, several inputs can *share* a single transmission line. An example of

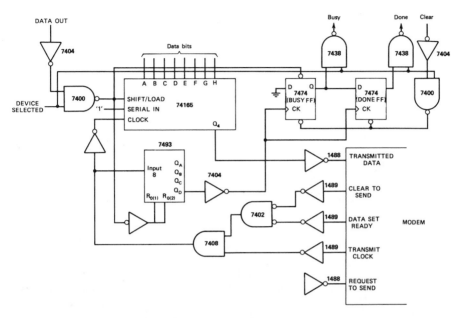

FIGURE 18-7 A controller using a high speed MODEM.

TDM has been presented in Ex. 11-1 where each of 8 lines had a 1-μs interval on a common output bus.

TDM can be economical where large quantities of data, which originate from many slow speed devices, must be sent to a remote location for processing. An example of this is a time-sharing computer company. Typically, such a company has many customers located in a large city and each customer has a teletype or some other low speed I-O device. Each customer's data must be transmitted to the time-sharing company's main computers, which could be many miles away. To handle this situation economically, the time-sharing company generally places a minicomputer or *data concentrator* in each city where it has a large number of customers. Each customer accesses the data concentrator using low speed MODEMS and local dial-up telephone lines. The data concentrator then TDMs the data onto a high speed MODEM that communicates with the computer via a leased line. The situation is shown in Fig. 18-8.

---

**EXAMPLE 18-6**

A company has 24 local customers, each having a 100 BPS data terminal. Design a circuit to TDM their data onto a single-leased transmission line using a 2400 BPS MODEM.

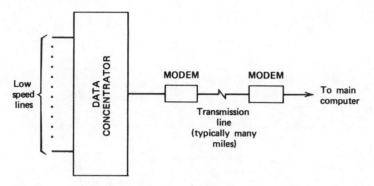

FIGURE 18-8   Data transmission using a data concentrator, MODEM and TDM.

## SOLUTION

A 24-line-to-1-line multiplexer is required, which must be under the control of the MODEM's TRANSMIT CLOCK. A solution is shown in Fig. 18-9. With DATA SET READY and CLEAR TO SEND both 0s, the TRANSMIT CLOCK increments the 5-bit counter consisting of the **74107** and the **7493**. The counter is wired to reset itself when the count reaches 24.

The output of the 5-bit counter is tied to the SELECT and STROBE inputs of the multiplexers. Each input to the multiplexers is tied to TRANSMITTED DATA for one count during a cycle. Because the output runs at 2400 BPS and samples 24 lines, each line is sampled 100 times a second, which matches its input speed.

The circuit of Fig. 18-9 is a simplification of a practical problem. While the circuit as shown can TDM 24 input lines onto a single output line, it makes no provision for error checking, data retransmission in the event of errors, line identification, and other problems associated with a practical transmission system. For these reasons, data concentrators or minicomputers are often used to handle multiplexing and high speed data transmission.

## 18-5.3   Synchronizing High Speed Lines

Since the data from high speed lines appear in a *steady bit stream*, special care must be taken to *synchronize the receiving end with the transmitting end so the receiving end can determine which bit in a stream is bit 0 of a character*. Generally this is done by preceding each transmission with three consecutive *sync* characters. A sync character carries no information. Its purpose is to synchronize the sending and receiving ends of the line. In addition, other special characters are used to synchronize data transmission.

Data are usually transmitted in groups of characters called *data blocks* and each block is a continuous stream of data characters. Each block is usually preceded by three sync characters and contains enough parity check-

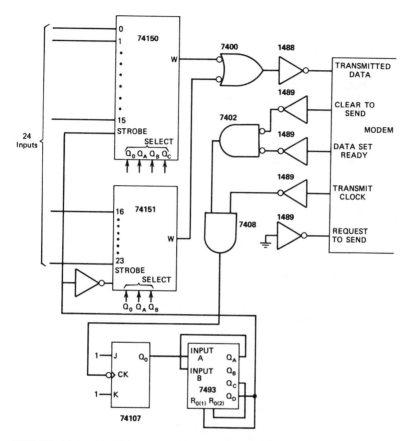

FIGURE 18-9   Multiplexing 24 slow speed inputs line onto a single high speed transmission line.

ing to allow the receiver to determine whether it has received the block correctly. If there has been an error, the receive side asks, via the return channel on a full duplex line, for a retransmission of the entire block. Very careful attention is paid to synchronizing protocol and conventions on a high speed data transmission line.

## 18-6   TELETYPES

The teletype (TTY) shown in Fig. 18-10 is the most common I-O device to be used with a computer. It operates on the 11-bit asynchronous code given in Sec. 18-4.1. The code requires:

- □   1 START bit
- □   7 data bits
- □   1 parity bit for even parity
- □   2 STOP bits

FIGURE 18-10 Teletype terminal. (From Louis Nashelsky, *Introduction to Digital Computer Technology*, 2nd ed. Copyright 1977, John Wiley & Sons, Inc. Reprinted with permission of John Wiley & Sons.)

---

**EXAMPLE 18-7**

Standard TTYs run at 10 characters per second. What is their bit rate?

**SOLUTION**

Each character requires 11 bits, so the bit rate is:

$$10 \text{ characters/sec} \times 11 \text{ bits/characters} = 110 \text{ BPS}$$

This means that the time duration of each bit is 9.1 ms.

---

A teletype consists of *two* devices, a *transmitter* and a *receiver*. The transmitter is activated by pushing a key on the teletype keyboard. It then produces an *n*-bit code corresponding to the key that was depressed.

The receiver must be sent an 11-bit asynchronous code. This causes the teletype to print the character. When connected to a computer in full duplex mode, it may appear to the user that he is actually typing when he depresses a key and sees the letter appear on his paper. Actually, the TTY transmits the key code to the computer or data concentrator, which then echoes or sends a print command back to the TTY. *The user does not activate the printer by depressing the key. A character is printed only at the command of the computer.*

## 18-6.1 Teletype I-O Characteristics

The basic teletype has only four I-O lines:

- A receive line and its ground.
- A transmit line and its ground.

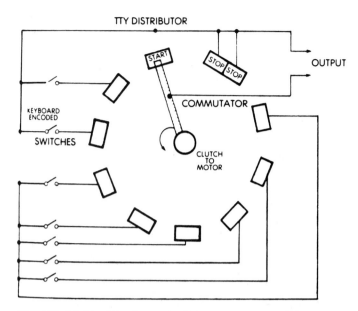

FIGURE 18-11 Distributor in Teletype. (Reprinted from *Microprocessor Interface Technique* by Rodney Zaks, by permission of Sybex, Inc. Copyright © 1977, 1978, 1979. World rights reserved.)

The most common TTY interface is the *20-mA current loop*. This is characterized by a steady current of 20 mA flowing in the keyboard and printer circuit when no data are being transmitted. This is the *mark* state. To print a character, the 0s of the character to be printed interrupt the 20-mA current loop. This is the *space* state. If the TTY receives no current for any appreciable period of time (constant space or BREAK), it will chatter, producing a clacking noise but not printing anything.

The TTY transmits when a key on its keyboard is struck. This causes a mechanical commutator (Fig. 18-11) to rotate and has the effect of opening and closing a switch at the 9.1-ms bit rate to identify the character being struck. When no keys are depressed, the switch is normally closed.

## 18-6.2 Design of a TTY Printer Interface Circuit

In many TTYs the input line to the TTY printer goes to the base of a *pnp* transistor within the TTY. The circuit of Fig. 18-12 was designed to transform TTL levels to the 20-mA current loop required by the TTY.[1] It works as follows.

[1] This circuit is in use at RIT.

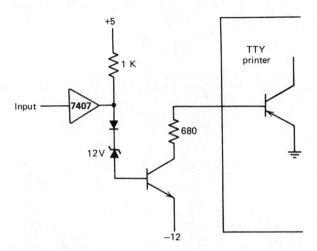

FIGURE 18-12   A TTL to 20-mA current loop converter connected to a TTY.

1.   In constant mark, or during the 1s of the character, a TTL high level must be present at the input to the 7407 (see Sec. 5-6.2). This turns the open-collector 7407 off and allows current to flow into the base of the transistor.

2.   This current saturates the transistor and drives approximately 20 mA through the base of the *pnp* transistor within the TTY.

3.   When a 0 appears at the input to the 7407, the open-collector transistor turns on driving the output to ground. Now there is not enough voltage to overcome the 12-V Zener diode, the switching diode, and the base to emitter drop. Consequently, the *npn* transistor receives no base current and remains off. This cuts off the input current to the TTY during the space bits.

### 18-6.3   Design of the Keyboard Interface

The design of a TTY receiver is shown in Fig. 18-13. As long as the commutator is closed, the input and output of the 7407 are at +5 V. When a key is depressed the commutator opens for the START bit and the 0s of the character. When the commutator is open the input of the 7407 drops to ground. This produces the asynchronous representation of the selected character.

### 18-6.4   Interfaces Using Optical Couplers

Circuits that have to operate in high noise environments (a circuit controlling an operation in a manufacturing plant, for example), or where isolation between input and output is necessary, often use optical couplers to achieve this isolation. The basic circuit of an optical coupler is shown in Fig. 18-14. It consists of a Light Emitting Diode (LED) and a photosensitive tran-

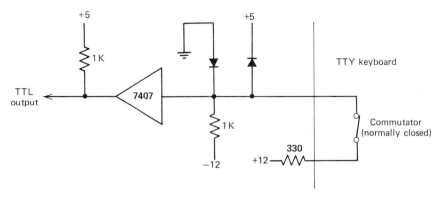

FIGURE 18-13   An interface circuit for receiving data from a TTY.

sistor, both encapsulated in the optical coupler package. If current is driven through the LED, the photosensitive transistor turns on.

Optical couplers are suggested for electrical signal isolation because the contacts in the TTY frequently produce serious noise spikes, and if the wire or cable to the terminal is very long, it can be exposed to interference from AC power lines, other TTYs, or radio signals. When such an isolation circuit is built, it is important to keep the wires associated with the digital side separated or shielded from those carrying the 20-mA currents to avoid electrostatic coupling of the noise spikes, which can defeat the purpose of the optical coupling. Common ground returns are to be avoided also, since extraneous current flowing can produce noise voltage differences that are effectively in series with the digital signals.

Note the use of the plus and minus 12 V in the TTY circuits. These voltages are desirable because it puts a higher voltage across the TTY keyboard or reader contacts. These contacts are designed to work with up to 130 V and will not work reliably with very low voltages (like 0.7 or even 5 V), since insulating oxides or even particles of dirt will form on the contacts

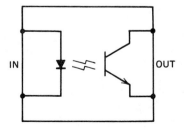

FIGURE 18-14   An optical coupler.

in time. The higher voltages will break through the oxides and provide reliable performance.

A circuit for connecting the TTY to TTL levels using optical couplers is shown in Fig. 18-15. It operates in the same manner as the circuits of Figs. 18-12 and 18-13. The transistors are turned on by light from the LEDs rather than base current.

## 18-7  UARTS

The circuits discussed in Sec. 18-6 shift TTL levels to 20-mA current loop. When a computer must write to a TTY, it generally provides an 8-bit character in ASCII code. A circuit is required between the computer and the 20-mA current loop to do the following.

1. Serialize the parallel data.
2. Produce the proper pulse width (9.1 ms per bit for a TTY).
3. Add START and STOP bits.
4. Leave the output in the mark state when no data are being transmitted.

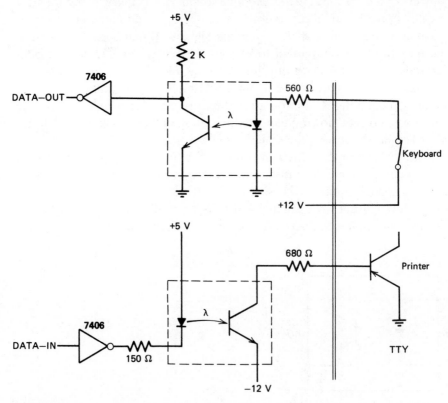

FIGURE 18-15  A TTY interface using optical couplers.

These circuits can be designed using clocks, shift registers, and counters (see Problems 18-10 and 18-11), but it is now common practice to use a UART.

The term UART stands for *Universal Asynchronous Receiver Transmitter.* Its purpose is to transform parallel data into asynchronous serial data, and vice versa, at a speed compatible with the receiving device. It performs all of the functions listed above.

UARTs are often used to communicate with printers and low speed modems as well as TTYs. These devices usually have RS-232 interfaces and 1488s and 1489s are used as level translators instead of 20-mA current loop converters.

## 18-7.1 The AY-5-1013

The AY-5-1013 is one of the most popular and commonly used UARTs. Its pinout and block diagram are shown in Fig. 18-16. For transmitting, the UART accepts 8 data bits on its DB lines and sends out the asynchronous character on its SO output. It receives data on its SI input and presents the data to the computer on its eight RD lines. All signal levels are TTL.

A list of the pins and the function of each pin is given in Fig. 18-17. The basic timing for the UART is generated by external clocks applied to pin 17 on the receive side and pin 40 on the transmit side. The frequency of these clocks must be 16 times the bit rate of the external device.

---

**EXAMPLE 18-8**

A UART must interface with a standard TTY. What frequency clock must be applied to pins 17 and 40?

**SOLUTION**

Since both the transmit and receive side of the TTY use the same frequency (110 BPS) the same clock can drive pins 17 and 40. Its frequency must be 16 × 110 BPS, or 1760 Hz.

---

## 18-7.2 The Receive Side of the AY-5-1013

On the receive side, besides the serial input and 8 data outputs, there are three error condition detectors (parity error, framing error, and over-run), a DATA AVAILABLE signal that goes HIGH when the UART has received a character and it is ready for the computer, and a $\overline{\text{RESET DATA AVAIL-ABLE}}$ line that will reset the DATA AVAILABLE (DAV) line when it goes LOW. There are also two 3-state enable lines. A LOW on the $\overline{\text{RDE}}$ line

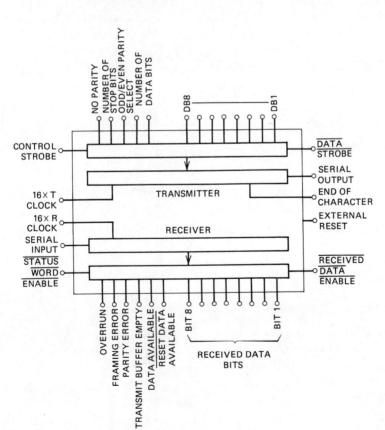

Pins

```
Vcc (+5V) ☐ 1        40 ☐ TCP
VGG (-12) ☐ 2        39 ☐ EPS
     GND ☐ 3         38 ☐ NB1
     RDE ☐ 4         37 ☐ NB2
     RD8 ☐ 5         36 ☐ TSB
     RD7 ☐ 6         35 ☐ NP
     RD6 ☐ 7         34 ☐ C5
     RD5 ☐ 8         33 ☐ DB8
     RD4 ☐ 9         32 ☐ DB7
     RD3 ☐ 10        31 ☐ DB6
     RD2 ☐ 11        30 ☐ DB5
     RD1 ☐ 12        29 ☐ DB4
      PE ☐ 13        28 ☐ DB3
      FE ☐ 14        27 ☐ DB2
      OR ☐ 15        26 ☐ DB1
     SWE ☐ 16        25 ☐ SO
     RCP ☐ 17        24 ☐ EOC
    RDAV ☐ 18        23 ☐ DS
     DAV ☐ 19        22 ☐ TBMT
      SI ☐ 20        21 ☐ XR
```

PINOUT FOR AY5-1013 UART

FIGURE 18-16  Pinout and block diagram of the **AY-5-1013**. (Courtesy of General Instruments Microelectronics.)

**PIN FUNCTIONS**

| Pin No. | Name (Symbol) | Function |
|---|---|---|
| 1 | V$_{CC}$ Power Supply (V$_{CC}$) | -5V Supply |
| 2 | V$_{GG}$ Power Supply (V$_{GG}$) | -12V Supply (Not connected for AY-3-1014A 1015) |
| 3 | Ground (V$_{GND}$) | Ground |
| 4 | Received Data Enable (RDE) | A logic "0" on the receiver enable line places the received data onto the output lines |
| 5-12 | Received Data Bits (RD8-RD1) | These are the 8 data output lines. Received characters are right justified the LSB always appears on RD1. These lines have tri-state outputs i e. they have the normal TTL ouput characteristics when RDE is "0" and a high impedance state when RDE is "1". Thus, the data output lines can be bus structure oriented |
| 13 | Parity Error (PE) | This line goes to a logic "1" if the received character parity does not agree with the selected parity. Tri-state |
| 14 | Framing Error (FE) | This line goes to a logic "1" if the received character has no valid stop bit. Tri-state. |
| 15 | Over-Run (OR) | This line goes to a logic "1" if the previously received character is not read (DAV line not reset) before the present character is transferred to the receiver holding register Tri-state. |
| 16 | Status Word Enable (SWE) | A logic "0" on this line places the status word bits (PE, FE, OR, DAV, TBMT) onto the output lines. Tri-state |
| 17 | Receiver Clock (RCP) | This line will contain a clock whose frequency is 16 times (16X) the desired receiver baud. |
| 18 | Reset Data Available (RDAV) | A logic "0" will reset the DAV line. The DAV F/F is only thing that is reset. Must be tied to logic "1" when not in use on the AY-3-1014A. |
| 19 | Data Available (DAV) | This line goes to a logic "1" when an entire character has been received and transferred to the receiver holding register. Tri-state. Fig.12,34. |
| 20 | Serial Input (SI) | This line accepts the serial bit input stream. A Marking (logic "1") to spacing (logic "0") transition is required for initiation of data reception. Fig.11,12,33,34. |
| 21 | External Reset (XR) | Resets all registers except the control bits register (the received data register is not reset in the AY-5-1013A and AY-6-1013). Sets SO, EOC, and TBMT to a logic "1". Resets DAV, and error flags to "0". Clears input data buffer. Must be tied to logic "0" when not in use. |
| 22 | Transmitter Buffer Empty (TBMT) | The transmitter buffer empty flag goes to a logic "1" when the data bits holding register may be loaded with another character. Tri-state. See Fig.18,20,40,42. |
| 23 | Data Strobe (DS) | A strobe on this line will enter the data bits into the data bits holding register. Initial data transmission is initiated by the rising edge of DS. Data must be stable during entire strobe. |
| 24 | End of Character (EOC) | This line goes to a logic "1" each time a full character is transmitted. It remains at this level until the start of transmission of the next character. See Fig.17,19,39,41. |
| 25 | Serial Output (SO) | This line will serially, by bit, provide the entire transmitted character. It will remain at a logic "1" when no data is being transmitted. See Fig.16. |
| 26-33 | Data Bit Inputs (DB1-DB8) | There are up to 8 data bit input lines available. |
| 34 | Control Strobe (CS) | A logic "1" on this lead will enter the control bits (EPS, NB1, NB2, TSB, NP) into the control bits holding register. This line can be strobed or hard wired to logic "1" level. |
| 35 | No Parity (NP) | A logic "1" on this lead will eliminate the parity bit from the transmitted and received character (no PE indication). The stop bit(s) will immediately follow the last data bit. If not used, this lead must be tied to a logic "0". |
| 36 | Number of Stop Bits (TSB) | This lead will select the number of stop bits, 1 or 2, to be appended immediately after the parity bit. A logic "0" will insert 1 stop bit and a logic "1" will insert 2 stop bits. For the AY-3-1014A/1015, the combined selection of 2 stop bits and 5 bits/character will produce 1½ stop bits. |
| 37-38 | Number of Bits/Character (NB2, NB1) | These two leads will be internally decoded to select either 5, 6, 7 or 8 data bits/character.<br><br>NB2 NB1 Bits/Character<br>0 0 5<br>0 1 6<br>1 0 7<br>1 1 8 |
| 39 | Odd/Even Parity Select (EPS) | The logic level on this pin selects the type of parity which will be appended immediately after the data bits. It also determines the parity that will be checked by the receiver. A logic "0" will insert odd parity and a logic "1" will insert even parity. |
| 40 | Transmitter Clock (TCP) | This line will contain a clock whose frequency is 16 times (16X) the desired transmitter baud. |

FIGURE 18-17 Pin functions of the **AY-5-1013**. (Courtesy of General Instruments Microelectronics.)

puts the received data on the lines while the $\overline{\text{SWE}}$ puts the error indicators and DAV on the lines.

---

### EXAMPLE 18-9

A computer must read both data and status from a UART over the same 8-bit data bus. How can this be done?

### SOLUTION

The error outputs and DAV can each be connected to one of the RECEIVED DATA BIT lines. The computer can request data by sending out a DATA REQUEST signal that causes $\overline{\text{RDE}}$ to go LOW, enabling the DATA outputs to be placed on the lines. If the computer sends out a STATUS REQUEST, $\overline{\text{SWE}}$ goes LOW and only the error status and DAV are placed on the output lines. Note how the use of 3-state outputs and separate enables allows data and status to be multiplexed onto the same lines.

---

When a character is being received by the UART, its bits are entered into a Shift Register at the clock rate. When it is fully received, and checked for parity and proper STOP bits (no FRAMING ERROR), it is transferred from the Shift Register to a Holding Register. The UART then raises DAV to inform the computer that a character is ready. The computer should then read the character and pulse $\overline{\text{RDAV}}$ LOW to acknowledge receipt of the data.

The character can remain in the Holding Register while the next character is being shifted into the Shift Register. This is known as *double-buffering* because one character is being held in the Holding Register, or buffer, while a second character is being shifted into the Shift Register (which is considered as the second buffer). If the second character gets completely in before receipt of $\overline{\text{RDAV}}$, the UART has more data than it can store since the Holding Register still contains the first character. In this case the UART raises the OVERRUN error to indicate that the computer did not accept the data fast enough, and the UART lost data that it could not store.

---

### EXAMPLE 18-10

A computer is connected to a TTY. In worst case, how much time does the computer have to respond to DAV?

### SOLUTION

The computer must read the data before the next character is completely entered.

Since a TTY runs at 10 characters per second (11 bits × 9.1 ms/bit), the computer has 100 ms to respond to DAV before an OVERRUN error occurs.

## 18-7.3 The Transmit Side of the Ay-5-1013

The transmit side of an **AY-5-1013** consists of pins 21-40. The characteristics of the asynchronous data for both transmission and reception are determined by the levels on pins 35–39. These include the number of bits in each character (pins 37–38), the number of STOP bits (pin 36), and parity (pins 35 and 39). These levels can be hard-wired in or entered from the computer using the CONTROL STROBE (CS), pin 34.

**EXAMPLE 18-11**

For a standard TTY, how would pins 34–40 be connected? Assume hard-wire connections.

**SOLUTION**

For hard-wire connections the CS input must be tied to 1 to allow the levels on pins 35–39 to enter the UART. The TTY takes 7-bit, even parity characters so the pins would be wired as follows:

**Pin 35-0.** Indicates parity is present. This is really for the receive side. The TTY does not check parity in the data sent to it, but the UART can now check parity on the data it receives from the TTY.

**Pin 39-1.** This is for even parity.

**Pin 36-1.** This inserts two STOP bits in each character.

**Pins 37 and 38.** They must be set up for a character comprised of 7 data bits and one parity bit. Figure 18-17 indicates that pin 37 should be connected to a 1 and pin 38 to a 0 for a 7 *data bit* character.

**Pin 40.** The transmit clock should be connected to a 1760-Hz clock for 110 BPS TTY.

When a computer is transmitting it must monitor either Transmitter Buffer Empty (TMBT, pin 22), or End of Character (EOC, pin 23). EOC is HIGH whenever no character is being transmitted. TMBT indicates the transmit holding register is empty and a character may be sent to the UART. Computers usually monitor TMBT, which is also enabled by the $\overline{SWE}$ strobe. If TMBT is HIGH, the computer sends a character to the UART and sends out $\overline{DATA\ STROBE}$ to inform the UART that a character is available on its transmit data input lines and should be strobed into its Transmit Holding

Register. The UART then serializes the character and sends it out via the SO line.

## 18-7.4 Connecting a TTY to a Computer Via a UART

For a TTY to communicate with a computer, the computer or microprocessor is connected to a UART. The UART is connected to a 20-mA loop converter (see Sec. 18-6) and then to the TTY. The proper frequency clocks must be supplied to pins 17 and 40 and pins 35–39 of the UART properly initialized (see Ex. 18-11).

The computer should next send out $\overline{\text{SWE}}$. If it finds that TMBT is a 1, it can send the next character to be transmitted, along with $\overline{\text{DATA STROBE}}$. On the receive side, the UART checks for DAV and no errors. If DAV is HIGH, it strobes in the received character and sends $\overline{\text{RDAV}}$ to the UART.

What the computer should do when an error occurs is a significant problem. Data may have been lost or changed and the received data should not be used. Somehow the computer must send a message back requesting a retransmission of the suspect data. Some systems allow a Negative Acknowledgement (NACK) to be sent back to indicate a transmission error.

### SUMMARY

This chapter introduced MODEMs, used to couple digital data onto a telephone line. The synchronous and asynchronous modes of data communication and the design of controllers for COMPUTER-TO-MODEM interfaces were presented.

The design of controllers to connect teletypes and other asynchronous devices to the I-O bus lines were considered. UARTs, which are ICs especially designed to handle these interface problems, were also introduced, and the **AY-5-1013** UART was discussed in detail.

### GLOSSARY

**MODEM.** Acronym for modulator-demodulator, generally used to convert digital signals to audio and vice versa.

**Mark.** A transmission line level corresponding to a logic 1.

**Space.** A transmission line level corresponding to a logic 0.

**FSK.** Frequency shift keying.

**BPS.** The rate of data transmission (bits per second).

**Asynchronous transmission.** Data transmission that is not synchronized by a clock, but uses START and STOP bits to synchronize the characters.

**Full duplex operation.** Simultaneous data transmission in both directions.

**Half duplex operation.** Data transmission in one direction at a time. The direction of data transmission alternates.

**Simplex operation.** Data transmission in one direction only.

**ASCII.** American Standard Code for Information Interchange.

**EIA.** Electronic Industries Association. This association issues standard specifications for the electronics industry.

**Time division multiplexing.** Multiplexing that assigns each input a particular interval or time slot on a transmission line.

**Data concentrator.** A minicomputer or other device that accepts many low speed data inputs and places the data on high speed output lines.

**Conditioned lines.** These are generally transmission lines leased from the telephone company and specially conditioned to pass high speed data.

**Sync character.** A character transmitted to synchronize a data block with the receiving end.

**UART.** Universal asynchronous receiver transmitter.

**Double buffering.** Using one Holding Buffer and one Shifting Buffer for temporary data storage.

### REFERENCES

Thomas R. Blakeslee, *Digital Design With Standard MSI and LSI*, Wiley, New York, 2nd ed., 1979.

Austin Lesea and Rodnay Zaks, *Microprocessor Interfacing Techniques*, 2nd ed., Sybex, Berkeley, CA, 1978.

### PROBLEMS

18-1.   Draw a MODEM transmission system using FSK and frequencies of 1000, 1100, 1200, and 1300 Hz.

18-2.   Find the bit pattern for the first two characters of Fig. P18-2. The 11-bit asynchronous code is used.

18-3.   Design a controller for a circuit that is to receive 8-bit words from a MODEM.

18-4.   A circuit receives 7-bit characters from a MODEM. For each seventh bit received, it is to append an eighth bit for odd parity and set DONE. Design the controller.

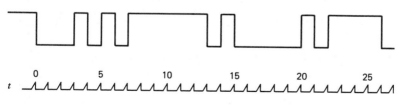

FIGURE P18-2

18-5.    In Problem 18-4, assume a 5000 BPS MODEM is used. How much time does the processor have to recognize DONE and take the character if:
   (a) Single buffering is used.
   (b) Double buffering is used.

18-6.    Design a double-buffered controller for Problem 18-5.

18-7.    If the users in a computer system all have 300 BPS I-O devices, design a system to multiplex them onto a 4800 BPS transmission line using 4800 BPS MODEMs.

18-8.    Consider the receive terminal that receives the data transmitted by the MODEM of Fig. 18-9. If the data are to be broken into 24 distinct lines to match the incoming data, design the required circuits.

18-9.    A synchronizing character has the code 00010110. Design a circuit to detect the presence of three consecutive sync characters anywhere within the data stream coming from a MODEM.

18-10.    Design a circuit to connect a computer to a TTY using the ICs discussed in the earlier chapters of the book. Assume the computer provides a 7-bit ASCII character and a DATA AVAILABLE pulse. The circuit must provide the START and STOP bits, proper timing, and add an eighth bit for even parity.

18-11.    Design a circuit to receive a character from a TTY and send it to a computer. The circuit must detect the START BIT, clock in the data bits, and provide a DATA READY signal.

18-12.    For your circuit of Problem 18-11, add an OVERRUN error detector. Assume the computer provides an ACK when it accepts a character.

18-13.    If there were no Holding Register in Problem 18-12, a received character could be read in during the stop bits. This is single buffering. For a TTY how long would the computer have to read a character if single buffering is used? How does this compare with the results of Ex. 18-10?

18-14.    A terminal is connected to an **AY-5-1013** and operates at 300 BPS.
   (a) What frequency must be supplied to the UARTs clock inputs?
   (b) How long does the computer have to respond to DAV?
   (c) If the character is 8 bits long, must have odd parity appended, and uses one STOP bit, how must pins 34–40 of the UART be wired?

After working the problems, the student should return to Sec. 18-2 and be sure he or she can answer all the self-evaluation questions. If any of them are still difficult, the student should review the appropriate sections of the chapter to find the answers.

# 19

# MICROPROCESSOR
# INTERFACING

## 19-1  INSTRUCTIONAL OBJECTIVES

This chapter introduces the methods of interfacing microprocessors ($\mu$Ps) to various peripherals. The most commonly used $\mu$Ps, the Intel **8085** and the Motorola **6800,** are considered as examples.

After reading the chapter, the student should be able to:

1.  List the signals on a $\mu$P bus and explain their function.
2.  Decode the $\mu$P address bus to access a particular IC.
3.  Design circuits to connect $\mu$Ps to memories.
4.  Design circuits and systems to exchange data with a $\mu$P using IN and OUT instructions.
5.  Generate an **8085** interrupt and properly respond to INTA.
6.  Connect a peripheral to a **6800** $\mu$P using the PIA.
7.  Control data transfers between the PIA and a peripheral using the control lines of the PIA.

## 19-2  SELF-EVALUATION QUESTIONS

Watch for the answers to the following questions as you read the chapter. They should help you understand the material presented.

1.  What is a multiplexed bus? What are its advantages and disadvantages?

2. Why must external signals to be applied to the data bus go through 3-state drivers?

3. What is the function of the IO/$\overline{\text{M}}$ line?

4. What is a vectored interrupt? What is its advantage?

5. What is the difference between a command or control register and a status register?

6. When should CA2 or CB2 be an input? When should it be an output?

## 19-3 INTRODUCTION TO MICROPROCESSOR INTERFACING

This chapter concentrates on interfacing a microprocessor ($\mu$P) to other devices, so that data and commands can be exchanged between the $\mu$P and the external world. It requires a full book to thoroughly cover a microprocessor, and several good ones have been written (see References). Many books on $\mu$Ps, however, tend to emphasize software or programming at the expense of interfacing. Successful $\mu$P interfacing requires a knowledge of both the instruction set and programming for the particular $\mu$P and digital hardware.

Interfacing to two of the most common $\mu$Ps, the Intel **8085** and the Motorola **6800**, will be discussed in this chapter. Most of the techniques for the **8085** also apply to the Intel **8080** and the Zilog **Z80**, while the techniques for interfacing the **6800** can be used with the Motorola **6809** and the **6502**. Knowledge of the instruction set and programming must be obtained elsewhere (see References), but interfacing uses digital gates, UARTs, D/A, and A/D converters, and many of the circuits and ICs discussed in the previous chapters. In this chapter we show how to connect a $\mu$P to these devices so it can monitor and control events that are external to the $\mu$P.

### 19-3.1 Microprocessor Busses

Interfacing to 8-bit $\mu$Ps is discussed here. Limitations do not permit a discussion of 16-bit $\mu$P interfacing, although many of the interfacing techniques are similar. A typical $\mu$P system is shown in Fig. 19-1. Most $\mu$Ps contain no memory within them and use memory ICs such as the **2114** (see Sec. 15-7.2). The $\mu$P is connected to the memory ICs and peripheral driver ICs via a 3-state bus that contains:

1. 8 bi-directional data lines.
2. 16 address lines (the $\mu$P controls the address except for DMAs).
3. 1 Read/Write line, also controlled by the $\mu$P.
4. Control lines, as required by the particular $\mu$P.

Most $\mu$P systems contain several memory ICs and several peripheral drivers. Of course only one IC on the 3-state bus may be selected at any

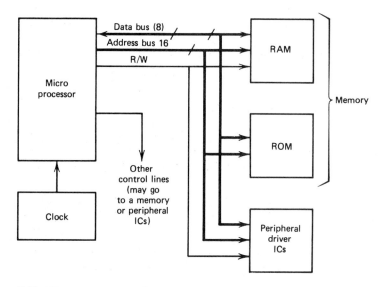

FIGURE 19-1 A typical μP system.

one time. Fortunately, all ICs on the bus have one or more CHIP SELECT (CS) lines. A particular IC is usually selected by using decoders on the high-order address lines.

---

**EXAMPLE 19-1**

The Motorola **6810** is a 128-word-by-8-bit memory.
  (a) What is the largest number of **6810**s that can be used in a system?
  (b) How can a particular **6810** be selected?

**SOLUTION**

The **6810** requires 7 address lines to select one of the 128 words in its memory. This leaves 9 address lines available for selecting a particular **6810** and 512 **6810**s could be used in a maximum system. A 9-line to 512-line decoder (see Sec. 11-5.1) would be needed to select a particular **6810**. A system of this size is impractical as it leaves no room for any other peripheral ICs.

**EXAMPLE 19-2**

Figure 19-2 shows the decoder circuitry inside the **SDK-85**, a μP kit for development and training using the Intel 8085 (see Sec. 19-11).
  (a) Explain what it is doing.
  (b) For which addresses will CS5 be LOW?

## SOLUTION

The **8205** is a 3-line to 8-line decoder, much like the **74155** (see Sec. 11-4.1). The major difference is that the **8205** has three ENABLE lines. The level on each of the ENABLE lines must be correct ($E_1$ and $E_2$ msut be 0, $E_3$ must be 1) or all eight outputs of the **8205** will be HIGH.

Figure 19-2 shows that $E_1$ and $E_2$ are connected to $A_{15}$ and $A_{14}$, the two highest addresses on the 16-line address bus. Thus the chip will only decode if the two high-order address bits are both 0.

The address inputs of the **8205**, $A_2$, $A_1$, and $A_0$, are connected to $A_{13}$, $A_{12}$, and $A_{11}$ of the $\mu$P address bus. CS5 will be LOW only if $A_{15}$ and $A_{14}$ are both 0, and $A_{13}$, $A_{12}$, and $A_{11}$ are 101 (5), respectively. Thus CS5 is LOW for any address on the $\mu$P bus that looks like 0010 1xxx xxxx xxxx or for any $\mu$P address between $(2800)_{16}$ and $(2FFF)_{16}$.

---

In addition to the basic $\mu$P bus discussed here, other busses for connecting components and systems, such as the S100 bus and the IEEE 488 bus, exist. They will be discussed in Sec. 19-12, after the reader is more familiar with $\mu$P interfacing.

## 19-3.2 Machine Cycles and Clock Cycles

Microprocessors are driven by an external *clock* that is used to *time and synchronize its internal operations*. The frequency of the clocks for the Motorola 6800 is about 1 MHz, while the clocks for the 8085 and **Z80** are two to four times faster. The 6800 requires less clock cycles to execute its instructions, however, so the speed that matters—the speed of instruction execution—is about the same for both $\mu$Ps.

In the 6800 there is one clock cycle for every memory fetch or store. An additional clock cycle or two may be required to give the $\mu$P extra time when complex instructions are being executed.

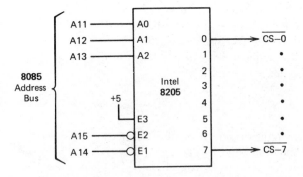

FIGURE 19-2 The Intel **8205** as used as a decoder in the **SDK-85** kit.

## EXAMPLE 19-3

In the 6800 $\mu$P an instruction to load the accumulator from location 0123 looks like this:

| Address Data | | Comments |
| --- | --- | --- |
| 20 | B8 | Op Code for a LOAD instruction in the extended mode. |
| 21 | 01 | MSByte of the address |
| 22 | 23 | LSByte of the address |

What is the least number of cycles required to execute this instruction?

### SOLUTION

This instruction requires four memory reads. All $\mu$P instructions start with a memory read called an *Op Code Fetch*, which reads the Op Code into the $\mu$P. When the 6800 has the Op Code and determines that it is a LOAD instruction in the extended mode, it knows it must go back to memory at the next two locations (21 and 22) to find the address. This requires two more memory read cycles. Now that the $\mu$P knows the required address (0123), it can put 0123 on the address bus and execute a fourth memory read cycle to get the data from 0123 and load it into the accumulator. An inspection of the 6800 instruction set reveals that this instruction actually does take four cycles, so no additional time is required.

The Intel 8085 divides its instructions into *Machine Cycles*. A Machine Cycle is required for each memory reference and some instructions also require extra Machine Cycles. For the 8080 and 8085, however, each Machine Cycle takes at least three clock cycles (or T states, as Intel calls them). The 8085 instruction equivalent to the LOAD EXTENDED is LOAD ACCUMULATOR DIRECT, which takes four Machine Cycles and 13 clock cycles.

## 19-3.3 The 8085 Bus

The microprocessor busses described in Sec. 19-3.1 pertain to the 6800, the 6502, and mainly to the 8080 $\mu$Ps. The 8085 bus, however, is a *multiplexed bus*. In a multiplexed bus *different signals appear on the same line at different times*.

For the 8085 the 8 data bits and the 8 LSBs of the address are multiplexed onto the same lines. During the *first T state* of each Machine Cycle the bus contains the *memory address* and during the *later T states* the bus contains the *memory data*, which may be bi-directional. The 8085 bus also contains an ALE (Address Latch Enable) signal, which goes HIGH when the bus contains addresses rather than data.

The 8085 bus also contains the following lines:

- $\overline{\text{RD}}$. A READ line that goes LOW when the $\mu$P is reading (from memory or a peripheral device).
- $\overline{\text{WR}}$. A WRITE line that goes LOW when the 8085 is writing to memory or a peripheral.
- IO/$\overline{\text{M}}$. This line distinguishes between a Memory Request (IO/$\overline{\text{M}}$ = 0) and an I-O Port Request (IO/$\overline{\text{M}}$ = 1). See Sec. 19-4 for further details.
- $S_1$, $S_0$. These two lines can be decoded to determine what type of memory cycle (I-O READ, MEMORY WRITE, OP CODE FETCH, etc.) is taking place. This was important for the 8080 $\mu$P, but for the 8085 the information is rarely required. The SDK-85 kit, for example, does not use these lines, although they are made available to the user on an external connector.
- READY. The READY line is an input to the 8085. During T2 of each Machine Cycle the status of the READY line is determined. If it is LOW the 8085 enters the WAIT state, and suspends operation until READY goes HIGH. This line is used to synchronize the 8085 with slow memories. If the memory's response is too slow for the 8085, the READY line can be held LOW until the memory had had sufficient time to READ or WRITE the byte.

The 8085 bus operation is shown in Fig. 19-3. Two READ cycles are shown: the first with no WAIT states (READY always HIGH) and the second with one WAIT state. Note that the addresses appear on AD0–AD7 (the

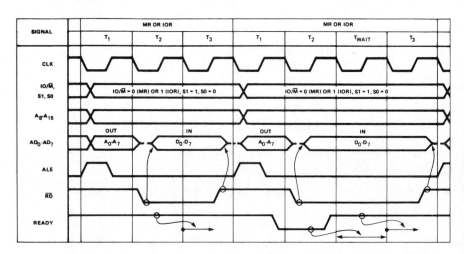

FIGURE 19-3 Memory read (OR I-O Read) Machine cycles. (Reprinted by permission of Intel Corporation, copyright 1978.)

multiplexed portion of the Address-Data bus) at T1 and the data appear on the same lines at the end of T2.

---

## EXAMPLE 19-4

If a slow memory is being used with the 8085, how can the READY line be connected so that it goes LOW on each memory cycle?

### SOLUTION

One way is to use ALE, which occurs at the beginning of each cycle, to trigger a one-shot. The output of the one-shot can be tied to READY and the time adjusted until the memory and $\mu$P speeds are compatible.

---

## EXAMPLE 19-5

Design a circuit to connect an 8085 to a 1K-byte memory using 2114s (see Sec. 15-7.2).

### SOLUTION

For a 1K-byte memory, two 2114s (which are 1K words by 4 bits) are required. The circuit is shown in Fig. 19-4. The multiplexed AD0–AD7 lines are connected to the data inputs of the 2114s and to the inputs of 7475 4-bit latches (see Sec. 15-7.2). The 7475s are clocked by ALE and supply the 8 LSBs of the address. The 2 MSBs of the address lines are tied directly to A8 and A9 coming from the 8085 because these lines are not multiplexed.

The 2114s are selected when either $\overline{RD}$ or $\overline{WR}$ go LOW. Note that the 2114s cannot always be selected because if they were, they would drive data onto the bus at the same time the 8085 was putting addresses on the bus.

---

## 19-4 INPUT AND OUTPUT PORTS

The 8080 and 8085 execute their I-O instructions primarily by using Input or Output *ports*. A port designates the *source* of data on an I-O input or I-O READ instruction. Thus input devices like UARTs, card readers, or cassettes would have to be connected to an input port. For an output or I-O WRITE instruction, the port designates which external device (line-printer, tape, etc.) is to *receive* information from the $\mu$P. The architecture of the 8080 and 8085 allows for 256 input and 256 output ports.

There are two special instructions for handling input and output: the IN instruction and the OUT instruction. An IN instruction causes data to be *read from the external device* into the accumulator, while an OUT instruction causes a byte from the accumulator to be *sent to an external device*.

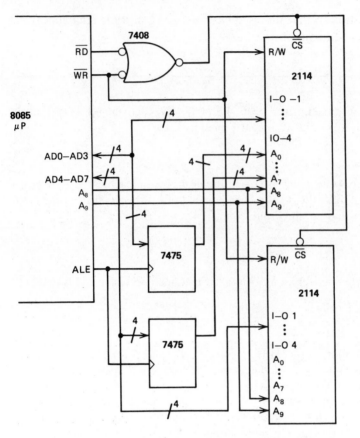

FIGURE 19-4 Connecting a **2114** memory to an **8085** $\mu$P.

Both use two bytes and three Machine Cycles. The first byte is the Op code (DB for an IN, D3 for an OUT). The second byte is the port address (1 of 256 ports may be selected). During the first Machine Cycle, the $\mu$P reads the IN or OUT Op code, during the second Machine Cycle it reads the port address.

Data transfer occurs during the third Machine Cycle. *The 8-bit port address from the second byte of the instruction is placed on the 16 address lines twice.* It occupies *both* bits 0–7 and bits 8–15 of the address bus. An IN instruction also causes $\overline{RD}$ to go LOW, and a byte is read from the external device to the accumulator. Thus an IN instruction is sometimes called an I-O READ (IOR). An OUT instruction causes $\overline{WR}$ to go LOW and a byte is written from the accumulator. This is also called an I-O WRITE (IOW).

Both IN and OUT instructions must inform all components connected to the $\mu$P that an I-O cycle is occurring instead of a memory cycle. In the

8080 it is necessary to decode the S1 and S2 lines to determine this. The 8085 simpliifes I-O by using the $\overline{IO/M}$ line. IN and OUT are the *only* two 8085 instructions that cause $\overline{IO/M}$ to go HIGH.

## 19-4.1  IN and OUT Timing

The timing for an OUT instruction for the 8085 is shown in Fig. 19-5. IN and OUT instructions take three Machine Cycles consisting of 10 clock cycles, or T states. Figure 19-5 shows the relationship of the clock, ALE, $\overline{RD}$, $\overline{WR}$, A8, and the lower 8 bits of the multiplexed address-data bus for an OUT 15 instruction. The instruction proceeds as follows:

1.  The first Machine Cycle is an Op code fetch. This requires four clock cycles. During the first clock cycle, the instruction address (N) is on the address-data bus and ALE goes HIGH. During the second and third cycles, $\overline{RD}$ goes LOW while the Op code (D3 for an OUT instruction) is placed on the bus. The fourth clock cycle gives the 8085 time to decode the instruction and prepare for the following cycles.

2.  The second cycle of the OUT instruction is a memory read of the second byte. The memory address is (N + 1), the location of the second byte, and the data read from memory is the port address (15 in this example).

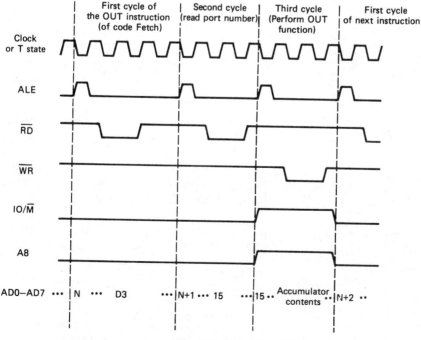

*Note:* N = address of the OUT instruction

FIGURE 19-5  Timing for an OUT instruction.

3. The data transfer takes place during the third Machine Cycle. During this cycle IO/M is HIGH. The port address is placed on the address-data bus, but is quickly replaced by the accumulator data. At this time the $\overline{\text{WR}}$ line goes LOW. Address bits 8–15 will also contain the port address (15), but they will remain there throughout the entire cycle as the behavior of A8 shows.

The timing for an **IN** instruction is identical except that an additional pulse occurs on the $\overline{\text{RD}}$ line instead of on the $\overline{\text{WR}}$ line. All timing can be checked on an oscilloscope by using the simple program:

$$
\begin{array}{llll}
2010 & \text{OUT} & 15 & \text{(or IN)} \\
2012 & \text{JMP} & 2012 &
\end{array}
$$

## 19-4.2  Execution of the OUT Instruction

The OUT instruction (I-O WRITE) can be executed by tying the address-data bus to the inputs of the various peripherals and then strobing the data in with a pulse that gates IO/M and the $\overline{\text{WR}}$ line.

---

**EXAMPLE 19-6**

An 8085 system has 16 output devices. Design the circuitry so the **8085** can send data to any of these devices.

**SOLUTION**

The output devices can be set up so they use ports 0–15. In this way only the 4 LSBs of the address lines need be used. The circuit is shown in Fig. 19-6. It operates as follows.

1. The AD0–AD7 lines go to all devices. These lines are shown going through a **74LS241**, 8 input, non-inverting, buffer/line driver. If the data input to all devices is CMOS, the **74LS241** might be omitted, but it has two advantages: it alleviates *fanout* or *loading problems*, and it *isolates* the $\mu$P from the external devices. Isolation is important. Otherwise, a miswire or other catastrophe on an external device could disable the $\mu$P. The buffer is enabled by IO/M so that it transmits data only during OUT or IN instructions.
2. The 4-bit port address is sent to a **74154** decoder (see Sec. 11-4.2). Addresses A8 through A11 are used (instead of ADO–AD3) because the higher address lines are not multiplexed and this eliminates the need for an address latch.
3. The decoder is enabled by the inverse of IO/M and by $\overline{\text{WR}}$. Thus it will only function during the write portion of an OUT instruction. When enabled the decoder provides a LOW input to the DEVICE-SELECTED line of the addressed peripheral. The peripheral must use this signal to *strobe* in the data. Note that the AD lines contain data at this time, not addresses, because DEVICE-SELECTED is synchronized with the $\overline{\text{WR}}$ signal.

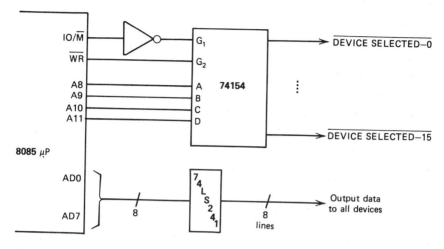

FIGURE 19-6    Sending data from an **8085** to 16 peripherals.

4.    More than 16 (up to 256) peripherals may be accommodated by enlarging the decoder and adding more drive capability if necessary.

## 19-4.3    Execution of the IN Instruction

The IN instruction (I-O READ) brings data in from the peripherals and enters them into the accumulator. The IN instruction can be executed by a circuit similar to Fig. 19-6. The major differences are:

1.    The $\overline{RD}$ line would enable the decoder instead of $\overline{WR}$.
2.    The peripherals must all put their data on the AD lines of the $\mu$P. If these data are present at the wrong time (when the $\mu$P is executing other instruction, for example), they will cause a conflict and cripple the $\mu$P. Therefore, all peripherals that are inputs must be tied to the address-data bus via 3-state gates that are only enabled during the third Machine Cycle of an IN instruction. Fortunately, the DEVICE-SELECTED outputs from the decoder only go LOW at this time, so DEVICE-SELECTED provides an ideal signal for enabling the 3-state drivers going from the peripherals to the $\mu$P.

## 19-5    INTERRUPTS ON THE 8080 AND 8085

*An interrupt is a high priority request for service made by a peripheral.* An interrupt causes the $\mu$P to *jump* out of the program it is currently running and go to a *service routine* for the peripheral. When it finishes the service routine it returns to its main program.

On the 8080 and 8085 an interrupt is initiated when the interrupting device places a HIGH on the INTR line. The $\mu$P responds by placing a LOW pulse on its $\overline{INTA}$ (Interrupt Acknowledge) line. While this pulse is

LOW the $\mu$P must receive the *Op code* of the next instruction it will execute on its AD0–AD7 lines. The interrupting device is responsible for placing the Op code on the AD0–AD7 lines.

The function of the instruction whose Op code is read in during $\overline{\text{INTA}}$ is to jump the program to the start of the interrupt service routine. It must also place the contents of the Program Counter (PC) on the $\mu$P's stack so the main program can be resumed after the interrupt routine is finished. For the 8080, there is only one instruction that performs these functions and should be placed on the AD0–AD7 lines. This is a RESTART. The 8085 can accept either a RESTART or a CALL.

### 19-5.1 The RESTART Instruction

The RESTART instruction looks like

$$11NNN111$$

During execution it first writes the PC onto the stack to preserve it, and then places NNN000 into the program counter, so the first instruction of the service routine is executed there. There are eight RESTARTS (RE-START-0 through RESTART-7), where the number of the RESTART is the binary value of NNN.

---

**EXAMPLE** 19-7

What does the Op code of a RESTART-5 look like? Where does the service routine start?

**SOLUTION**

A RESTART-5 looks like

$$111\underbrace{01}_{5}111$$

The service routine will start at 0000 0000 0010 1000 or $(0028)_{16}$.

$$\underbrace{0010\ 1000}_{5}$$

**EXAMPLE** 19-8

Design a circuit to connect a RESTART to the AD0–AD7 lines.

## SOLUTION

Since the RESTART Op code should only be placed on the AD lines when $\overline{\text{INTA}}$ is LOW, it must come in through 3-state gates that are enabled by $\overline{\text{INTA}}$. The circuit is shown in Fig. 19-7. It can use either two **74125**s (see Sec. 5-7.1) or a **74LS244**. The **74LS244** is shown. The switches can be set to give any desired

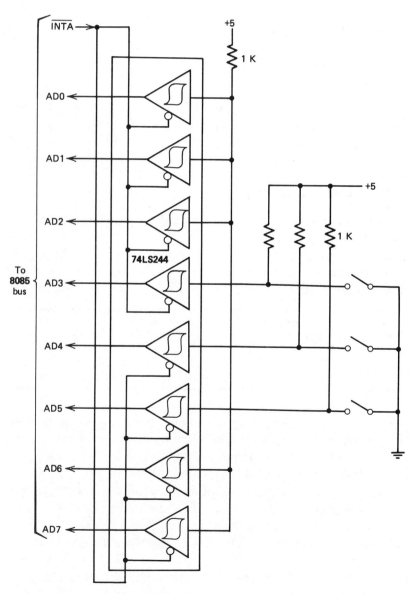

FIGURE 19-7 Connecting a RESTART to an **8085**.

RESTART (0–7). Of course peripherals that generate RESTARTs will probably use gates rather than switches to drive the AD lines during INTA (see Problem 19-6).

The possibility of eight different RESTART addresses allows the service routines for the various interrupting devices to start at (or vector to) eight different addresses. This is called *Vectored Interrupts* by Intel since each interrupting device can use a different RESTART that *vectors the $\mu P$ to that device's service routine.* There are only eight locations between the RESTART vectors, however, so programs cannot really be written there. Most RESTART addresses contain a JUMP to the actual start of the service routine.

The 8080 and 8085 also contain a RESET IN pin. A LOW on this input causes the $\mu P$ to go to location 0. Location 0 is thus reserved for the start of the main program. Turning power on also causes a RESET. Note that both a RESET and a RESTART-0 vector the $\mu P$ to location 0.

## 19-5.2 The CALL Instruction

The 8085 will also accept an unconditional CALL instruction. The CALL instruction can be considered as a Jump to Subroutine. It places the address of the next instruction (the contents of the PC) on the stack and jumps to the address specified by the two bytes following the Op code. For example, a CALL instruction to location 2050, that is in 2010, looks like the following:

| Address | Data | Function |
|---------|------|----------|
| 2010 | CD | Op code for an unconditional CALL |
| 2011 | 50 | LS byte of destination address |
| 2012 | 20 | MS byte of destination address |

This instruction causes the program to jump to the subroutine at **2050**, and it places the address of the next instruction, **2013**, on the stack so the program can resume after the subroutine finishes and RETURNs.

The advantage of using a CALL in response to an interrupt is that the 8085 can be vectored to a service routine *anywhere* in memory. Of course, it requires more hardware to do this. The Interrupting Peripheral must place the CALL Op code (CD) on the AD0–AD7 lines in response to INTA. It will then receive two more INTAs, and it must place the address on the bus.

The timing for a CALL response to an interrupt is shown in Fig. 19-8. It requires five Machine Cycles. During the first three cycles INTA goes

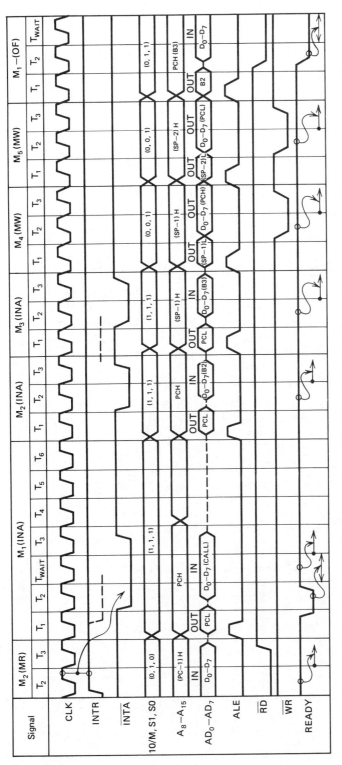

FIGURE 19-8 Interrupt acknowledge machine cycles with CALL instructions in response to INTR. (Reprinted by permission of Intel Corporation, copyright 1978.)

LOW and the 8085 receives the CALL Op code and the two bytes of the address (B2 and B3). During cycles 4 and 5 $\overline{WR}$ goes LOW and the PC is written to the stack. Figure 19-8 also shows the Op code fetch for the next instruction. Note that addresses A8–A15 contain B3 throughout the Machine Cycle, and AD0–AD7 contain B2 during ALE.

---

**EXAMPLE 19-9**

A peripheral is to interrupt and place a CALL 2050 on the AD0–AD7 lines when the 8085 responds. Design the circuitry.

**SOLUTION**

The circuit is shown in Fig. 19-9. It was designed as follows:

1. The outputs to the AD0–AD7 lines must only be active when INTA is LOW. This suggests a **74LS244** (or two **74125s**) enabled by $\overline{INTA}$.
2. Three different outputs are required in response to an interrupt. This suggests a 3s counter (see Sec. 8-7) triggered by $\overline{INTA}$. Assuming the 3s counter is cleared prior to the start, the leading edge of the first INTA will set it to a count of 1, so for each interrupt it counts 1, 2, 0.
3. The bits to be placed on the line are shown in Table 19-1.
4. By examining the table we see that bits 0, 2, 3, and 7 should only be 1s during the first pulse, when $Q_1$ is 1. Therefore, $Q_1$ is connected to the inputs of these gates.
5. Bit 1 is always 0. Therefore, the input to that gate is grounded.
6. By looking at the logic required for the other bits we can complete the circuit as shown in Fig. 19-9.

---

## 19-5.3 Other 8085 Interrupts

The 8085 has four interrupts that come in on pins of the $\mu$P. When an interrupt is received on one of the pins, the 8085 executes a RESTART, where the RESTART addresses are placed between the normal RESTART

TABLE 19-1 BIT CONFIGURATION FOR A CALL 2050

| INTA | $Q_1$ | $Q_2$ | 7 | 6 | 5 | 4 | 3 | 2 | 1 | 0 | | |
|------|-------|-------|---|---|---|---|---|---|---|---|-----|------|
| 1 | 1 | 0 | 1 | 1 | 0 | 0 | 1 | 1 | 0 | 1 | CD | CALL |
| 2 | 0 | 1 | 0 | 1 | 0 | 1 | 0 | 0 | 0 | 0 | 50 | 50 |
| 3 | 0 | 0 | 0 | 0 | 1 | 0 | 0 | 0 | 0 | 0 | 20 | 20 |

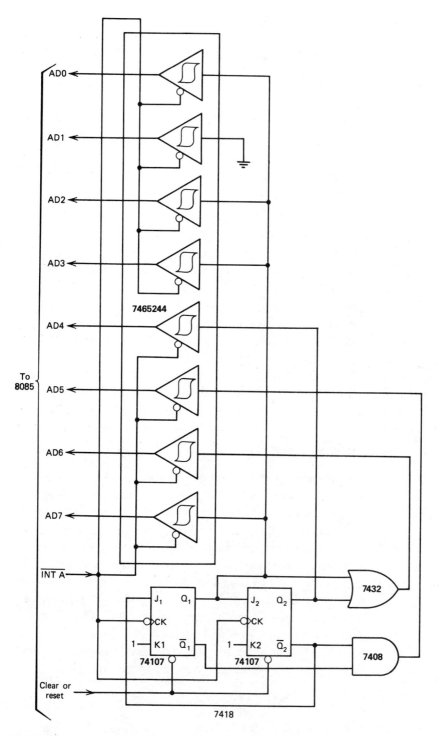

FIGURE 19-9 A circuit of generate a CALL 2050 in response to an interrupt for an **8085**.

TABLE 19-2   RESTART   ADDRESSES
FOR THE 8085

| Name | RESTART Address |
|------|-----------------|
| TRAP | $(24)_{16}$ |
| RST 5.5 | $(2C)_{16}$ |
| RST 6.5 | $(34)_{16}$ |
| RST 7.5 | $(3C)_{16}$ |

addresses. This gives each RESTART four locations before it impinges on another RESTART. This is long enough, however, to accommodate a JUMP instruction to the start of the appropriate service routine.

The four interrupts and their RESTART addresses are shown in Table 19-2. Note that the TRAP address is between RESTART 4 and RESTART 5, RST 5.5 is between RESTART 5 and RESTART 6, and so forth.

TRAP is a non-maskable interrupt that that should be used only for catastrophic events, such as a power failure somewhere in the system. The other inputs are maskable by the SIM (Set Input Mask) instruction. Refer to the 8085 literature (see Sec. 9-15) for the details on how these operate.

## 19-6   MEMORY MAPPED I-O

While Intel uses primarily port I-O, Motorola uses *memory mapped I-O* exclusively. In memory mapped I-O *all peripheral interface ICs occupy locations in the address field*. The most commonly used Motorola Peripheral is the 6821 Peripheral Interface Adapter (PIA). Of course, addresses used by PIAs or other I-O ICs cannot also be used by memory. This reduces the available memory space. Furthermore, each peripheral IC generally occupies only a few addresses (a PIA takes four addresses), so in a large system, considerable decoding may be needed to access a particular IC. The advantages of memory mapped I-O are that the $\mu$P treats the I-O ICs just like memory; thus any instructions that access memory (LOADs, STOREs, SHIFTS[1], etc.) can also access the I-O ICs. No special instructions such as IN and OUT are needed and this simplifies programming.

---

[1]The 6800 has instructions that allow the user to shift, increment, decrement, and perform other operations on a memory location. The 8080/8085 does not have this capability.

## 19-6.1 Command and Status Registers

Most ICs that use memory mapped I-O have several registers in them. The main registers are:

1. **The Data Register.** This holds the data being transferred from the I-O ICs to the $\mu$P on an I-O READ and from the $\mu$P on an I-O WRITE.

2. **The Command Register.** Most I-O ICs have several *modes* of operation, or ways they operate. The particular mode they use at any time is determined by the bits in a Command Register. The Command Register is written by the $\mu$P. Thus the $\mu$P can command the I-O IC to operate in a given mode, and the $\mu$P can change these commands as the program progresses. Often, however, the appropriate mode of operation is selected at the beginning of the program and does not change.

3. **The Status Register.** The Status Register of the I-O IC is written into by the external device. It contains information about the status of the external device such as: is the printer out of paper, does the tape drive need another character, and so on. The $\mu$P can read the Status Register to monitor the progress of data transfers and take appropriate action when required.

4. **Command-Status Register.** Sometimes the Command and Status Registers are combined. The PIA is an example.

5. **The Data Direction Register (DDR).** This register determines whether each line of the data register is input or output. Generally a 0 written into a bit in the DDR forces the corresponding bit in the Data Register to be an input and a 1 in the DDR forces the Data Register bit to be an output. The DDR is written to by the $\mu$P. This allows the $\mu$P to control the direction of data flow.

---

**EXAMPLE 19-10**

A DDR contains F0. What is the direction of data flow?

**SOLUTION**

Because the 4 LSBs of the DDR are 0, the corresponding four LSBs of the Data Register are inputs. The 4 MSBs of the Data Register are outputs. Thus this IC can take data in on its 4 LSBs and send data to the peripheral on its 4 MSBs. This situation will continue until the $\mu$P commands a change in the DDR.

---

In many ICs the Command Register is WRITE-ONLY (the $\mu$P cannot read it) and the Status Register is READ-ONLY (the $\mu$P cannot write it).

## 19-7 THE MOTOROLA PIA

The Motorola Peripheral Interface Adapter (PIA) is the **MC6820** or **MC6821**. The 6821 is the newer and preferred part. The difference between the two parts is minimal and not essential for this discussion. The PIA is too complex an IC to be thoroughly discussed here, but the salient points and some I-O examples will be presented.[2]

The PIA is divided into two 8-bit I-O ports, called the A side and the B side. Each side has eight I-O lines to go to external devices.

There are three registers that control the data flow on each side:

1. The Control-Status Register, which uses 6 bits for control and has two read-only status bits.
2. The 8-bit Data Register.
3. An 8-bit DDR that controls the direction of data flow. The $\mu$P uses address lines A0 and A1 to select the proper register, then communicates with the PIA via the bi-directional data bus and R/W. The PIA is enabled by Chip Select (CS) lines. The address of the PIA is decoded from higher address bits and the decoder outputs drive the CS lines.

To read a character from a peripheral device into the $\mu$P system, the peripheral's data output lines are connected to the PIA's data lines (PA0–PA7 on the A side and PB0–PB7 on the B side). The DDR is set at 00 (all input). Now a LOAD from the Data Register's memory address loads the input data into one of the 6800's two accumulators.

When a character is written to a peripheral, it usually is first stored in memory. The DDR is set for output (FF) and the character is LOADED from memory into an accumulator and then stored into the PIA's data register. A PIA port does not have to be all input or all output. The direction of each data bit is individually controlled by the corresponding DDR bit.

---

### EXAMPLE 19-11

Write a program to read the state of four switches and send the data out to a hexadecimal display. Assume the address of the PIA's Data Register is 8008.

### SOLUTION

One solution is shown in the circuit of Fig. 19-10 where the four switches are connected to lines PA0 through PA3, and the display inputs are connected to peripheral lines PA4 to PA7. The program for the solution proceeds.

1. During initialization, F0 is written into the Direction Register. This configures the 4 LSBs as inputs and the 4 MSBs as outputs.
2. The Control Register is then rewritten to make bit 2 a 1. The instruction LOAD A (from) 8008 reads the switch contents into the 4 LSBs of the accumulator.

---

[2]For a complete discussion of the PIA, see Greenfield and Wray, *Using Micropro-cessors and Microcomputers: The 6800 Family,* Wiley, New York, 1981.

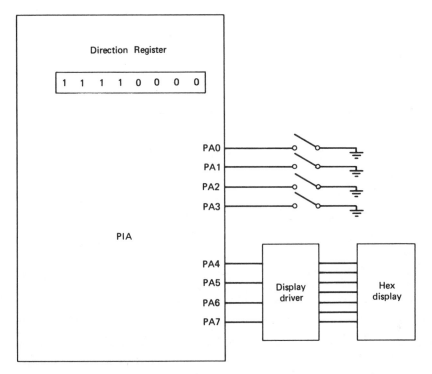

FIGURE 19-10   The PIA and its connections for Example 19-11. (From Greenfield and Wray. *Using Microprocessors and Microcomputers: The* **6800** *Family.* Copyright John Wiley & Sons, Inc. 1981. Reprinted by permission of John Wiley & Sons, Inc.)

3.   Four LEFT SHIFT instructions shift the switch settings into the MSBs of the accumulator.

4.   Now a STORE A at 8008 places the switch settings on the output lines (PA4–PA7) where they can drive the display.

## 19-7.1   CA1 and CB1

Each side of the PIA has two lines that are used to *control the data transfer.* These lines are labeled CA1 and CA2 for the A side, and CB1 and CB2 for the B side. The CA1 and CB1 lines are input only. They can be used by the peripheral to synchronize data transfers.

When a peripheral has a character to input, for example, it causes a *transition* on the CA1 or CB1 line. This transition causes a status bit in the Control Register to set (CRA-7 or CRB-7). The $\mu$P program must periodically examine this bit to determine if the peripheral has data. When the $\mu$P sees the bit SET, it reads the data. Reading the byte from the Data

Register also clears the bit in the Control Register. This bit will not go HIGH again until the peripheral has another character for the $\mu$P and again causes a transition on CA1 or CB1.

---

**EXAMPLE 19-12**

Show how a TTY can be connected to a $\mu$P via an **AY5-1013** UART (see Sec. 18-7.1) on the receive side.

**SOLUTION**

The connection is shown in Fig. 19-11. The Received Data bits are connected to PA0–PA7 and DAV is connected to CA1. When the UART has a character from the TTY, it raises DAV causing CA1 and CRA-7, the bit in the Control Register, to go HIGH. When the $\mu$P reads the PIA's Control Register, it finds the bit SET, so it reads the Data Register to accept the character. This also resets the bit in the Control Register.

---

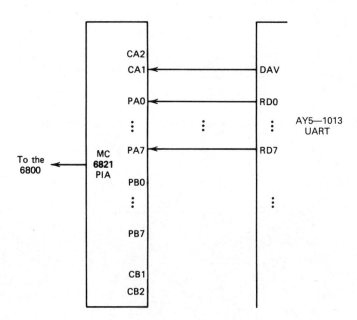

FIGURE 19-11   Connecting a UART to a **6800** system using a PIA.

## 19-8   CA2   AND   CB2

CA2 and CB2 can operate in any one of four modes. The modes are:

1.   Input
2.   Output level
3.   Output pulse
4.   Handshaking

The mode of operation of these lines is determined by the bits in the Control Register, which are set by the $\mu$P.

### 19-8.1   Input Mode

In input mode CA2 and CB2 are very similar to CA1 and CB1. A transition on the CA2 or CB2 line causes bit 6 of the corresponding Control Register to be SET. In this mode CA1 and CA2 can be connected to two different peripherals that require service. The first peripheral can cause a transition on CA1, setting CRA-7 and the second peripheral can cause a transition on CA2, setting CRA-6. Of course, the $\mu$P program must now examine both bits to determine if either peripherals is requesting service (see Problem 19-11).

### 19-8.2   Level Mode

In the output level mode CA2 (CB2) assumes a level of 1 or 0 as determined by the $\mu$P. In this mode the $\mu$P writes a bit into the Control Register that sets the level, so CA2 and CB2 can be used as two additional output lines, provided they are not needed for other uses.

---

EXAMPLE  19-13

How can the level mode of the PIA be used to connect a $\mu$P to an A/D converter? (see Sec. 17-7.5).

SOLUTION

Every time the $\mu$P needs to read the A/D converter it must generate a Start-Of-Conversion (SOC) pulse. The $\mu$P can do this by first lowering CA2 or CB2 and then raising it. The End-Of-Conversion (EOC) line can be connected to CA1 or CB1. After generating SOC the $\mu$P monitors the control register to see when EOC goes HIGH. It can then read the digital equivalent of the analog input voltage.

---

### 19-8.3  Pulse Mode

In the pulse mode, a negative pulse is produced for one $\mu$P clock cycle each time the PIA data register is read or written to. Here the A and B sides of the PIA differ. The A side pulses CA2 each time the A data Register is read, but the B side pulses CB2 each time the B Data Register is written to. In this mode, and in the handshaking mode, the A side is designed primarily for input and B side for output because of this difference.

The pulse mode can be used to connect the $\mu$P to a UART or memory (see Problem 19-9). In Example 19-12, the reader may have noticed a problem. While the $\mu$P is set up to respond to DAV, it must also provide RESET DATA AVAILABLE ($\overline{\text{RDAV}}$) when a character has been received. This can be done simply by connecting CA2 to $\overline{\text{RDAV}}$ and setting the PIA into pulse mode. Then each time the $\mu$P reads a character from the UART, the LOW pulse on CA2 serves as $\overline{\text{RDAV}}$.

### 19-8.4  The Handshaking Mode

The *handshaking mode* provides for total coordination of the $\mu$P and a peripheral during a data transfer. It is equivalent to the sending device saying "I have a byte of data for you," and then waiting until the receiving device replies with "I acknowledge receipt of the data byte. Now you may send another byte."

For handshaking, the A side of the PIA is used for input from a peripheral to the $\mu$P. The connection is shown in Fig. 19-12.

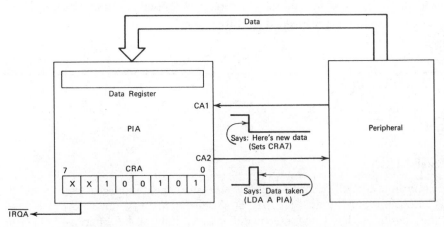

FIGURE 19-12  Handshaking with a Peripheral on the "A" side. (From Greenfield and Wray. *Using Microprocessors and Microcomputers: The* **6800** *Family*. Copyright John Wiley & Sons, Inc. 1981. Reprinted by permission of John Wiley & Sons, Inc.)

When the peripheral has a byte for the $\mu$P it places a transition on CA1. This sets bit 7 of Control Register A to inform the $\mu$P that a byte is available. The transition also causes CA2 to go HIGH. The peripheral must not send another byte when CA2 is HIGH because this indicates that the $\mu$P has not accepted the first byte.

The $\mu$P accepts the byte by reading the PIA's Data Register. In this mode the read of the data register also resets the control bit to indicate to the $\mu$P that the data have been taken, and it lowers CA2. When the peripheral sees that CA2 has gone LOW, it knows it can send another byte, which it must synchronize by placing another transition on CA1.

The B side of the PIA is used for writing data to the peripheral. The connection is shown in Fig. 19-13. The sequence of events for writing follows.

1. The control word to the B side is written as shown. Note that CRB 5, 4, and 3 are 100, respectively.
2. When the system must send data out it writes the data to Data Register B. This causes CB2 to go LOW.
3. The peripheral device acknowledges receipt of the data by placing an acknowledge pulse on CB1. The negative transition of this pulse causes CB2 to return HIGH and raises the CRB7 flag.

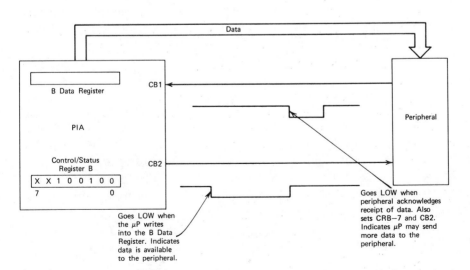

FIGURE 19-13 Handshaking with the "B" side of the PIA. (From Greenfield and Wray. *Using Microprocessors and Microcomputers: The* **6800** *Family.* Copyright John Wiley & Sons, Inc. 1981. Reprinted by permission of John Wiley & Sons, Inc.)

4. The **6800** system responds to a HIGH on CRB7 as an indication that the data have been accepted by the peripheral, and the next byte can be written to the PIA.
5. Before the next output byte can be written to the PIA, Data Register B must be read to reset CRB7.

---

**EXAMPLE 19-14**

The data between memory locations 1000 and 104F must be sent to a peripheral using the system of Fig. 19-13. Draw a flowchart to show how this is accomplished.

**SOLUTION**

The solution is shown in Fig. 19-14 and proceeds in the following way.

1. The Control/Status Register is set to 00 (or the PIA is RESET).
2. Direction Register B is initialized to FF (all bits to be outputs) and then the Control/Status Register is set to 24 (hex) to initialize the handshaking procedure described above.
3. The index register is set equal to 1000 and the first memory word to be transmitted is loaded from that location into the accumulator and stored in Data Register B. This causes CB2 to go LOW.
4. The program goes into a loop until the peripheral accepts the data and causes a negative transition ( ↓ ) on CB1. This SETS CRB7 and CB2. Before the transition, the result is positive (MSB = 0) and the program continues to loop. After the transition the result is negative and the program exits from the loop.
5. The program reads the B data register to reset CRB7. It then tests to determine if it has exhausted its data area by comparing the index register with 1050, each time after it has been incremented. If it has, it stops; if it has not, it loads the next data word causing CB2 to go LOW again and repeats the above steps.

---

## 19-9  INTERRUPTS ON THE 6800

The **6800** $\mu$P has two pins available to cause it to interrupt. Normal interrupts are made using the $\overline{\text{IRQ}}$ input. This is a *level-sensitive* input that causes an interrupt when it goes LOW. On an IRQ interrupt the **6800** jumps to an address that is stored in locations FFF8 and FFF9. These locations must contain the *starting address of the interrupt service routine*. An interrupt also causes all the **6800** registers to be stored on the stack and inhibits further interrupts by setting the *I bit* of the **6800**'s *Condition Code Register*, which disables interrupts.

### 19-9.1  Interrupts Via the PIA

Most **6800** interrupts come in through the PIA. It has two interrupt outputs, $\overline{\text{IRQ}}$-A and $\overline{\text{IRQ}}$-B, which drive the $\overline{\text{IRQ}}$ input to the **6800**. The PIA can

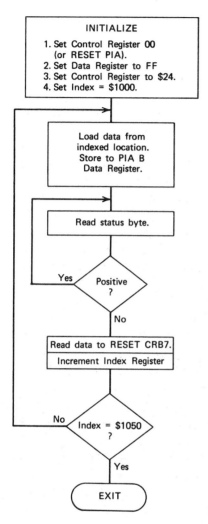

FIGURE 19-14 Flowchart for Example 19-14. (From Greenfield and Wray. *Using Microprocessors and Microcomputers: The* **6800** *Family*. Copyright John Wiley & Sons, Inc. 1981. Reprinted by permission of John Wiley & Sons, Inc.)

be set to enable or disable interrupts by properly setting the Control Register. If enabled, interrupts will occur in response to the same transitions that cause Control Register bit 7 (or 6 when CA2 or CB2 is used in the input mode) to go HIGH. Interrupts that come in via the PIA are *edge or transition sensitive* rather than level sensitive. The interrupt service routine must read the PIA's Data Register to clear bit 7. This also clears the IRQ request and the PIA then waits for another transition from the peripheral to cause another interrupt.

### 19-9.2 The NMI

The 6800 can also be interrupted by a LOW on its $\overline{\text{NMI}}$ (Non-Maskable Interrupt) input. The starting address of the service routine for an $\overline{\text{NMI}}$ must be stored in FFFC and FFFD. As its name implies, this interrupt cannot be masked by the DISABLE INTERRUPT flag, so the $\mu$P loses some control over it. The $\overline{\text{NMI}}$ should be reserved for disaster situations where top priority action must be taken immediately.

### 19-9.3 Comparison of 8085 and 6800 Interrupts

On interrupt, the 8085 can be sent to one of many service routines by properly using RESTARTs or CALLs. The 6800 service routine always starts at the address in FFF8 and FFF9. If many peripherals can interrupt the 6800, it must *poll* to identify the interrupting device. Additional hardware (an Interrupt Controller IC) can be added to the 6800 to vector the $\mu$P to any one of eight service routines.

The 6800, however, has the advantage of stacking all its registers when an interrupt occurs. The 8085 only stacks the PC. Other registers that must be preserved for the main program must be stacked by the interrupt routine using PUSHes and POPs.

### 19-10 OTHER INTERFACE ICs

Both Intel and Motorola manufacture a variety of interface ICs to augment their $\mu$Ps. The PIA is an example. These ICs are designed to perform specific interfacing functions.

Space limitations preclude a thorough discussion of these ICs (that would take a whole book). They will be described briefly, however. The reader is referred to the references or the manufacturer's literature for further information.

### 19-10.1 Intel Peripheral ICs

Intel has recently come out with a new series of ICs: the 8251, 8253, and 8255 are the most important of these. They can be used either in memory mapped or port I-O. In Intel systems they are usually given port addresses. All the Intel ICs have their data transfers synchronized with the $\mu$P by pulses on the $\overline{\text{RD}}$ or $\overline{\text{WR}}$ lines.

The 8251 is a *communications interface*. It serializes data and sends them out to synchronous or asynchronous peripherals, such as a TTY, lineprinter or MODEM. In asynchronous mode it functions much like a UART. It also has pins to connect directly to a MODEM, such as REQUEST TO SEND ($\overline{\text{RTS}}$) and CLEAR TO SEND ($\overline{\text{CTS}}$).

The 8251 has two addresses or ports that are selected by the level on its C/$\overline{\text{D}}$ input. One port is for data, both in and out, depending on whether the $\overline{\text{RD}}$ or $\overline{\text{WR}}$ line goes LOW, and the other port is for the Control Register. The Control Register determines whether the IC operates in synchronous or asynchronous mode, the asynchronous character length, the number of STOP bits, and so on.

The 8253 is a *counter/timer*. It contains three 16-bit counters. The counters can count events. If they are clocked by a fixed frequency clock, they become a clock and can be used to give a reading of real time to the $\mu$P. The counters within an 8253 can be cascaded to provide a 32- or 48-bit counter.

The 8255 is an *interface* IC whose functions are similar to the Motorola PIA. It contains 24 lines to peripheral ICs that can be used as inputs or outputs. It also can operate in a mode where four or eight of the output lines are used to synchronize the data transfer. This allows it to operate in a handshake mode.

Other Intel ICs are the 8257 DMA controller, the 8259 programmable interrupt controller, and the 8279 keyboard display interface. These ICs are used in larger systems and not as often as those discussed above.

## 19-10.2   Motorola ICs

Like Intel, Motorola has developed a variety of peripheral ICs. The PIA has already been discussed. The most important of the others are the 6850 ACIA and the 6840 PTM.

The 6850 ACIA (Asynchronous Communications Interface Adapter) is Motorola's equivalent of a UART, and similar to Intel's 8251. It takes parallel characters from the $\mu$P, adds START and STOP bits, and sends them out to asynchronous receivers. It also receives asynchronous characters, and checks for errors and sends them to the $\mu$P. Like a UART and the 8251, the 6850 requires an external clock to regulate its data transmission or bit rate.

The 6840 PTM (Programmable Timer Module) is a *counter/timer* similar to Intel's 8253. It also has three internal 16-bit registers and can be operated as a counter or as a real time clock.

## 19-11   THE SDK-85 KIT

Intel has produced the **SDK-85** kit primarily for educational and training purposes. The kit is an 8085 system with a keyboard for data entry and 7-segment displays of the address and data lines. It allows the user to read and write memory and run or single step a program. In single step mode the user can examine the contents of all of the $\mu$P's internal registers after each step; this can be very helpful when trying to debug complicated programs.

The **SDK-85** uses **8355** ROMs to store its monitor program, which supervises the keyboard and controls execution of the commands entered by the user. The ROM memories occupy the lower address bytes. The kit uses **8155** RAMs. These are at locations $(2000)_{16}$ and $(2800)_{16}$. There is also an **8279** in the kit that controls the keyboard and the display panel.

I-O is not done by connecting external devices directly to the data bus. Instead each **8355** has two I-O ports and the **8155**s each have 22 I-O lines and a timer built in. Data transfer between the SDK-85 and external devices is accomplished via these ICs. Their ports are accessed by IN and OUT instructions from the **8085**. The **8155** and **8355** memories are designed specifically to work with the **8085**. In addition to the normal signals they accommodate $\overline{RD}$, $\overline{WR}$, ALE (they have internal registers to latch the addresses on the multiplexed bus), and IO/$\overline{M}$, to distinguish between memory and port addresses.

The **SDK-85** makes its data bus available externally via **8216** transceivers. Interrupts are allowed. During INTA the transceivers turn around, allowing an external CALL or RESTART to enter the system. RESTARTs cause the system to go to the ROM. RESTARTs 1–4 are disabled because the monitor uses those locations for the program. Each of the other RESTART locations contains a jump to a three-address segment in RAM where the user can place another jump to the start of his service routine.

---

### EXAMPLE 19-15

An interrupt service routine on the **SDK-85** must start at 2810. In response to INTA the user places a RESTART-5 on the data bus. How can the $\mu$P be directed to 2810?

### SOLUTION

The RESTART-5 vectors the user to $(0028)_{16}$, which is in ROM. Because more than one version of the monitor exists, the user should use the **SDK-85** to read this location to find out where the program will be sent. In one version we find the following code:

| Address | Data |
| --- | --- |
| 0028 | C3 JMP (Op code) |
| 0029 | C8 |
| 002A | 20 |

This is a JUMP to 20C8 which is in RAM. In 20C8 the user should write:

| Address | Data |
|---------|------|
| 20C8 | C3 |
| 20C9 | 10 |
| 20CA | 28 |

This directs the program to 2810 where it can start. Note that 20C8, 20C9, and 20CA are reserved for the RESTART-5, and a RESTART-6 vectors the user to 20CB, so this location should not be used in the RESTART-5 routine.

### 19-11.1   The MEK-6800-D2 Kit

Motorola manufactures an **MEK-6800-D2** kit. This D2 kit is similar in function to a **SDK-85,** but of course it uses the **6800,** the PIA, and other Motorola parts.

## 19-12   OTHER MICROPROCESSOR BUSSES

There are two *busses* in widespread use to interconnect microprocesssor systems to peripherals. The S-100 bus is used for system interconnections, and the IEEE 488 bus is often used for instrumentation. Both of these busses have had books written on them, so they will only be introduced and described briefly here.

### 19-12.1   The S-100 Bus

As its name implies, the S-100 bus uses a standard 100-pin connector. Its 100 connections include 16 address lines, 8 data input lines, 8 data output lines, and many control lines, including some directly compatible with the 8080 and 8085, such as WAIT, INTA, $\overline{WR}$, and so on. All these signals have standard locations on the bus.

The bus is physically constructed on a backplane. Standard cards (10″ by 5.3″) plug into the bus. Various manufacturers are building a variety of cards to plug into the S-100 bus and perform specific functions. The bus distributes $+8$, $+16$, and $-16$ V and ground on its pins. Each card must have a regulator to reduce these voltages to the voltages needed on the boards (the $+8$ V is often reduced to $+5$ V for TTL ICs).

## 19-12.2 The IEEE 488 Bus

The IEEE 488 bus is designed to control a network of instruments (scopes, digital, voltmeters, etc.). It consists of 8 data lines and 8 control lines. One device on the line is designated as a talker (originator of data) and one or more devices may be designated as listeners (receivers of data). The control lines include signals to designate the talker and the listener, whether the listener is ready to receive data, and so on. A complete list of the signals and their functions can be obtained from the IEEE (see References).

### SUMMARY

This chapter contains several examples of interfacing the Intel 8085 and the Motorola 6800 to various peripherals. These peripherals include UARTs, A/D converters, and memories that were discussed in the earlier chapters of this book. The principles involved can also be applied to lineprinters, disks, and other devices commonly connected to computers.

Interfacing to the 8085 using IN and OUT instructions, ports and Interrupt responses were described, as was interfacing to the 6800 using the Motorola PIA. Most of these interfaces used smaller scale ICs, discussed in the earlier chapters of this book, working in conjunction with the new large scale special purpose ICs to complete the interconnection.

Finally, various special purpose interface ICs were introduced and described briefly. The **SDK-85** $\mu$P kit, the S-100 bus, and IEEE 488 bus were also described briefly.

### GLOSSARY

**ALE.** Address Latch Enable. An 8085 signal that is present when addresses are on the multiplexed bus.

**CALL.** An 8080/8085 instruction that is essentially a JUMP-TO-SUB-ROUTINE.

**Control Register.** A register in an I-O IC that determines the mode or way that IC will function.

**DDR.** Data Direction Register. A register that determines the direction of data flow of each bit in an I-O interface IC.

**I bit.** A bit in the Condition Code Register that enables or disables interrupts.

**INTA.** Interrupt Acknowledge. A signal from the 8085 acknowledging interrupts and requiring the interrupting device to provide an instruction Op code.

**INTR.** An interrupt request to an 8085.

**IO/M.** An 8085 generated signal to determine whether the address is for a memory location or an I-O port.

**IRQ.** A 6800 input used for interrupts.

**Multiplexed bus.** A $\mu$P bus that contains different signals at different times.

**PIA.** Peripheral Interface Adapter. Motorola's IC for performing I-O.

**RD.** A low-going READ pulse generated by the 8085.

**RESTART.** An 8080/8085 instruction that causes the $\mu$P to start at a specific location in memory.

**Status register.** A register in an I-O IC that holds signals pertaining to the status of the peripheral device.

**WR.** A low-going WRITE pulse generated by the 8085.

### REFERENCES

W.M. Goble, *"Introducing the S-100: Standard Small Computer Bus Structure,"* *Interface Age*, June 1977.

J. Greenfield and W. Wray, *Using Microprocessors and Microcomputers: The 6800 Family*, Wiley, New York, 1981.

Lance A. Leventhal, *Introduction to Microprocessors: Software, Hardware, Programming*, Prentice-Hall, Inc., Englewood Cliffs, N.J., 1978.

Peter R. Rony, *"Interfacing Fundamentals: Bused Flags,"* *Computer Design*, July 1981.

Kenneth L. Short, *Microprocessors and Programmed Logic*, Prentice-Hall, Inc., Englewood Cliffs, N.J., 1981.

*MCS-85 User's Manual*, Intel Corp., Santa Clara, Ca., 1977.

*SDK-85 User's Manual*, Intel Corp., Santa Clara, Ca., 1977.

*The 8080/8085 Microprocessor Book*, Intel Corp., Wiley-Interscience, New York, 1980.

*IEEE Standard Digital Interface for Programmable Instrumentation*, IEEE, Inc., New York, 1978.

### PROBLEMS

19-1.    A system has 32 peripherals that must send data to an 8085 $\mu$P using the IN instruction. Design the circuitry to tie the $\mu$P to the peripheral devices.

19-2.    Design a 16K-byte memory using 2114s to work with an 8085.

19-3.    A tape controller has both a Status Register and a Data Register. When an 8085 needs to read the Status Register, it executes an IN 1 instruction, but it reads the Data Register with an IN 2 instruction. Design the circuitry needed to place the proper register on the AD lines.

19-4.    The speed of an 8085 memory is such that each cycle requires exactly one WAIT state. Design a circuit to produce this state using FFs instead of a one-shot.

19-5.    A slow memory has an access or cycle time of two $\mu$s. Show how it can be connected to an 8085 that has a clock frequency of 2 MHz.

19-6.    A tape controller has a Status Problem line and a Data Ready line. If the Status Problem line goes HIGH, it must interrupt the $\mu$P and send out a RESTART-

1 in response to INTA. If Data Ready goes HIGH, it must interrupt and send out a RESTART-2. If they both go HIGH, Status Problem has priority. Design the circuitry.

19-7.    A circuit is to respond to an **8085** interrupt with a CALL 2859. Design the circuit using **74125**s.

19-8.    What character should be in the DDR for the following data directions:
 (a) All lines input.
 (b) All lines output.
 (c) Lines 0–5 output, lines 6 and 7 input.

19-9.    A 256-byte memory external to a $\mu$P is to be connected to a PIA. A 256 byte section of the $\mu$P's memory is to be written to the external memory. Show the circuit and the hardware connections. Describe the program. Assume the R/W line to the external memory is LOW on WRITE. (*Hint:* Write to the memory using the pulse mode.)

19-10.   An **AY5-1013** UART has data and status connected to the same lines. Which one is read depends on whether RDE or SWI is LOW. How can a $\mu$P read the word it requires?
 (a) Assume an **8085** using IN instructions.
 (b) Assume a PIA. (*Hint:* Use the level mode and an inverter.)

19-11.   A peripheral has two control lines A and B. If it sets A, it wants to READ the data in location 100. If it sets B, it wants to READ the data in 200. Explain how this can be done using a PIA:
 (a) Using polling.
 (b) Using interrupts.

After working the problems, the student should return to Sec. 19-2 and be sure all the self-evaluation questions can be answered. If any of them are still difficult, the student should review the appropriate sections of the chapter to find the answers.

# 20

# DISPLAY GENERATORS

## 20-1 INSTRUCTIONAL OBJECTIVES

A *display terminal* is an I-O device commonly used to read and enter data to a computer. A display terminal consists of a *cathode ray tube* (CRT) display and a *keyboard*. The computer operator or user can cause the contents of the computer's memory to be displayed on the CRT, and can enter data into memory via the keyboard.

The *display generator* is that portion of the display terminal that *receives the characters from the computer and causes them to be displayed on the CRT.* This chapter explains the operation of a Display Generator. After reading it, the student should be able to:

1. List and explain the required functions of a display generator.
2. List the memories required in a display generator and explain the relationship between them.
3. Draw timing charts and design timing circuits for a display generator.
4. Design a CHARACTER LOCATION COUNTER and a CURSOR ADDRESS REGISTER.
5. Design a circuit for writing the data into the REFRESH MEMORY at the CURSOR LOCATION.

## 20-2   SELF-EVALUATION QUESTIONS

Watch for the answers to the following questions as you read the chapter. They should help you understand the material presented:

1. What are *Vertical* and *Horizontal Sync pulses?* Why are they required?
2. What is the relationship between the frequency response of the video circuits and the number of characters that can be displayed?
3. What is the difference between rows and lines? What limits the number of rows on a CRT display?
4. What are the inputs and outputs of the character generator ROM? Where do they come from?
5. What counters are used in a display generator? What is the function of each counter?
6. What is the function of the CHARACTER LOCATION COUNTER and of the CURSOR ADDRESS REGISTER?

## 20-3   INTRODUCTION TO THE DISPLAY GENERATOR

*The function of a display generator is to display alphanumeric characters on a cathode ray tube* (CRT) *screen.* The CRT screen resembles the face of an ordinary television set, and amateur computer hobbyists have used commercial TV sets for this purpose.

Most display generators work with a keyboard, and become a *display terminal* that interfaces to a computer. Data are entered into the computer via the keyboard. Output information from the computer appears on the CRT screen. Generally, the normal set of *alphanumeric characters* (letters, numbers, and punctuation) are used. The display generator can interface to the computer either directly or remotely via MODEMs.

A typical display terminal is shown in Fig. 20-1. Display terminals usually operate at a variety of speeds or bit rates, and are often less expensive and more mobile than a TTY. They can also operate at higher speeds than a TTY. Unfortunately, CRT terminals do not produce "hard copy" (letters printed on paper), so a permanent record of the computer output is not available unless additional equipment to print the output is used.

Modern small computers such as Apple, Atari, TRS-80, and so on, all use display terminals. Some of them, like the TRS-80, use dedicated CRT terminals whose only purpose is to display the alphanumeric characters on its screen. These are called *monitors.* Other computers such as the Apple can use either a monitor or a commercial TV set to display its information.

FIGURE 20-1 A display terminal. (From Thomas Blakeslee, *Digital Design With Standard MSI & LSI.* Copyright John Wiley & Sons, Inc. 1975. Reprinted by permission of John Wiley & Sons, Inc.)

## 20-3.1 The Function of the Display Generator

The display generator must drive the CRT so that information displayed on the screen. To accomplish this it must:

1.   Obtain the information from the computer. This information generally consists of 8-bit characters that form the code for the letter or number to be displayed. The ASCII code (see Appendix G) is most commonly used.
2.   Present the information to the CRT circuits in a form suitable for display.
3.   Remember the information so that it can continue to be displayed as long as it is needed.
4.   Generate the proper synchronizing signals required by the CRT.

CRT-keyboard combinations are available as commercially manufactured display terminals for about $500 (and higher). These perform the functions listed above, and the user need only interface the terminal with the computer. The principles of operation of a display generator are covered in this chapter for the reader who must design his own display generator, or who wants to understand how display generators work. Their design is an interesting example of the use of digital ICs and memories to build a practical computer I-O terminal.

## 20-4 THE RASTER-SCAN DISPLAY GENERATOR

There are two types of display generators available: *raster-scan* displays that have a continuous beam sweeping the screen as in an ordinary TV set, and produce their characters by modulating the beam, and *stroke generators*, which produce their characters by a series of strokes of predetermined length and direction.

The raster-scan display terminals are far more common and are discussed in this book. They produce their characters using a *Character Generator* that is actually a Read Only Memory (ROM). The Signetics **2513** is a typical Character Generator ROM that produces the set of characters shown in Fig. 20-2. It has a set of 64 characters and can only display upper-case characters. More elaborate character generators such as the **MCM 6571**[1] are available. It produces both lower-case and upper-case characters and also produces Greek letters.

### 20-4.1 Timing the Raster-Scan

In a raster-scan display, such as a TV set, an electron beam starts at the upper left-hand corner and sweeps horizontally across the screen drawing the top line. The beam illuminates the portion of the screen (the dot) that it strikes. By *modulating* the beam only some dots are illuminated, creating the desired display. When the beam reaches the right end of the screen, it quickly returns to the left end of the screen at a point just below the top line and proceeds to trace out the second line of the picture, from left to right. The scanning is shown in Fig. 20-3.

The horizontal motion of the beam is controlled by a *horizontal oscillator*. For a commercial TV set, the frequency of the horizontal oscillator is 15,750 Hz and that many lines are written on the screen every second. The time when the beam is returning from the right end of the screen to the left end is called the *retrace* time. At this time the beam is blanked, so that the retrace lines are not visible, and *horizontal sync pulses* are added to the beam. These pulses synchronize the beam with the desired display.

As the beam is *swept* across the screen by the *horizontal oscillator*, it is slowly lowered by the *vertical oscillator*, which controls the vertical position of the beam on the screen. Vertical oscillators typically run at 60 Hz. Therefore, a *typical screen has 60 complete pictures, or fields, drawn on it per second*. This is the *refresh rate* of a display terminal.

Information and specifications for both the horizontal and vertical synchronizing pulses are available in EIA (Electronic Industries Association) standard No. RS-170. These specifications indicate that the horizontal re-

[1]Manufactured by Motorola, Inc.

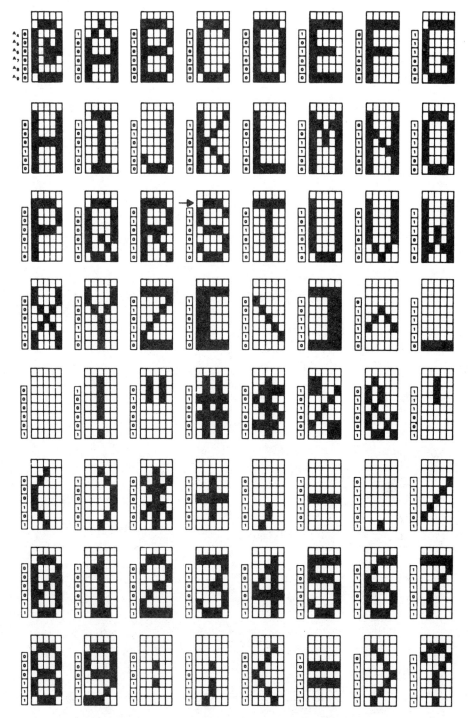

FIGURE 20-2 The character set of the **2513** Character Generator. (Permission to reprint granted by Signetics Corp., a subsidiary of U.S. Philips Corp., 811 E. Arques Avenue, P. O. Box 409, Sunnydale, Calif. 94086.)

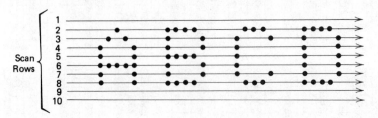

FIGURE 20-3 The scan rows of a display. Characters from the **2513** Character Generator are shown.

trace time takes about 17 percent of the time required to write a horizontal line.

The time of an entire horizontal line, including retrace, is 1/15,750 Hz, or 63.9 $\mu$s. Experience with monitors, especially those used with commercial TV sets, indicates that the beam can only be displaying dots for about 42 $\mu$s (this allows for reasonably margins). Attempts to use the beam longer than 42 $\mu$s cause the displayed line to be too wide for the screen and characters disappear off the edges.

---

**EXAMPLE 20-1**

How many horizontal lines comprise each field in an ordinary display?

**SOLUTION**

The answer is obtained by dividing the number of horizontal lines per second by the number of fields:

$$\frac{15,750 \text{ lines/second}}{60 \text{ fields/second}} = \textbf{262.5} \text{ lines/per field}$$

---

In an ordinary television set the horizontal lines are *interlaced*, which means the alternate lines start at *different* points on the screen (the left-hand corner and the middle), as shown in Fig. 20-4. It therefore requires two

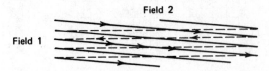

FIGURE 20-4 Interlaced scanning. *Note:* The dashed lines indicate the retrace lines. The arrows indicate the scan path of field 1.

fields to make up a complete picture. These *two fields are called a frame,* and TV sets present 30 *frames per second.* Display generators rarely use interlaced scanning so that fields and frames become synonomous. The refresh rate or number of fields for most display generators is 60 per second.

## 20-4.2 Vertical Retrace

VERTICAL RETRACE is the time required for the beam to return to the top of the screen from the bottom. With EIA RS-170 timing, this requires between 18 and 21 horizontal line times. During this time the beam is blanked and *vertical synchronizing pulses* control the CRT timing.

## 20-4.3 Character Size and Screen Capacity

Each character in the set of Fig. 20-2 is produced by illuminating the proper dots in a *dot matrix* 5 bits wide by 7 bits high. This is the most popular size for a dot matrix, although 5 × 5 is sometimes used and 5 × 9 is used when it is necessary to display both capital and lowercase letters.

To allow some space between characters, the *dot matrix is imbedded in a larger matrix or grid where the additional rows are always blanked.* Note that the 5 × 7 characters of the **2513** ROM are in a 5 × 8 matrix, and the top line is always blanked. Additional blanking is usually required between characters. If the character is described as a 5 × 7 dot matrix embedded in the 8 × 10 field, it means that each character occupies 8 horizontal dots by 10 vertical dots. The three rightmost columns and three rows are always blanked so that the characters have ample space between them.

Figure 20-3 shows the characters A, B, C, D in a 5 × 7 matrix embedded in an 8 × 10 grid as they would appear using a **2513**. The top line and the two bottom lines are blanked, and there are three blank dots between the characters. Note that the horizontal scan lines in Fig. 20-3 actually have a very slight downward tilt because the beam is slowly being lowered by the vertical oscillator as it sweeps across the screen.

---

**EXAMPLE 20-2**

Using a 5 × 7 character imbedded in an 8 × 10 field, how many character lines can be written on a CRT screen if standard timing is used? Assume the characters are not interlaced.

**SOLUTION**

At this point, we must make a distinction between *rows* and *lines.* We define a *row* as a *single horizontal sweep of the beam.* A *line* is a *line of horizontal characters* that requires several rows for its presentation.

Figure 20-3 shows 10 rows, or sweeps, of the beam that comprise one character line. Example 20-1 showed that there are 262.5 rows per field on an ordinary TV screen. Subtracting the time required for Vertical Sync and Retrace and for leaving adequate margins on the top and bottom of the display leaves about 200 usable rows per frame. Therefore, if 10 rows are used for each character, 20 lines of characters can be displayed.

## 20-4.4 The Dot Rate

The time required to write a single dot on the screen is called the *dot rate*. It is the highest frequency used in the character generator. The maximum dot rate is about twice the bandwidth or upper frequency response of the video circuitry driving the CRT. The bandwidth of a commercial TV set is about 3.5 MHz, which limits the number of dots that can be displayed to about 7 MHz. Monitors, which are CRT drivers built specifically for the purpose of displaying alphanumeric information, have higher bandwidths and can display more characters on a line.

### EXAMPLE 20-3

If the dot rate for a display is 5 MHz, how many characters can be displayed on a CRT? Assume the character is 8 dots wide (including blanks) and 40 $\mu$s is available for each line.

### SOLUTION

Each dot requires 0.2 $\mu$s for display. The number of dots that can be displayed on a single line is therefore 40 $\mu$s/0.2 $\mu$s per dot = 200 dots. The number of characters per line is 200 dots/8 dots per character = 25 characters.

Example 20-3 explains why CRT circuits that use commercial TVs usually do not have a large number of characters per line. If something like 80 characters per line are required, a monitor should be used rather than a TV set. A commercial TV set is the least expensive display one can buy. Hobbyist or homebuilt CRT drivers using a TV set usually display 32 characters per line.

The Apple computer uses a **2513** character generator and produces a 5 $\times$ 7 dot matrix embedded in a 7 $\times$ 8 grid. The dot clock of an Apple is 7.159 MHz, or 140 ns per dot.

---

EXAMPLE 20-4

An Apple computer in text mode uses 24 lines of 40 characters each.
  (a) How long does it take to display each line?
  (b) How long does it take to display each frame?


SOLUTION

  (a) The time each line is displayed is:

  140 ns/dot × 7 dots/character × 40 characters/line = 39.1 μs/line

  (b) The Apple uses 24 lines times 8 rows per line, or 192 rows. It therefore requires 192 rows × 63.5 μs/row or 12.2 ms. Thus the Apple display is on for 39.1 of the 63.5 μs on each horizontal line and for 12.2 ms of the 16.7 ms for each frame. The remaining time is used for vertical and horizontal retrace. Note that an alphanumeric display does not use the entire available time because it cannot completely fill up the screen the way a TV picture does. Alphanumeric displays require margins.

---

## 20-5   COMPONENTS OF THE DISPLAY GENERATOR

The components and data flow of a typical Display Generator are shown in Fig. 20-5. It shows that the Refresh Memory and Row Counter feed the Character Generator. The output of the Character Generator is loaded into a Shift Register, which is shifted out at the dot rate of the system. Finally, the Vertical and Horizontal Sync signals are added to the output of the Shift Register to produce the final video output that is fed to the CRT.

### 20-5.1   The Dot Counter

The dot counter contains the highest frequency in the Display Generator, and is used to clock the Shift Register that shifts the bits to the video output. Dot clocks are often generated by crystal oscillators, although some hobbyist Display Generators use a simple and inexpensive one-shot oscillator (see Sec. 7-6.2). Some sophisticated Display Generators use a dot clock that is synchronized with the 60-Hz power line frequency by a *phase-locked loop*. The dot clock frequency is then counted down to provide character counters, row counters, line counters, etc. With the method, this Horizontal and Vertical synchronizing circuits are *precisely synchronized* with the power lines. Failure to synchronize the 60-Hz Vertical Sync signal with the 60-Hz power line can cause interference between the signals and result in a weaving or "seasick" display.

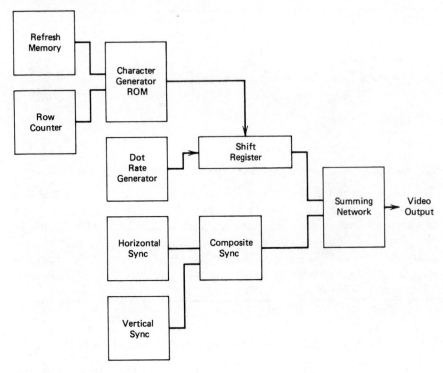

FIGURE 20-5   Data flow in a typical Display Generator.

## 20-5.2   The REFRESH Memory

The REFRESH MEMORY is a RAM memory that contains a code for every character on the CRT screen. It remembers the characters so that the same character is displayed in every position until it is changed on command from the computer.

Character Generators like the **2513** (see Sec. 20-5.4) can only display 64 different characters and therefore require only a 6-bit code from the RE-FRESH MEMORY to identify a particular code. The character code that identifies each character displayed by the **2513** is a truncated form of ASCII.

---

EXAMPLE 20-5

A CRT displays 24 lines and 60 characters per line and uses a **2513**. What are the dimensions of its REFRESH MEMORY?

SOLUTION

The REFRESH MEMORY must remember 1440 characters, each identified by a 6-bit word. Therefore, the memory dimensions are **1440** words by **6** bits per word.

---

REFRESH MEMORIES can use core, semiconductor RAMs, or circulating shift registers. A typical circulating shift register memory is the **2524**, manufactured by Signetics, Inc. This is a 512-bit shift register that requires a 2-phase clock similar to the clock of the **2521** (Sec. 9-8.1).

---

**EXAMPLE 20-6**

(a) Design a circulating shift register memory for the 1440-word memory of Ex. 20-5 using **2524**s.

(b) What is the clock frequency of the shift register?

**SOLUTION**

(a) Since 1440 6-bit words are required, 18 shift registers must be used, as shown in Fig. 20-6. The shift register must be counted around once for each field. Since the **2524**s are arranged as in groups of 3, they effectively become a 1536-bit shift register. Because only 1440 words are required, an additional 96 pulses must be applied once each field to completely recycle the shift registers so that they start at word 0 each time a field is repeated. Generally the additional pulses can be applied during vertical retrace.

(b) If the refresh rate is 60 Hz, the clock frequency is:

$$60 \text{ fields/sec} \times 1536 \text{ clocks/field} = \textbf{92,160 Hz}$$

This is well below the maximum shift frequency and well above the minimum shift frequency of a **2524**. Note that the clocks are not applied at a constant rate (see Problem 20-9).

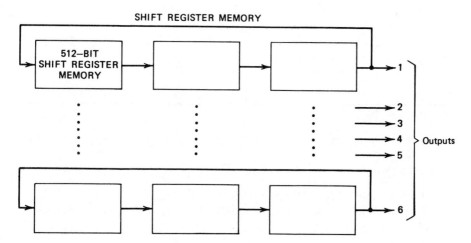

FIGURE 20-6  A 1536-bit shift register. *Note:* All ICs are **2524**s with READ and WRITE inputs tied to $V_{CC}$.

## EXAMPLE 20-7

If a core or semiconductor memory is used for the REFRESH MEMORY of Ex. 20-5, design the MAR.

## SOLUTION

With 1440 words, an 11-bit MAR is required, but only the first 1440 of the 2048 locations are used. Since the memory accesses are always sequential, the MAR can be a counter that increments each time a character is required. It can be reset during vertical retrace so that it restarts for each field.

---

## 20-5.3  The ROW COUNTER

Because the beam sweeps continuously across the screen, *the top row of each character must be presented followed by the second row for each character, and so on.* Each line of characters being displayed consists of at least eight rows of dots. The ROW COUNTER contains the number of the row that is currently being displayed. It is incremented at the end of each row (generally during Horizontal Sync) and is reset when one line of characters has been displayed and the next line is to start.

## 20-5.4  The Character Generator

At any particular time the Character Generator must provide the dots for the row of the character currently being displayed. The **2513**, for example, is a 512-word-by-5-bit ROM and has nine inputs. The REFRESH MEMORY provides six inputs to determine which of the 64 characters is being displayed and the ROW COUNTER provides three more inputs to select one of the eight rows of the character. The 6 bits shown beside each character in Fig. 20-2 ($A_4$–$A_9$) designate the character. The 3 other address bits designate one of the eight lines in each character.

---

## EXAMPLE 20-8

What are the five outputs of a 2513 if the nine inputs are $A_1 \cdots A_9 = 001110010$?

## SOLUTION

The 6 higher bits (110010) designate the letter S. The 3 LSBs, 001, designate line 1. The outputs taken from Fig. 20-2 are 01110, where a dark square represents a 1 and a light square represents a 0. The arrow on Fig. 20-2 shows the selected line.

---

### 20-5.5 The Shift Register

The 5-bit output of the Character Generator (the five dots for the row of the character to be displayed) is loaded into the Shift Register at the appropriate time. Extra 0s to blank the display between characters are also loaded into the Shift Register at this time. They are then clocked out serially at the dot rate to form the light and dark dots on the screen.

### 20-5.6 The Synchronizing Circuits

The Horizontal and Vertical synchronizing signals (15,750 Hz and 60 Hz, respectively) are generated by a separate oscillator or derived from the dot rate. They should be synchronized to the 60-Hz line frequency. They are often mixed together to form a signal called COMPOSITE SYNC.

### 20-5.7 The Summing Network

The summing network then combines the dot output of the Shift Register and the sync signals together to produce the final video output. One function of the synchronizing signals is to blank the display during retrace time. The summing network is often only a transistor whose base current is controlled by the Shift Register and sync circuit outputs.

The summing network for a Display Generator kit built at RIT is shown in Fig. 20-7. Vertical and Horizontal Sync are combined to form Composite

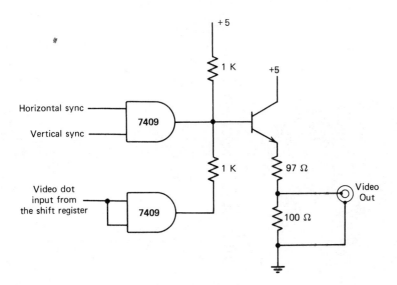

FIGURE 20-7 The summing network for a hobbyist Display Generator.

Sync by one **7409** open-collector gate (see Sec. 5-6.1). Composite Sync is wire ANDed to the dot output from the Shift Register via a second **7409** gate. These outputs form the input to a transistor emitter follower that drives the final video output to the CRT.

Note that Composite Sync is directly connected to the base of the transistor, but the dot output is connected to the base via a resistive voltage divider. When Composite Sync is LOW the transistor is cut off, but when the dot output is LOW the base of the transistor is about 2.5 V and there is about 0.9 V on the video output (see Problem 20-5).

## 20-5.8 The Timing Waveforms

Timing waveforms for the Display Generator are shown in Figs. 20-8 and 20-9. They were taken from the Display Generator kit built at RIT.

Figure 20-8 shows Composite Sync. The wider pulses are Vertical Sync. There are 10 Vertical Sync pulses per frame. The rest of the display lines only receive Horizontal Sync, which are the narrower pulses following the Vertical Sync pulses. In this Display Generator, Vertical Sync is LOW for 22 of the 63.4 $\mu$s for each row, and Horizontal Sync is LOW for 5 $\mu$s of each 63.4 $\mu$s. These signals do not conform exactly to the RS-170 specification, but they are sufficient to drive a CRT display.

Figure 20-9a is a photograph of the video output. It shows the 16 lines being sent to the display. The lowest line on the figure is the sync pulses (the photo does not contain sufficient resolution to separate them). If the lower line is examined closely, one can see the slightly heavier area where Vertical Sync occurs.

Figure 20-9b is a more detailed view of Figure 20-9a, where the visual level is shown. The 1s in the video output drive the signal above the visual

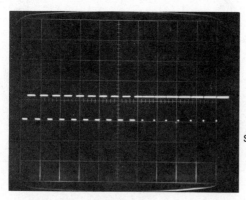

Scales:
2 Volts/cm vertical
.1 ms/cm horizontal

FIGURE 20-8  Composite Sync.

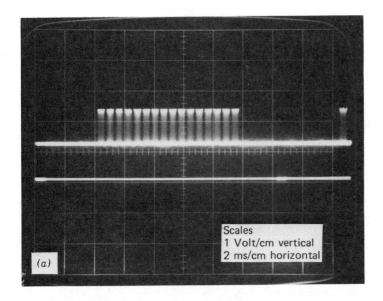

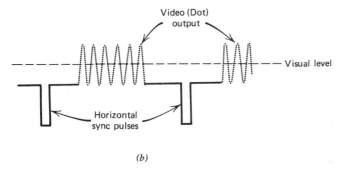

FIGURE 20-9   Video output.

level and appear on the screen. The 0s remain below the visual level and leave the screen dark. The sync pulses turn the output transistor OFF (see Fig. 20-7), resulting in 0 V at the output (the lowest line on Fig. 20-9*a*) and surely blank the screen. (See also Problem 20-5). From the photograph (Fig. 20-9*a*) the visual level can be seen to be about 1.8 V.

## 20-6   TIMING THE DISPLAY GENERATOR

It is not possible to understand and appreciate the function of each component of a Display Generator without a detailed examination of the timing. Therefore, the timing involved in a hypothetical Display Generator that is similar to the Display Generator in the Apple computer is described in this section. It has the following specifications:

1. **Dot rate**—7 MHz or 143 ns/dot. This is the frequency of the Dot Rate Generator, the highest frequency in the display generator, and is limited by the bandwidth of the CRT video circuits.
2. **Character size**—5 × 7 dot matrix imbedded in a 7 × 8 grid.
3. **Horizontal lines per frame or field**—262.5.
4. **Refresh rate**—60 fields per second.

## 20-6.1  Horizontal Character Timing

As with the Apple, we will assume a raster size of 40 characters by 24 lines. The time for each character is 7 dots/character × 143 ns/dot = 1 $\mu$s/character. In this time the following events must occur:

1. The Refresh RAM must be read.
2. The output of the Refresh RAM must be sent to the character generator.
3. The character generator must have time to produce the proper output.
4. The shift register must be loaded with this output.

The access time for a **2513** character generator, after all addresses are stable, is 400 ns. If we allow 200 ns for loading the shift register and other delays, this allows 400 ns for the access time of the RAM. Figure 20-10 is a timing chart for this circuit.

This discussion shows that there are problems to overcome if more characters per line are to be displayed:

1. The dot rate must be higher. This means we need video circuitry with a higher bandwidth and probably precludes the use of an ordinary TV set.
2. The Refresh RAM and character generator must be faster. If 80 characters have to be displayed on a line, there are still only about 40 $\mu$s of display time available. Consequently, there is now only 500 ns between characters and this is

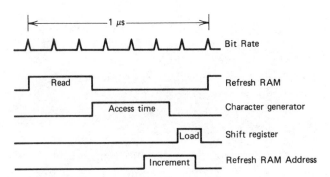

FIGURE 20-10  Horizontal character timing for a Display Generator.

not enough time for the Refresh RAM and the Character Generator to settle. Either faster ICs or a more sophisticated scheme, such as using two memories and character generators for alternate characters, must be used (see Problem 20-8).

## 20-6.2   Horizontal Row Timing

Because each character in the display generator under discussion consists of 7 dots, the Dot Rate Generator must drive a divide-by-7 counter (see Chapter 8). It rolls over every 1 $\mu$s. During this time the Refresh Address must be incremented and the shift register loaded, as shown in Fig. 20-10.

The divide-by-7 Dot Counter must also drive a divide-by-40 Character Counter that counts the number of characters on the line. This counter counts only during 40 of the 63.9 $\mu$s used to display a line. During the rest of the time the display should be blanked. The divide-by-40 counter can be cleared by the HORIZONTAL SYNC pulse so that it is clear at the start of each horizontal line. The display should be blanked during the two dots used to separate characters and the 23.9 $\mu$s when it is not active. Blanking the dots between characters is accomplished by loading the five outputs of the character generator and two 0s into the shift register. It shifts 7 times per character and always inserts the two 0s (or "undots") between the characters to provide the required separation. Blanking during the 23.9 $\mu$s when the display is inactive is often accomplished by gating the shift register so that its output is 0 during this time. HORIZONTAL SYNC always occurs during this interval and also contributes to blanking the display.

Some display generators have WIDTH and CENTERING controls. A WIDTH control can make a small adjustment in the dot rate. By increasing or decreasing the dot rate, each line becomes narrower or wider. A WIDTH control allows the user to adjust the line width to properly fit the CRT screen. Circuits using crystal oscillators for their dot rate generators do not have WIDTH controls because the dot frequency is fixed by the crystal.

A CENTERING control varies the time between HORIZONTAL SYNC and the start of the display. It can be used to move the line left or right on the display until it is centered properly.

## 20-6.3   Horizontal Line Timing

In our hypothetical display generator eight horizontal rows or scans make up a single text line. A ROW COUNTER is required to keep track of the current row of the display. The ROW COUNTER is incremented each time the CHARACTER COUNTER rolls over. The outputs of the ROW COUNTER provide the ROW inputs to the character generator so it can provide the dot pattern for the proper row of the character to be displayed.

The timing for the Display Generator is shown in Fig. 20-11. It shows how the 143-ns dot rate is applied to the Dot Counter. The Dot Counter's output is applied to the Character Counter. Each time the Character Counter rolls over it increments the Row Counter.

In this display there is only one blank row separating the lines. The blank row can be obtained directly from the 2513 Character Generator. The top line of each character of the 2513 is all 0s (see Fig. 20-2). If more than one line is to be used to separate the characters, the Row Counter would divide by 9 or 10 and additional gating might have to be included to blank the display while the separating rows occur.

The Refresh RAM must be cycled through the same addresses for each row of the line. After the line is complete the number of characters in the line are added to the memory address and the next set of addresses is scanned.

---

### EXAMPLE 20-9

If we assume that address 0 in the Refresh RAM holds the character at the upper left corner of the screen, how does the Refresh RAM addressing progress?

### SOLUTION

For the first row the top row of the 40 characters must be displayed. Therefore, memory addresses 0–39 are accessed sequentially. Then the second row of the same 40 characters must be displayed, so memory addresses 0–39 must be accessed

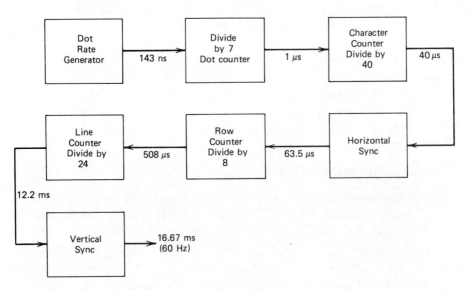

FIGURE 20-11   Display Generator timing.

again. This continues, with the memory address going from 39 at the end of one row to 0 for the start of the next row, until all rows of the first character line are written. The Display Generator must then add 40 to the memory address and access addresses 40–79 for each row of the second line of characters.

In microprocessor systems like the Apple, the Refresh RAM is part of the computer's memory. The area of memory reserved for the Refresh RAM is called the *screen image*. In the Apple, the primary screen image is between locations 1024 and 2047, but it is not sequential and not all locations in the area are used.

## EXAMPLE 20-10

Assume a screen must display 240 lines (24 rows) of 40 characters each. Design the addressing circuits for the Refresh RAM.

## SOLUTION

The Refresh RAM must contain 40 × 24 or 960 locations. The circuit to control its addresses is shown in Fig. 20-12. The addresses require 10 bits that are taken from a counter made up of a **7493** (only the divide-by-8-section is used) and two **74193**s. The circuit operates as follows:

1.   The character counter increments the 10 bit counter 40 times for each row displayed. Circuitry to blank the display after the 40 characters have been displayed is not shown.
2.   The Horizontal Sync pulse clears the **7493** so each line starts with a memory address whose 3 LSBs are 0.
3.   Horizontal Sync also loads the **74193**s. The output of these counters are the MSBs of the address. The trailing edge of Horizontal Sync clocks the outputs of the **74193**s into the **74174**s where it is retained while the row is displayed.
4.   Row Counter-N is a signal that is HIGH only when the row N, the highest row in the line, is being displayed.
   The **7483** adders (see Sec. 14-7) receive an input of 5 when row N is displayed. The 5 is displaced three bits from the LSB of the address, however, so the adder effectively adds 5 × 8 or 40 to the address. For all other rows, the inputs to the adder are zero and the original start-of-line location.
5.   The outputs of the **7483** adder are loaded into the **74193**s during Horizontal Sync. When Row N is being displayed, the adder adds 40 to the memory address. For all other rows the **7483**s add 0 to the row address. Thus each row except ROW N starts at the same memory address as the previous row. This is necessary so that all rows of the same 40 characters are displayed on a single line before advancing to the next line (see Problem 20-6).
6.   Vertical Sync is used to clear the **74193**s so the memory address is always 0 when the trace returns to the top of the screen.

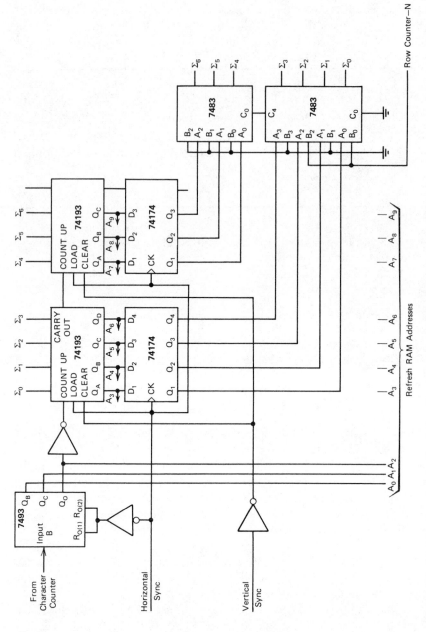

FIGURE 20-12  A control circuit for the Refresh RAM addresses in a Display Generator.

Many hobbyist displays use 32 characters per line. This greatly simplifies the addressing problems (see Problem 20-7).

### 20-6.4 The Line Counter

The line counter is the last counter shown in Fig. 20-10. It simply counts the number of lines in a frame, and can be used to blank the display at other times. The Apple computer and our Display Generator provide 24 lines of 8 rows each, or 192 rows. For the remaining 70 rows of the frame (remember there are 262 rows per frame) the display must be blanked while Vertical Sync occurs.

In many display generators the line counter is synchronized to Vertical Sync, which, in turn, is synchronized to the 60-Hz AC power line. The line counter keeps the display blanked after Vertical Sync until the trace is at the proper position near the top of the screen. The line counter then counts the number of lines to be displayed and blanks the screen until the next Vertical Sync pulse occurs.

### 20-7 THE CURSOR

The *cursor* is a symbol on the CRT screen that *indicates where the next character will be entered.* The cursor is usually identified by a blinking box, but sometimes by an underscore or overscore. To allow the operator to manipulate the cursor, most keyboards have a HOME key that moves the cursor to the upper left-hand corner of the display (the location of the first character in the field) and keys to shift the cursor UP, DOWN, RIGHT, or LEFT.

*Display generators must know the cursor position and the position of each character that is being displayed.* This requires a CURSOR ADDRESS REGISTER, which contains the address of the cursor, and a CHARACTER LOCATION COUNTER, which contains the position of the character currently being displayed. The CHARACTER LOCATION COUNTER consists of two parts: the CHARACTER COUNTER, discussed in Sec. 20-6.2 and the VERTICAL LINE COUNTER, discussed in Sec. 20-6.4 The former contains the horizontal position and the latter the vertical position of the present character.

If the REFRESH MEMORY is a RAM, there is a one-to-one correspondence between the MAR and the CHARACTER LOCATION COUNTER. They can actually be the same counter.

The CURSOR ADDRESS REGISTER is the same size as the CHARACTER LOCATION COUNTER. But *it is only changed in response to a cursor command* from the keyboard *or when a new character is written to the screen.*

## EXAMPLE 20-11

In a 24 line by 60 column display, the operator depresses the ↑ key on his keyboard. What must happen to the CURSOR ADDRESS REGISTER?

## SOLUTION

The cursor must move up one line when the ↑ key is depressed. To accomplish this, 60 must be subtracted from the CURSOR ADDRESS REGISTER because there are 60 characters in the line.

## EXAMPLE 20-12

Design the CURSOR ADDRESS REGISTER. Make provision for the HOME, →, and ← keyboard commands that move the cursor.

## SOLUTION

The design is shown in Fig. 20-13. The CURSOR ADDRESS REGISTER consists of four **74193** up-down counters (Sec. 8-10.1). The two upper **74193**s form the 6-

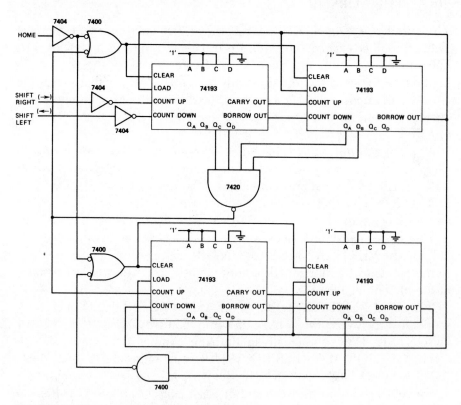

FIGURE 20-13   A cursor address register.

bit HORIZONTAL CHARACTER COUNTER. The two lower **74193**s form the 5-bit VERTICAL LINE COUNTER.

The HOME command clears all counters, thereby placing the cursor at location 0, or the upper left-hand corner of the screen.

The SHIFT RIGHT ($\rightarrow$) command is connected to the count-up input of the **74193** and increments the CURSOR ADDRESS REGISTER, thereby moving the cursor one position to the right. For a 60 character display, the cursor is at the extreme right end of the line when the count is 59. A SHIFT RIGHT command at this time causes the count to go to 60. This is decoded by the **7420**, which clears the horizontal portion of the CURSOR ADDRESS COUNTER and increments the VERTICAL COUNTER. The effect is that the cursor moves down to the leftmost (first) position on the next line.

The SHIFT LEFT ($\leftarrow$) command decrements the CURSOR ADDRESS COUNTER, thereby moving the cursor one place to the left. If the cursor is at the leftmost position of the screen, a SHIFT LEFT command causes a BORROW-OUT which loads a 59 into the HORIZOINTAL COUNTER and decrements the vertical counter, placing the cursor at the end of the row above. If the cursor is HOME (the upper leftmost position of the screen), a SHIFT LEFT command loads a 59 into the horizontal portion of the counter and a 23 into the vertical portion of the counter. This moves the cursor to the rightmost position on the bottom line of the screen.

## 20-7.1 Displaying the Cursor

Additional circuitry is required to display the blinking cursor. The Character Counter and Line Counter that indicate which character is being displayed are compared to the Cursor Address Register. Comparators such as the **7485** (see Sec. 13-3.3) are used. When the addresses compare, the blinking box is placed on the screen so the cursor location is apparent to the operator.

## 20-7.2 Interfacing the Display Terminal with a Computer

Display terminals are interfaced to computers and behave like TTYs. Depending on the display generator, synchronous (RS-232-C) and asynchronous interfaces are available.

To write to the computer's memory, the operator depresses a key on the keyboard. The code for the character is sent to the computer via the display terminal controller.

When the computer sends a character to the display terminal, it is written on the CRT *at the cursor location* and the *cursor is then advanced*. When the display generator receives a character from the computer, it compares to the CURSOR ADDRESS REGISTER to the CHARACTER LOCA-TION COUNTER. When equality occurs, the REFRESH MEMORY receives a WRITE command rather than a READ command, and writes the characters to the CRT. The CURSOR ADDRESS REGISTER is then advanced to be ready for the next character.

When a character is received from the computer, the display generator waits until the REFRESH MEMORY address agrees with the CURSOR ADDRESS REGISTER before writing it. Therefore, the CRT screen cannot be updated faster than once per field or 60 characters per second, but this is faster than the maximum bit rate for most display generators.

## 20-8   CRT CONTROLLERS

Microprocessor manufacturers are now producing ICs called CRT controllers (CRTCs). These ICs are designed to allow a $\mu$P to drive a CRT display and contain within them many of the functions required in a Display Generator.

One CRT controller is the Motorola **6845,** whose functional block diagram is shown in Fig. 20-14. The **6845** performs the following functions:

1.   Generating the Refresh RAM addresses.
2.   Generating the row selects (the row counter) for the character generator.
3.   Generating Horizontal and Vertical Sync.

The **6845** is designed to work with the **6800** and other Motorola $\mu$Ps. It contains 18 internal registers that must be initialized by the $\mu$P. The $\mu$P communicates with the **6845** registers through a buffered 8-bit data bus. These **6845** registers hold the basic system parameters such as the number of horizontal lines for each character row, number of lines per field, the

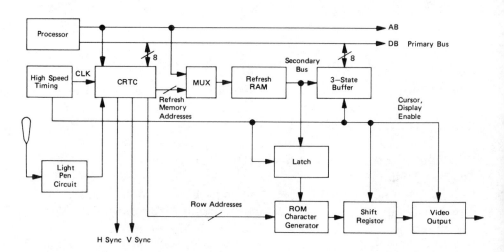

FIGURE 20-14   Logic functions provided by **6845** CRT controller. (Courtesy of Motorola Integrated Circuits Division.)

cursor location, and whether interlaced or noninterlaced scan is desired. These parameters are written into the CRTC registers by the $\mu$P and control the operation of the CRTC, making it produce the desired screen image.

The use of the **6845** CRT controller (CRTC) in a display generator is shown in Fig. 20-15. The CRTC receives its inputs from the $\mu$P, which sends it commands and sets its internal registers, and from a clock and Light Pen (if used). The output of the **6845** drives the Refresh RAM and the row addresses. The Refresh RAM also receives inputs from the $\mu$P so that the $\mu$P can change the screen image as required. Figure 20-15 also shows that the output of the RAM is then latched and fed to the character generator. The character generator's output is sent to the shift register and then becomes the video output.

Other manufacturers also produce CRT controllers. Intel's is the **8275**, which is designed to work with the **8080** or **8085**. CRT controllers are too complex to be discussed in detail here. The reader is referred to Gerry Kane's book (see References) or to the manufacturer's specifications.

### SUMMARY

This chapter considers the use of display terminals as I-O devices for computers. The emphasis is on the display generator that controls the characters written to the CRT screen.

The progression of a character from a 6-bit code in REFRESH MEMORY to an image on the screen is explained. The various memories required and the important timing of the display generator are discussed in detail. Finally, the use of the cursor and the method of writing a character on the CRT screen are considered.

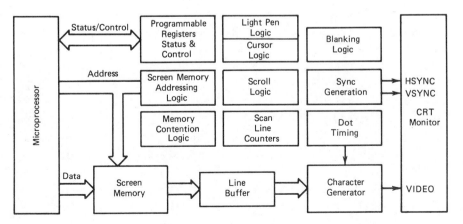

FIGURE 20-15  Typical CRT controller application. (Courtesy of Motorola Integrated Circuits Division.)

## GLOSSARY

**Display terminal.** A computer terminal consisting of a CRT screen and a keyboard.

**Display generator.** That part of a display terminal that presents the data to the CRT screen.

**Alphanumeric.** The set of characters consisting of alphabetics, numbers, and common punctuation.

**Hard copy.** Data output in typed or printed form.

**Raster-scan.** The method of displaying a CRT image by constantly scanning the screen and modulating the scanning beam to produce light and dark areas in the display.

**Sync pulses.** Pulses added to the CRT information to synchronize the horizontal and vertical oscillators.

**Retrace time.** The time when a CRT beam is blanked and returning to its starting position.

**Horizontal retrace time.** The time the beam takes to return from the right end of the screen to the left end in order to start a new line.

**Vertical retrace time.** The time required for the beam to return to the top of the screen from the bottom to start a new field.

**Field.** One complete CRT picture, drawn when the beam traverses the screen once.

**Frame.** Two fields, generally interlaced.

**Refresh rate.** The rate at which fields are placed on a CRT screen.

**Interlacing.** Starting alternate fields at different points on the CRT screen.

**Dot matrix.** The matrix of dots used to display a single character.

**CHARACTER GENERATOR.** A ROM that produces the character configuration for display on a CRT.

**REFRESH MEMORY.** The memory that contains all the characters currently being displayed on the CRT screen.

**SYNC GENERATOR.** An IC that provides the vertical and horizontal pulses needed to synchronize a display generator otuput with a CRT.

**Cursor.** An underscore or overscore on a CRT display that indicates where the next position is to be written.

**CHARACTER LOCATION COUNTER.** Contains the location of the character (vertical and horizontal coordinates) currently being displayed.

**CURSOR ADDRESS REGISTER.** The register that contains the cursor address.

**HOME.** The upper left-hand corner of the display.

**SHIFT RIGHT** ($\rightarrow$). A command to move the cursor one space to the right.

**SHIFT LEFT** (←). A command to move the cursor one space to the left.
**SHIFT UP** (↑). A command to move the cursor one line up.
**SHIFT DOWN** (↓). A command to move the cursor one line down.
**CRT Controller (CRTC).** An IC designed to provide many of the functions
   of a display generator.

## REFERENCES

**Joseph D. Greenfield** and **William C. Wray,** *Using Microprocessors and
   Microcomputers: The 6800 Family*, Wiley, New York, 1981.
**Gerry Kane,** *CRT Controller Handbook*, Osborne/McGraw-Hill, Berkeley,
   Ca., 1980.
**Don Lancaster,** *TV Typewriter Cookbook*, Howard W. Sams & Co., Inc.,
   Indianapolis, Ind., 1976.
*Signetics Digital Linear MOS Data Book*, Signetics Corporation, Menlo
   Park, Ca., 1976.
***Apple II Reference Manual,*** Apply Computer, Inc., Cupertino, Cal., 1979.

## PROBLEMS

20-1.    A 5 × 5 dot matrix is imbedded in a 6 × 6 grid. How many character
lines can be written to the screen if:
   (a) The refresh rate is 60 Hz.
   (b) The refresh rate is 50 Hz.
Assume the horizontal oscillator is 15.750 Hz in both cases and that the display
can show 240 lines. Also assume that 80% of the rows are available for display.
20-2.    Repeat Problem 20-1 for a 5 × 7 dot matrix imbedded in a 6 × 8 grid.
20-3.    What is the maximum number of horizontal characters that can be displayed if the frequency response of the CRT limits the input to $5 \times 10^6$ BPS. The
horizontal oscillator runs at 15,750 Hz. Assume:
   (a) The characters are 8 bits wide.
   (b) The characters are 6 bits wide.
   (c) 42 $\mu$s are available for displaying each line.
20-4.    A commercial terminal displays 24 lines of 80 characters.
   (a) What are the dimensions of its REFRESH MEMORY.
   (b) If the REFRESH MEMORY is built of **2524** shift registers, how many **2524**s
are required? How many extra pulses must be injected at the end of each frame?
   (c) If the REFRESH MEMORY is built using core or semiconductor RAM,
how many bits are required in the MAR? How many unused locations are there
in the RAM?

TABLE P20-6

| | 74174 outputs during H sync | B inputs to the 7483 | 7483 outputs during H sync | 74174 outputs after H sync | memory address at start of row | memory address at end of row |
|---|---|---|---|---|---|---|
| Line 0 | | | | | | |
| Row 0 | 0 | 5 | 5 | 5 | 40 | 79 |
| 1 | | | | | | |
| 2 | | | | | | |
| 3 | | | | | | |
| 4 | | | | | | |
| 5 | | | | | | |
| 6 | | | | | | |
| 7 | | | | | | |
| Line 1 | | | | | | |
| Row 0 | | | | | | |
| 1 | | | | | | |
| 2 | | | | | | |

20-5.  For the circuit of Fig. 20-7, complete the table:

| Composite Sync | Dot Output | Video Output (volts) |
|---|---|---|
| 0 | 0 | |
| 0 | 1 | |
| 1 | 0 | |
| 1 | 1 | |

Where is the visual level for this display?

20-6.  Table P20-6 applies to Fig. 20-12. Fill in the table. Note that the memory actually starts at location 40 for this circuit.

20-7.  Design an address controller for the memory similar to Fig. 20-11 if the raster size is:

(a) 32 characters by 24 lines.

(b) 48 characters by 24 lines.

20-8.  A display generator uses a semiconductor RAM as a REFRESH MEMORY and no LINE MEMORY. Both the CHARACTER GENERATOR and the RAM have an access time of 500 ns. A character must be displayed every 800 ns. Design a buffered circuit to make this possible.

20-9.  For use with an ordinary TV set, one author recommends using only 32 characters, each 6 bits wide. The memory is then a single **2518** and the CHARACTER GENERATOR is a **2513**. Draw a timing chart for this display.

20-10.  If the REFRESH MEMORY for Problem 20-9 is a **2524** that is read during a single blank line between the characters, draw a timing chart for the reading of the **2524**.

20-11.  What happens to the CURSOR ADDRESS REGISTER if the operator depresses the ↓ key, which causes the cursor to move down one line.

20-12.  Assuming a RAM is being used for the REFRESH MEMORY, design a circuit to write a character on the screen when it is received from the computer.

20-13.  Modify the circuit of Fig. 20-13 to permit ↑ and ↓ keyboard commands.

20-14.  Design a CURSOR LOCATION REGISTER for an 80 character by 48 line display.

20-15.  Design a display generator to display 24 lines of 80 characters each. Assume each character is $5 \times 7$ dots embedded in a $7 \times 9$ matrix.

After attempting the problems, return to Sec. 20-2 and reread the self-evaluation questions. If any of them are still difficult, review the appropriate sections of the chapter to find the answers.

# APPENDIX

Table of powers of 2.

| $2^n$ | n | $2^{-n}$ |
|---|---|---|
| 1 | 0 | 1.0 |
| 2 | 1 | 0.5 |
| 4 | 2 | 0.25 |
| 8 | 3 | 0.125 |
| 16 | 4 | 0.062 5 |
| 32 | 5 | 0.031 25 |
| 64 | 6 | 0.015 625 |
| 129 | 7 | 0.007 812 5 |
| 256 | 8 | 0.003 906 25 |
| 512 | 9 | 0.001 953 125 |
| 1 024 | 10 | 0.000 976 562 5 |
| 2 048 | 11 | 0.000 488 281 25 |
| 4 096 | 12 | 0.000 244 140 625 |
| 8 192 | 13 | 0.000 122 070 312 5 |
| 16 384 | 14 | 0.000 061 035 156 25 |
| 32 768 | 15 | 0.000 030 517 578 125 |
| 65 536 | 16 | 0.000 015 258 789 062 5 |
| 131 072 | 17 | 0.000 007 629 394 531 25 |
| 262 144 | 18 | 0.000 003 814 697 265 625 |
| 524 288 | 19 | 0.000 001 907 348 632 812 5 |
| 1 048 576 | 20 | 0.000 000 953 674 316 406 25 |
| 2 097 152 | 21 | 0.000 000 476 837 153 203 125 |
| 4 194 304 | 22 | 0.000 000 238 418 579 101 562 5 |
| 8 388 608 | 23 | 0.000 000 119 209 289 550 781 25 |
| 16 777 216 | 24 | 0.000 000 059 604 644 775 390 625 |
| 33 554 432 | 25 | 0.000 000 029 802 322 387 695 312 5 |
| 67 108 864 | 26 | 0.000 000 014 901 161 193 817 656 25 |
| 134 217 728 | 27 | 0.000 000 007 450 580 596 923 828 125 |
| 268 435 456 | 28 | 0.000 000 003 725 290 298 461 914 062 5 |
| 536 870 912 | 29 | 0.000 000 001 862 645 149 230 957 031 25 |
| 1 073 741 824 | 30 | 0.000 000 000 931 322 574 615 478 515 625 |
| 2 147 483 648 | 31 | 0.000 000 000 465 661 287 307 739 257 812 5 |
| 4 294 967 296 | 32 | 0.000 000 000 232 830 613 653 869 628 906 25 |
| 8 589 934 592 | 33 | 0.000 000 000 116 415 321 826 934 814 453 125 |
| 17 179 869 184 | 34 | 0.000 000 000 058 207 660 913 467 407 226 562 5 |
| 34 359 738 368 | 35 | 0.000 000 000 029 103 830 456 733 703 613 281 25 |
| 68 719 476 736 | 36 | 0.000 000 000 014 551 228 366 851 806 640 625 |
| 137 438 953 472 | 37 | 0.000 000 000 007 275 957 614 183 425 903 320 312 5 |
| 274 877 906 944 | 38 | 0.000 000 000 003 637 978 807 091 712 951 660 156 25 |
| 549 755 813 888 | 39 | 0.000 000 000 001 818 898 403 545 856 475 830 078 125 |
| 1 099 511 627 776 | 40 | 0.000 000 000 000 909 494 701 772 928 237 915 039 062 5 |
| 2 199 023 255 552 | 41 | 0.000 000 000 000 454 747 350 886 464 118 957 519 531 25 |
| 4 398 046 511 104 | 42 | 0.000 000 000 000 227 373 675 443 232 059 478 759 765 625 |
| 8 796 093 022 208 | 43 | 0.000 000 000 000 113 686 837 721 616 029 739 379 882 812 5 |
| 17 592 186 044 416 | 44 | 0.000 000 000 000 056 843 418 860 808 014 869 698 941 406 25 |
| 35 184 372 088 832 | 45 | 0.000 000 000 000 028 421 709 431 404 007 434 844 970 703 125 |
| 70 368 744 177 664 | 46 | 0.000 000 000 000 014 210 854 715 202 003 717 422 485 351 562 5 |
| 140 737 488 355 328 | 47 | 0.000 000 000 000 007 105 427 357 601 001 858 711 242 675 781 25 |
| 281 474 976 710 656 | 48 | 0.000 000 000 000 003 552 713 678 800 500 929 355 621 337 890 625 |
| 562 949 953 421 312 | 49 | 0.000 000 000 000 001 776 356 839 400 250 464 677 810 668 945 312 5 |
| 1 125 899 906 843 624 | 50 | 0.000 000 000 000 000 888 178 419 700 125 232 338 905 334 472 656 25 |
| 2 251 799 813 685 248 | 51 | 0.000 000 000 000 000 444 089 209 850 062 616 169 452 667 236 328 125 |
| 4 503 599 627 370 496 | 52 | 0.000 000 000 000 000 222 044 804 925 031 308 084 726 333 618 164 062 5 |
| 9 007 199 254 740 992 | 53 | 0.000 000 000 000 000 111 022 302 462 515 654 042 363 166 809 082 031 25 |
| 18 014 398 509 481 984 | 54 | 0.000 000 000 000 000 055 511 151 231 257 827 021 181 583 404 541 015 625 |
| 36 028 797 018 963 968 | 55 | 0.000 000 000 000 000 027 755 575 615 628 913 510 590 791 702 270 507 812 5 |
| 72 057 594 037 927 936 | 56 | 0.000 000 000 000 000 013 877 787 807 814 456 755 295 395 851 135 253 906 25 |
| 144 115 188 075 855 872 | 57 | 0.000 000 000 000 000 006 938 893 903 907 228 377 647 697 925 567 626 953 125 |
| 288 230 376 151 711 744 | 58 | 0.000 000 000 000 000 003 469 446 951 953 614 188 823 848 962 783 813 476 562 5 |
| 576 460 752 303 423 488 | 59 | 0.000 000 000 000 000 001 734 723 475 976 807 094 411 924 481 391 906 738 281 25 |
| 1 152 921 504 606 846 976 | 60 | 0.000 000 000 000 000 000 867 361 737 988 403 547 205 962 240 695 953 369 140 625 |
| 2 305 843 009 213 693 952 | 61 | 0.000 000 000 000 000 000 433 680 868 994 201 773 602 981 120 347 976 684 570 312 5 |
| 4 611 686 018 427 387 904 | 62 | 0.000 000 000 000 000 000 216 840 434 497 100 886 801 490 560 173 988 342 285 156 25 |
| 9 223 372 036 854 775 808 | 63 | 0.000 000 000 000 000 000 108 420 217 248 550 443 400 745 280 086 994 171 142 578 125 |
| 18 446 744 073 709 551 616 | 64 | 0.000 000 000 000 000 000 054 210 108 624 275 221 700 372 640 043 497 085 571 289 062 5 |
| 36 893 488 147 419 103 232 | 65 | 0.000 000 000 000 000 000 027 105 054 312 137 610 850 186 320 021 748 542 785 644 531 25 |
| 73 786 976 294 838 206 464 | 66 | 0.000 000 000 000 000 000 013 552 527 156 068 805 425 093 160 010 874 271 392 822 265 625 |
| 147 573 952 589 676 412 928 | 67 | 0.000 000 000 000 000 000 006 776 263 578 034 402 712 546 580 005 437 135 696 411 132 812 5 |
| 295 147 905 179 352 825 856 | 68 | 0.000 000 000 000 000 000 003 388 131 789 017 201 356 273 290 002 718 567 848 205 566 406 25 |
| 590 295 810 358 705 651 712 | 69 | 0.000 000 000 000 000 000 001 694 065 894 508 600 678 136 645 001 359 283 924 102 783 203 125 |
| 1 180 591 620 717 411 303 432 | 70 | 0.000 000 000 000 000 000 000 847 032 947 254 300 339 068 322 500 679 641 962 051 391 601 562 5 |
| 2 361 183 241 434 822 606 848 | 71 | 0.000 000 000 000 000 000 000 423 516 473 627 150 169 534 161 250 339 820 981 025 695 800 781 25 |
| 4 722 366 482 869 645 213 696 | 72 | 0.000 000 000 000 000 000 000 211 758 236 813 575 084 767 080 625 169 910 490 512 847 900 390 625 |

# APPENDIX

# B

Calculation of resistances for open collector ICs.

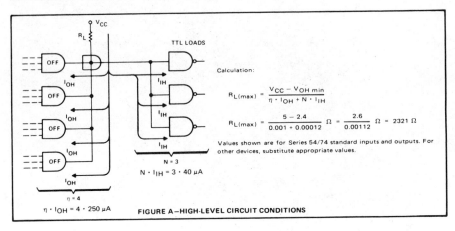

Calculation:

$$R_{L(max)} = \frac{V_{CC} - V_{OH\,min}}{\eta \cdot I_{OH} + N \cdot I_{IH}}$$

$$R_{L(max)} = \frac{5 - 2.4}{0.001 + 0.00012}\,\Omega = \frac{2.6}{0.00112}\,\Omega = 2321\,\Omega$$

Values shown are for Series 54/74 standard inputs and outputs. For other devices, substitute appropriate values.

**FIGURE A—HIGH-LEVEL CIRCUIT CONDITIONS**

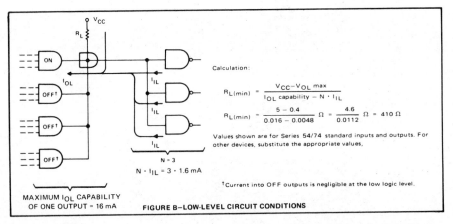

Calculation:

$$R_{L(min)} = \frac{V_{CC} - V_{OL\,max}}{I_{OL}\text{ capability} - N \cdot I_{IL}}$$

$$R_{L(min)} = \frac{5 - 0.4}{0.016 - 0.0048}\,\Omega = \frac{4.6}{0.0112}\,\Omega = 410\,\Omega$$

Values shown are for Series 54/74 standard inputs and outputs. For other devices, substitute the appropriate values.

†Current into OFF outputs is negligible at the low logic level.

**FIGURE B—LOW-LEVEL CIRCUIT CONDITIONS**

Reduction of Eq. (12-2) to EXCLUSIVE ORs.

$$
\begin{aligned}
f(A,B,C,D) &= \overline{A}\overline{B}\overline{C}D + \overline{A}\overline{B}C\overline{D} + \overline{A}B\overline{C}\overline{D} + \overline{A}BCD \\
&\quad + A\overline{B}\overline{C}\overline{D} + A\overline{B}CD + AB\overline{C}D + ABC\overline{D} \\
&= \overline{A}\overline{B}(\overline{C}D + C\overline{D}) + \overline{A}B(\overline{C}\overline{D} + CD) \\
&\quad + A\overline{B}(\overline{C}\overline{D} + CD) + (AB)(\overline{C}D + C\overline{D}) \\
&= \overline{A}\overline{B}(C \oplus D) + \overline{A}B\overline{(C \oplus D)} \\
&\quad A\overline{B}\overline{(C \oplus D)} + AB(C \oplus D) \\
&= (C \oplus D)(AB + \overline{A}\overline{B}) + \overline{(C \oplus D)}(A\overline{B} + \overline{A}B) \\
&= (C \oplus D)\overline{(A \oplus B)} + \overline{(C \oplus D)}(A \oplus B) \\
&= A \oplus B \oplus C \oplus D
\end{aligned}
$$

To show that if all the outputs of a register are XORed together the resulting output is HIGH on odd parity, we use an induction proof.

Assume the statement is true for an $n$ bit register. We proceed to show that it must be true for an $n + 1$ bit register.

1.  If there is an odd number of 1s in the $n$-bit register, there will be an odd number of 1s in the $n + 1$ bit register only if the $n + 1$ bit is 0. By XORing the $n$-bit output (HIGH) with the $n + 1$ bit (0) the results will be HIGH.

2.  If there are an even number of 1s in the $n$-bit register, the output of the XOR circuits for the $n$-bit register is LOW. The $n + 1$ bit register will have an odd number of 1s if the $n + 1$ bit is a 1. By XORing the $n + 1$ bit with the output of the $n$-bit register, we obtain the correct results.

3.  Since the statement is true by inspection for a 2-bit register, it must be true for a register of any size.

## Characteristics of TTL NAND gates

### recommended operating conditions

| | 54 FAMILY / 74 FAMILY | SERIES 54 / SERIES 74 '00, '04, '10, '20, '30 | | | SERIES 54H / SERIES 74H 'H00, 'H04, 'H10, 'H20, 'H30 | | | SERIES 54L / SERIES 74L 'L00, 'L04, 'L10, 'L20, 'L30 | | | SERIES 54LS / SERIES 74LS 'LS00, 'LS04, 'LS10, 'LS20, 'LS30 | | | SERIES 54S / SERIES 74S 'S00, 'S04, 'S10, 'S20, 'S30, 'S133 | | | UNIT |
|---|---|---|---|---|---|---|---|---|---|---|---|---|---|---|---|---|---|
| | | MIN | NOM | MAX | MIN | NOM | MAX | MIN | NOM | MAX | MIN | NOM | MAX | MIN | NOM | MAX | |
| Supply voltage, $V_{CC}$ | 54 Family | 4.5 | 5 | 5.5 | 4.5 | 5 | 5.5 | 4.5 | 5 | 5.5 | 4.5 | 5 | 5.5 | 4.5 | 5 | 5.5 | V |
| | 74 Family | 4.75 | 5 | 5.25 | 4.75 | 5 | 5.25 | 4.75 | 5 | 5.25 | 4.75 | 5 | 5.25 | 4.75 | 5 | 5.25 | V |
| High-level output current, $I_{OH}$ | 54 Family | | | -400 | | | -500 | | | -100 | | | -400 | | | -1000 | µA |
| | 74 Family | | | -400 | | | -500 | | | -200 | | | -400 | | | -1000 | µA |
| Low-level output current, $I_{OL}$ | 54 Family | | | 16 | | | 20 | | | 2 | | | 4 | | | 20 | mA |
| | 74 Family | | | 16 | | | 20 | | | 3.6 | | | 8 | | | 20 | mA |
| Operating free-air temperature, $T_A$ | 54 Family | -55 | | 125 | -55 | | 125 | -55 | | 125 | -55 | | 125 | -55 | | 125 | °C |
| | 74 Family | 0 | | 70 | 0 | | 70 | 0 | | 70 | 0 | | 70 | 0 | | 70 | °C |

### electrical characteristics over recommended operating free-air temperature range (unless otherwise noted)

| PARAMETER | TEST FIGURE | TEST CONDITIONS† | | SERIES 54 / SERIES 74 '00, '04, '10, '20, '30 | | | SERIES 54H / SERIES 74H 'H00, 'H04, 'H10, 'H20, 'H30 | | | SERIES 54L / SERIES 74L 'L00, 'L04, 'L10, 'L20, 'L30 | | | SERIES 54LS / SERIES 74LS 'LS00, 'LS04, 'LS10, 'LS20, 'LS30 | | | SERIES 54S / SERIES 74S 'S00, 'S04, 'S10, 'S20, 'S30, 'S133 | | | UNIT |
|---|---|---|---|---|---|---|---|---|---|---|---|---|---|---|---|---|---|---|---|
| | | | | MIN | TYP‡ | MAX | MIN | TYP‡ | MAX | MIN | TYP‡ | MAX | MIN | TYP‡ | MAX | MIN | TYP‡ | MAX | |
| $V_{IH}$ High-level input voltage | 1, 2 | | | 2 | | | 2 | | | 2 | | | 2 | | | 2 | | | V |
| $V_{IL}$ Low-level input voltage | 1, 2 | | 54 Family | | | 0.8 | | | 0.8 | | | 0.7 | | | 0.7 | | | 0.8 | V |
| | | | 74 Family | | | 0.8 | | | 0.8 | | | 0.7 | | | 0.8 | | | 0.8 | V |
| $V_{IK}$ Input clamp voltage | 3 | $V_{CC}$ = MIN, $I_I$ = § | | | | -1.5 | | | -1.5 | | | | | | -1.5 | | | -1.2 | V |
| $V_{OH}$ High-level output voltage | 1 | $V_{CC}$ = MIN, $V_{IL}$ = $V_{IL}$ max, $I_{OH}$ = MAX | 54 Family | 2.4 | 3.4 | | 2.4 | 3.5 | | 2.4 | 3.3 | | 2.5 | 3.4 | | 2.5 | 3.4 | | V |
| | | $V_{IH}$ = 2 V | 74 Family | 2.4 | 3.4 | | 2.4 | 3.5 | | 2.4 | 3.2 | | 2.7 | 3.4 | | 2.7 | 3.4 | | V |
| $V_{OL}$ Low-level output voltage | 2 | $V_{CC}$ = MIN, $I_{OL}$ = MAX | 54 Family | | 0.2 | 0.4 | | 0.2 | 0.4 | | 0.15 | 0.3 | | 0.25 | 0.4 | | | 0.5 | V |
| | | $V_{IH}$ = 2 V | 74 Family | | 0.2 | 0.4 | | 0.2 | 0.4 | | 0.2 | 0.4 | | 0.25 | 0.5 | | | 0.5 | V |
| | | $I_{OL}$ = 4 mA | Series 74LS | | | | | | | | | | | | 0.4 | | | | V |
| $I_I$ Input current at maximum input voltage | 4 | $V_{CC}$ = MAX, $V_I$ = 5.5 V / $V_I$ = 7 V | | | | 1 | | | 1 | | | 0.1 | | | 0.1 | | | 1 | mA |
| $I_{IH}$ High-level input current | 4 | $V_{CC}$ = MAX, $V_{IH}$ = 2.4 V / $V_{IH}$ = 2.7 V | | | | 40 | | | 50 | | | 10 | | | 20 | | | 50 | µA |
| $I_{IL}$ Low-level input current | 5 | $V_{CC}$ = MAX, $V_{IL}$ = 0.3 V / 0.4 V / 0.5 V | | | | -1.6 | | | -2 | | | -0.18 | | | -0.4 | | | -2 | mA |
| $I_{OS}$ Short-circuit output current♦ | 6 | $V_{CC}$ = MAX | 54 Family | -20 | | -55 | -40 | | -100 | -3 | | -15 | -20 | | -100 | -40 | | -100 | mA |
| | | | 74 Family | -18 | | -55 | -40 | | -100 | -3 | | -15 | -20 | | -100 | -40 | | -100 | mA |
| $I_{CC}$ Supply current | 7 | $V_{CC}$ = MAX | | | | | | | | | | | | | | | | | See table on next page | mA |

† For conditions shown as MIN or MAX, use the appropriate value specified under recommended operating conditions.
‡ All typical values are at $V_{CC}$ = 5 V, $T_A$ = 25°C.
§ $I_I$ = -12 mA for SN54'/SN74', -8 mA for SN54H'/SN74H', and -18 mA for SN54LS'/SN74LS' and SN54S'/SN74S'.
♦ Not more than one output should be shorted at a time, and for SN54H'/SN74H', SN54LS'/SN74LS', and SN54S'/SN74S', duration of short-circuit should not exceed 1 second.

*From the TTL Data Book for Design Engineers. Texas Instruments 1976. Courtesy of Texas Instruments, Inc.

# F

# ANSWERS TO SELECTED PROBLEMS

## CHAPTER 1

1-1    7
1-5    (c)  395
       (d)  27.71875
       (f)  399.453125
1-6    (c)  1011010100011
       (g)  0.110100011
1-7    (c)  $A + B = 111101100$
$A - B = 110100010$
1-8    (b)  $A + B = 11011111$, $A - B = 0100011$
1-10   The numbers 0 through 1023

## CHAPTER 2

2-4    $P = M + F\overline{A}$ (A policy will be issued to anyone who is married or to a female under 25 years of age.)
2-5    (c)  $X + Z$
       (d)  1
2-7    (c)  The equation is false for $a = 1$, $b = 1$, $c = 0$.
2-8    (a)  $Y = \overline{C}(AB + CD) = \overline{C}AB$

2-9 (a)

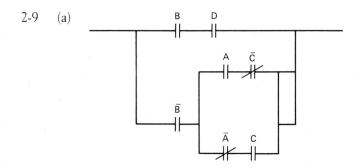

2-10 (a) $(a + \bar{d} + \bar{b}\bar{c})\,(a + d + \bar{b}\,\bar{c})\,bc$

2-12 (b)

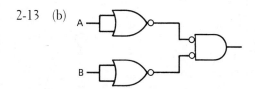

2-13 (b)

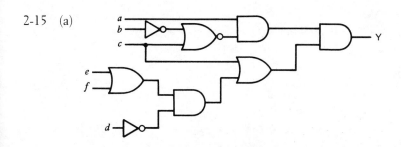

2-15 (a)

2-15 (d)

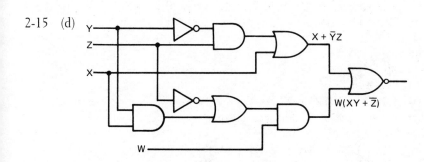

2-15 (f)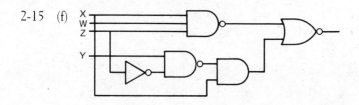

2-16 (d) $\overline{Y} = ABC + A\overline{B} + \overline{A}BC$;

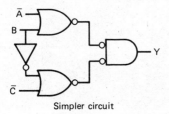

Simpler circuit

(e) $Y = AB \cdot (\overline{\overline{A}\overline{B} + \overline{C})\overline{D}} \cdot (\overline{C} + \overline{D})$; simpler circuit

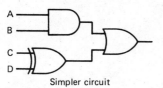

Simpler circuit

(f) $\overline{Y} = (\overline{A} + \overline{B}\overline{C})(CD)\overline{D}$; simpler circuit; $Y = 1$

# CHAPTER 3

3-1 (b) $w\bar{x}\bar{y}\bar{z} + w\bar{x}\bar{y}z + w\bar{x}y\bar{z} + w\bar{x}yz + wxyz + \bar{w}xyz + \bar{w}\bar{x}\bar{y}z + \bar{w}x\bar{y}z$

3-2 (b) $(\bar{w} + \bar{x} + \bar{y} + z)(\bar{w} + \bar{x} + y + \bar{z})(\bar{w} + \bar{x} + y + z)$
$(w + \bar{x} + \bar{y} + z)(w + \bar{x} + y + z)(w + x + \bar{y} + \bar{z})$
$(w + x + \bar{y} + z)(w + x + y + z)$

3-3 (b) $f(w,x,y,z) = \Sigma\ (8,9,10,11,7,15,1,5) = \pi\ (0,2,3,4,6,12,13,14)$

3-4 (b) 8,9,10,11
    (f) 2,3,6,7

3-5 Subcube 1 $\overline{X}\ \overline{Y}\ \overline{Z}$
        5 $B\ D$
        6 $W + \overline{X} + \overline{Z}$

3-6 $f(W,X,Y,Z) = \overline{W}\overline{Y}\overline{Z} + \overline{X}Y + WYZ + W\overline{X}$
The subcube composed of the four corner squares is not essential.

3-10 $F(A,B,C,D) = \overline{B}\overline{C} + \overline{A}C + AB$

3-13 (b) $F(W,X,Y,Z) = \overline{X}Z + W\overline{Y}Z + W\overline{X}Y$

3-14 (b) $F(W,X,Y,Z) = (W + \overline{X})\ (Y + Z)\ (W + Z)\ (\overline{X} + \overline{Y})$

3-16 (b)

| YZ \ WX | 00 | 01 | 11 | 10 |
|---------|----|----|----|----|
| 00 |  |  | 1 | d |
| 01 | 1 |  | 1 |  |
| 11 |  |  |  |  |
| 10 |  |  | 1 | 1 |

$$f(W,X,Y,Z) = W\bar{Z} + WX\bar{Y} + \bar{W}X\bar{Y}Z$$
$$f(W,X,Y,Z) = (W + \bar{X})(\bar{Y} + \bar{Z})(W + Z)(\bar{W} + X + \bar{Z})$$

(c)

| cd \ ab | 00 | 01 | 11 | 10 |
|---------|----|----|----|----|
| 00 |  | 0 |  |  |
| 01 |  |  |  | 0 |
| 11 | 0 | d | 0 | d |
| 10 | 0 |  |  | 0 |

$$f(a,b,c,d) = (\bar{c} + \bar{d})(b + \bar{c})(\bar{a} + \bar{b} + c + d)(\bar{a} + b + \bar{d})$$
$$f(a,b,c,d) = \bar{a}\,\bar{c} + \bar{b}\bar{c}\bar{d} + bc\bar{d} + b\bar{c}d$$

3-17  $f(V,W,X,Y,Z) = \bar{X}\bar{Y} + \bar{V}X\bar{Z} + WXY + VWX\bar{Z} + VXY\bar{Z}$

3-18  Segment $f = A + B\bar{C} + B\bar{D} + \bar{C}\bar{D}$

3-22 (a)                                              (b)

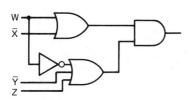

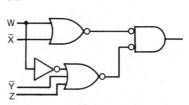

# CHAPTER 5

5-2 (a) Circuit

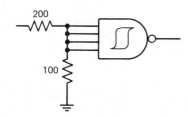

(b) For minimum pulse width,

$$\sin \theta = \frac{0.9}{1.7} = 0.53$$
$$\theta = 148° - 90° = 58°$$
$$\text{Time} = \frac{58°}{360°} \times 1 \ \mu s = 161 \ \text{ns}$$

5-4 (a) Yes
  (b) No
  (c) No
  (d) Yes

5-8

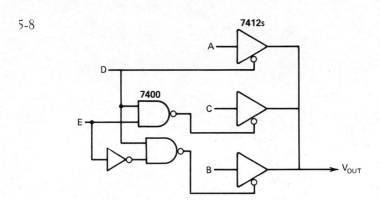

5-10 (a)

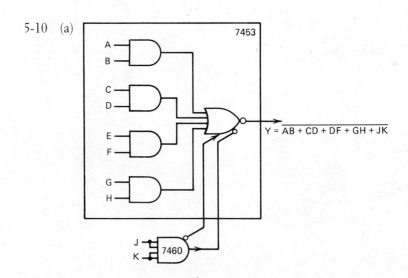

$Y = \overline{AB + CD + DF + GH + JK}$

5-13   (a)

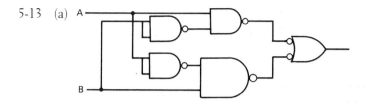

## CHAPTER 6

6-3     Both outputs would be LOW.

6-5

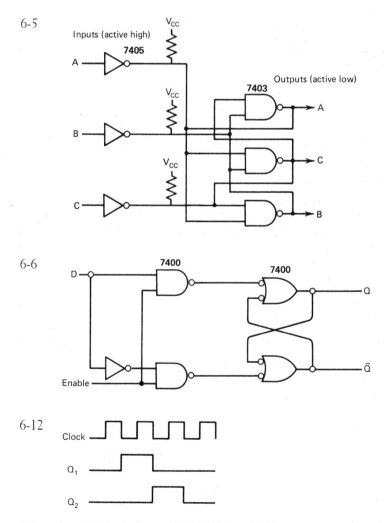

6-6

6-12

When $Q_2$ SETS, it direct CLEARS $Q_1$. $Q_1$ does not toggle on the next pulse because $\overline{Q}_2$ is LOW.

6-17   (a)  Using 7474s: Note additional gating to provide the required drive capability. CLEAR inputs also require additional gating.

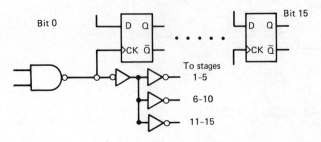

6-17   (b) Using 74174s

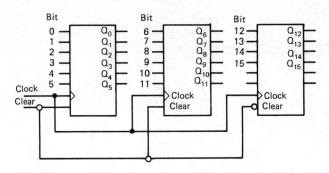

6-18

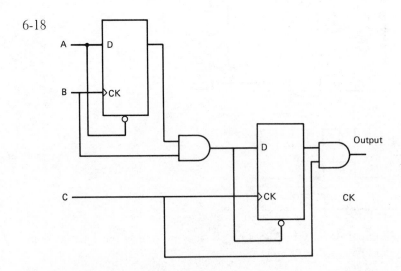

6-21

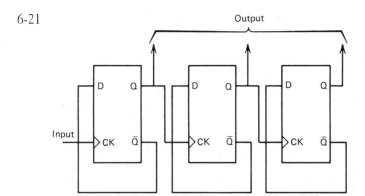

6-22

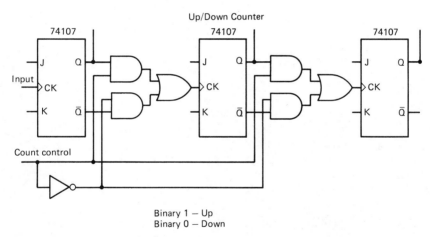

6-27

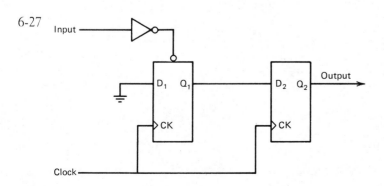

6-30 (a)

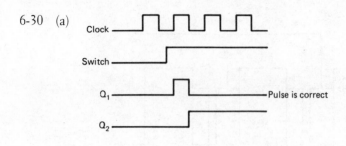

(b)

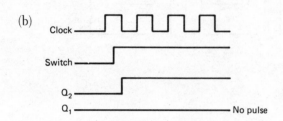

## CHAPTER 7

7-2 (a) $C = 3.57 \ \mu f$. Time between pulses—2.5 $\mu s$.

(b) $C = 0.175 \ \mu f$ Time between pulses—0.5 $\mu s$.

7-3 (a) 350 $\mu s$

(b) 1.75 ms

(c) The fixed resistor eliminates the possibility of connecting the $R_{EXT}$ input directly to $V_{CC}$.

7-4c C for a 10K resistor, $C = 130$ pf

7-5c C for a 10K resistor, $C = .166 \ \mu f$

7-6 (a) 3.5 ms

(b) 1.7 ms

(c) 1.5 ms

7-11 (a) Each **74121** can be set for 10 $\mu s$ using the 2K internal resistor $C = 0.00715 \ \mu f$

(b) Using two halves of a **74123** with a 10K external resistor $C = 0.00333 \ \mu f$

(c) For the Schmitt trigger, $R = 330$, $C = 0.0256 \ \mu f$.

(d) For the 555, let $C = 10^{-9}$; then we can choose $R_A = 12K$ and $R_3 = 8.2K$.

7-13 From the chart $f \approx 8$ Hz.

From the formula, $f = \dfrac{1.44}{300 \times 10^3 \times 10^{-6}} = \dfrac{1.44}{0.3} = 4.8$ Hz

Note that the charts give a very approximate reading.

7-15

Circuit shown using all **74123**s. Timing resistors and capacitors not shown for clarity.

7-16 (a) Yes
　　(b) Yes
　　(c) No

　Switches generally bounce longer than 500 $\mu$s, so bounces could cause additional triggers in parts a and b. Switches never bounce for more than 500 ms, so additional triggers will not be generated in part c. (Note that if a retriggerable one-shot and an undebounced switch are used, the pulse time will be the one-shot time plus the switch bounce time.)

7-19

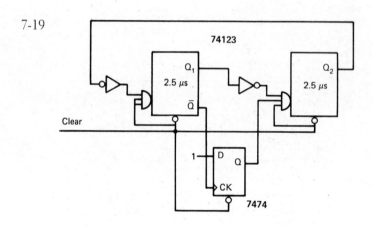

# CHAPTER 8

8-1　Use an ordinary five-stage ripple counter.

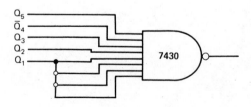

8-3　The XOR gate acts as a complementer circuit and complements $Q_3$ only when $Q_1$ and $Q_2$ are both HIGH. Otherwise $Q_3 = D_3$ and the FF does not change. When $Q_1$ and $Q_2$ are both 1s, $D_3 = \bar{Q}_3$ and the FF toggles.

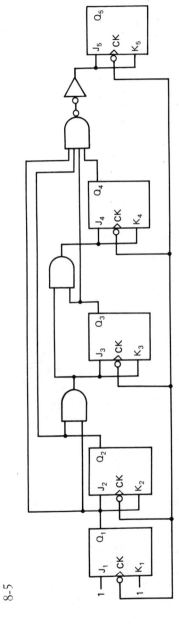

8-5

(a) Minimum time delay (2 gates max)

8-10

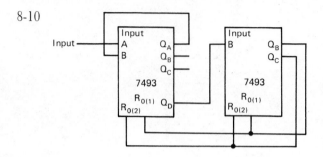

8-14 (a)

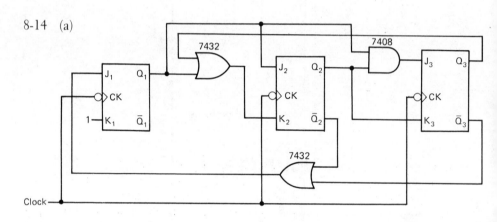

8-15 (b)

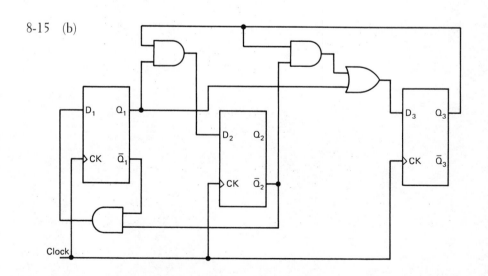

8-16   The count sequence is $0,1,3,5,7,0,\ldots$

8-18

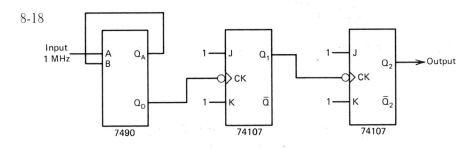

8-24   (a)

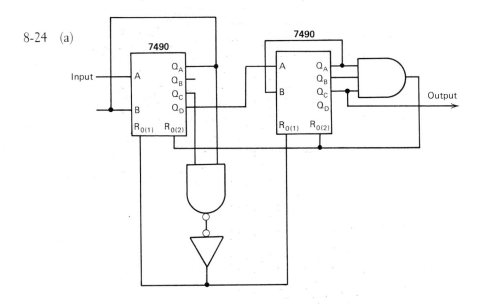

8-25   **7490**s — 95
       **7492**s — 89
       **7493**s — 149

8-27 (a)

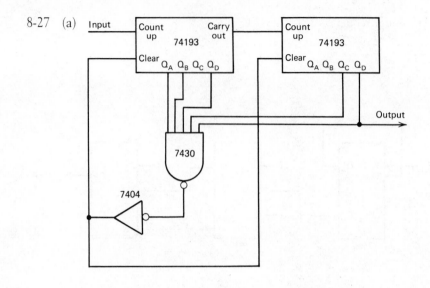

(b)

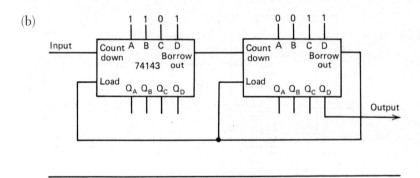

8-28 (c) For the **7490**s, the output should be HIGH for 830 μs and LOW for 830 μs. For the **74193**s the output should be HIGH for 385 μs and LOW for 1275 μs.

8-30

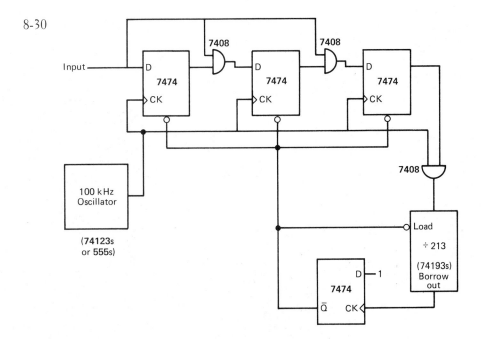

# CHAPTER 9

9-1    (a)  000100101101
       (b)  011011010000

9-2

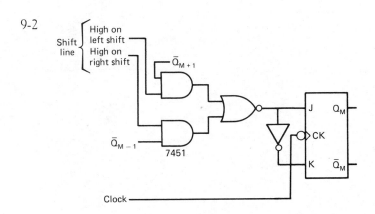

9-6

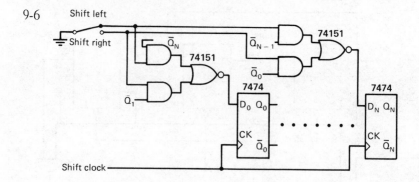

9-8

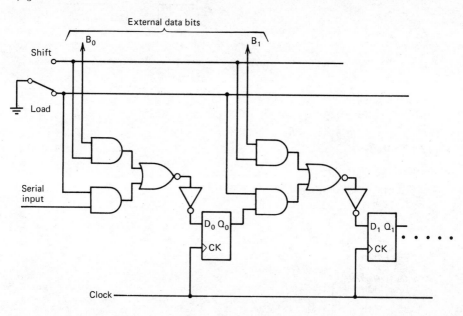

9-13

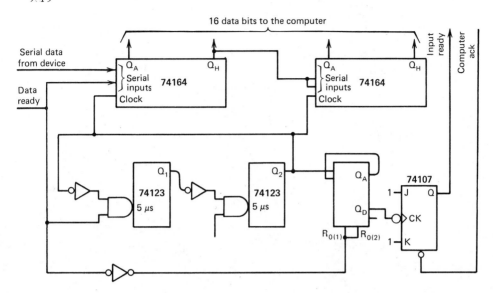

9-15  (a)

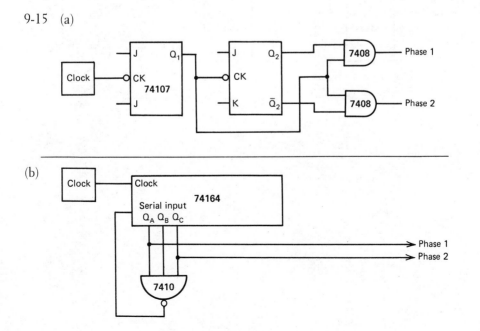

(b)

9-18

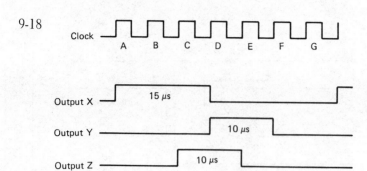

9-20

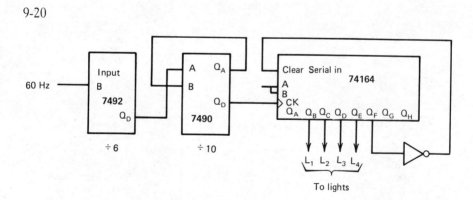

To lights

# CHAPTER 11

11-1

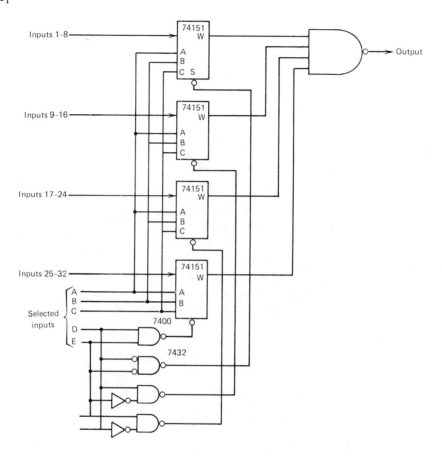

11-2    Delete one IC from the circuit of Problem 11-1 or use a **74150** and a **74151**.

11-4    Use a 100-KHz oscillator to drive a **7493** binary counter. Connect the four counter outputs to the select inputs of the **74150**.

11-8

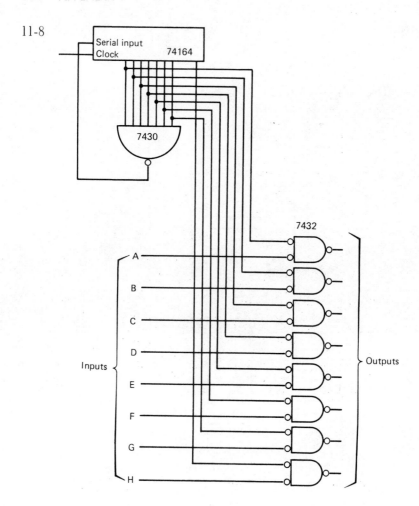

11-10 (a) The decoder requires four **74154**s. Use the select lines as follows:

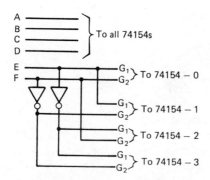

11-11   The circuit requires 11 **7442** 4-line-to-10-line decoders. The select lines of one decoder must be connected to the most significant BCD decade and the select lines of the other decoders must be connected to the least significant BCD decade. Because **7442s** have no strobe, the following circuit should be used to deselect 9 of the multiplexers:

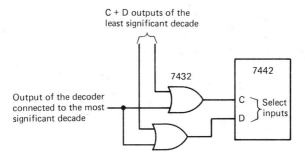

C + D outputs of the least significant decade

7432

7442

Output of the decoder connected to the most significant decade

C ⎫ Select
D ⎭ inputs

The outputs of the decoder connected to the most significant decade consists of nine 1s and a 0. The 1s cause the C and D inputs of all the corresponding decoders to be HIGH, which causes them to present all HIGH outputs. Only the least significant decoder connected to the LOW output of the most significant decoder receives a genuine BCD input and produces a LOW output.

11-12

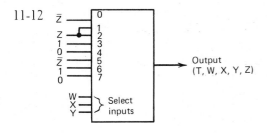

Output
(T, W, X, Y, Z)

Select inputs

11-14   (a)

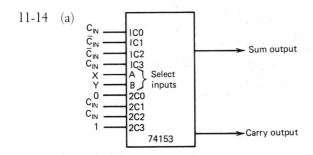

Sum output

Carry output

74153

# CHAPTER 12

12-1    (b)  010100001001
12-2    (b)  2905
12-3    (a)  799
        (b)  Not BCD—Most significant decade is a 10.
        (c)  Not BCD—Second decade is a 12.
12-4B   (c)
12-6B   (a)

|  |  | Place position |
|---|---|:---:|
| 1 0 1 1 1 0 0 1 1 |  |  |
| 1 1 0 0 1 0 0 0 |  | 10 |
| 1 0 1 0 1 0 1 1 |  |  |
| 1 1 0 0 1 0 0 |  | 9 |
| 1 0 0 0 1 1 1 |  |  |
| 1 0 1 0 0 0 |  | 7 |
| 1 1 1 1 1 |  |  |
| 1 0 1 0 0 |  | 6 |
| 1 0 1 1 |  |  |
| 1 0 1 0 |  | 5 |
| 1 |  |  |
| 1 |  | 1 |
| 0 |  |  |

The result has 1s in positions 1, 5, 6, 7, 9 10

$$1 1 0 1 1 1 0 0 0 1 = 371$$

12-4B   (c)

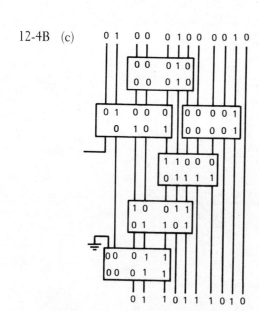

12-7    (a)  The circuit attempts to convert 698 to binary. The left IC in the second column is not working. Its output should be 1000. This throws the rest of the calculation off. The bottom IC is also not working. Its output should be 00110.

    (b)  The wire between ground and pin F of the lowest **74185** could be broken. The IC would then see inputs of 11001 and in the original circuit produce the output shown.

12-9

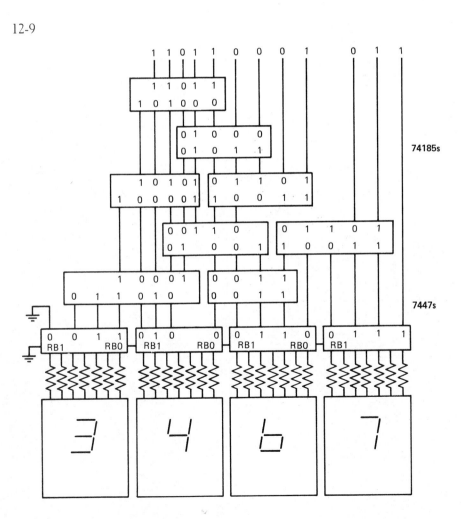

12-12

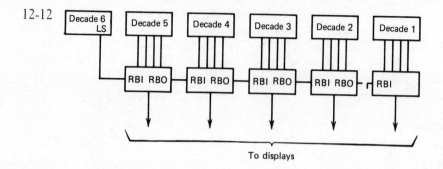

To displays

Note the number 50 will cause the 3 most significant displays to be blanked and will appear as 50, but 100050 will not cause any blanking because the leading 1 is tied to RBI of the MSD. Therefore 100050 appears as 00050, which is distinguishable from 50.

# CHAPTER  13

13-3

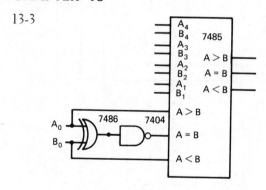

13-5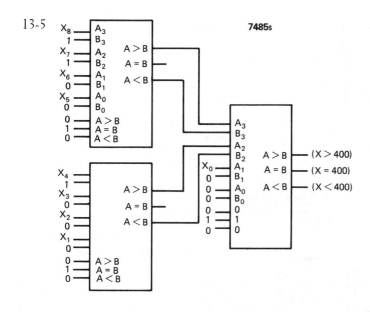

13-8 (a)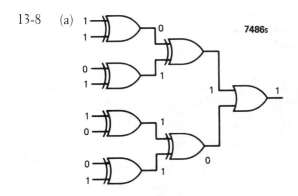

13-10 (b)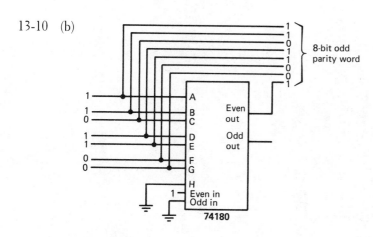

The circuit is shown as an 8-bit odd parity generator. As a checker the odd output is HIGH for any odd parity input.

13-13 (b)

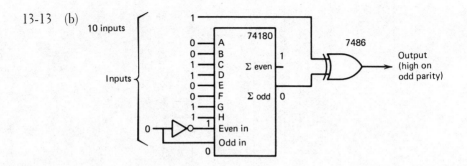

13-16

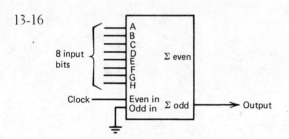

13-18 The second bit of the third word is wrong.

13-21 $G_0 = B_1 \oplus B_0$

$G_1 = B_2 \oplus B_1$

$G_2 = B_3 \oplus B_2$

$G_3 = B_3$

# CHAPTER 14

14-3 (b)
$$\begin{array}{r} 8193 \\ 3904 \\ \hline 12097 \\ \searrow 1 \\ \hline 2098 \end{array}$$

14-4 (c) 10011010

14-5   Only numbers $b$, $c$, and $f$ have 2 LSBs of 0 and are divisible by 4.

14-7   (b)
$$\begin{array}{rl} 001010101 & (85) \\ +\ \underline{111011011} & (-37) \\ 000110000 & (+48) \end{array}$$

14-8   (b)
$$\begin{array}{rl} 835 = & 01101000011 \\ 214 = & 00011010110 \\ -214 = & 11100101010 \end{array}$$

$$\begin{array}{ll} 01101000011 & 835 \\ \underline{11100101010} & \underline{-214} \\ 01001101101 & 621 \end{array}$$

14-10   (c)  20FC

14-11   (c)  0101 1100 1111 0000 0011 0101

14-12   (c)  6090805

14-13   (b)  205

14-14   (c)  19B40

14-15   (c)  A0A

14-16   (c)  B00
        (d)  E0CFE

14-18   (a)  $(-55) - (-45)$
        (b)  Gates 3 and 10 are defective

14-21   Refer to Fig. 14-10 and start at $D_0$, where the carry in is 1.

| Original numbers | $D_3$ | $D_2$ | $D_1$ | $D_0$ | |
|---|---|---|---|---|---|
| A | 2 | 4 | 3 | 2 | |
| B | 0 | 6 | 7 | 6 | |
| Carry in | 0 | 0 | 0 | 1 | |
| $\bar{B}$ | 15 | 9 | 8 | 9 | |
| Sum of upper 7483 and an input to lower 7483 | 1 | 13 | 11 | 12 | |
| Carry out | 1 | 0 | 0 | 0 | |
| B input to lower 7483 | 0 | 10 | 10 | 10 | |
| Difference digit | 1 | 7 | 5 | 6 | Answer |

14-23   (a)

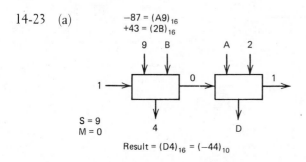

$$\begin{array}{l} -87 = (A9)_{16} \\ +43 = (2B)_{16} \end{array}$$

S = 9
M = 0

Result = $(D4)_{16} = (-44)_{10}$

(d)

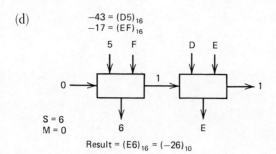

$-43 = (D5)_{16}$
$-17 = (EF)_{16}$

S = 6
M = 0

Result = $(E6)_{16} = (-26)_{10}$

(f)

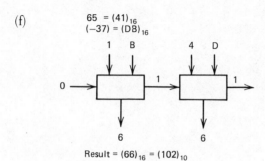

$65 = (41)_{16}$
$(-37) = (DB)_{16}$

Result = $(66)_{16} = (102)_{10}$

14-25  Use S = 15 and tie $C_{in}$ of the least significant stage HIGH.

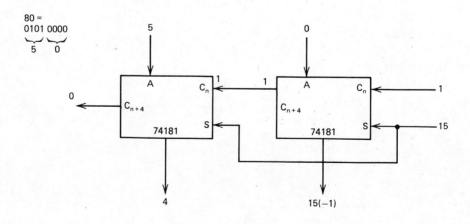

$80 =$
0101 0000
5    0

Result is $\overbrace{0100}^{4} \ \overbrace{1111}^{15} = 79$

14-28

| | $A_0$ | $B_0$ | $F_0$ | $C_0$ | $A_1$ | $B_1$ | $F_1$ | $C_1$ |
|---|---|---|---|---|---|---|---|---|
| $S = 12$ <br> $M = 0$ <br> $C_{in} = 0$ | B | C | 7 | 0 | 5 | 3 | B | 1 |
| $S = 6$ <br> $M = 0$ <br> $C_{in} = 0$ | A | D | D | 1 | 6 | A | B | 1 |
| $S = 9$ <br> $M = 1$ <br> $C_{in} = 1$ | A | D | 8 | X | 3 | 9 | 5 | X |
| $S = 15$ <br> $M = 0$ <br> $C_{in} = 1$ | 0 | 7 | F | 1 | 5 | 7 | 4 | 0 |

14-29  If $A = 0$, the carry out of the second stage will be HIGH.

14-31  Multiplier = $45 = 101101$

Multiplicand = $37 =$     1 0 0 1 0 1

```
                                            LSB
              0 0 0 0 0 0
              1 0 0 1 0 1                 1   Add
              0 1 0 0 1 0     1               Shift
              0 0 1 0 0 1     0 1           0   Shift
              1 0 1 1 1 0     0 1           1   Add
              0 1 0 1 1 1     0 0 1             Shift
              1 1 1 1 0 0     0 0 1         1   Add
              0 1 1 1 1 0     0 0 0 1           Shift
              0 0 1 1 1 1     0 0 0 0       0   Shift
              1 1 0 1 0 0     0 0 0 0 1     1   Add
              0 1 1 0 1 0     0 0 0 0 0 1       Shift
```

(Final answer = 1665)

# CHAPTER 15

15-1  (a)  14
      (b)  32
      (c)  32
      (d)  $2^{19}$
      (e)  32
      (f)  16,384
      (g)  32
      (h)  16,384
      (i)  128 each
      (j)  4096
      (k)  512
      (l)  96

15-3

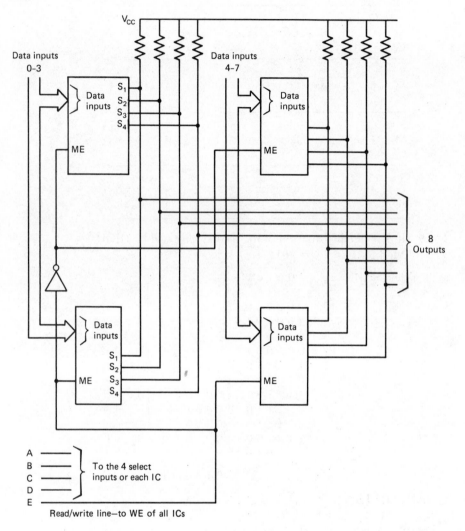

15-6   (a)  12
       (b)  12
       (c)  13
       (d)  1
       (e)  8

15-6    (f)

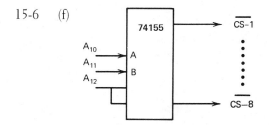

15-9    (a)  12—They come from the computer system.
        (b)  6—They go to the Dynamic RAM.
        (c)  The lower 6 address bits
        (d)  The upper 6 address bits
        (e)  The 6 bits of the Refresh Counter
15-12   (a)  16
        (b)  CSA is selected when $A_{15} = 0$, $A_{14} = 1$, $A_{13} = 0$, and $A_{12} = 1$.
        (c)  The **8216**s are transceivers.
        (d)  They are synchronizing FFs.
        (e)  The **3242** can accommodate a 16K memory IC but is being used only
for 4K memory ICs.
        (f)  Left for student to draw.

15-14   (a)

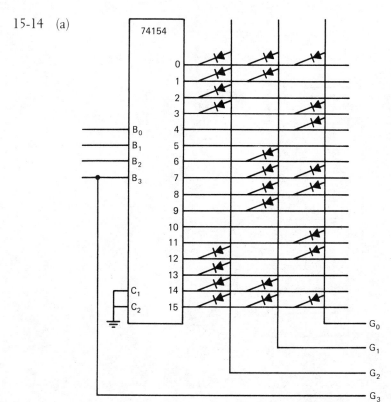

# CHAPTER 16

16-2

| 10 | LOAD | 100 |
|----|------|-----|
| 11 | SUBTRACT | 22 |
| 12 | BRZ | 25 |
| 13 | LOAD | 10 |
| 14 | ADD | 50 |
| 15 | STORE | 10 |
| 16 | LOAD | 51 |
| 17 | SUB | 50 |
| 18 | STORE | 50 |
| 19 | BPL | 10 |
| 20 | HALT | |
| 25 | LOAD | 52 |
| 26 | ADD | 50 |
| 27 | STORE | 52 |
| 28 | BRA | 13 |

At the start of the program:

50 contains 1
51 contains 200
52 contains 0

It retains the number of 22s in the program.

16-6

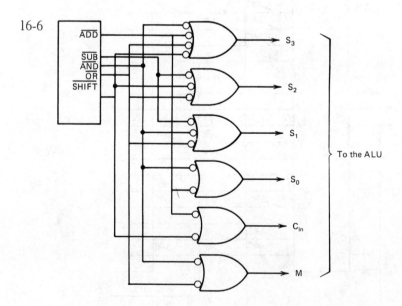

16-8    (a)  0—The INCREMENT ACCUMULATOR instruction uses the F = A + 1 function of the 181s.

(b) 1—The SKIP ON 0 ACCUMULATOR instruction uses the $F = A$ minus 1 function of the 181s and checks $C_{n + y}$ for a 1, which only occurs if the accumulator is 0.

# CHAPTER 17

17-1　(a)　179 ohms
　　　(b)　2.7 V and $\approx 0$ ($V_{CE(SAT)}$)
　　　(c)　30 mA
　　　(d)　Yes

17-2　$\dfrac{R_1 R_2}{R_1 + R_2} = 150 \dfrac{R_1}{R_1 + R_2} = \dfrac{3}{5}$

Results $R_1 = 375$ ohms, $R_2 = 250$ ohms

17-6

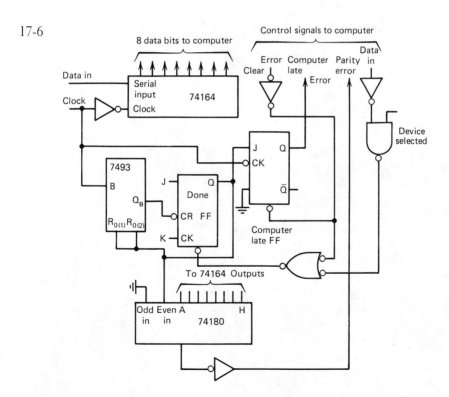

17-8　　1　Memory request FF
　　　　1　READ/WRITE FF
　　　 12　FFs for the MDR
　　　 13　FFs for the MAR
　　　 $\overline{27}$　FFs are required

17-10

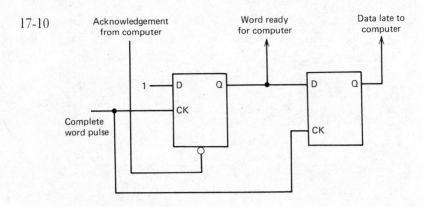

17-12   5 volts ÷ 256 steps = 0.0195 volts/step

17-14 At least 2000 steps are required. Therefore, 11 bits are required on the digital input since $2^{11} = 2048$.

17-15   (a)  37 = 100101

Assuming 1 = 5 volts,

$$I = \frac{5}{R} + \frac{5}{R/4} + \frac{5}{R/32} = \frac{185}{R} \quad R = 10\text{K, so } I = 18.5 \text{ mA}$$

17-17   (a)  13 mA
        (b)  19.5 mA
        (c)  32.5 mA

17-19

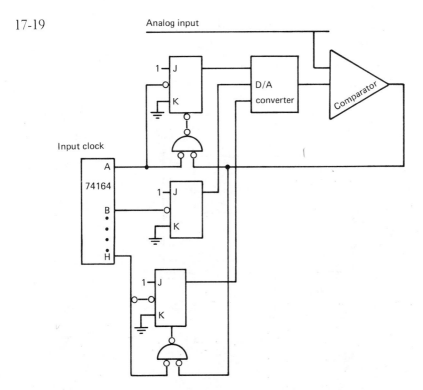

A single negative pulse is passed down the shift register and sets each FF. If the D/A converter output is too high, the FF just set resets. A worst case comparison requires N steps for an N-bit output.

17-20   1.9 volts

# CHAPTER 18

18-1

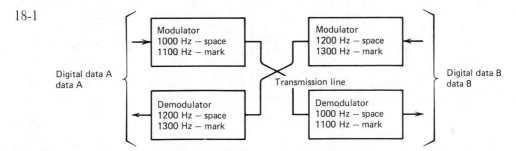

18-2     First character = $11010100 = (D4)_{16}$
         Second character = $01000001 = (41)_{16}$

18-3

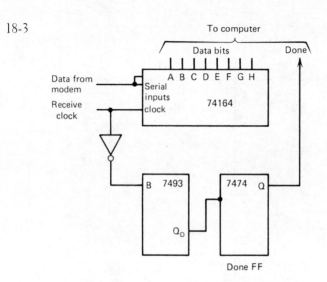

18-7

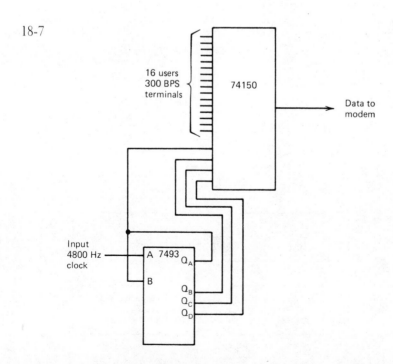

18-9

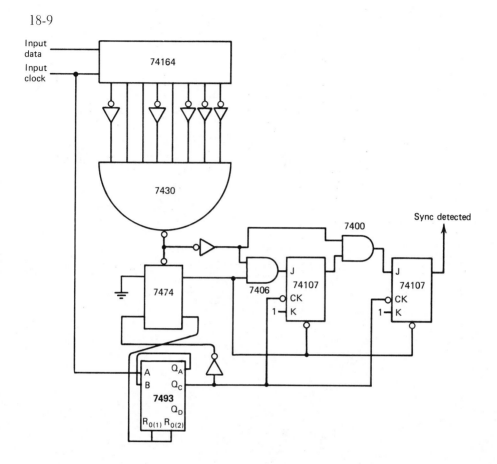

18-13   18.2 ms
18-14   (a)  4800Hz
        (b)  36.63 ms
        (c)  Pin 34–1
             Pin 35—0
             Pin 36—0
             Pin 37—1
             Pin 38—1
             Pin 39—0
             Pin 40—4800 Hz

# CHAPTER 19

19-2

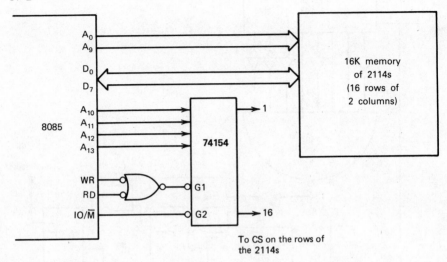

To CS on the rows of
the 2114s

19-3

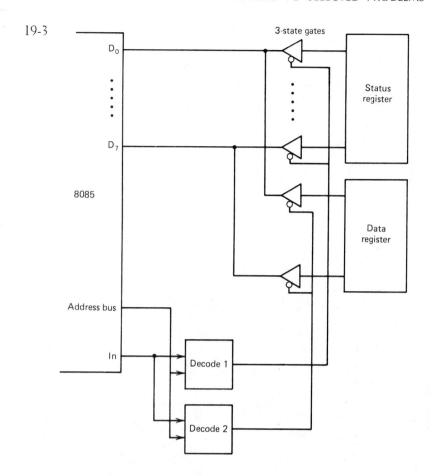

19-4

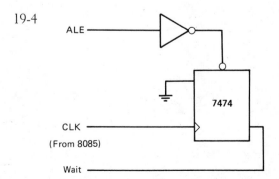

19-7 CD 1 1 0 0 1 1 0 1
     59 0 1 0 1 1 0 0 1
     28 0 0 1 0 1 0 0 0

Use Fig. 19-9 connect:

AD0  to  $\overline{Q1} + \overline{Q2}$
AD1  to  0
AD2  to  Q1
AD3  to  1
AD4  to  $\overline{Q2}$
AD5  to  $\overline{Q1} + \overline{Q2}$
AD6  to  Q1 + Q2
AD7  to  Q1

19-8    (a)  All 0's
         (b)  All 1s
         (c)  Lines 0–5 1's Lines 6 and 7 0s (Hex. 3F).
19.10.  (a)  Use an IN 1 to enable $\overline{RDE}$ and an IN 2 to enable $\overline{SWE}$.
        (b)  Use CA2 to enable $\overline{RDE}$ if it is a 0 and $\overline{SWE}$ if it is a 1 by connecting
it to the UART via an inverter.

# CHAPTER 20

20-1    (a)  40-character lines can be written at a refresh rate of 60 Hz.
        (b)  48-character lines can be written at a refresh rate of 50 Hz.
20-3    Assume 42 $\mu$s for nonblanked line time. Each dot requires 0.2 $\mu$s; therefore,
210 dots can be displayed.

$$(a)\frac{210}{8} = 26 \text{ characters per line}$$

$$(b)\frac{210}{6} = 35 \text{ characters per line}$$

20-5

| COMPOSITE SYNC | DOT OUTPUT | VIDEO OUTPUT |
|:---:|:---:|:---:|
| 0 | 0 | 0 |
| 0 | 1 | 0 |
| 1 | 0 | 1.2 |
| 1 | 1 | 3.2 |

20-8

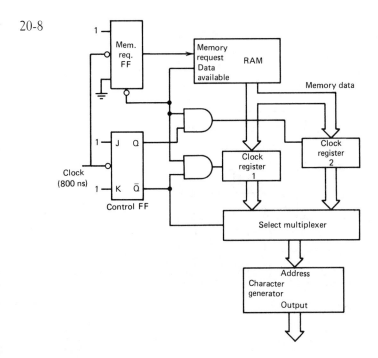

20-11   The number of characters in a line is added to the cursor address register.

20-12

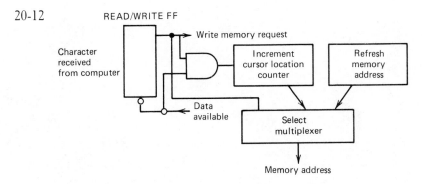

Cursor manipulation circuitry not shown for clarity.

20-14

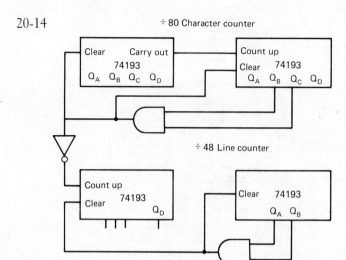

÷ 80 Character counter

÷ 48 Line counter

# G

# ASCII CONVERSION CHART

The conversion chart listed below is helpful in converting from a two-digit (two-byte) hexadecimal number to an ASCII character or from an ASCII character to a two-digit hexadecimal number. The example provided below shows the method of using this conversion chart.

EXAMPLE

| | | Bits | | | | | | |
| | | MSB ← | | | | | → LSB | |
| ASCII | Hex # | 6 | 5 | 4 | 3 | 2 | 1 | 0 |
|---|---|---|---|---|---|---|---|---|
| T | 54 | 1 | 0 | 1 | 0 | 1 | 0 | 0 |
| ? | 3F | 0 | 1 | 1 | 1 | 1 | 1 | 1 |
| + | 2B | 0 | 1 | 0 | 1 | 0 | 0 | 1 |

| Bits 0 to 3 Second Hex Digit (LSB) | Bits 4 to 6 First Hex Digit (MSB) | | | | | | | |
|---|---|---|---|---|---|---|---|---|
| | 0 | 1 | 2 | 3 | 4 | 5 | 6 | 7 |
| 0 | NUL | DLE | SP | 0 | @ | P | | p |
| 1 | SOH | DC1 | ! | 1 | A | Q | a | q |
| 2 | STX | DC2 | '' | 2 | B | R | b | r |
| 3 | ETX | DC3 | # | 3 | C | S | c | s |
| 4 | EOT | DC4 | $ | 4 | D | T | d | t |
| 5 | ENQ | NAK | % | 5 | E | U | e | u |
| 6 | ACK | SYN | & | 6 | F | V | f | v |
| 7 | BEL | ETB | ' | 7 | G | W | g | w |
| 8 | BS | CAN | ( | 8 | H | X | h | x |
| 9 | HT | EM | ) | 9 | I | Y | i | y |
| A | LF | SUB | * | : | J | Z | j | z |
| B | VT | ESC | + | ; | K | [ | k | { |
| C | FF | FS | , | < | L | / | l | / |
| D | CR | GS | – | = | M | ] | m | } |
| E | SO | RS | . | > | N | ∧ | n | ≈ |
| F | SI | US | / | ? | O | — | o | DEL |

# INDEX

Acceptance test, 309
Access time, 461, 469, 476, 550
Accumulator, 492–495, 507, 509, 510, 512–514, 527, 528
ADD instruction, 493, 495
Adder, 97, 98, 403–406
  BCD, 427, 428
  full adder, 404–406
  half adder, 403, 404
  **7483,** 420–425, 427–431, 436
Adder/subtracter, 422–425
Addition:
  binary, 14, 15
  Boolean, 23
  hexadecimal, 417–418
  instruction, 493–495, 513
  theorems, 25–28
  two's complement, 412–413
Address multiplexer, 514
Algorithms, 349–355, 438–440
  add algorithm for BCD-to-binary conversion, 349, 350
  multiplication, 438–440
  shift and add algorithm, 354, 355

shift and subtract algorithm, 351, 352
subtract algorithm for binary-to-BCD conversion, 352, 353
Analog-to-digital (A/D) converters, 558–565, 617
AND:
  logical, 24
  gates, 33, 34
  instruction, 513, 514
AND-OR, 69, 70, 288
AND-OR-INVERT gates, 147–151, 270, 273
Apple computer, 1, 630, 636, 637, 647
Arithmetic/logic units (ALUs), 431–435, 507, 512–514
  **74181**, 431–435, 527, 528
Arithmetic processing units, 444, 445
ASCII code, 573–575, 631, Appendix G
Asynchronous data transmission, 573–575, 581–592

Binary:
    binary to decimal conversion, 5–8
    boat, 6, Appendix A
    numbers, 516, 65
    system, 5, 6
Binary coded decimal (BCD), 91, 92,
    347–370
    arithmetic, 427–431
Bit, 3–5, 6, 7–9
Boolean algebra, 19–32
    for relays, 53–55
    theorems, 25–28
Branch (instruction), 498–505,
    516–517
Bubble, 35–37, 165
    memory, *see* memories
    placement of, 48–52
Buffer drivers, 136–139, 604, 605
Bus:
    Bidirectional, 537
    IEEE-488, 626
    I-O, 534–545
    memory, 546–552
    microprocessor, 598, 601
    multiplexed (8085), 599–601
    S-100, 625
BUSY, 535, 541–545
Byte, 274, 414

CALL instruction, 608–611
Carry FF, 274, 275, 527, 528
Cathode ray oscilloscope (CRO), *see*
    Oscilloscope
Character generator, 632–638, 640,
    644, 646
    **2513**, 632–638, 640, 644, 646
Check table, 81, 94
Clamp circuit, 132
Clock, 186, 231, 598, 599. *See also*
    Oscillators
CMOS, 118–122
Code conversion, 100, 101, 347–362
    Gray to binary, 395–397
    **74184**, 355–359
    **74185**, 359–362
Code wheel, 394, 395
Comparator, 559–562
Comparison circuits, 151, 375–380, 651
    **7485,** 377–380, 651

Complementation, 23
    of functions, 37–41
Control unit, 505–509
Controller, 533, 534, 541–545, 556,
    577, 579
    CRT, 652–653
    memory, 546–552
    priority interrupt (74148),
        549–552
Cores, magnetic, 455–459
Counters, 182–185, 231–262,
    645–649
    down counter, 184, 254, 255
    ripple, 182–185, 232–234
    synchronous, 234–239
    3s counter, 238, 239, 610
    truncated, 241
    **7490** decade counter, 247, 248,
        257–259
    **7492** divide by 12, 247, 249,
        250, 286
    **7493** 4-bit binary counter,
        250–252
    **74192** decade up-down counter,
        257
    **74193** up-down counter,
        252–257, 260–262
Current loop, 583–584, 586, 592
Cursor, 648–652
Cycle time, 461, 477
Cyclical redundancy checking, 390,
    392

Debouncing circuits, 220–224
Debugging, 310–316
Decimal numbers, 5, 10
    decimal to binary conversion,
        8–14
Decoder, 232–234, 337, 469, 597,
    598, 604, 605
    keyboard, 340–344
    **7442**, 337
DeMorgan's theorem, 38–41, 45, 46,
    73, 74
Demultiplexer, 332–337
    **74154**, 334–336, 480, 482, 604
    **74155**, 332–334, 341, 466, 467
Device select lines, 538–540, 545
Digital to analog converters, 558, 565

Diode matrices, 480–482

Direct memory access (DMA), 546–552, 557

Display drivers, 138, 363–370
    **7447,** 363–367
    **7448,** 367–370

Display generator, 630–653

Divide-by-N circuits, 231, 232, 257–262

Discriminator, 224, 225

Domain, 62, 65

DONE, 535, 541–545, 552

Don't cares, 91–93, 100, 101

Duty cycle, 196, 197, 200, 209

Emitter coupled logic (ECL), 116, 117

Equivalence, logical, 31, 32

Excess-3 code, 100, 101

Exclusive NOR (equality) gate, 375, 377
    **74LS266**, 375, 377

Exclusive-OR gates, 151, 152, 285, 375, 422–425
    **7486**, 151, 152, 375

Execute, 505, 508–511, 516–517, 524, 526

Expandable gates, expanders, 145–147

Fanout, 108, 114, 132

Fetch, 505–508, 511, 514–516, 524, 526

Flip flops, 159–189, 453
    data lookout, 172, 173, 174, 175
    D type, 165–167, 174, 220, 268, 269
    edge triggered, 172, 173, 174, 175
    J-K, 167–175, 269
    latch type, 167, 168, 174, 463, 464
    master-slave, 167–172, 174, 175
    NAND gate, 162–164
    NOR gate, 160–162
    SET-RESET, 159–160

Flow chart, 8, 9, 10, 11, 12,13, 496, 503, 620, 621

Fractions, 7, 8, 10–14

Frequency shift keying (FSK), 225, 572, 573

Gates, 32–37, 41–52

Gray code, 393–397

Glitch, 179, 233, 234, 257, 258, 262, 314, 315, 321
    controlled, 187–189

HALT, 526
    FF, 526

Hexadecimal:
    arithmetic, 414–419
    system, 5, 414

Instruction (IN), 528, 601–605

Interrupt, 535, 552–557, 605–612, 624
    FF, 535
    NMI, 622
    **6800,** 620–622
    vectored, 553–555

Interrupt Acknowledge (INTA), 605–612, 624

Inverter, 23, 35–37, 221

IO/$\overline{M}$, 600, 603–605

Irregular count sequence, 230–241, 244, 245

Jitter, 197, 200

Karnaugh maps, 74–101, 383, 404, 405
    five variable maps, 93–95
    POS maps, 84–87

Keyboards, 340–344

Light and switch panels, 313

Light emitting diodes (LEDs), 363, 367, 368, 584, 585

Literal, 61, 62, 65, 68

Liquid crystal displays, 370, 371, 398, 399

LOAD instruction, 494, 495, 614

Logicals (logic diagrams), 304–305

Logic analyzers, 316–321

Logic probe, 315–316

Longitudinal redundancy checking, 390–392

Look-ahead carry, 435, 438
    **74182**, 438, 439

Loop, 499–504, 620

Manuals, 309, 310
Mark, 573, 574, 583, 584, 566
Mask, 552, 553
Memories, 317, 453–489, 505, 511, 522, 546
    bubble, 292, 293
    core, 455–459
    refresh, 638–640
    semiconductor, 459
    *see also* RAM, ROM
Memory address register (MAR), 454, 455, 506–509, 511, 512, 516, 548, 557
Memory data register (MDR), 454, 455, 506, 507, 512, 548, 557
Microprocessor, 471, 472, 596–625
    parity, 386, 387
    shifting, 274, 275
MODEM, 571–581
Module chart, 306, 307
MOS, 118
    shift registers, 288–292
Multiple outputs, 89–91
Multiplexed displays, 369, 370
Multiplexer logic, 337–340
Multiplexers, 95, 324–332, 522, 580, 581
    **74LS257,** 327, 328, 522
    **74LS258,** 327, 328
    **74150,** 328–332, 581
    **74151,** 328–329, 581
    **74152,** 328–329, 338–342
    **74153,** 326, 327, 369, 370
    **74157,** 327
Multiplication, 272, 273
    Binary, 438–444
    Boolean, 24, 25, 26
    theorems, 25

NAND gates, 41–44, 114, 115, 220
    implementation of logic functions, 69, 70
Negation, 419
Nibble, 414–415
Noise margin, 112, 113
NOR gates, 41, 43
    implementation, 71

Octal, 419, 420
One shot, 196–217, 222–225, 313

    retriggerable, 201–217
    **74121,** 197–201, 207–209
    **74122,** 203–206
    **74123,** 206–207, 210–211, 222–225
Op code, 493, 498, 505, 512, 526, 599, 602, 606, 608, 609
Open collector gates, 135–138, 375, 376, 584
    memories, 462, 466
Optical couplers, 584–586
OR:
    gate, 35, 44
    logical, 23, 24
OR-AND implementation, 71
Oscillators, 207–216
    crystal, 231
    horizontal, 632–635, 637, 641
    one shot, 207–211
    Schmitt trigger, 211, 212
    vertical, 632–635, 641
    **555,** 212–215
Oscilloscope traces (CRO), 258, 312–315
    shadows, 259, 260
OUT instruction, 528, 529, 601–605
Overflow, 425–426

Page register, 526, 527
Panel, IC, 298–301, 306
Parallel-to-serial conversion, 285–287
Parity, 380–392
    **74180,** 387–390
Peripheral interface adapter (PIA), 612–622
Peripherals, 533, 534, 556, 557
Port (I/O), 529, 601–605, 622, 624
Power dissipation, 107, 112
    in CMOS, 120
Printed circuits, 301–303
Printers, 290–292
Product of sums (POS), 62, 64, 66–74
Product term, 61–66
Program counter, 506, 514, 516, 526, 606, 609
Programmable logic arrays, 95
Programmable read only memories (PROMs), 485–487
    **2708,** 485, 486

Programs, 492–505, 552
    self-modifying, 504, 505
Propagation delay, 106, 107, 112
Pull-up resistors, 131, 132, 135

Race condition, 176–178
Random access memory (RAM),
   460–476, 653
    bipolar, 461–468
    dynamic, 472–480
    MOS, 468–476
    static, 460–472
    **74LS215**, 468
    **2102**, 468–470, 511
    **2114**, 470–472, 522, 601, 602
    **4116**, 473–476
    **7489,** 464–467
    **74170**, 462–464
Read only memory (ROM), 95,
   480–487
    diode matrix, 480–481
    MOS IC, 483
    NAND gate, 481–483
    PROMs, 484–485
Refresh controller, 477–480
    **i3242**, 478–480
Refresh cycles, 476–480
Register, 147, 151, 152, 180, 326,
   327, 512–514
    command, 613, 614
    control, 613–620, 623
    cursor address, 649–651
    data, 613, 614, 620
    data direction, 613, 614, 620
    status, 613, 614
Relays, 52–55
Resistors:
    pull-up, 131–132, 135
    terminating, 534–537
Resolution, 558, 561, 562
RESTART instruction, 606–608, 610,
   612, 624
Ripple blanking, 364–367
Rotate, 273
RS-232-C, 575–578, 587

Schmitt trigger, 126–130, 200, 201,
   211, 212
    debouncing, 222

SDK-85, 597, 598, 623–625
Serial-to-parallel conversion, 285
Seven segment displays, 99, 100,
   363–370, 527, 541, 542
Shift register, 268–293
    dynamic, 288–290
    left-right, 270–272, 281–283
    MOS, 288–292
    parallel-loading, 275–276,
      279–281
    static, 290–292
    **7494**, 283
    **7495**, 283
    **74164**, 276–278, 283–285, 577
    **74165**, 279–281, 286, 287, 578,
      579
    **74166**, 279, 281
    **74198**, 281–283
Sign-magnitude, 409
Skip instruction, 525, 528
Sockets, IC, 299–301
Space, 573, 574, 583, 584
Specifications, 304
Speed-power product, 113, 114
State checkout table, 239, 242, 246
State table, 242–246
STORE instruction, 494, 495, 615
Strobe, 143–144, 151, 326, 327, 329,
   332
    strobed gates, 143–144, 151
Subcubes, 76–94
    essential subcubes, 81–84, 86,
      90–92
Subtraction:
    BCD, 428–431
    binary, 15, 16, 406, 407
    by complementation, 407, 408
    hexadecimal, 418
    two's complement, 413, 414
Sum of products (SOP), 62–66, 72–74
Sum term, 62, 64, 67–69
Switch bounce, 217–224, 342, 343
Synchronous circuits, 186–187
Synchronous data transmission,
   575–581

Teletype, 349, 581–592, 616
Three state ICs, 139–143, 607
    **74LS257, 74LS258**, 327–328
    **2102**, 468–470

Three state ICs (*continued*)
**74125, 74126**, 142
**74173**, 180
**74365, 74366, 74367, 74368,**
142, 143
Time division multiplexing, 324, 325,
578–581
Timing charts, 173–175, 236, 256,
644
for logic analyzers, 3, 9, 320
Timing generation circuits, 510, 511
using one shots, 215–217
using shift registers, 284–285
Transceivers, 143, 144, 537–539, 624
Trigger, 196–198, 201, 202, 319
Truth tables, 21–23, 72, 96, 97, 101,
241, 404, 405, 407, 480, 481
TTL:
characteristics of, 111–115
high speed, 110
low power, 110
low power Schottky, 111
Schottky, 111
Standard, 109, 110
Twisted pair wires, 535–537, 547
Two's complement arithmetic,
409–414

Universal asynchronous
receiver-transmitter (UART),
586–592, 616, 618
**AY5-1013**, 587–592, 616
**6850**, 623
**8251**, 622, 623
Unused inputs, 130–132

V-flag, 425
Volatility, 457, 459

Wire-ANDing, 132–135, 312, 376
Wire list, 307–309
Wire wrapping, 298–301, 303–309
Word, 7, 453

# SUPPLEMENTARY INDEX

This is an index of ICs discussed in this book. The numbers are the pages where the ICs are *first* discussed.

*7400* series ICs
**7400** 2-input NAND gate, 41, 42, 114, 115
**7401** open collector 2-input NAND gate, 136
**7402** 2-input NOR gate, 43
**7403** open collector 2-input NAND gate, 136
**7404** inverter, 36
**7405** open collector inverter, 136
**7406** open collector inverter buffer driver, 136–138
**7407** open collector buffer driver, 136, 137
**7408** AND gate, 33, 34
**7409** open collector AND gate, 136
**7410** 3-input NAND gate, 41, 42
**7412** open collector 3-input NAND gate, 136
**7413** Schmitt trigger, 127, 128

**7414** Schmitt trigger, 127
**7420** 4-input NAND gate, 41, 42
**7422** open collector 4-input NAND gate, 136
**7423** expandable gate, 145
**7425** 4-input NOR gate, 43, 143–145
**7427** 3-input NOR gate, 43
**7430** 8-input NAND gate, 41, 42
**7432** OR gate, 35
**7433** open collector NOR gate, 136
**7442** BCD decoder, 337
**7447** 7-segment display driver, 364–367
**7448** common cathode 7-segment display driver, 367–370
**7450, 7451** AND-OR INVERT gates, 147, 148, 150
**7453, 7454** AND-OR INVERT gates, 147, 149
**7460** expander, 145, 146

**7474** D flip flop, 164, 165
**7475** QUAD latch flip flop, 167, 168
**7476** master-slave flip flop, 172
**7483** 4-bit adder, 420–425
**7485** 4-bit comparator, 377–381
**7486** EXCLUSIVE OR gate, 151, 152, 375
**7489** 64-bit memory, 464–467
**7490** decade counter, 247, 248
**7492** divide-by-12 counter, 247–250
**7493** binary counter, 250, 251
**7494, 7495** shift registers, 283
**74107** J-K master-slave flip flop, 169–172
**74111** J-K flip flop with data lockout, 173, 174
**74113, 74114** J-K edge-triggered flip flop, 172
**74121** one shot, 197–201
**74122** retriggerable one shot, 203–206
**74123** dual retriggerable one shot, 206, 207
**74125** 3-state driver, 142
**74126** 3-state driver, 142
**74148** priority interrupt controller, 549–551
**74150, 74151, 74152** multiplexers, 328–332
**74153** 4-line to 1-line multiplexer, 326, 327
**74154** demultiplexer, 334–336
**74155** demultiplexer, 332–334
**74157** QUAD 2-line to 1-line multiplexer, 327
**74164** shift register, 276–278
**74165, 74166** shift registers, 279–281
**74170** 16-bit memory, 462–464
**74173** QUAD D flip flop with 3-state outputs, 180, 181
**74174** hex D flip flop, 166, 167, 180
**74180** parity generator/checker, 387–390
**74181** arithmetic logic unit (ALU), 431–435

**74182** look-ahead carry generator, 438–439
**74184** BCD-to-binary converter, 355–359
**74185** binary-to-BCD converter, 359–362
**74192** decade up-down counter, 257
**74193** up-down counter, 252–257
**74198** shift register, 281–283
**74LS241, 74244** 3-state drivers, 142
**74LS245** 3-state transceiver, 143
**74LS257, 74LS258** multiplexers with 3-state outputs, 327, 328
**74LS266** EXCLUSIVE-NOR, 375–377
**74365, 74366, 74367, 74368** 3-state drivers, 142
**75138** transceiver, 537–539

*Other ICs*
**555** timer, 212–216
**ADC 0800** A/D converter, 564, 565
**DAC 0800** D/A converter, 563, 564
**AY-5-1013** UART, 587–592
**MC1488/1489** TTL to RS232 level converters, 577, 579, 581
**2102** 1K memory, 468–470
**2114** 4K memory, 470–472
**2513** character generator (ROM), 484, 632–638
**2521, 2522** static shift registers, 290–292
**2524, 2525** dynamic shift registers, 288, 289, 639
**2708** EPROM, 485, 486
**i3242** dynamic RAM controller, 478–480
**4116** dynamic RAM, 472–476
**6800** microprocessor, 598, 599, 612–622

**6810** memory, 597
**6821/6820** PIA, 612–622
**6840** programmable timer
   module, 623
**6845** CRT controller, 652–653
**6850** ACIA, 623

**8085** microprocessor, 599–612
**8205** decoder, 598
**8251** communications interface,
   622, 623
**8253** counter/timer, 623
**8255** interface IC, 623